ANNALS OF THE NEW YORK ACADEMY OF SCIENCES

Volume 492

The New York Academy of Sciences
2 East 63rd Street
New York, New York 10021

ALCOHOL AND THE CELL

ANNALS OF THE NEW YORK ACADEMY OF SCIENCES
Volume 492

ALCOHOL AND THE CELL

Edited by Emanuel Rubin

The New York Academy of Sciences
New York, New York
1987

Library of Congress Cataloging-in-Publication Data

Alcohol and the cell.

(Annals of the New York Academy of Sciences; v. 492)
Based on papers presented at a conference held by the New York Academy of Sciences and the National Institute on Alcohol Abuse and Alcoholism in Philadelphia, Apr. 30 to May 2, 1986.
Includes bibliographies and index.
1. Alcohol—Physiological effect—Congresses. 2. Cell physiology—Congresses. I. Rubin, Emanuel, 1928– . II. New York Academy of Sciences III. National Institute on Alcohol Abuse and Alcoholism (U.S.) IV. Series. [DNLM: 1. Alcohol, Ethyl—metabolism—congresses. 2. Alcohol Oxidoreductases—metabolism—congresses. 3. Cells—enzymology—congresses. W1 AN626YL v.492 / QV 84 A3538 1986]
Q11.N5 vol. 492 500 s 87-5799
[QP801.A3] [615′.7828]

SP
Printed in the United States of America
ISBN 0-89766-379-9 (cloth)
ISBN 0-89766-380-2 (paper)

ANNALS OF THE NEW YORK ACADEMY OF SCIENCES

Volume 492
April 30, 1987

ALCOHOL AND THE CELL[a]

Editor
EMANUEL RUBIN

CONTENTS

[a]The papers in this volume were presented at a conference entitled Alcohol and the Cell, which was held by the New York Academy of Sciences and the National Institute on Alcohol Abuse and Alcoholism in Philadelphia, Pennsylvania on April 30 to May 2, 1986.

Poster Papers

Part III. Effects of Ethanol on Biological Membranes: Chemical Studies

Poster Papers

Part IV. Effects of Ethanol on Transport and Protein Synthesis

Poster Papers

Part V. Effects of Ethanol on Cell Organization and Function

Poster Papers

Major funding for this conference was provided by:

- NATIONAL INSTITUTE ON ALCOHOL ABUSE AND ALCOHOLISM

Financial assistance was received from:

- ALCOHOLIC BEVERAGE MEDICAL RESEARCH FOUNDATION
- ELI LILLY AND COMPANY
- HAHNEMANN FOUNDATION OF PATHOLOGY AND LABORATORY MEDICINE
- OLYMPUS CORPORATION
- SANDOZ CORPORATION
- SMITHKLINE BIO-SCIENCE LABORATORIES

Preface

EMANUEL RUBIN

Department of Pathology
Jefferson Medical College of Thomas Jefferson University
Philadelphia, Pennsylvania 19107

Like most of the things that give pleasure to life, for example food and sex, alcohol taken in moderation is an adjunct to civilized behavior. When abused, there ensue complications that are the subject of this volume. The dimensions of the public health problem presented by alcoholism are not commonly recognized by the public. The total costs of alcoholism are actually over $100 billion a year. The recognition of alcohol-related problems has an interesting history. Acute intoxication was described in the biblical account of Noah, and chronic illnesses were noted in ancient Indian texts and by Hippocrates and Galen. Yet the scientific study of problems related to alcohol consumption is relatively new.

Interest in the scientific aspects of alcoholism has paralleled the revolution in cell biology. The papers in this volume provide ample evidence of the impact of biochemistry, biophysics, ultrastructural techniques, electrophysiology, and many other contemporary methods. Another aspect of alcohol studies emphasized by some of the contributions in this volume is the fact that alcohol can be a valuable tool for elucidating certain features of normal physiology, in the same way that many processes are studied by the use of specific inhibitors.

Chronic alcohol abuse is responsible for damage to a bewildering variety of organs, including the liver, pancreas, brain, heart, skeletal muscle, gastrointestinal tract, endocrine system, and hematopoietic system. Despite a substantial research effort for more than a quarter of a century, we do not understand the pathogenesis of any chronic disease caused by long-term alcohol intake, nor do we perceive the mechanisms underlying many of its acute effects. The only generally agreed upon principle is that alcohol is toxic, and even that was not accepted by all until a few years ago. What is the relationship of irreversible chronic diseases, such as those of the liver, heart, and brain, to the easily reversible acute effects? Why are irreversible changes seen only in noncycling cells (liver, pancreas, muscle, and brain), whereas all cycling cells (gastrointestinal and hematopoietic cells) revert to a normal condition upon discontinuation of alcohol intake? Are there any irreversible changes in a living cell, or is cell death a *sine qua non* for irreversible disease? What is the role of alcohol and aldehyde metabolism? Do the different alcohol-related maladies reflect a separate pathogenesis for each organ, or is there a common effect underlying all of them? These and other questions pose difficult problems for which no easy solutions are in sight. It seems that a global approach at the level of organ physiology is unlikely to provide the fundamental insights that are necessary 1) to explain the pathogenesis of diseases caused by chronic alcoholism and 2) to devise preventive measures and rational therapy. For such insights we require a reductionist approach wherein we may study the disposition of alcohol and its acute and chronic effects in the smallest self-contained living unit, namely the cell.

Cell biology can be thought of as "organized complexity," whereas classical physics has been described as "organized simplicity." The effective study of the cell requires an overall strategy to resolve the complex relationships between various organelles and metabolic pathways. In the context of an alcohol-induced disease it appears at this time to be a task of staggering magnitude. Although an understanding of biophysical and biochemical alterations at the cellular and subcellular levels is most likely to elucidate the basic mechanisms of alcohol-induced injury, to consider the organism merely as a chemical machine is to take a simplistic approach, in which life itself can be explained only in thermodynamic and kinetic terms. These methodologies, while yielding highly relevant information from closed systems in equilibrium (such as physics), do not apply strictly to systems not in equilibrium. Since cells are irreversible systems, the necessary methodologies (nonequilibrium thermodynamics) have yet to acquire the techniques or sophistication necessary to describe their inherent complexity. Thus, a completely unitary approach to the problems of alcoholism, while desirable, is not in sight. However, the nearest approach, and the one likely to yield most information, is the elucidation of the spatial and temporal parameters of regulatory systems in the cell.

The contributors to this book, who have utilized this approach in their own work, are in the forefront of research in this area of immense public health concern.

Distribution and Properties of Human Alcohol Dehydrogenase Isoenzymes

TING-KAI LI AND WILLIAM F. BOSRON

Indiana University School of Medicine
and
Veterans Administration Medical Center
Indianapolis, Indiana 46223

INTRODUCTION

Mammalian alcohol dehydrogenase (ADH) is a zinc metalloenzyme composed of two protein subunits of 40,000 daltons. Most animal species examined thus far have shown multiple forms of the enzyme in liver that differ in molecular charge. Multiple forms of human ADH were first recognized more than twenty years ago during the early attempts at its purification.[1] Subsequent studies revealed a degree of complexity that seemed to have no counterpart in lower animal species.[2] This heterogeneity appeared to arise from genetic variance within and among sample populations as well as from the existence of a larger number of different ADH subunits in humans, as compared with other animal species. A genetic model was postulated in 1971 that assumed three human ADH structural genes encoding three different subunits that combine randomly to form active dimeric isoenzymes.[3] Allelic variants involving two of the three gene loci were proposed.[4] This model could account for all the enzyme forms then identified, but additional enzyme forms discovered subsequently[5–7] have necessitated extension of this basic model.

In the last ten years, efficient affinity and ion-exchange chromatographic methods have been developed for the purification and separation of all the enzyme forms now known to exist.[8,9] As a result, the kinetic properties of the major enzyme forms and the structural interrelations among them have been characterized. These studies as well as recent reports of the amino acid and cDNA sequences of different ADH subunits support the existence of at least five different structural genes for human ADH, three of which encode very closely related protein subunits. From this viewpoint, it is remarkable that the different isoenzymes of human ADH, some of them being alleloenzymes, differ so strikingly in catalytic properties. This paper summarizes some of the recent advances in this area of our knowledge about this enzyme.

BIOCHEMICAL GENETICS OF HUMAN ADH ISOENZYMES

The isoenzymes of human ADH have been most clearly identified by means of electrophoresis in starch gels (FIG. 1) or isoelectric focusing in polyacrylamide or agarose gels, followed by the staining of the gels for alcohol oxidizing capacity.[2,4,10,11] The number and the electrophoretic mobility of the individual ADH molecular forms in human livers vary from specimen to specimen depending upon the genetic background and the state of health of the donors. The nature of this variability has been elucidated in the last fifteen years through a variety of studies:

1. The systematic comparison of the molecular forms present in liver and other

tissues of different population groups by means of starch gel electrophoresis and isoelectric focusing.[2,3,10,11]

2. The isolation of the individual molecular forms and the establishment of their identity as homo- or heterodimers through dissociation-recombination studies.[9,11]
3. The examination of subunit-relatedness through hybridization or dissociation-recombination studies,[9,12,13] studies of immunochemical cross-reactivity,[14] and studies of structural differences and similarities through HPLC peptide mapping[15] and amino acid sequence analysis.[16–18]

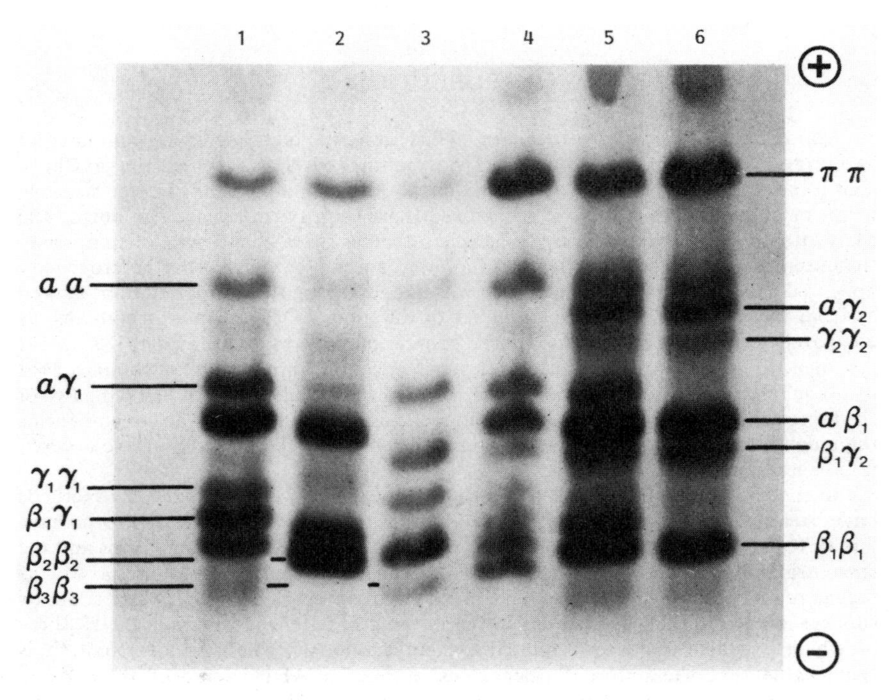

FIGURE 1. Starch gel electrophoresis of ADH in human liver homogenate-supernatants. Electrophoresis of 6 samples with different ADH phenotypes was performed according to Bosron *et al.*[9] The ADH phenotype of the liver in *lane 1* is ADH_2 1-1, ADH_3 1-1; *lane 2* is ADH_2 2-2, ADH_3 1-1; *lane 3* is ADH_2 3-3, ADH_3 1-1; *lane 4* is ADH_2 3-1, ADH_3 1-1; *lane 5* is ADH_2 1-1, ADH_3 2-1; *lane 6* is ADH_2 1-1, ADH_3 2-2.

4. Comparison of the catalytic properties of the different forms, *e.g.*, substrate specificity, sensitivity to inhibition by the pyrazole class of compounds, and pH-optima for ethanol oxidation.[9,11,19]

The findings have been consistent with the notion that there are five structural genes for ADH in humans, designated ADH_1 through ADH_5,[2,3,20] which encode five different polypeptide subunits—α, β, γ, π, and χ, respectively (TABLE 1). Three of these subunits—α, β, and γ—can combine randomly to form active homodimeric, *e.g.*, $\beta\beta$ or

TABLE 1. Genetic Model of Human ADH

Class	Gene	Allele	Subunit
I	ADH_1	ADH_1	α
I	ADH_2	ADH_2^1	β_1
I		ADH_2^2	β_2
I		ADH_2^3	β_3
I	ADH_3	ADH_3^1	γ_1
I		ADH_3^2	γ_2
II	ADH_4	ADH_4	π
III	ADH_5	ADH_5	χ

$\gamma\gamma$, and heterodimeric, *e.g.*, $\alpha\beta$ or $\beta\gamma$, isoenzymes.[9] The heterodimers have mobilities that are intermediate between the respective homodimers. The π and χ subunits have not been found to form heterodimers with each other or with the α, β, and γ subunits. Based upon the similarities and differences in each of the isoenzymes thus studied, those composed of the α, β, and γ subunits have been designated class I isoenzymes; π-ADH or $\pi\pi$, class II ADH; and χ-ADH, class III ADH.[20]

The class I and class II ADHs primarily are liver-specific isoenzymes. The class I forms appear at different stages of development. Only the $\alpha\alpha$ form is seen in fetal liver before the first trimester. Isoenzymes containing β subunits appear later in fetal life, followed by those containing γ subunits after birth.[3] Class I isoenzymes are found in small amounts in other tissues; however, not all the forms are expressed. For example, only isoenzymes with γ subunits have been detected in the upper gastrointestinal tract and the $\beta\beta$ form predominates in fetal lung, hair root, and skin.[4] π-ADH appears to be present also in other tissues such as the gastrointestinal tract, kidney, and lung, but its tissue distribution has not been systematically studied. Class III χ-ADH has been detected in all tissues examined, including white blood cells and brain.[21,22]

A number of large studies of the electrophoretic patterns of ADH isoenzymes in adult liver autopsy specimens have been conducted in different racial populations.[2-4,6,10,13,23-26] The differences in isoenzyme patterns within and among the sample populations are best explained by genetic polymorphism at the ADH_2 and ADH_3 gene loci (FIG. 1). Polymorphism has not been observed for the ADH_1, ADH_4, or ADH_5 genes by electrophoretic methods in any racial population thus far examined.

Three different electrophoretic patterns of liver ADH are observed most frequently in European and White-American populations (FIG. 1, lanes 1, 5, and 6). They arise from genetic polymorphism at ADH_3, with two alleles—ADH_3^1 and ADH_3^2—giving rise to γ_1 and γ_2 subunits, respectively.[4] The distribution of alleles in White-Americans and White-Europeans (British and German)is 0.5–0.6 for ADH_3^1 and 0.4–0.5 for ADH_3^2 (TABLE 2). The distribution of these alleles is substantially different in Japanese, Chinese, Black-Americans, and Brazilians, where the frequency of ADH_3^1 is 0.85 or greater (TABLE 2). The assignment of γ_1 and γ_2 subunits as polymorphic variants

TABLE 2. Frequency of ADH Alleles in Racial Populations[a]

	ADH_2^1	ADH_2^2	ADH_2^3	ADH_3^1	ADH_3^2
White-American	>95%	<5%	<5%	50%	50%
White-European	85%	15%	<5%	60%	40%
Japanese	15%	85%	<5%	95%	5%
Black-American	85%	<5%	15%	85%	15%

[a]Allele frequencies are taken from REFERENCES 2–4, 6, 10, 13, 23–26.

produced by ADH_3 is based on the goodness of fit of their observed distribution in different populations to the Hardy-Weinberg binomial equation for an autosomal gene whose alleles are codominantly expressed.[2]

Examination of ADH isoenzyme patterns in different populations by isoelectric focusing[10,24] and starch gel electrophoresis[2,3,6] have shown that isoenzyme variants other than those encoded by ADH_3 are present, and these have been assigned to polymorphism at the ADH_2 gene locus (FIG. 1, lanes 2, 3, and 4). The predominant subunit produced by the ADH_2 gene, which appears in European[3,10] or White-American[2] populations, is the β_1 subunit (TABLE 2). All homo- and heterodimeric isoenzymes containing α, β_1, γ_1, and γ_2 subunits exhibit pH-optima for ethanol oxidation at 10.5. On the other hand, about 85% of Japanese and Chinese[23,26] have isoenzymes containing the variant β_2 subunit encoded by the ADH_2^2 allele. Isoenzymes with β_2 subunits have been found in only 5% or less of European and White-American populations,[2,3] although they were detected initially in as high as 15% of a Swiss population.[27] The presence of isoenzymes with β_2 subunits in liver homogenate supernatants can be identified by examination of the pH-dependence of ethanol oxidation. Livers with the homozygous ADH_2 1–1 phenotype (β_1-type isoenzymes only) exhibit an optimum for activity at pH 10.5, whereas livers with the homozygous ADH_2 2–2 phenotype (β_2-type isoenzymes only) and those with the heterozygous ADH_2 2–1 phenotype (both β_1 and β_2 types of isoenzymes) exhibit pH-optima at 8.5.[2,27] All homo- and heterodimeric isoenzymes containing β_2 exhibit a pH-optimum for ethanol oxidation at 8.5.[11]

In 25% of Black-Americans, isoenzymes with a pH-optimum for alcohol oxidation at 7.0 or dual pH-optima at 7.0 and 10.5 have been identified,[6] and they exhibited electrophoretic mobilities unlike isoenzymes with β_1 or β_2 subunits, (FIG. 1, lanes 3 and 4). This variant subunit was initially called $\beta_{Indianapolis}$, because it was first identified in Indianapolis, Indiana; however, since this is the third type of β subunit that has been characterized, it is now called β_3 (TABLE 1). The homodimeric $\beta_3\beta_3$ exhibits a single pH-optimum for ethanol oxidation at 7.0, whereas heterodimeric forms containing β_3 exhibit dual pH-optima at 7.0 and 10.5.[6] Thus a characteristic feature of the allelic variants of the ADH_2 gene locus is the distinctive difference in pH-optima for ethanol oxidation: pH 10.5 for β_1, pH 8.5 for β_2, and pH 7.0 for β_3. As is true for the two types of γ subunits, the occurrences of β_1, β_2, and β_3 in the different racial groups studied fit the Hardy-Weinberg distribution.[2]

STRUCTURAL STUDIES OF CLASS I ALCOHOL DEHYDROGENASE ISOENZYMES AND GENES

Further evidence substantiating the genetic model for human ADH shown in TABLE 1 has been obtained by protein sequence analysis of different ADH subunits and by sequencing the cDNAs of the ADH subunits. Jornvall and colleagues recently sequenced the entire β_1 and γ_1 subunits and reported partial sequences for the α and β_2 subunits.[16–18,28] The subunits are 374 amino acids in length. A 94% homology was found between β_1 and γ_1, but the two fragments of the α subunit that were sequenced differed from β_1 in 7 out of 64 positions and from γ_1 in 11 out of 64. The β_1 and β_2 subunits differ only in one residue, which is the substitution of Arg-47 in the NAD(H) pyrophosphate binding site of β_1 by His-47 in β_2.[28] Thus the functional differences between the allelic β_1 and β_2 forms derive solely from this single amino acid exchange. Interestingly, the α, β, and γ_1 subunits also show structural dissimilarities in this active site region. Amino acid replacements affect both position 47 that is thought to be involved in coenzyme binding (Gly in α, Arg in β_1 and γ_1, and His in β_2) and position 48 that is involved in determining substrate specificity (Thr in α, β_1, and β_2, and Ser in γ_1).

The cDNAs for the α, β_1, and γ_1 subunits were recently isolated and sequenced.[29,30] The data show 93–96% identity in the deduced amino acid sequences of the three subunits. The cDNA for the γ subunit was presumed to be γ_2, because the only difference in the deduced amino acid sequence for this subunit from the determined amino acid sequence of γ_1 was at position 276,[17] where there is a Val (γ_1) to Met (γ_2) exchange. Both α and β_1 have Met at this position. As judged from a consensus sequence for the human ADH subunits, which might represent an archetypal human ADH subunit, the number of amino acid changes in the subunits would be: β_1, 10; β_2, 11; γ_2, 12; γ_1, 13; and α, 15.[30]

A recent publication has reported the isolation of three full-length human ADH genes encoding the α, β, and γ subunits.[31] Hybridization studies indicated that the α, β, and γ ADH genes form a closely related gene family, but that the genes encoding the π and χ subunits share a more distant evolutionary relation. Nucleotide sequence analysis of the β ADH gene showed that the coding region is interrupted by 8 introns and spans approximately 15 kilobases.

From studies of ADH in somatic cell hybrids, Smith *et al.*[32] have deduced that *ADH*$_1$, *ADH*$_2$, and *ADH*$_3$ genes that encode α, β, and γ subunits, respectively, are clustered together on the long arm of chromosome 4. The gene for χ-ADH is apparently also situated on this chromosome.

TABLE 3. Kinetic Constants for Human Liver ADH at pH 7.5[a]

Constant	Value for Isoenzyme						
	$\alpha\alpha$	$\beta_1\beta_1$	$\beta_2\beta_2$	$\beta_3\beta_3$	$\gamma_1\gamma_1$	$\gamma_2\gamma_2$	$\pi\pi$
K_m NAD$^+$, μM	13	7.4	180	530	7.9	8.7	14
K_m ethanol, mM	4.2	0.049	0.94	34	1	0.63	34
K_i 4-methylpyrazole, μM	1.1	0.13	—	2.1	0.1	—	2000
V_{max}, min^{-1}	27	9.2	400	—	87	35	20
pH-optimum	10.5	10.5	8.5	7.0	10.5	10.5	10.5

[a]Data was taken from REFERENCES 9, 11, 33, and 38.

SUBSTRATE SPECIFICITY AND THE THREE CLASSES OF HUMAN ADH ISOENZYMES

Studies of purified human liver ADH isoenzymes have shown that some of them are strikingly different in substrate specificity and other catalytic properties as well as in molecular charge. In fact, the differences in electrophoretic mobility in starch gels (FIG. 1), the K_m values for ethanol (TABLE 3), and the K_i values for pyrazole and its 4-substituted derivatives have served as the principal basis for dividing the isoenzymes into the three distinct classes.[20] Class I ADH molecular forms comprise the homo- and heterodimeric isoenzymes with α, β, and γ subunits (TABLE 1). They exhibit cathodic mobility in starch gels with electrophoresis at pH 7 to 8, and all but $\beta_3\beta_3$ have K_m values for ethanol between 0.05 and 5 mM[9,11] at pH 7.5. That for $\beta_3\beta_3$ is 34 mM.[6] The K_i values for 4-methylpyrazole of the class I isoenzymes are between 0.1 and 10 μM.

The class II π-ADH isoenzyme migrates anodically to the class I isoenzymes on starch gel electrophoresis (FIG. 1). This isoenzyme differs from the class I isoenzymes in that it does not oxidize methanol and it has a high K_i value for 4-methylpyrazole—2 mM at pH 7.5.[33] The amount of π-ADH activity in postmortem livers varies substantially depending on the state of health of the donor and the duration of storage of the tissue after death.[5,34] Recent studies of purified π-ADH indicate that only a single molecular form is found in human liver[35] and that it is homodimeric (ππ in FIG. 1).[36]

The class III isoenzyme χ-ADH migrates to the anode on starch gel electrophoresis and is seen only after staining for alcohol oxidizing activity with long chain alcohols like 1-pentanol or 1-hexanol as substrate.[7] χ-ADH is not active with methanol or cyclohexanol as substrate, and activity cannot be saturated with ethanol even at concentrations as high as 2.5 M.[7] This suggests that χ-ADH does not participate in the oxidation of ethanol in liver under physiological conditions. Only long chain alcohols like 1-pentanol or 16-OH-hexadecanoic acid or aromatic alcohols like cinnamyl alcohol are efficiently oxidized by χ-ADH.[20] χ-ADH is not inhibited by 12 mM 4-methylpyrazole.[7] Interestingly, χ-ADH has been the only isoenzyme isolated from human brain and testis. Its functional role in these tissues has not been determined.[22,37]

KINETICS OF ETHANOL OXIDATION BY HUMAN LIVER ADH ISOENZYMES

The steady-state kinetic properties of the ethanol-active class I and II isoenzymes have been examined at pH 7.5.[9,11,33,38] All of these isoenzymes obey an ordered BiBi reaction mechanism with NAD^+ combining first with ADH to form the binary ADH·NAD^+ complex and ethanol then combining with the binary complex to form the active ternary complex. Comparison of the NADH dissociation rate constant for the class I isoenzymes calculated from steady-state kinetic constants with the V_{max} values for ethanol oxidation indicates that NADH dissociation is partially rate-limiting for the reaction.[9,11]

The steady-state kinetic constant for six homodimeric class I isoenzymes and the class II ππ isoenzyme are shown in TABLE 3. All of them exhibit simple Michaelis-Menten kinetics for ethanol saturation except for the $\gamma_1\gamma_1$ and $\gamma_2\gamma_2$ alloenzymes. The two γγ alloenzymes exhibit negative cooperativity for ethanol saturation with Hill coefficients of 0.5–0.6.[9]

K_m values for ethanol vary about seven hundredfold among the seven isoenzymes. The $\beta_3\beta_3$ and ππ isoenzymes exhibit very high K_m for ethanol—34 mM. K_m values for NAD^+ are similar for the αα, $\beta_1\beta_1$, $\gamma_1\gamma_1$, $\gamma_2\gamma_2$, and ππ isoenzymes (TABLE 3). The $\beta_2\beta_2$ and $\beta_3\beta_3$ alloenzymes have K_m values for NAD^+ that are about 10 and 50 times higher, respectively, than the values for the other five isoenzymes. Since NADH dissociation is thought to be partially rate-limiting for ethanol oxidation, the high V_{max} of $\beta_2\beta_2$—400 min^{-1} (TABLE 3)—as compared with the other isoenzymes is consistent with its weaker affinity for NADH. According to the X ray crystallographic model of horse liver ADH, Arg-47 is one of the residues that participates in the binding of the coenzyme pyrophosphate group at the enzyme active site.[39] The difference between $\beta_1\beta_1$ and $\beta_2\beta_2$ in V_{max} and affinity for coenzymes can be explained by the substitution of the bulky, weak base His-47 in the active site of β_2 for the strong base Arg-47 in β_1. The amino acid substitution in $\beta_3\beta_3$, which gives rise to the even weaker affinity for coenzymes, is not known at this time.

It has been proposed that the above-mentioned differences in catalytic properties in the polymorphic ADH isoenzymes may relate to individual differences in alcohol metabolic rate.[40] Recent studies comparing alcohol metabolic rates in mono- and dizygotic twins suggest that the two to threefold difference in ethanol elimination rate among individuals is in part genetically determined.[41,42] The large differences in kinetic constants among the three types of β subunits encoded by ADH_2 (TABLE 3) and the variation in expression of these alloenzymes in different racial groups (TABLE 2) are particularly interesting in this regard. Unfortunately, it is not possible at this time to compare alcohol elimination rates with ADH_2 genotypes, because the liver is the only tissue in which all of the ADH genes are expressed and in which the isoenzymes are present in sufficient quantities for electrophoretic analysis; this tissue is not available for examination except by biopsy. However, with the recent cloning of ADH genes,[30,31] specific DNA probes should become available for the study of ADH genotypes by restriction fragment length polymorphism analysis of DNA in extrahepatic tissues and cells such as leukocytes.[32]

REFERENCES

1. BLAIR, A. H. & B. L. VALLEE. 1966. Some catalytic properties of human liver alcohol dehydrogenase. Biochemistry **5:** 2026–2034.
2. BOSRON, W. F. & T. -K. LI. 1981. Genetic determinants of alcohol and aldehyde dehydrogenases and alcohol metabolism. Seminars in Liver Dis. **1:** 179–188.
3. SMITH, M., D. A. HOPKINSON & H. HARRIS. 1971. Developmental changes and polymorphism in human alcohol dehydrogenase. Ann. Hum. Genet. Lond. **34:** 251–271.
4. SMITH, M., D. A. HOPKINSON & H. HARRIS. 1972. Alcohol dehydrogenase isozymes in adult human stomach and liver: evidence for activity of the ADH_3 locus. Ann. Hum. Genet. Lond. **35:** 243–253.
5. LI, T. -K. & L. J. MAGNES. 1975. Identification of a distinctive molecular form of alcohol dehydrogenase in human livers with high activity. Biochem. Biophys. Res. Commun. **63:** 202–208.
6. BOSRON, W. F., T. -K. LI & B. L. VALLEE. 1980. New molecular forms of human liver alcohol dehydrogenase: isolation and characterization of $ADH_{Indianapolis}$. Proc. Natl. Acad. Sci. **77:** 5784–5788.
7. PARÉS, X. & B. L. VALLEE. 1981. New human liver alcohol dehydrogenase forms with unique kinetic characteristics. Biochem. Biophys. Res. Commun. **98:** 122–130.
8. LANGE, L. G. & B. L. VALLEE. 1976. Double-ternary complex affinity chromatography: preparation of alcohol dehydrogenases. Biochemistry **15:** 4681–4686.
9. BOSRON, W. F., L. J. MAGNES & T. -K. LI. 1983. Kinetic and electrophoretic properties of native and recombined isoenzymes of human liver alcohol dehydrogenase. Biochemistry **22:** 1852–1857.
10. HARADA, S., D. P. AGARWAL & H. W. GOEDDE. 1978. Human liver alcohol dehydrogenase isoenzyme variations. Improved separation methods using prolonged high voltage starch gel electrophoresis and isoelectric focusing. Hum. Genet. **40:** 215–220.
11. YIN, S. -J., W. F. BOSRON, L. J. MAGNES & T. -K. LI. 1984. Human liver alcohol dehydrogenase: Purification and kinetic characterization of the $\beta_2\beta_2$, $\beta_2\beta_1$, $\alpha\beta_2$, and $\beta_2\tau_1$ "Oriental" isoenzymes. Biochemistry **23:** 5847–5853.
12. SMITH, M., D. A. HOPKINSON & H. HARRIS. 1973. Studies on the subunit structure and molecular size of human alcohol dehydrogenase isoenzymes determined by the different loci, ADH_1, ADH_2 and ADH_3. Ann. Hum. Genet. **37:** 49–67.
13. BOSRON, W. F., L. J. MAGNES & T. -K. LI. 1983. Human liver alcohol dehydrogenase: $ADH_{Indianapolis}$ results from genetic polymorphism at the ADH_2 gene locus. Biochem. Genet. **21:** 735–744.
14. ADINOLFI, A., M. ADINOLFI, D. A. HOPKINSON & H. HARRIS. 1978. Immunological properties of the human alcohol dehydrogenase isozymes. J. Immunogenet. **5:** 283–296.
15. STRYDOM, D. J. & B. L. VALLEE. 1982. Characterization of human alcohol dehydrogenase isoenzymes by high-performance liquid chromatographic peptide mapping. Anal. Biochem. **123:** 422–429.

16. HEMPEL, J., R. BÜHLER, R. KAISER, B. HOLMQUIST, C. DE ZALENSKI, J. P. VON WARTBURG, B. VALLEE & H. JORNVALL. 1984. Human liver alcohol dehydrogenase. 1. The primary structure of the $\beta_1\beta_1$ isoenzyme. Eur. J. Biochem. **145:** 437–445.
17. BÜHLER, R., J. HEMPEL, R. KAISER, C. DE ZALENSKI, J. -P. VON WARTBURG & H. JORNVALL. 1984. Human liver alcohol dehydrogenase. 2. The primary structure of the τ_1 protein chain. Eur. J. Biochem. **145:** 447–453.
18. HEMPEL, J., B. HOLMQUIST, L. FLEETWOOD, R. KAISER, J. BARROS-SODERLING, R. BÜHLER, B. L. VALLEE & H. JORNVALL. 1985. Structural relationships among class I isozymes of human liver alcohol dehydrogenase. Biochemistry **24:** 5303–5307.
19. WAGNER, F. W., A. R. BURGER & B. L. VALLEE. 1983. Kinetic properties of human liver alcohol dehydrogenase: oxidation of alcohols by class I isoenzymes. Biochemistry **22:** 1857–1863.
20. VALLEE, B. L. & T. J. BAZZONE. 1983. Isozymes of human liver alcohol dehydrogenase. Isozymes: Current Topics in Biological and Medical Research, Cellular Localization, Metabolism, and Physiology **8:** 219–244.
21. ADINOLFI, A., M. ADINOLFI & D. A. HOPKINSON. 1984. Immunological and biochemical characterization of the human alcohol dehydrogenase chi-ADH isozyme. Ann. Hum. Genet. **48:** 1–10.
22. BEISSWENGER, T. B., B. HOLMQUIST & B. L. VALLEE. 1985. χ-ADH is the sole alcohol dehydrogenase isozyme of mammalian brains: implications and inferences. Proc. Natl. Acad. Sci. **82:** 8368–8373.
23. TENG, Y. -S., S. JEHAN & L. E. LEI-INJO. 1979. Human alcohol dehydrogenase ADH_2 and ADH_3 polymorphisms in ethnic Chinese and Indians of West Malaysia. Hum. Genet. **53:** 87–90.
24. YIN, S. -J., W. F. BOSRON, T. -K. LI, K. OHNISHI, K. OKUDA, H. ISHII & M. TSUCHIYA. 1984. Polymorphism of human liver alcohol dehydrogenase: identification of ADH_2 2-1 and ADH_2 2-2 phenotypes in the Japanese by isoelectric focusing. Biochem. Genet. **22:** 169–180.
25. AZEVEDO, E. S., C. B. O. DA SILVA & J. TAVARES-NETO. 1975. Human alcohol dehydrogenase ADH_1, ADH_2 and ADH_3 loci in a mixed population of Bahia, Brazil. Ann. Hum. Genet. **39:** 321–327.
26. HARADA, S., S. MISAWA, D. P. AGARWAL & H. W. GOEDDE. 1980. Liver alcohol dehydrogenase and aldehyde dehydrogenase in the Japanese: isozyme variation and its possible role in alcohol intoxication. Am. J. Hum. Genet. **32:** 8–15.
27. VON WARTBURG, J. -P. & P. M. SCHURCH. 1968. Atypical human liver alcohol dehydrogenase. Ann. N. Y. Acad. Sci. **151:** 936–946.
28. JÖRNVALL, H., J. HEMPEL, B. L. VALLEE, W. F. BOSRON & T. -K. LI. 1984. Human liver alcohol dehydrogenase: amino acid substitution in the $\beta_2\beta_2$ Oriental isozyme explains functional properties, establishes an active site structure, and parallels mutational exchanges in the yeast enzyme. Proc. Natl. Acad. Sci. **81:** 3024–3028.
29. IKUTA, T., T. FUJIYOSHI, K. KURACHI & A. YOSHIDA. 1985. Molecular cloning of full-length cDNA for human alcohol dehydrogenase. Proc. Natl. Acad. Sci. **82:** 2703–2707.
30. IKUTA, T., S. SZETO & A. YOSHIDA. 1986. Three human alcohol dehydrogenase subunits: cDNA structure and molecular and evolutionary divergence. Proc. Natl. Acad. Sci. **83:** 634–638.
31. DEUSTER, G., M. SMITH, V. BLANCHIONE & G. W. HATFIELD. 1986. Molecular analysis of the human class I alcohol dehydrogenase gene family and nucleotide sequence of the gene encoding the β subunit. J. Biol. Chem. **162:** 2027–2033.
32. SMITH, M. 1986. Genetics of human alcohol and aldehyde dehydrogenase. Adv. Hum. Genet. **15:** 249–290.
33. BOSRON, W. F., T. -K. LI, W. P. DAFELDECKER & B. L. VALLEE. 1979. Human liver π-alcohol dehydrogenase: kinetic and molecular properties. Biochemistry **18:** 1101–1105.
34. LI, T. -K., W. F. BOSRON, W. P. DAFELDECKER, L. G. LANGE & B. L. VALLEE. 1977. Isolation of π-alcohol dehydrogenase of human liver: Is it a determinant of alcoholism? Proc. Natl. Acad. Sci. **74:** 4378–4381.

35. DITLOW, C. C., B. HOLMQUIST, M. M. MORELOCK & B. L. VALLEE. 1984. Physical and enzymatic properties of a class II alcohol dehydrogenase isozyme of human liver: π-ADH. Biochemistry **23:** 6363–6368.
36. KEUNG, W. M., C. C. DITLOW & B. L. VALLEE. 1985. Identification of human alcohol dehydrogenase isozymes by disc polyacrylamide gel electrophoresis in 7 M urea. Anal. Biochem. **151:** 92–96.
37. DAFELDECKER, W. P. & B. L. VALLEE. 1986. Organ-specific human alcohol dehydrogenase: isolation and characterization of isozymes from testis. Biochem. Biophys. Res. Commun. **134:** 1056–1063.
38. BURNELL, J. C., W. F. BOSRON & T. -K. LI. 1986. Kinetic characterization of human liver $\beta_3\beta_3$ ($\beta_{Indianapolis}$) alcohol dehydrogenase. Fed. Proc. **45:** 1504.
39. EKLUND, H., J. -P. SAMAMA & T. A. JONES. 1984. Crystallographic investigations of nicotinamide adenine dinucleotide binding to horse liver alcohol dehydrogenase. Biochemistry **23:** 5982–5996.
40. LI, T. -K. 1983. The absorption, distribution and metabolism of ethanol and its effects on nutrition and hepatic function. *In* Medical and Social Aspects of Alcohol Abuse. B. Tabakoff, P. B. Sutker, and C. L. Randall, Eds. 47–77. Plenum. New York, NY.
41. KOPUN, M. & P. PROPPING. 1977. The kinetics of ethanol absorption and elimination in twins and supplementary repetitive experiments in singleton subjects. Eur. J. Clin. Pharmacol. **11:** 337–344.
42. MARTIN, N. G., J. PERL, J. G. OAKESHOTT, J. B. GIBSON, G. A. STARMER & A. V. WILKS. 1985. A twin study of ethanol metabolism. Behav. Genet. **15:** 93–109.

DISCUSSION OF THE PAPER

L. T. CLARK (*Brooklyn Veterans Administration Medical Center, Jamaica Estates, NY*): Do the different forms of ADH that you have described correlate with any specific disease processes? In particular, are there any differences in ethnic manifestations of disease that might be explained by these differences?

LI: There are no known associations of various isozymes of ADH with different diseases. If one compares livers from normal populations, say from accidental deaths, with those from the hospital population, one will find that certain forms are missing in the hospitalized patients. The main difference is a falling out of the π- and αα-ADH. This finding explains why π-ADH was missed initially. The large population that Moyra Smith examined was composed of hospitalized patients, more specifically cancer patients. The absence of these isoenzymes may simply reflect malnutrition.

Aldehyde dehydrogenase seems to be necessary for the alcohol flush reaction. It is not known how the ADH polymorphism alters that reaction. I would imagine that there might be some relationship, because the β_2 forms are so much more active than the β_1.

R. PIETRUSZKO (*Rutgers University Center for Alcohol Studies, Piscataway, NJ*): Is anything known about the function of extreme enzyme multiplicity?

LI: There is no information other than the fact that class I forms appear to be a polygene family, and through evolution there may have been gene duplication.

PIETRUSZKO: Why do you find enzymes in so many molecular forms? Is it an adaptation to evolutionary changes?

LI: I believe they are due to evolutionary changes. Many high K_m forms appear, and the γ forms are thought to be closer to the lower species than the α and β forms.

M. J. COON (*University of Michigan, Ann Arbor, MI*): I assume that your pH-optima curves represent enzyme conformational changes. Have you ruled out the possibility that binding constants for dimer formation are influenced by pH?

LI: No, we haven't. We have unpublished data that show that we can drastically alter the pH-optima, say of the β form, simply with chloride.

QUESTION: You mentioned that there is a $\gamma_1\gamma_1$ form in about 80% of Orientals. Do you have data on Native American populations, such as Navajos or Summis? Do they resemble the Orientals or the Anglos?

LI: The American Indians we have examined are from the Southwest. Both alcohol-dehydrogenase and aldehyde-dehydrogenase isoenzyme patterns are similar to European populations.

QUESTION: Would you speculate on the relation of the polymorphic forms of ADH to endogenous substrates?

LI: I am unaware of any reliable data on that subject.

C. S. LIEBER (*Veterans Administration Medical Center, Bronx, NY*): Would you discuss other substrates and substrate specificity?

LI: Substrate specificity is very broad. Long chain alcohols are active toward sterols, a phenomenon that Dr. Pietruszko has investigated. ADH can be thought of as a detoxifying enzyme. As a general group, enzymes of detoxification typically have a broad substrate specificity.

The Microsomal Ethanol Oxidizing System and Its Interaction with Other Drugs, Carcinogens, and Vitamins

C. S. LIEBER, J. M. LASKER, J. ALDERMAN,
AND M. A. LEO

Alcohol Research and Treatment Center
Bronx Veterans Administration Medical Center
Bronx, New York 10468
and
Section of Liver Disease and Nutrition
Mount Sinai School of Medicine
City University of New York
New York, New York 10029

INTRODUCTION

The hepatocyte contains three main pathways for ethanol metabolism, each located in a different subcellular compartment: the main alcohol dehydrogenase (ADH) pathway of the cytosol or the soluble fraction of the cell, the microsomal ethanol oxidizing system (MEOS) located in the endoplasmic reticulum, and catalase located in the peroxisomes. The latter pathway appears to be of limited quantitative importance, but activation of the second pathway, namely MEOS, may be associated with significant metabolic and hepatic disorders.

METABOLISM OF ETHANOL VIA THE MICROSOMAL PATHWAY AND ASSOCIATED INTERACTIONS WITH XENOBIOTICS

Pharmacologic Interactions

The interactions between ethanol and other drugs are extremely complex. Indeed, alcoholic patients have an increased tolerance to a variety of drugs when sober but, paradoxically, display an increased susceptibility to these same agents when intoxicated. Such changes in drug response are a result of alterations in rates of drug metabolism as well as of adapative and synergistic effects of the drug at the primary site of action.

The interplay between the pharmacologic effects of ethanol and that of other drugs at the primary site of action, the central nervous system, is significant; they have been discussed elsewhere[1] and will not be reviewed here in detail so as not to exceed the scope of this article. Alcohol consumption also can alter drug absorption from the alimentary tract in a number of ways, most notably by interfering with gastric emptying. The effective therapeutic level of certain drugs depends on the degree of plasma protein binding. Hypoproteinemia may result in reduced plasma protein binding. Therefore, patients with alcoholic liver disease may be prone to develop toxicity from some drugs although their plasma concentrations are within the usual "therapeutic" range. A major site of ethanol-drug interaction is the liver, either through interferences with hepatic blood flow, uptake, or metabolism. Hepatic metabolism can be affected in different ways, as outlined below.

The Effect of Ethanol on Cytochrome P-450-Dependent Drug Metabolism in Microsomes

A major advance towards our understanding of the ways both acute and chronic ethanol consumption modify drug disposition has resulted from the discovery of an additional pathway for hepatic ethanol metabolism. This auxiliary enzymatic pathway is commonly known as the microsomal ethanol oxidizing system (MEOS).

Microsomal Ethanol Oxidizing System

Until recently, it was commonly believed that the primary pathway for hepatic ethanol metabolism involved cytosolic alcohol dehydrogenase (ADH), with minor contributions from catalase in the peroxisomes. Non-ADH ethanol oxidation occurring in other subcellular fractions isolated from liver was usually attributed to a hydrogen peroxide-dependent reaction mediated by the supposed catalase contamination of these fractions. Indeed, this oxidative reaction not mediated by ADH showed substrate specificity for methanol rather than ethanol and other higher aliphatic alcohols (*e.g.*, butanol), and was extremely sensitive to inhibitors of catalase activity.[2] In the 1960s, however, it was determined that liver endoplasmic reticulum (*i.e.*, microsomes) contained enzymes, collectively known as cytochromes P-450, which oxidized drugs and other xenobiotics. Furthermore, it was also found that the liver was capable of an adaptive response: upon prior drug administration, more of these hemeproteins were produced. The morphological counterpart to this was proliferation of endoplasmic reticulum membranes.

The observation in rats[3] as well as in man[4] that chronic ethanol consumption was associated with such proliferation suggested that liver microsomes might also be a site of ethanol metabolism, as was subsequently demonstrated *in vitro*.[5–7] Because of the cofactor requirements, inducibility after chronic ethanol intake, and response to inhibitors, we concluded that cytochromes P-450 were an integral component of MEOS. This thesis stirred up a decade of extensive experimentation with lively and, at times, acrimonious debate.[3] The issue was finally settled after: a) resolution of a cytochrome P-450-containing fraction from liver microsomes, which, although devoid of any ADH and catalase activity, could still oxidize ethanol as well as higher aliphatic alcohols; the latter alcohols are not substrates for catalase[8] (FIG. 1); b) reconstitution of ethanol-oxidizing activity using NADPH:cytochrome P-450 reductase, phospholipid, and either partially-purified or highly-purified microsomal cytochrome P-450 isolated from untreated or ethanol-treated[9] and phenobarbital-treated rats,[10] respectively; c) isolation of an ethanol-inducible cytochrome P-450 isozyme that oxidized ethanol at rates much higher than phenobarbital-inducible or polycyclic aromatic hydrocarbon-inducible isozymes of liver microsomal cytochrome P-450[9,11] (FIG. 2).

More recent studies in our laboratory have focused on characterization of an ethanol-inducible form of cytochrome P-450, termed $P\text{-}450_{alc}$, that has been isolated from hamster liver microsomes. The hamster was chosen as a source of the hemeprotein because this animal species is ethanol-preferring,[12] enabling ethanol to be chronically administered in drinking water, and because liver microsomes from ethanol-treated hamsters oxidize ethanol and aniline at very high rates,[13] especially when compared to liver microsomes from ethanol-treated rabbits. In addition, neither catalase nor oxygen radicals appear to play an important role in ethanol oxidation by liver microsomes from ethanol-treated hamsters, in contrast to what has been reported with rats.[14] Aniline is mentioned here since this compound, among several others (see below), is a good substrate for ethanol-inducible cytochrome P-450. When reconstituted with highly purified NADPH:cytochrome P-450 reductase and dilauroyl phos-

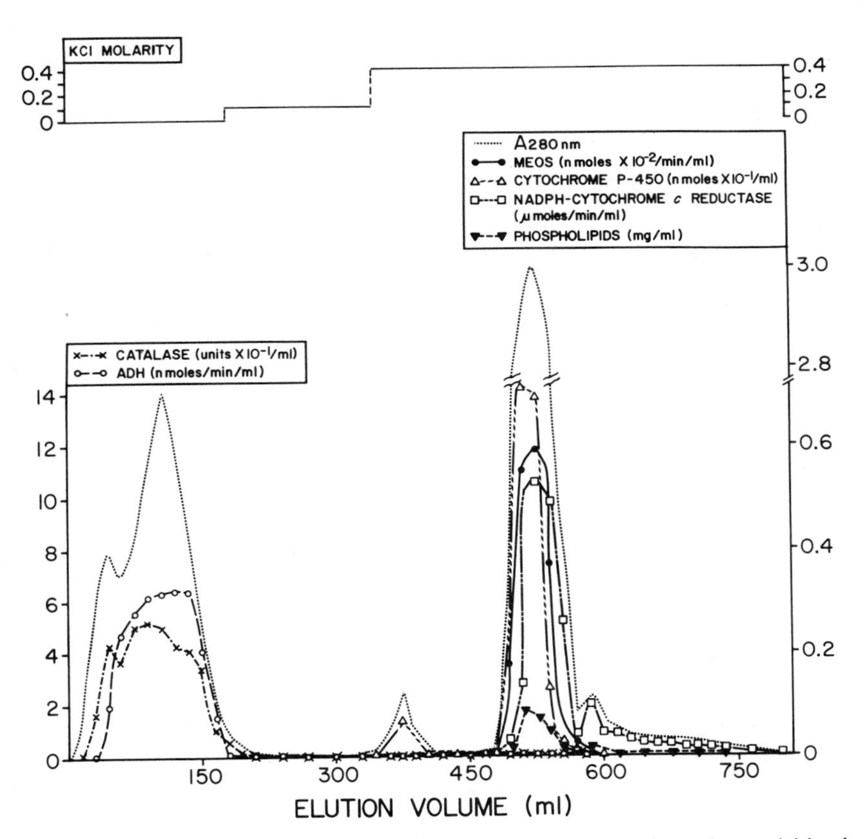

FIGURE 1. Separation of MEOS from alcohol dehydrogenase- and catalase-activities by ion-exchange column chromatography on DEAE-cellulose. Sonicated microsomes from rats fed laboratory chow were further solubilized by treatment with sodium deoxycholate and put onto a DEAE-cellulose column (2.5 × 45 cm). The separation of the enzyme activities was achieved by a stepwise increase of the salt gradient.[79]

phatidylcholine, cytochrome P-450_{alc} catalyzes ethanol oxidation and aniline *p*-hydroxylation at rates much higher than cytochromes P-450_{PB-1}, P-450_{PB-2}, or P-450_{PB-3}; the latter P-450 isozymes have been purified from phenobarbital-treated hamsters.[15,16]

Using a polyclonal antibody prepared in rabbits against ethanol-inducible hamster P-450_{alc}, we have examined whether immunochemically-related forms of cytochrome

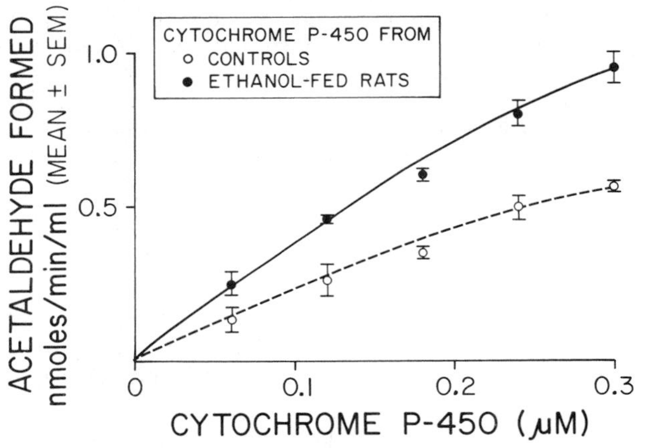

FIGURE 2. Ethanol oxidation by the microsomal ethanol-oxidizing system reconstituted with cytochrome P-450 from ethanol-fed or control rats. The acetaldehyde produced by reductase plus lecithin (0.36 ± 0.01 nmol/ml) was substracted from each value. The values represent means (± SE) of six experiments.[9]

P-450 exist in other species. Liver microsomes from naive rats, rabbits, ADH^+ and ADH^- deermice, and baboons were all found to contain a single protein that cross-reacted with anti-P-450_{alc}, as determined by Western blotting. The molecular weights of these P-450_{alc}-related proteins were, depending on the source of microsomes, in the 50,000–55,000 range. Moreover, the amounts of these P-450_{alc}-related proteins were significantly higher in liver microsomes from ethanol-treated animals.[17] Based upon their molecular weights and inducibility by ethanol, we are putatively identifying these cross-reacting proteins as forms of cytochrome P-450. Most importantly, we have determined that a protein (mol. wt. = 54,000) immunochemically homologous to hamster P-450_{alc} also exists in liver microsomes from nonalcoholic and alcoholic humans, and preliminary observations indicate that levels of this protein are higher in liver microsomes from alcoholics.

As described above, aniline, like ethanol, is an excellent substrate for hamster ethanol-inducible cytochrome P-450_{alc}. The administration of xenobiotics other than ethanol that increase MEOS activity, such as isoniazid and pyrazole, invariably increases microsomal aniline *p*-hydroxylase activity as well.[13] Anti-P-450_{alc}, when added to liver microsomes from ethanol-treated hamsters, inhibited aniline hydroxylation by 70%. By contrast, only 30% of the total aniline hydroxylase activity of control hamster liver microsomes was inhibited by this antibody. Since rates of aniline *p*-hydroxylation are threefold higher in liver microsomes from ethanol-treated hamsters (as is MEOS activity), cytochrome P-450_{alc} may represent a significant percentage of the total liver microsomal P-450 in these animals. The exact percentage is being determined using immunochemical techniques, which should also provide confirmatory evidence for our initial observation that ethanol-inducible P-450_{alc} may have the shortest *in-vivo* half-life of any liver microsomal P-450 isozyme yet described.[14]

Several lines of evidence support the role of cytochrome P-450 in hepatic ethanol metabolism *in vivo*. These include the incomplete inhibition of ethanol metabolism using ADH inhibitors,[6,18] the pattern of labeling of metabolites derived from stereospecifically-labeled ethanol,[19] the increased rate of ethanol metabolism at high ethanol concentrations (well above those needed to saturate alcohol dehydrogenase),[18,20] and the isotopic effects observed with deuterated ethanol.[21] Furthermore, the persistence of a substantial rate of ethanol metabolism in deermice genetically devoid of liver ADH, in association with high MEOS activity,[22,23] illustrated most elegantly the *in-vivo* importance of cytochrome P-450-dependent ethanol metabolism. Finally, the contribution of cytochrome P-450 to ethanol metabolism has now also been determined in the presence of ADH. To quantitate contributions of the various metabolic pathways, ethanol oxidation rates were initially examined in isolated deermouse hepatocytes with or without added 4-methylpyrazole;[21,24] 4-methylpyrazole significantly reduced rates of ethanol oxidation in both ADH^+ and ADH^- hepatocytes. This decrease seen in ADH^- cells was used to correct for the inhibitory effect of 4-methylpyrazole on cytochrome P-450-catalyzed ethanol oxidation in ADH^+ deermouse hepatocytes. After such correction, cytochrome P-450-dependent oxidation was found to catalyze 42% of the total ethanol metabolism at 10 mM substrate and 62% at 50 mM substrate. Similar results (FIG. 3) were derived *in vivo* with the ADH inhibitor 4-methylpyrazole and with the catalase inhibitor aminotriazole, the latter used at a dose that inhibited H_2O_2-dependent ethanol peroxidation.[24] By a different approach (involving measurement of isotope effects), cytochrome P-450 was calculated to account for 35% of the total substrate oxidation at low ethanol concentrations and about 70% at high ethanol concentrations (FIG. 4).[21] With a third approach, namely by determining the fate of ^3H from [1-^{3}H] ethanol, the contribution of non-ADH pathways to ethanol oxidation was found in rats to approach 50%.[25] Thus, very different experimental approaches yielded similar results, namely, that cytochrome P-450 plays a significant role in hepatic ethanol oxidation even in the presence of ADH.

Elucidation of this new cytochrome P-450-dependent pathway of ethanol oxidation has improved not only our understanding of the metabolic adaptation to ethanol but also of a number of interactions of ethanol with xenobiotics. This new knowledge offers

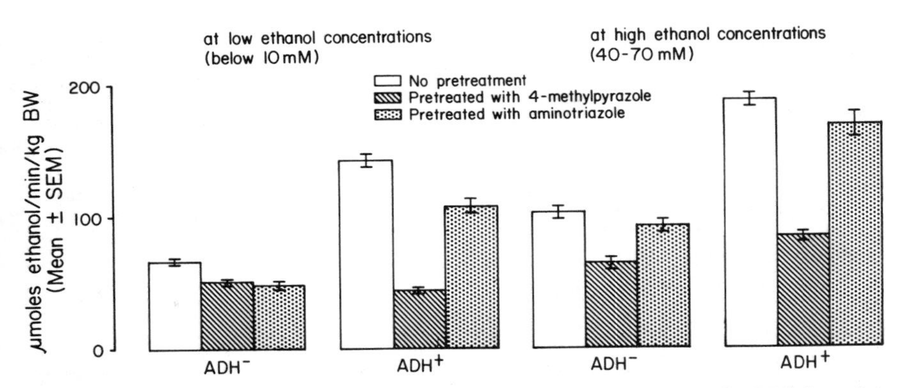

FIGURE 3. *In-vivo* ethanol clearance with or without inhibitors in ADH^- and ADH^+ deermice: animals were given 0.5 mmol 4-MP/kg BW 15 min prior to ethanol, or 1 g AT/kg BW 3 hr prior to ethanol. 4-MP inhibited ($p < 0.01$) ethanol clearance *in vivo* in ADH^- deermice at all concentrations of ethanol tested, whereas after AT, there was a small inhibition at low ethanol concentrations only. Corrected for the nonspecific effect of 4-MP *in vivo*, non-ADH pathways in ADH^+ deermice were calculated to contribute about 40% of total ethanol metabolism below 10 mM and 60% at ethanol concentrations of 40 to 70 mM.[24]

us better insight into acute and chronic alcohol-drug interactions, the altered drug metabolism in the alcoholic, and its possible pathologic consequences.

Acute Interactions

The main effect of the acute presence of ethanol is inhibition of drug metabolism. The primary mechanism involved appears to be competition for a common microsomal oxidation pathway, as reviewed by Lieber.[26] It was reported more recently that inhibition of drug oxidation by ethanol may be due to the displacement of substrates

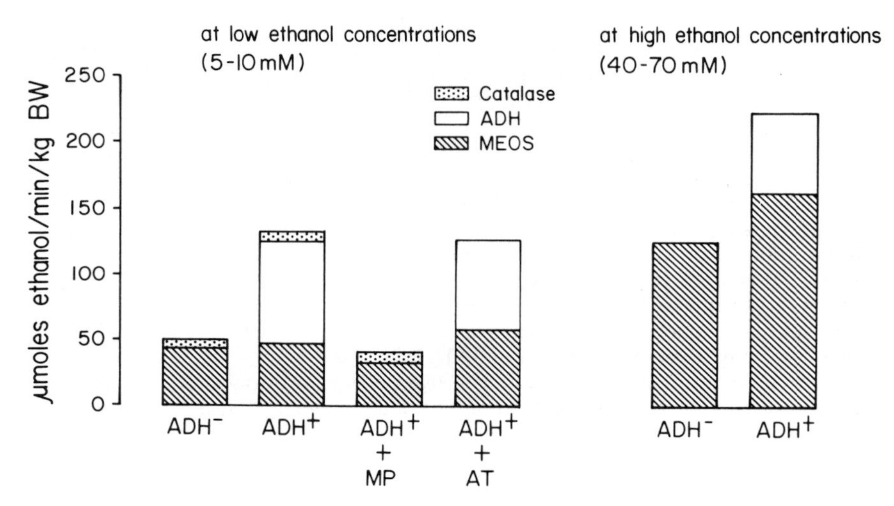

FIGURE 4. *In-vivo* contributions of ADH, catalase, and MEOS in ADH^- and ADH^+ deermice calculated from the observed isotope effect *in vivo*. The ethanol administered to these deermice contained 1R (1, 2-^{14}C, 1-^{2}H) and 2-^{3}H ethanol diluted with unlabeled ethanol. 0.3 g/kg BW of ethanol was injected IP in the low dose experiments (5–10 mM in blood) and 2.5 g/kg BW at the high dose (40–70 mM). Fifteen min after ethanol injection, animals were sacrificed and acetate was isolated from deproteinized blood by diffusion. The isolated acetate solution was evaporated at alkaline pH and the residue was washed with unlabeled ethanol and counted. 0.5 mmol/kg BW 4-methylpyrazole (MP) or 1 g/kg BW 3-amino-1,2,4-triazole (AT) was injected prior to ethanol administration to inhibit one pathway. The MEOS accounted for almost all of ethanol oxidation in ADH^- deermice. The MEOS contribution in ADH^+ deermice was calculated to be about 35% at low and about 70% at high ethanol concentration.[21]

from cytochrome P-450 and/or to the inhibition of reduction of cytochrome P-450 by NADPH:cytochrome P-450 reductase.[27] Ethanol may also interfere with oxidative drug metabolism in microsomes by affecting the supply of NADPH via an indirect mechanism that involves ADH.[28] However, a recent study revealed that, even in the absence of ADH, ethanol inhibited the demethylation of aminopyrine *in vivo* and *in vitro*. This suggests that a non-ADH-related mechanism is responsible for the inhibition of metabolism of this drug by ethanol.[29] Phase II microsomal drug metabolism such as glucuronidation has also been found to be inhibited by ethanol, in some cases by 50% or more.[30,31] By contrast, other drug-metabolizing reactions occurring in microsomes, including acetylation[32] and sulfation,[31] are unaltered by ethanol.

The acute interaction of ethanol with drug metabolism may have some important practical consequences. As reviewed elsewhere,[26] it has been shown that acute oral administration of ethanol results in prolongation of the clearance of meprobamate from the blood. Alcohol acts synergistically with meprobamate to depress performance tasks and driving-related skills such as time estimation, attention, reaction time, body steadiness, oculomotor control, and alertness. Benzodiazepines are other frequently used minor tranquilizers. Although earlier studies showed no synergism between alcohol and diazepam, the majority of recent investigations give clear indication that the combination is dangerous.

Ethanol also interacts with narcotics. Several epidemiologic studies suggest that combined use of morphine and alcohol potentiates the effects of both drugs and increases the probability of death. Experimentally, acute administration of ethanol results in increased brain and liver concentrations of methadone.[33]

Ethanol also interacts with industrial solvents. Occupational exposure to xylene is widespread in the manufacture and application of xylene-containing chemicals, notably paints, glues, printing inks, pesticides, etc. Alcoholic beverages are extensively consumed by some populations, occasionally in the course of the workday during lunch time and particularly after work hours. Since the excretion of xylene is delayed by its high solubility and storage in lipid-rich tissues, the simultaneous presence of xylene and ethanol in the body is probably not uncommon. Indeed, after ethanol intake, blood xylene levels were found to rise about 1.5–2.0-fold, while urinary excretion of methylhippuric acid (a xylene metabolite) declined by about 50%, suggesting that ethanol decreases the metabolic clearance of xylene by about one-half during xylene inhalation.[34] Such alterations in xylene pharmacokinetics are likely to be caused primarily by ethanol-mediated inhibition of hepatic xylene oxidation. Acute interactions of this type have also been described for other industrial solvents. Besides enhancing the toxicity of the parent compound, the presence of ethanol can also decrease the toxicity of certain solvents by inhibition of the hepatic metabolism of a relatively innocuous parent compound to a toxic product. Such is the case for carbon tetrachloride.[35]

Chronic Interactions

Tolerance to Ethanol and Other Drugs. It is well known that chronic alcohol abusers develop tolerance to ethanol. Such tolerance is due, in part, to central nervous system adaptation (pharmacodynamic tolerance). In addition, there is also metabolic tolerance that results from an increased capacity to oxidize ethanol. This occurs because of the adaptive increase in activity of non-ADH pathways involved in ethanol metabolism, most likely cytochromes P-450 (see above).

In addition to tolerance to ethanol, the alcoholic also displays tolerance to various other drugs, which again has been usually attributed to central nervous system adaptation. However, metabolic adaptation must also be considered. Indeed, the increased levels of cytochrome P-450 observed after chronic alcohol consumption not only result in enhanced ethanol oxidation but also in the increased microsomal oxidation of other drugs as well. As a consequence, administration of ethanol to man (under metabolic ward conditions) resulted in a striking increase in the rate of blood clearance of meprobamate and pentobarbital.[36] Similarly, increases in the metabolism of aminopyrine, tolbutamide, propranolol, and rifampicin have been described. Experimentally, this effect of chronic ethanol consumption can be modulated, in part, by the dietary content in carbohydrates,[37] lipid,[38] and proteins.[39] It is likely that genetic factors also play a role.

Thus, a recovered alcoholic's previous history of alcohol abuse can be a key factor in prescribing medication, because even abstaining alcoholics need doses different from those required by nondrinkers to achieve therapeutic levels of certain drugs, such as warfarin, phenytoin, tolbutamide, and isoniazid. The half-life of each of these drugs can be 50% shorter in abstaining alcoholics than in nondrinkers. This metabolic tolerance may persist several days to weeks after cessation of alcohol abuse, and the duration of recovery will probably vary depending on the particular drug, as has been shown experimentally.[40]

Enhanced Susceptibility to Hepatotoxic Agents. The general increase in activities of microsomal cytochrome P-450-dependent monooxygenases also applies to those that convert certain exogenous compounds to toxic products. For instance, carbon tetrachloride (CCl_4) exerts its toxicity only after conversion by cytochrome P-450 to an active metabolite; alcohol pretreatment markedly enhances CCl_4 hepatotoxicity.[41] Thus, clinical observations of the enhanced susceptibility of alcoholics to the hepatotoxic effect of CCl_4 may be due at least in part to increased metabolic activation of this compound. Bromobenzene liver toxicity was also found to be increased following chronic alcohol consumption.[42] It is likely that a number of other toxic agents will be found to display a selective injurious action in the alcoholic. This pertains not only to industrial solvents but also to a variety of prescribed drugs. For instance, the increased hepatotoxicity of isoniazid observed in alcoholics may well be due to increased microsomal production of a reactive metabolite of this drug.[43] Similarly, chronic ethanol administration enhances phenylbutazone hepatotoxicity, possibly also because of increased biotransformation.[44]

The same mechanism of hepatotoxicity also pertains to some over-the-counter medications. Acetaminophen (paracetamol, N-acetyl-p-aminophenol), widely used as an analgesic and an antipyretic, is considered safe when taken in recommended doses. However, large doses of acetaminophen, mostly taken in suicide attempts, have been shown to produce fulminant hepatic failure. It is now clear that in addition to glucuronidation and sulfation, acetaminophen is also metabolized in the liver by microsomal cytochrome P-450-dependent monooxygenases; biotransformation by the latter pathway yields a reactive intermediate highly toxic to the liver. The practical implications are obvious, especially in view of the widespread use of drugs known to induce microsomal drug metabolism. Since alcohol can also induce microsomal drug-metabolizing activities, it was expected that a history of ethanol consumption might favor the hepatotoxicity of acetaminophen. This has been suggested in various case reports.[45] Experimentally, after chronic ethanol feeding, enhanced covalent binding of a reactive metabolite(s) of acetaminophen to liver microsomes from ethanol-fed rats was found to be associated with enhanced hepatotoxicity.[46] Furthermore, in a reconstituted system, an ethanol-inducible form of cytochrome P-450 was found to have a high capacity to oxidize acetaminophen to a reactive intermediate that readily formed a conjugate with reduced glutathione.[47] For these reasons, it is likely that the enhanced hepatotoxicity of acetaminophen observed after chronic ethanol consumption is due, at least in part, to increased microsomal production of N-acetyl-*p*-benzoquinoneimine, the purported reactive metabolite of acetaminophen.[48] In contrast to pretreatment with alcohol, which accentuates toxicity, the presence of ethanol may prevent acetaminophen-induced hepatotoxicity, most likely because of inhibition of acetaminophen oxidation to a reactive metabolite.[49–51] This type of interaction is similar to that described for other drugs (see above). To what extent acute ethanol intake competitively inhibits the microsomal metabolism of acetaminophen remains to be evaluated clinically.

Enhanced Carcinogenicity. Alcohol abuse is associated with an increased incidence of upper alimentary and respiratory tract cancers. Many factors have been incriminated in the co-carcinogenic effect of ethanol.[52] One of these factors is believed to be the effect of ethanol on enzyme systems involved in carcinogen activation. Ethanol's enhancement of cytochrome P-450-dependent carcinogen activation has been demonstrated using microsomes derived from a variety of tissues including the liver (the principal site of xenobiotic metabolism), the lungs and intestines (the major portals of entry for tobacco smoke and dietary carcinogens, respectively), and the esophagus (where ethanol consumption is a major risk factor in cancer development). The general approach used in these studies has been to isolate microsomes from tissues of ethanol-exposed rats, hamsters, or mice, and then assay these preparations for their ability either to metabolically activate procarcinogens to mutagens detectable in the Ames/Salmonella mutagenesis assay or to produce defined metabolites.[53–55] Ethanol also has a unique effect concerning bioactivation of the chemical carcinogen dimethylnitrosamine (DMN): it induces a liver microsomal DMN N-demethylase which has high activity at low DMN concentrations, which may be physiologically relevant.[56] This contrasts with other inducers of cytochromes P-450 such as phenobarbital, 3-methylcholanthrene, or polychlorinated biphenyls, which increase the activity of DMN N-demethylases (whose activity is detectable only at relatively high DMN concentrations) while at the same time repressing the activity of low K_m DMN N-demethylases.[57–59] Some of the effects described above may be due to the induction of a unique species of cytochrome P-450 by ethanol (see above)[9,11] that differentially affects the activation of various carcinogens; the ethanol-inducible form of rabbit liver cytochrome P-450 oxidizes DMN at rates higher than any other cytochrome P-450 isozyme; furthermore, the high catalytic activity of this hemeprotein has been demonstrated at low DMN concentrations.[60]

Interactions with Steroids and Vitamins. Ethanol also affects microsomal metabolism of exogenous and endogenous steroids, as discussed in detail elsewhere;[26] included are enhanced hepatic microsomal testosterone degradation as well as decreased testicular synthesis and increased conversion to estrogens, in part because of enhanced aromatase activity.[61] Furthermore, ethanol alters the metabolism of structurally related vitamins such as vitamin D.[62,63] Since vitamin D and other micronutrients are metabolized by liver microsomes, the induction of microsomal cytochrome P-450-dependent oxidative activities by ethanol may alter vitamin requirements and could even affect the integrity of liver and other tissues. It is known that plasma vitamin A[64] as well as retinol-binding protein (RBP) levels[65] are decreased in patients with alcoholic cirrhosis. These complications have usually been attributed to malnutrition or to hepatic injury, because decreased plasma vitamin A and RBP levels have also been reported in nonalcoholic patients with liver disease.[66] Inadequate dietary intake of zinc, increased urinary zinc losses, and depressed serum zinc levels are common in alcoholics with or without liver disease.[67] Alcoholic cirrhosis is often associated with zinc deficiency, which has also been postulated to decrease plasma RBP through impaired mobilization of this protein from the liver.[68]

In addition, it was found recently that even at the early fatty liver stage, alcoholics commonly have very low hepatic vitamin A levels despite normal circulating vitamin A and the absence of obvious dietary vitamin A deficiency.[69] In experimental animals, ethanol administration was shown to depress hepatic vitamin A levels even when administered with diets containing adequate amounts of this vitamin.[70] When dietary vitamin A was virtually eliminated, the depletion rate of hepatic vitamin A stores was two to threefold faster in ethanol-fed rats than in controls, possibly because of accelerated microsomal degradation of vitamin A.[71] Furthermore, it has been shown in reconstituted systems that retinoic acid is oxidized by several purified forms of cytochrome P-450 isolated from rat liver microsomes.[72] In comparison to retinoic acid, rates of microsomal retinol metabolism were found to be much greater. Indeed, a new pathway of microsomal retinol oxidation has been discovered in the liver, inducible by ethanol and other xenobiotics, that is capable of degrading an amount of retinol comparable to the daily intake.[73]

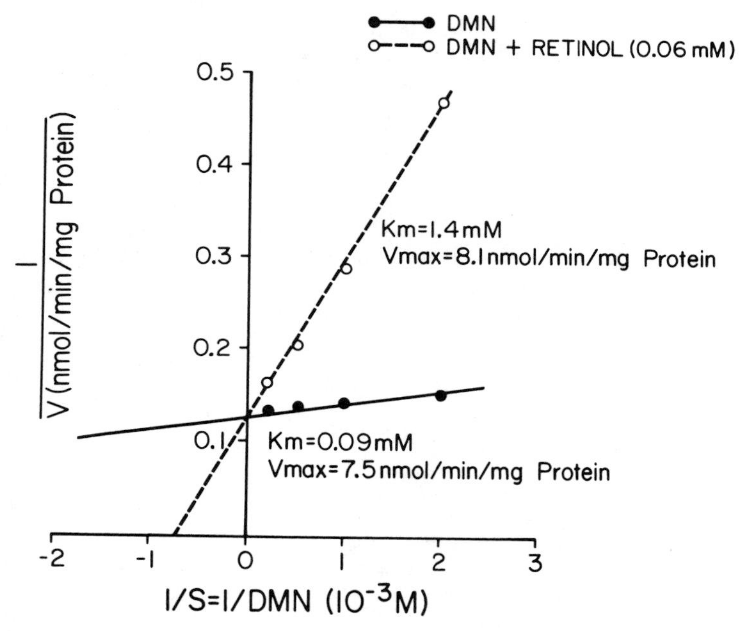

FIGURE 5. Effect of retinol on hepatic microsomal DMN demethylation. Liver microsomes of rats fed ethanol were incubated with various concentrations of DMN, with or without retinol. Competitive inhibition was observed.[76]

For these reasons, not only should vitamin A supplementation be given to correct the problems of night blindness and sexual dysfunction of the alcoholic, but also such therapy might be useful with regard to the liver pathology. Indeed, low hepatic vitamin A levels might be responsible for some hepatic functional or structural abnormalities, including giant lysosomes.[74] Vitamin A depletion can also result in mucosal alterations, thus indirectly favoring carcinogenesis.[75] Retinol may also compete with microsomal activation of carcinogens[76] (FIG. 5); therefore, retinol depletion may have adverse consequences in terms of chemical carcinogenesis. For all of these reasons, vitamin A repletion should be considered. The therapeutic usage of vitamin A, however, is complicated by the fact that excessive amounts of this vitamin are known to be hepatotoxic and that the alcoholic has an enhanced susceptibility to this effect.[77,78] In control rats, amounts of vitamin A equivalent to those commonly used for the treatment of alcoholics were found to be without significant toxic effects on the liver, while in animals chronically fed alcohol, signs of hepatotoxicity developed, such as marked morphologic and functional alterations of liver mitochondria,[77] together with necrosis and fibrosis.[78] It should be emphasized here that this increased hepatotoxicity was not necessarily associated with increased vitamin A levels in the liver. In fact, because alcohol administration tends to decrease vitamin A levels in the liver (as mentioned above), even after vitamin A supplementation, some ethanol-fed animals had vitamin A levels in the liver that were not higher than normal;[77] nevertheless, toxicity developed. One possible explanation, still to be proved, is that vitamin A toxicity may be mediated at least in part by the enhanced production of a toxic metabolite, as in the case of certain xenobiotic agents. In any event, further studies will be needed to more clearly define the narrowed "therapeutic window" for vitamin A in the alcoholic.

SUMMARY

The interaction of ethanol with the oxidative drug-metabolizing enzymes present in liver microsomes results in a number of clinically significant side effects in the alcoholic. Following chronic ethanol consumption, the activity of the microsomal ethanol oxidizing system (MEOS) increases. This enhancement of MEOS activity is primarily due to the induction of a unique microsomal cytochrome P-450 isozyme, which has a high capacity for ethanol oxidation, as shown in reconstituted systems. Normally present in liver microsomes at low levels, this form of cytochrome P-450 increases dramatically after chronic ethanol intake in many species, including baboons. The *in-vivo* role of cytochrome P-450 in hepatic ethanol oxidation, especially following chronic ethanol consumption, has been conclusively demonstrated in deer-mice lacking liver ADH. Induction of microsomal cytochrome P-450 by ethanol is associated with the enhanced oxidation of other drugs as well, resulting in metabolic tolerance to these agents. There is also increased cytochrome P-450-dependent activation of known hepatotoxins such as carbon tetrachloride and acetaminophen, which may explain the greater susceptibility of alcoholics to the toxicity of industrial solvents and commonplace analgesics. In addition, the ethanol-inducible form of cytochrome P-450 has the highest capacity of all known P-450 isozymes for the activation of dimethylnitrosamine, a potent (and ubiquitous) carcinogen. Moreover, cytochrome P-450-catalyzed oxidation of retinol is accelerated in liver microsomes, which may contribute to the hepatic vitamin A depletion seen in alcoholics. In contrast to chronic ethanol consumption, acute ethanol intake inhibits the metabolism of other drugs via competition for shared microsomal oxidation pathways. Thus, the interplay between ethanol and liver microsomes has a profound impact on the way heavy drinkers respond to drugs, solvents, vitamins, and carcinogens.

REFERENCES

1. PIROLA, R. C. 1978. Drug Metabolism and Alcohol: A Survey of Alcohol-Drug Reactions-Mechanisms, Clinical Aspects, Experimental Studies. Adis Press.
2. ZIEGLER, D. M., R. THURMAN, T. R. TEPHLY & C. S. LIEBER. 1972. *In* Microsomes and Drug Oxidation. R. W. Estabrook, J. R. Gillette & K. C. Liebman, Eds. 458–460. Williams and Wilkins. Baltimore, MD.
3. ISERI, O. A., L. S. GOTTLIEB & C. S. LIEBER. 1964. Fed. Proc. **23:** 579.

4. LANE, B. P. & C. S. LIEBER. 1966. Am. J. Pathol. **49:** 593–603.
5. LIEBER, C. S. & L. M. DECARLI. 1968. Science **162:** 690–691.
6. LIEBER, C. S. & L. M. DECARLI. 1970. J. Biol. Chem. **245:** 2505–2512.
7. ISHII, H., J.- G. JOLY & C. S. LIEBER. 1973. Biochim. Biophys. Acta **291:** 411–420.
8. TESCHKE, R., Y. HASUMURA & C. S. LIEBER. 1975. J. Biol. Chem. **250:** 7397–7404.
9. OHNISHI, K. & C. S. LIEBER. 1977. J. Biol. Chem. **252:** 7124–7131.
10. MIWA, G. T., W. LEVIN, P. E. THOMAS & A. Y. H. LU. 1978. Arch. Biochem. Biophys. **187:** 464–475.
11. KOOP, D. R., E. T. MORGAN, G. E. TARR & M. J. COON. 1982. J. Biol. Chem. **257:** 8472–8480.
12. KULKOSKY, P. J. & N. W. CORNELL. 1977. Pharmacol. Biochem. Behav. **11:** 439–444.
13. LASKER, J. M., C. M. ARDIES & C. S. LIEBER. 1985. Fed. Proc. **44:** 1449.
14. LASKER, J. M. & C. S. LIEBER. 1984. Alcoholism: Clin. Exp. Res. **8:** 102.
15. LASKER, J. M., B. P. BLOSWICK & C. S. LIEBER. 1984. Fed. Proc. **43:** 2033.
16. LASKER, J. M., C. M. ARDIES, B. P. BLOSWICK & C. S. LIEBER. 1986. Fed. Proc. **45:** 1665.
17. LASKER, J. M., C. M. ARDIES, B. P. BLOSWICK & C. S. LIEBER. 1986. Hepatology **4:** 1108.
18. LIEBER, C. S. & L. M. DECARLI. 1972. J. Pharmacol. Exp. Ther. **181:** 279–287.
19. ROGNSTAD, R. & D. G. CLARK. 1974. Eur. J. Biochem. **42:** 51–60.
20. MATSUZAKI, S., E. GORDON & C. S. LIEBER. 1981. J. Pharmacol. Exp. Ther. **217:** 133–137.
21. TAKAGI, T., J. ALDERMAN & C. S. LIEBER. 1985. Alcohol **2:** 9–12.
22. BURNETT, K. G. & M. R. FELDER. 1980. Biochem. Pharmacol. **28:** 1–8.
23. SHIGETA, Y., S. IIDA, M. A. LEO, F. NOMURA, M. R. FELDER & C. S. LIEBER. 1984. Biochem. Pharmacol. **33:** 807–814.
24. TAKAGI, T., J. ALDERMAN, J. GELLERT & C. S. LIEBER. 1986. Biochem. Pharmacol. **35:** 3601–3606.
25. VIND, C. & N. GRUNNET. 1985. Biochem. Pharmacol. **34:** 655–661.
26. LIEBER, C. S. 1982. Medical Disorders of Alcoholism: Pathogenesis and Treatment. 589. W.B. Saunders. Philadelphia, PA.
27. HAYASHI, N., T. KAMADA, N. SATO, A. KASAHARA & J. A. PETERSON. 1985. Dig. Dis. Sci. **30:** 334–339.
28. REINKE, L. A., F. C. KAUFFMAN, S. A. BELINSKY & R. G. THURMAN. 1980. J. Pharmacol. Exp. Ther. **201:** 125.
29. GELLERT, J., J. ALDERMAN & C. S. LIEBER. 1986. Biochem. Pharmacol. **35:** 1037–1041.
30. MOLDEUS, P., B. ANDERSSON & A. NORLING. 1978. Biochem. Pharmacol. **27:** 2583–2588.
31. SUNDHEIMER, D. W. & K. BRENDEL. 1984. Life Sci. **34:** 23–29.
32. HUTCHINGS, A., R. D. MONIE, B. SPRAGG & P. A. ROUTLEDGE. 1984. Br. J. Clin. Pharmacol. **18:** 98–100.
33. BOROWSKY, S. A. & C. S. LIEBER. 1978. J. Pharmacol. Exp. Ther. **207:** 123–129.
34. RIIHIMAKI, V., K. SAVOLAINEN, P. PFAFFLI, K. PEKARI, H. W. SIPPEL & A. LAINE. 1982. Arch. Toxicol. **49:** 253–263.
35. TESCHKE, R., K. -H. HAUPTMEIER & H. FRENZEL. 1983. Liver **3:** 100.
36. MISRA, P. S., A. LEFEVRE, H. ISHII, E. RUBIN & C. S. LIEBER. 1971. Am. J. Med. **51:** 346–351.
37. TESCHKE, R., F. MORENO & A. S. PETRIDES. 1981. Biochem. Pharmacol. **30:** 1745–1751.
38. JOLY, J. -G. & C. HETU. 1975. Biochem. Pharmacol. **24:** 1475–1480.
39. MITCHELL, M. C., E. MEZEY & C. MADDREY. 1981. Hepatology **1:** 336–340.
40. HETU, C. & J. -G. JOLY. 1985. Biochem. Pharmacol. **34:** 1211–1216.
41. HASUMURA, Y., R. TESCHKE & C. S. LIEBER. 1974. Gastroenterology **66:** 415–422.
42. HETU, C., A. DUMONT & J. -G. JOLY. 1983. Toxicol. Appl. Pharmacol. **67:** 166–177.
43. TIMBRELL, J. A., J. R. MITCHELL, W. R. SNODGRASS & S. D. NELSON. 1980. J. Pharmacol. Exp. Ther. **23:** 364–369.
44. BESKID, M., J. BIALEK, J. DZIENISZEWSKI, J. SADOWSKI & J. TLALKA. 1980. Exp. Path. **18:** 487–491.
45. SEEF, L. B., B. A. CUCCHERINI, M. P. H. HYMAN, J. ZIMMERMAN, E. ADLER & S. B. BENJAMIN. 1986. Ann. Int. Med. **104:** 399–404.
46. SATO, C., Y. MATSUDA & C. S. LIEBER. 1981. Gastroenterology **80:** 140–148.

47. MORGAN, E. T., D. R. KOOP & M. J. COON. 1983. Biochem. Biophys. Res. Commun. **112:** 8–13.
48. CORCORAN, G. B., J. R. MITCHELL, Y. N. VAISHNAV & E. C. HORNING. 1980. Mol. Pharmacol. **18:** 536–542.
49. SATO, C. & C. S. LIEBER. 1981. J. Pharmacol. Exp. Ther. **218:** 811–815.
50. ALTOMARE, E., M. A. LEO & C. S. LIEBER. 1984. Alcoholism: Clin. Exp. Res. **8:** 405–408.
51. ALTOMARE, E., M. A. LEO, C. SATO, G. VENDEMIALE & C. S. LIEBER. 1984. Biochem. Pharmacol. **33:** 2207–2212.
52. LIEBER, C. S., H. K. SEITZ, A. J. GARRO & T. M. WORNER. 1981. *In* Frontiers in Liver Disease. P. D. Berk & T. C. Chalmers, Eds. 320–335. Thieme-Stratton. New York, NY.
53. SEITZ, H. K., A. J. GARRO & C. S. LIEBER. 1981a. Eur. J. Clin. Invest. **11:** 33–38.
54. SEITZ, H. K., A. J. GARRO & C. S. LIEBER. 1981b. Cancer Letters **13:** 97–102.
55. FARINATI, F., Z. C. ZHOU, J. BELLAH, C. S. LIEBER & A. J. GARRO. 1985. Drug Metab. Dispos. **13:** 210–214.
56. GARRO, A. J., H. J. SEITZ & C. S. LIEBER. 1981. Cancer Res. **41:** 120–124.
57. VENKATESAN, N., J. C. ARCOS & M. F. ARGUS. 1968. Life Sci. **7:** 1111–1119.
58. CZYGAN, P., H. GREIM, A. J. GARRO, F. HUTTERER, F. SCHAFFNER, H. POPPER, O. ROSENTHAL & D. Y. COOPER. 1973. Cancer Res. **33:** 2983–2986.
59. GUTTENPLAN, J. B. & A. J. GARRO. 1977. Cancer Res. **37:** 329–330.
60. YANG, C. S., Y. Y. TU, D. R. KOOP & M. J. COON. 1985. Cancer Res. **45:** 1140–1145.
61. GORDON, G. G., A. L. SOUTHREN, J. VITTEK, C. S. LIEBER. 1979. Metabolism. **28:** 20–24.
62. GASCON-BARRE, M. & J. -G. JOLY. 1981. Life Sci. **28:** 279–286.
63. GASCON-BARRE, M. 1982. Metabolism **31:** 67–72.
64. SMITH, J. C., E. D. BROWN, S. C. WHITE & J. D. FINKELSON. 1975. Lancet **1:** 1251–1252.
65. MCCLAIN, C. J., D. H., VAN THIEL, S. PARKER, L. K. BADZIN & H. GILBERT. 1979. Alcoholism: Clin. Exp. Res. **3:** 135–141.
66. SMITH, F. R. & D. S. GOODMAN. 1971. J. Clin. Invest. **50:** 2426–2436.
67. MCCLAIN, C. J. & L. C. SU. 1983. Alcoholism: Clin. Exp. Res. **7:** 5–10.
68. SMITH, J. C., E. G. MCDANIEL, F. F. FAN & J. A. HALSTED. 1973. Science **181:** 954–955.
69. LEO, M. A. & C. S. LIEBER. 1982. N. Engl. J. Med. **37:** 597–601.
70. SATO, M. & C. S. LIEBER. 1981. J. Nutr. **111:** 2015–2023.
71. SATO, M. & C. S. LIEBER. 1982. Arch. Biochem. Biophys. **213:** 557–564.
72. LEO, M. A., S. IIDA & C. S. LIEBER. 1984. Arch. Biochem. Biophys. **234:** 305–312.
73. LEO, M. A. & C. S. LIEBER. 1985. J. Biol. Chem. **260:** 5228–5231.
74. LEO, M. A., M. SATO & C. S. LIEBER. 1983. Gastroenterology **84:** 562–572.
75. MAK, K. M., M. A. LEO & C. S. LIEBER. 1984. Trans. Assoc. Am. Physicians **98:** 210–221.
76. LEO, M. A., N. LOWE & C. S. LIEBER. 1986. Biochem. Pharmacol. **35:** 3949–3953.
77. LEO, M. A., M. ARAI, M. SATO & C. S. LIEBER. 1982. Gastroenterology **82:** 194–205.
78. LEO, M. A. & C. S. LIEBER. 1983. Hepatology **2:** 1–11.
79. TESCHKE, R., Y. HASUMURA & C. S. LIEBER. 1974. Arch. Biochem. Biophys. **163:** 404–415.

DISCUSSION OF THE PAPER

M. J. COON (*University of Michigan, Ann Arbor, MI*): Whereas, there are a number of hetero- and homodimers, as we heard about with ADH, with P-450 we have a very different system of some 200 enzymes, ten of which have been purified from the liver. There is some sequence homology, and they vary in their ability to bind. They also vary in the rates at which they turn over the various substrates. You have alluded to the huge number of substrates, carcinogens, xenobiotics of all kinds, and lipids. We know that if we administer ethanol or other foreign compounds, a specific P-450, which is ethanol-inducible, is a primary catalyst. However, we should also emphasize that, in addition, all of the other P-450s will catalyze the same reactions, although they are slower and have different binding constants.

One of the most puzzling aspects of the P-450 oxygenase field is the breadth of the inducers. We have shown by immunological techniques, using poly- and monoclonal antibodies, that the inducers include ethanol. Acetone, the diabetic state, isoniazide, various other heterocyclic compounds, proline, and imitazol are also excellent inducers.

One of the most challenging questions is how these compounds, with very different structures, induce a specific cytochrome P-450. Are you working on this fundamental question?

LIEBER: I've certainly thought about it, but unfortunately we don't have a good handle on this fundamental question. There is no obvious common factor that has been identified.

COON: We should probably be thinking about common hormonal mechanism, in view of the influence of the diabetic state.

LIEBER: If evidence that induction may occur in cell cultures is verified, hormonal mechanisms may not be operative.

E. MEZEY (*Johns Hopkins School of Medicine, Baltimore, MD*): What is the rate-limiting step in microsomal oxidation of ethanol?

LIEBER: Activity depends on the amount of cytochrome P-450 present, but other factors can also affect the rate of reaction. It is not clear what the most important factors might be *in vivo*. Obviously, cofactors, which are always optimal in our *in-vitro* system, may become limiting *in vivo*. There are probably a number of rate-limiting factors.

Subcellular Localization of Acetaldehyde Oxidation in Liver[a]

HENRY WEINER

Department of Biochemistry
Purdue University
West Lafayette, Indiana 47907

INTRODUCTION

In the past few years there has been an increased interest in the role of acetaldehyde in alcohol-related problems. The enzymes responsible for the metabolism of the compounds have been studied in detail only for the past decade, compared to the alcohol metabolizing enzymes, which have been investigated since the 1950s. Three older review articles are recommended for general reference to acetaldehyde-related problems.[1-3] This article will not focus on any of the physiological or pharmacological effects of acetaldehyde, but only on its subcellular metabolism; an article in 1985 reviews many of those aspects of the problem.[4]

While ethanol oxidizing enzymes are primarily localized in liver, those involved in acetaldehyde oxidation are ubiquitous to the body. Inasmuch as blood acetaldehyde levels are extremely low (μM) even in the presence of relatively high blood ethanol levels, the bulk of the acetaldehyde oxidation must then be confined to liver, where ethanol is primarily converted to the aldehyde. In fact, it has been estimated that, at least for rats, *ca* 95% of acetaldehyde oxidation occurs in liver.[5,3] A number of enzymes are capable of oxidizing acetaldehyde to acetate, the most important one being aldehyde dehydrogenase (ALDH), an NAD-dependent enzyme found in cytosol, endoplasmic reticulum, and mitochondria. Other dehydrogenases such as succinic semialdehyde and glyceraldehyde-3-phosphate dehydrogenase[6] can, *in vitro,* oxidize acetaldehyde. Furthermore, an enzyme misnamed "aldehyde oxidase" can also catalyze the *in-vitro* oxidation of acetaldehyde.[7] This enzyme, like the others, has a very high K_m for the compound in contrast to mitochondrial aldehyde dehydrogenase, which has a K_m value typically in the μM range.

In rodents, many isozymes of liver aldehyde dehydrogenase exist. In man, it appears that there is only one form in each subcellular organelle. Unlike human liver alcohol dehydrogenase, where numerous polymorphic forms of the enzyme exist, only two forms of human mitochondrial ALDH have been described thus far.

All people examined have the same cytoplasmic isozyme of ALDH. This enzyme, called "E_1" by some investigators and "E-II" by others,[8] has a K_m for acetaldehyde of *ca* 200 μM and is very susceptible to inhibition by the antialcohol drug disulfiram (Antabuse), which is often prescribed to deter alcoholics from drinking.[9,10]

It is mitochondrial ALDH that has an altered primary sequence in different populations. Basically, the two forms of the enzyme differ from each other by just one amino acid resulting from a single base change in DNA. A guanine being mutated to

[a]Supported in part by Public Health Service Grant AA05812. Author is recipient of National Institute on Alcohol Abuse and Alcoholism Research Scientist Development Award K05 AA00028.

an adenine caused the glutamate residue to become a lysine at the 14th position from the C-terminal of the enzyme.[11] Many orientals have this altered form of the enzyme compared to very few caucasians.[12]

The "oriental" form of ALDH has been reported to have essentially no or very little catalytic activity when acetaldehyde or other common aldehydes are used as substrate.[11,13] Thus far, no one has reported the enzyme to be a good catalyst for the oxidation of any aldehyde, nor do the people who possess it suffer in any way, provided they do not consume alcoholic beverages.

The fact that individuals possessing the inactive form of ALDH have high blood acetaldehyde levels[12,14] after consuming ethanol leads one to conclude that mitochondrial ALDH is important in the *in-vivo* oxidation of aldehyde. It has been suggested, however, that the cytoplasmic ALDH may be the major enzyme responsible for acetaldehyde metabolism. Evidence for this conclusion is indirect. First, is the well documented report that disulfiram reacts much more rapidly with cytoplasmic than with mitochondrial ALDH.[9] The second supporting evidence is a report that shows that the level of cytoplasmic ALDH decreases in the alcoholic.[15] Since it has been reported that alcoholics obtain higher blood levels of acetaldehyde after being given a dose of ethanol than do nonalcoholics,[16] it was concluded that the cytoplasmic enzyme was most important in maintaining low blood levels of acetaldehyde. One finds then two conflicting views as to the subcellular localization of acetaldehyde metabolism in liver. Thus far, no investigator has implicated the importance of the high K_m microsomal ALDH in maintaining low blood acetaldehyde levels.

The rate of acetaldehyde oxidation in liver depends upon three factors: enzyme activity, NAD/NADH ratio (NADH is a good inhibitor of the enzyme), and substrate availability. It is difficult to state unequivocally which step is actually rate-limiting for acetaldehyde oxidation independent of where in the cell the oxidation occurs. As the enzyme requires NAD as a cofactor, if the ratio of NAD/NADH were drastically lowered, then the rate of oxidation of acetaldehyde would decrease. Thus, independent of whether cytoplasmic or mitochondrial ALDH was the major enzyme involved, the redox potential of the cell is also of prime importance in maintaining low blood acetaldehyde levels.

Barring drastic alterations in NAD levels it is worthwhile to ascertain which subcellular organelle is dominant in acetaldehyde oxidation. Knowing where in the cell acetaldehyde is oxidized could help in our understanding of some of the complications produced by this reactive intermediate. It is equally desirable to know which isozyme is involved, so that pharmacologists could possibly design alternate drugs that inhibit the specific isozyme required for acetaldehyde oxidation. Pharmacological intervention thus far is the major way to deter alcoholics from drinking.

Elegant experiments performed using labelled acetaldehyde showed that the bulk of the isotope was found in water and not in lactate, suggesting that mitochondria was the major site of oxidation.[17–19] This conclusion was based on the fact that in mitochondria the labelled ^{3}H-NADH would pass its electrons to the electron transport system releasing a proton (^{3}H), producing water. If the cytosol was the principal site of oxidation then the ^{3}H-NADH would have transferred the hydride ion to compounds such as pyruvate, producing labelled lactate, or to a component of the coenzyme shuttle system such as oxaloacetate, producing labelled malate.

We attempted to answer the question as to which subcellular organelle was most important in acetaldehyde oxidation by utilizing a pharmacological approach to specifically inhibit one form of ALDH. The study was based on the work of others who found that cyanamide preferentially inhibited mitochondrial ALDH[20,21] and that animals treated with the compound had elevated blood acetaldehyde after being given ethanol.[22] Our plan then was to use tissue slices and inhibit the mitochondrial enzyme.

The ability of this modified liver slice to oxidize acetaldehyde would then be correlated with the remaining enzymatic actvity of the mitochondrial isozyme. Tissue from rat livers was employed in our initial studies, since in that case the "answer" was already known: mitochondria was the site of oxidation. The study was then extended to use tissue from beef and pig livers. No attempts were made by us to use hepatocytes, but the protocol to be described could easily be adopted for studies using isolated liver cells from any species, including human. The use of cells would be necessary, for in all probability it would be virtually impossible to obtain human tissue, except perhaps from biopsy, that would not possess damaged mitochondria membranes.

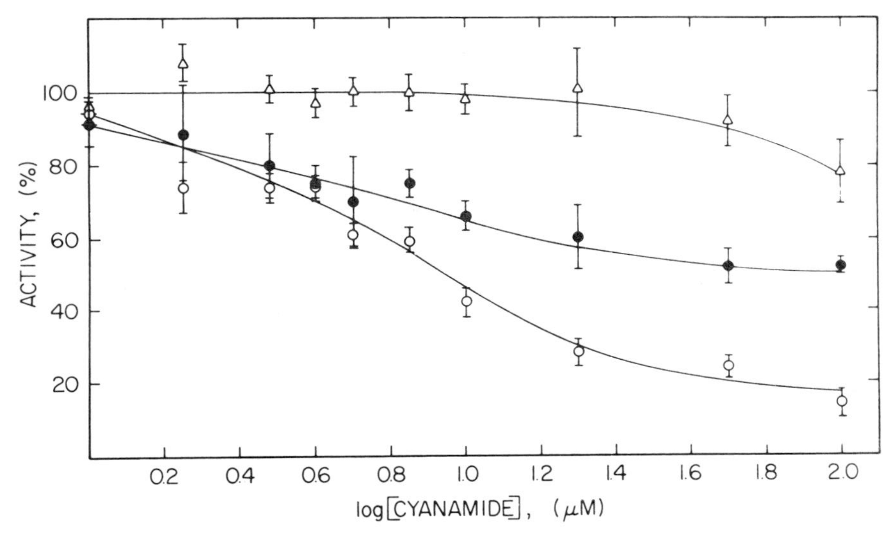

FIGURE 1. Effect of cyanamide on the activities of high-K_m (Δ) and low-K_m ALDH (O) and on acetaldehyde metabolism (●) in slice incubations. Rat liver slices were treated with various concentrations of cyanamide, and their ability to metabolize acetaldehyde was measured. The slices were then homogenized in order to determine enzyme activities. Data are reported as a percentage of the activity in the absence of inhibitor and are the mean ± SE for 2–12 animals. At least 3 incubations were performed for each cyanamide concentration with each animal. (This FIG. and FIGS. 2–5 from Svanas and Weiner.[23] Reprinted by permission from *Archives of Biochemistry and Biophysics.*)

EXPERIMENTAL

Detailed protocols employed in this study were based on those published by Svanas and Weiner.[23] These are briefly summarized here. Thin slices of fresh tissue were bathed in Krebs-Ringer buffer for 10 minutes in the presence of varying doses of cyanamide. The slices, which weighed approximately 50 mg, prepared with the aid of a Staddie-Riggs tissue slicer, were then transferred to a vial containing fresh Krebs-Ringer buffer and acetaldehyde. After a 10-minute incubation at 37° C the slice was transferred to cold Krebs-Ringer buffer. The acetaldehyde remaining in the incubation mixture was measured by either head space gas chromatography or by a new high-performance liquid-chromatography method described by Lieber's group.[24] To determine the enzymatic activity remaining, the slice was first homogenized with buffer. The activity of the high- and the low-K_m enzymes were then determined by assaying with 10 μM or 10 mM acetaldehyde; the total protein content was also determined. Thus, in any one slice it was possible to determine the rate of acetaldehyde oxidation and the amount of high- and low-K_m ALDH present in the tissue.

In some experiments disulfiram was added to the incubation mixture, while in others chloralhydrate was employed. In still other experiments, ethanol was used to generate acetaldehyde *in situ*. Wistar rats were obtained from the Purdue University Biochemistry Department animal breeding facilities and fresh beef and pig livers from a local abattoir.

RESULTS

Inhibition of ALDH by Cyanamide

Cyanamide is an *in-vivo* inhibitor of ALDH. The compound is enzymatically converted into an unknown intermediate, which in turn is the actual inhibitor of the enzyme. We[25] as well as others[20,21] have done extensive studies to verify that catalase is the required enzyme. It was shown that rat mitochondrial ALDH as found by others is much more susceptible to inhibition than is the cytosol enzyme.

Incubating the slice with cyanamide caused the same inhibition pattern to occur as did incubations with homogenates. Mitochondrial ALDH can be substantially inhibited before any appreciable inhibition of the cytosolic enzyme occurs. This is illustrated in FIGURE 1. Similar experiments were also performed with tissue slices obtained from beef and pig liver.

Acetaldehyde Metabolism in Rat Liver Slices

Isolated rat mitochondria oxidizes acetaldehyde. The rate of this oxidation was defined as 100% with any particular batch of organelles obtained from one rat liver. To aliquots of organelles were added various doses of cyanamide. The enzymatic activity and the rate of acetaldehyde oxidation in these treated organelles were measured; the results are presented in FIGURE 2. It can be observed that any decrease in the low-K_m mitochondrial ALDH activity resulted in a decreased rate of acetaldehyde oxidation. Thus, the amount of enzyme and not the reoxidation of NADH is rate-limiting. It was previously concluded in other studies utilizing different techniques that the rate of oxidation of acetaldehyde in intact mitochondria was limited by the level of enzyme and not by regeneration of coenzyme.[26]

Having established that it was possible to retard acetaldehyde oxidation when inhibiting ALDH in the intact mitochondria, we performed a similar study in the isolated slice. Conditions were chosen such that less than 5% of the cytosolic enzyme was inhibited, while up to 80% of the low-K_m mitochondrial enzyme was inhibited. Presented in FIGURE 3 are the results of such an experiment. It can be noted that a linear decrease in rates of acetaldehyde metabolism occurred as the low-K_m enzyme was inhibited. Most surprising, though, was the fact that the extrapolation of the data did not intercept at the origin representing 0% metabolism, but was at *ca* 40%. This observation revealed that some metabolism of acetaldehyde could occur even if there were no active mitochondrial ALDH.

The data in FIGURE 3 were obtained using a high concentration of acetaldehyde—200 μM. The same experiment was repeated using 100 μM acetaldehyde, and the results are presented in FIGURE 4. The intercept is at 20% rate of metabolism, when complete inhibition of the mitochondrial enzyme occurred. Presumably with a lower level of acetaldehyde less is being oxidized by the higher-K_m cytoplasmic enzyme.

The systems responsible for this nonmitochondrial ALDH metabolism were not unequivocally determined. We found that including chloralhydrate (200 mM, a concentration that totally inhibits both cytosol and mitochondrial ALDH) reduced the extrapolation from 40% to intercept at 20% when 200 μM acetaldehyde was employed in the incubation. Thus even in the absence of functioning ALDH, acetaldehyde

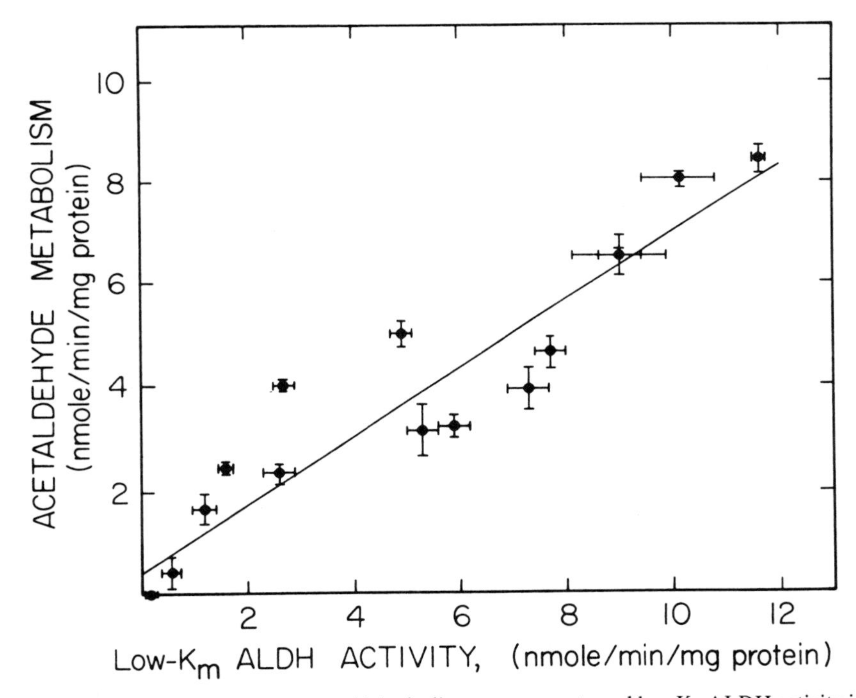

FIGURE 2. Correlation between acetaldehyde disappearance rate and low-K_m ALDH activity in incubations with intact mitochondria. Mitochondria from 3 animals were treated with 2 μM cyanamide for various periods of time to partially inhibit low-K_m ALDH activity. Their ability to metabolize 200 μM acetaldehyde and remaining low-K_m ALDH activity was measured in parallel experiments. Each point represents the mean ± SE of both variables for 3–4 incubations using mitochondria from a single animal.

disappearance still occurred. It appears then that the loss of acetaldehyde in the slice is due to the action of three systems: mitochondrial ALDH (60%), cytoplasmic ALDH (20%), and a yet to be determined system (20%). Of this remaining 20% it is felt that some is due to binding of acetaldehyde to proteins. Though no experiments were performed to test this hypothesis, it is consistent with the findings by others that hemoglobin will bind acetaldehyde[27] as will other proteins[28,29] and with our previous studies, which showed that aldehydes derived from biogenic amines bind to proteins.[30] Independent of the system responsible for this 20% loss of acetaldehyde, the studies

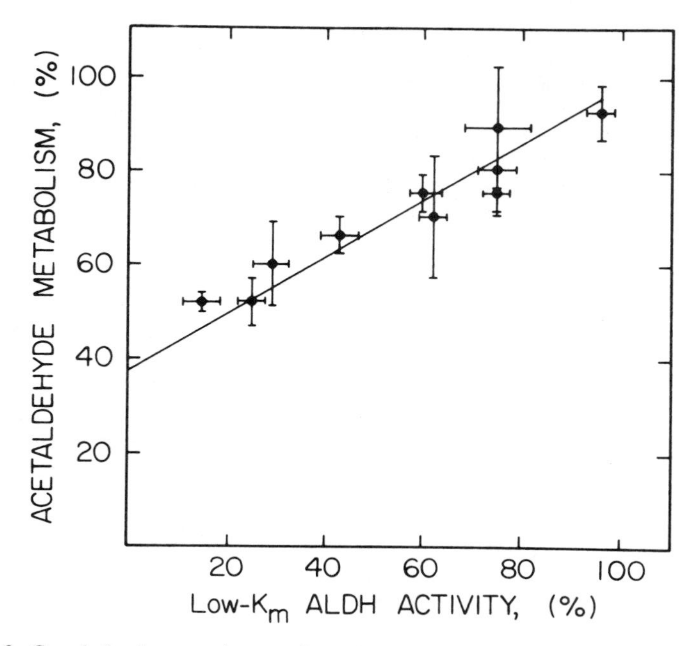

FIGURE 3. Correlation between the rate of metabolism of 200 μM acetaldehyde and the fraction of the remaining low-K_m ALDH activity. Data from FIGURE 1 were replotted to directly compare acetaldehyde metabolism and low-K_m ALDH activity. Data are reported as the mean ± SE.

with rat tissue basically replicate what has been found previously using the radio-labelling method: mitochondria is the major organelle responsible for acetaldehyde oxidation in rat liver.

In all the above described experiments, acetaldehyde was added to the slice. It was necessary to verify that inhibition of mitochondrial ALDH would result in an increased accumulation of acetaldehyde produced during ethanol oxidation. Slices were treated

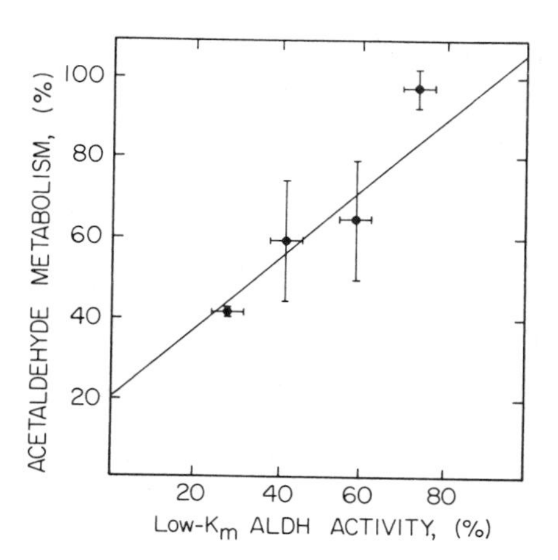

FIGURE 4. Correlation between the rate of metabolism of 100 μM acetaldehyde and the fraction of the remaining low-K_m ALDH activity. Experimental procedure was the same as in FIGURE 1, except for substrate concentration. Presentation of data is the same as in FIGURE 3.

with the inhibitor and then incubated with 8.5 mM ethanol for 30 minutes. The acetaldehyde concentration was then determined. The results, presented in FIGURE 5, show that any inhibition of low-K_m mitochondrial enzyme resulted in an accumulation of acetaldehyde.

Acetaldehyde Metabolism in Beef and Pig Liver Slices

Essentially similar experiments as were described for the rat tissue were repeated in tissue obtained from beef and pig livers. In both cases it was found that cyanamide

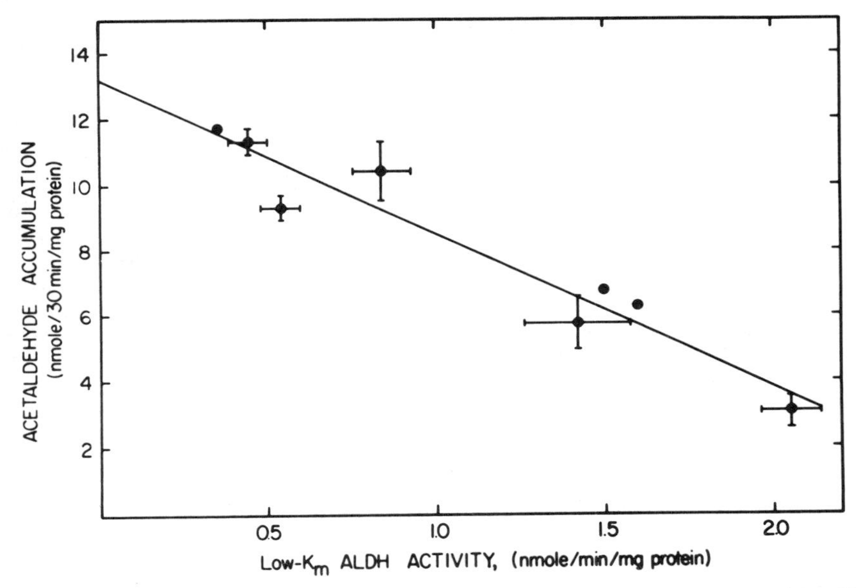

FIGURE 5. Acetaldehyde generation during ethanol metabolism after inhibition of low-K_m ALDH. Slices were treated with various concentrations of cyanamide to partially inhibit low-K_m ALDH, and then acetaldehyde appearance in the incubation medium was measured after administration of 8.5 mM ethanol. Data were combined from experiments with 3–4 animals for each data point, except for the ones without error bars, in which single animals were used. Data from each animal are a combination of 3 incubations and are presented at the mean ± SE.

was an effective inhibitor of the low-K_m mitochondrial enzyme, while hardly inhibiting the high-K_m cytosol enzyme. It was necessary for these studies to isolate the enzyme from tissue and determine the K_ms of the enzyme from the different subcellular organelles. These data will be presented elsewhere.

The metabolic experiments performed with tissue slices once again proved that inhibition of the mitochondrial enzyme caused a decreased rate of acetaldehyde oxidation. The data obtained with all three animals are essentially the same. Complete inhibition of the low-K_m enzyme results in approximately a 60% decrease in the tissue's ability to oxidize acetaldehyde.

To assess the role of the high-K_m cytosolic enzyme, beef liver slices were incubated with disulfiram prior to the addition of acetaldehyde. It was possible by incubating the slice with 10 μM disulfiram to achieve 100% inhibition of the high-K_m enzyme while barely decreasing the activity of the low-K_m mitochondrial enzyme. Under these conditions the rate of oxidation of acetaldehyde was 75% of the control rate. Thus, this additional evidence supports the conclusion that mitochondria is the major site of metabolism of acetaldehyde in liver.

DISCUSSION

The observation that there is virtually no detectable blood acetaldehyde during active ethanol oxidation in man leads to the conclusion that liver has the ability to oxidize acetaldehyde at least as rapidly as it is produced. Our report that any decrease in the activity of the mitochondrial ALDH will cause the tissue to accumulate acetaldehyde is supportive of the conclusion reached by many that mitochondria is the subcellular site of the oxidation and that the level of ALDH is the rate-determining step for the metabolism of acetaldehyde.

Supporting evidence for the conclusion that the level of mitochondrial ALDH is rate-limiting in the metabolism comes from two separate observations. First, the rate of an electron transfer is faster than is the rate of acetaldehyde oxidation. That is, the cell can regenerate NAD at a rate faster than it can oxidize acetaldehyde. Second, any decrease in the activity of the enzyme results in a decreased rate of acetaldehyde oxidation. If some other process were rate-limiting, it would have been possible to inhibit a portion of the ALDH without causing a decrease in the rate of acetaldehyde oxidation. It is obvious that under conditions that cause a drastic decrease in the rate of electron transport system will make it appear that the rate of oxidation of acetaldehyde is governed not by ALDH activity but by the regeneration of NADH.

It is possible to increase up to twofold the activity of mitochondrial ALDH.[31] Divalent ions such as Mg^{2+} and Ca^{2+} will increase the specific activity of the enzyme. The exact mechanism for the activation appears to be species-dependent. Whether or not this activation occurs *in vivo* has yet to be explored. It is possible that the enzyme exists normally in the cell in its activated form, and in the alcoholic it functions as the less active enzyme. If the latter be the case, it could explain why the alcoholic has a somewhat elevated acetaldehyde level after receiving a dose of ethanol. This question as to whether enzymes level changes could be addressed by the use of an ELISA assay, so that one could determine the actual concentration of enzyme in the liver of the normal and alcoholic patient. A measure of the catalytic activity alone will not indicate whether or not there has been an increased or decreased rate of synthesis of the enzyme.

In addition to being oxidized to acetate or reduced to ethanol, acetaldehyde can interact with biological material. The aldehyde can condense with biogenic amines such as dopamine to form salsolinol. The importance of the amine-aldehyde adducts in alcohol-related problems has been a well debated issue.[32] The aldehyde can also bind to proteins, including hemoglobin, though the latter has recently been challenged.[33] How important these nonenzymatic reactions are to the *in-vivo* removal of acetaldehyde is not known. It appears, though, that the enzymatic oxidation in liver will be sufficient for allowing the organ to maintain a very low cellular, and hence blood level of acetaldehyde, especially in the nonalcohol abuser.

ACKNOWLEDGMENT

The unpublished data from the author's laboratory was obtained with Q-N Cao.

REFERENCES

1. LINDROS, K. O. 1978. Acetaldehyde—its metabolism and rate in the actions of alcohol. *In* Research Advances in Alcohol and Drug Problems. Y. Israel, F. B. Glaser, H. Kalant, R. E. Popham, W. Schmidt & R. G. Smalt, Eds. Vol. 4: 111–176. Plenum. New York, NY.
2. WEINER, H. 1979. Aldehyde dehydrogenase: mechanism of action and possible physiological roles. *In* Biochemistry and Pharmacology of Ethanol. E. Majchrowicz & E. P. Nobel, Eds. Vol. 1: 107–124. Plenum. New York, NY.
3. WEINER, H. 1979. Acetaldehyde metabolism. *In* Biochemistry and Pharmacology of Ethanol. E. Majchrowicz & E. P. Nobel, Eds. Vol. 1: 125–144. Plenum. New York, NY.
4. TOPEL, H. 1985. Alcohol **2:** 711–788.
5. ERIKSSON, C. S. P. 1977. Acetaldehyde metabolism *in vitro* during ethanol oxidation. *In* Alcohol Intoxication and Withdrawal-Experimental Studies. M. M. Gross, Ed. Vol. 3: 319–341. Plenum. New York, NY.
6. FORTE-MCROBBIE, C. M. & R. PIETRUSZKO. 1986. J. Biol. Chem. **261:** 2154–2163.
7. WEINER, H. 1980. Aldehyde oxidizing enzymes. *In* Enzymatic Basis of Detoxication. W. B. Jakoby, Ed. 261–280. Academic Press. New York, NY.
8. WEINER, H. 1985. Alcoholism: Clin. Exp. Res. **9:** 541–542.
9. GREENFIELD, N. J. & R. PIETRUSZKO. 1977. Biochem. Biophys. Acta **483:** 35–45.
10. HEHEHAN, G. T. M., K. WARD, N. P. KENNEDY, D. G. WEIR & K. F. TIPTON. 1985. Alcohol **2:** 107–110.
11. YOSHIDA, A., M. IKAWA, L. C. HSU & K. TANI. 1985. Alcohol **2:** 103–106.
12. HARADA, S., D. P. AWARWAL & H. W. GOEDDE. 1985. Further characterization of aldehyde dehydrogenase isozyme variant in Japanese. *In* Enzymology of Carbonyl Metabolism 2: Aldehyde Dehydrogenase, Aldo-Keto Reductase and Alcohol Dehydrogenase. T. G. Flynn & H. Weiner, Eds. 129–135. Alan Liss. New York, NY.
13. FERENCZ-BIRO, K. & R. PIETRUSZKO. 1984. Biochem. Biophys. Res. Commun. **118:** 97–102.
14. INOUE, K., M. FUKUNAGA, T. KIRIYAMA & S. KOMURA. 1984. Alcoholism: Clin. Exp. Res. **8:** 319–322.
15. THOMAS, M., S. HAISALL & T. J. PETERS. 1982. Lancet **2:** 1057–1059.
16. KORSTEIN, M. A., S. MATSUZAKI, L. FEINMAN & C. S. LIEBER. 1975. N. Engl. J. Med. **292:** 386–389.
17. CORRALL, R. J., P. HAVRE, J. M. MARGOLIS, M. KONG & B. R. LANDAU. 1976. Biochem. Pharmacol. **25:** 17–20.
18. GRUNNET, N., H. I. D. THIEDEN & B. QUISTORFF. 1976. Acta Chem. Scand. **30:** 345–352.
19. HAVRE, P., J. M. MARGOLIS & M. A. ABRAMS. 1976. Biochem. Pharmacol. **25:** 2757–2758.
20. DEMASTER, E. G., F. N. SHIROTA & H. T. NAGASAWA. 1984. Biochem. Biophys. Res. Commun. **122:** 358–366.
21. DEMASTER, E. G., F. N. SHIROTA & H. T. NAGASAWA. 1985. Alcohol **2:** 117–121.
22. MARCHNER, H. & O. TOTTMAR. 1976. Acta. Pharmacol. Toxicol. **39:** 331–339.
23. SVANAS, G. W. & H. WEINER. 1985. Arch. Biochem. Biophys. **236:** 36–46.
24. DIPADOVA, C., J. ALDERMAN & C. S. LIEBER. 1986. Alcoholism: Clin. Exp. Res. **10:** 86–89.
25. SVANAS, G. W. & H. WEINER. 1985. Biochem. Pharmacol. **34:** 1197–1204.
26. TANK, A. W., H. WEINER & J. A. THURMAN. 1981. Biochem. Pharmacol. **30:** 3265–3275.
27. HOBERMAN, H. D. & S. M. CHIODO. 1982. Alcoholism: Clin. Exp. Res. **6:** 260–266.
28. DONOHUE, T. M., D. J. TUMA & M. F. SORRELL. 1983. Arch. Biochem. Biophys. **220:** 239–246.
29. COLLINS, M. A., Ed. 1985. Aldehyde Adducts in Alcoholism. Alan Liss. New York, NY.
30. BERGER, D. & H. WEINER. 1977. Biochem. Pharmacol. **26:** 741–747.
31. WEINER, H. & K. TAKAHASHI. 1983. Pharmacol. Biochem. Behav. **18:** 109–112.
32. BLOOM, F., J BARCHAS, M. SANDLER & E. USDIN. Eds. 1985. Beta Carbolines and Tetrahydroisoquinolines. Alan Liss. New York, NY.
33. PETERSON, C. M., C. M. POLIZZI & P. J. FRAWLEY. 1986. Alcoholism: Clin. Exp. Res. **10:** 219–220.

DISCUSSION OF THE PAPER

D. McCarthy (*University of New Mexico School of Medicine, Albuquerque, NM*): Although the K_ms are very different, do we know anything about the saturation conditions in the various fractions? In other words, at high blood ethanol levels, is acetaldehyde in the mitochondria saturating?

Weiner: Yes, because the concentrations of acetaldehyde are micromolar and the K_m for the mitochondrial aldehyde dehydrogenase is micromolar. The latter enzyme will be saturated. The more important question is how saturated is the cytoplasmic enzyme, its K_m being in the 100-μM range. If the liver acetaldehyde concentration is about 100 μM, the enzyme would be 50% saturated. The true concentration of acetaldehyde in the cells during metabolism has not been unequivocally determined.

McCarthy: There is extensive mitochondrial injury associated with the excess consumption of alcohol. Is there any evidence that kinetic parameters, such as K_m, V_{max}, etc., for the mitochondrial enzyme are altered by chronic exposure to alcohol?

C. S. Lieber (*Veterans Administration Medical Center, Bronx, NY*): Perhaps Dr. Pietruszko would answer.

R. Pietruszko (*Rutgers University Center for Alcohol Studies, Piscataway, NJ*): In preliminary experiments we found such parameters altered. The mitochondrial enzyme in alcoholics seems to have a lower specific activity.

E. Mezey (*Johns Hopkins School of Medicine, Baltimore, MD*): What is the mechanism for the effect of magnesium in activating the mitochondrial aldehyde dehydrogenase?

Weiner: In the horse liver the enzyme is a tetramer that functions with half-site reactivity. Thus, in the four identical sub-units, only two active sites function. In the presence of magnesium the horse liver enzyme dissociates into dimeric forms, and each subunit functions.

The human and sheep enzymes show a different mechanism for activation. It appears that the rate-limiting steps for those enzymes are different, and they seem to be more involved with coenzyme dissociation. It is interesting that in all species there is approximately a twofold activation.

Microsomal Generation of Hydroxyl Radicals: Its Role in Microsomal Ethanol Oxidizing System (MEOS) Activity and Requirement for Iron[a]

ARTHUR I. CEDERBAUM

Department of Biochemistry
Mount Sinai School of Medicine
New York, New York 10029

INTRODUCTION

Hydroxyl radicals (·OH) can be produced in biological systems from a number of sources including the interaction of ionizing radiation with water, autoxidation of chemical agents (*e.g.*, ascorbate or 6-hydroxydopamine), or during enzyme-catalyzed reactions (*e.g.*, the xanthine-xanthine-oxidase reaction). Ethanol is a potent scavenger of ·OH, the product of the reaction being acetaldehyde.[1,2] Evidence has been presented that acetaldehyde may be generated from ethanol during the xanthine-oxidase reaction or during the autoxidation of ascorbate.[3] Inhibition of oxidative reactions by ethanol has been used to invoke the participation of ·OH in various reactions. For example, Beauchamp and Fridovich[4] provided evidence for formation of ·OH during the xanthine-oxidase reaction (oxidation of reduced cytochrome *c*; generation of ethylene gas from methional). These reactions were blocked by various ·OH scavengers, including ethanol. Similarly, the disruption of lysosomes that occurred during the oxidation of NADPH by liver microsomes was shown to be mediated by ·OH, since the addition of known scavengers of ·OH, including ethanol, provided significant protection against lysis.[5]

Besides these *in-vitro* experiments, there is evidence that ethanol is an effective ·OH scavenger *in vivo*. The administration of alloxan results in the destruction of the beta cells of the pancreas and hyperglycemia. The diabetogenic action of alloxan is believed to be mediated via the reaction of the reduced metabolite, dialuric acid, with oxygen.[6] *In-vitro* studies have shown that dialuric acid can give rise to ·OH.[7] 6-Hydroxydopamine is a neurotoxic agent that causes destruction of adrenergic nerve terminals.[8] This compound is capable of spontaneous autoxidation; during this autoxidation, ·OH is generated.[7] The specific cytotoxic action of alloxan and 6-hydroxydopamine is believed to be due to the generation of ·OH during the autoxidation of these agents. Ethanol, as well as other known ·OH scavengers, provided *in-vivo* protection against the toxic actions of alloxan and 6-hydroxydopamine,[6–8] *e.g.*, ethanol-intoxicated mice failed to develop diabetes after the injection of alloxan.[6] The ability of thiourea to block the artifactual production of acetaldehyde from ethanol in tissue extracts[9] may reflect the ·OH scavenging properties of thiourea.

The following reactions may be relevant to various aspects concerning the

[a]Supported by the National Institute on Alcohol Abuse and Alcoholism.

generation of $\cdot OH$:

$$H_2O_2 + O_2^{\cdot -} \rightarrow \cdot OH + OH^- + O_2 \quad (1)$$

$$Fe^{3+} + O_2^{\cdot -} \rightarrow Fe^{2+} + O_2 \quad (2)$$

$$Fe^{2+} + H_2O_2 \rightarrow \cdot OH + Fe^{3+} + OH^- \quad (3)$$

Reaction 1 is the Haber-Weiss[10] reaction, in which the superoxide anion radical ($O_2^{\cdot -}$) reacts with hydrogen peroxide (H_2O_2) to form $\cdot OH$ as one of the products. This reaction requires catalysis by heavy metals (Reactions 2 and 3). The superoxide anion radical arises as the result of the transfer of one electron to oxygen:

$$O_2 + e^- \rightarrow O_2^{\cdot -} \quad (4)$$

Superoxide is produced as an intermediate in several reactions, *e.g.*, the reactions catalyzed by xanthine oxidase, dihydroorotic acid dehydrogenase, diamine oxidase, tryptophan dioxygenase, during active phagocytosis of leukocytes, during illumination of chloroplasts, and during mitochondrial electron transport (for review, see REFS. 11, 12). Of great interest are the reports that $O_2^{\cdot -}$ can be generated by microsomal electron transport associated with NADPH-cytochrome *c* reductase and cytochrome P-450 of hepatic microsomes.[13–16] Superoxide radicals can dismute spontaneously to yield (H_2O_2):

$$O_2^{\cdot -} + O_2^{\cdot -} + 2H^+ \rightarrow O_2 + H_2O_2 \quad (5)$$

The dismutation of $O_2^{\cdot -}$ is catalyzed by the enzyme superoxide dismutase. Consequently, conditions that result in the production of $O_2^{\cdot -}$ also lead to the accumulation of H_2O_2 via Reaction 5. These conditions can facilitate the Haber-Weiss reaction (Reaction 1).

There are numerous reports that have described systems involving the participation of $\cdot OH$ (for review, see REFS. 17–19). The Haber-Weiss reaction has generally been invoked in these cases as the reaction mechanism whereby $\cdot OH$ was generated. However, the Haber-Weiss reaction has become a subject of debate, with recent reports indicating that the reaction of $O_2^{\cdot -}$ with H_2O_2 in pure solution is too slow. Reaction 1 can be catalyzed by heavy metals, especially iron.[20,21] Reaction 3 is the well-known Fenton reaction.[22] Thus, in biological systems, iron salts, iron chelates, or iron-containing compounds may participate in the generation of $\cdot OH$ by catalyzing Reaction 1 or by participating directly by forming complexes with $O_2^{\cdot -}$ or H_2O_2 and thereby accelerating the formation of $\cdot OH$.[23] In biological systems, iron is present in various chelated forms; hence, an understanding of the structure-function relationship of various iron chelates as it concerns oxygen activation mechanisms and the interaction of such chelates with biological reductants such as superoxide radical, ascorbate, and flavoprotein dehydrogenases is important.

The ability of isolated liver microsomal fractions to oxidize ethanol to acetaldehyde has been well characterized by Lieber and co-workers.[24,25] In view of a) the ability of hepatic microsomes to generate $O_2^{\cdot -}$ and H_2O_2 (and thus the potential to generate $\cdot OH$ via Reactions 1, 2, and 3) and b) the ability of ethanol to react with $\cdot OH$, it seemed possible that the microsomal ethanol oxidizing system (MEOS) could result from the interaction of ethanol with $\cdot OH$. We, therefore, evaluated the ability of rat liver microsomes to generate $\cdot OH$, the role of iron in the production of $\cdot OH$ by microsomes, the enzymes involved in this production, and the role of $\cdot OH$ in the molecular mechanism of MEOS. This article summarizes and reviews previous results.

RESULTS

Methods to Detect ·OH

The techniques that were employed were to evaluate the production of products known to be generated when certain potent ·OH scavengers reacted with radiolytically generated ·OH. This serves as chemical evidence for the production of ·OH or, more correctly, potent oxidants with the oxidizing power of ·OH. The scavengers routinely utilized were dimethyl sulfoxide (DMSO), 2-keto-4-thiomethylbutyric acid (KMB or KTBA, a transamination product of the amino acid methionine), benzoic acid, and primary, secondary, and tertiary alcohols such as ethanol, and 1-, 2-, 3-butyl alcohol.

DMSO reacts with ·OH to yield the methyl radical, which has been detected by ESR spectroscopy.[26] Methyl radicals can produce methane gas via hydrogen abstraction or can dimerize to yield ethane gas. We found that under most conditions, the predominant reaction of the methyl radical was to react with molecular oxygen to generate the methylperoxy radical, which upon dismutation (Russell-type mechanism), produced formaldehyde.[27] KMB reacts with ·OH to produce ethylene gas (from carbon atoms 3 and 4 of KMB), which can easily be detected via head space gas chromatography. This agent has been used in numerous studies to detect ·OH-like products (reviewed in REF. 28). Benzoic acid reacts with ·OH to produce several hydroxylated benzoate derivatives.[29] In addition, the reaction with ·OH can also result in the decarboxylation of benzoate to produce CO_2 plus phenol.[30] The production of $^{14}CO_2$ from [7-^{14}C]benzoate via ·OH-mediated decarboxylation makes this scavenger very useful for the detection of ·OH. Of greater importance may be the recent observations that benzoate, relative to other ·OH scavengers such as KMB or ethanol, appears to be more specific for reaction with ·OH than with other potent oxidants such as alkoxyl or peroxyl radicals.[31]

Primary or secondary aliphatic alcohols react with ·OH mainly at the alpha position to produce hydroxyalkyl radicals.[2] These radicals give rise to the aldehyde or ketone product upon dismutation or loss of an electron, *e.g.*, to a ferric chelate. An alcohol that is a ·OH scavenger but does not contain an alpha hydrogen is *tertiary*-butyl alcohol. The reaction of *t*-butyl alcohol with ·OH produces either the α-hydroxyalkyl radical or the *t*-butoxy radical. The latter can decompose to produce acetone plus the methyl radical[32] with $\cdot CH_3$ ultimately producing methane, ethane, and formaldehyde.[33] Of considerable interest is the observation by Baker *et al.* that ^{14}C- or ^{13}C-labeled acetone was produced in the breath of mice treated *in vivo* with ^{14}C- or ^{13}C-labeled *t*-butyl alcohol.[34] TABLE 1 is a brief summary of these reactions. Full details as to methodologies to assay these reactions are provided elsewhere.[35,36]

Microsomal Oxidation of ·OH Scavengers[37–38]

TABLE 2 shows that isolated rat liver microsomes catalyze an NADPH-dependent oxidation of DMSO to formaldehyde, of KMB to ethylene, of *t*-butyl alcohol to acetone plus formaldehyde, and of benzoate to $^{14}CO_2$. Product formation is markedly reduced when NADH replaces NADPH or when boiled microsomes are utilized. Reactions are generally linear for 15 min with KMB as substrate, and for at least 30 min with the other three scavengers. The addition of 1 mM azide to inhibit the activity of contaminating catalase results in a three- to fivefold stimulation of product formation. This increase probably reflects the accumulation of H_2O_2 in the reaction system when catalase activity is blocked by azide. The stimulation by azide suggests that H_2O_2 is the precursor of the oxidant (·OH) responsible for the oxidation of the ·OH scavengers.

TABLE 1. Interaction of Hydroxyl Radical Scavengers with ·OH

Dimethylsulfoxide
$(CH_3)_2 - SO + \cdot OH \rightarrow CH_3SOOH + \cdot CH_3$
$\cdot CH_3 \rightarrow CH_4, CH_3CH_3, CH_2O$

Methional
$CH_3S - CH_2CH_2CHO + \cdot OH \rightarrow \frac{1}{2}(CH_3S)_2 + HCOOH + CH_2 = CH_2$

KTBA
$CH_3S - CH_2CH_2 - CO - COOH + \cdot OH \rightarrow \frac{1}{2}(CH_3S)_2 + 2\,CO_2 + CH_2 = CH_2$

Benzoic acid
$C_6H_5\,^{14}COOH + \cdot OH \rightarrow {}^{14}CO_2 + C_6H_5OH$

Ethanol
$CH_3CH_2OH + \cdot OH \rightarrow CH_3 \cdot CHOH \rightarrow CH_3CHO$

Isopropanol
$(CH_3)_2CHOH + \cdot OH \rightarrow (CH_3)_2 \cdot OH \rightarrow (CH_3)_2CO$

t-Butyl alcohol
$(CH_3)_3 - COH + \cdot OH \rightarrow \cdot CH_2 - C(OH) - (CH_3)_2 \rightarrow$ dimerization
↘
$(CH_3)_3 - CO \cdot \rightarrow (CH_3)_2CO + \cdot CH_3$
$\cdot CH_3 \rightarrow CH_4, CH_3CH_3, CH_2O$

H_2O_2 alone, in the absence of NADPH, is not effective in supporting the oxidation of the scavengers; therefore, microsomal electron transfer is also required. Other studies have shown that there is cross-competition among the various scavengers for oxidation, *e.g.*, benzoate inhibits DMSO oxidation and DMSO inhibits benzoate oxidation. This suggests competition for the generated oxidant.

Role of Iron in the Oxidation of ·OH Scavengers

It is becoming more apparent that biological systems shown to produce ·OH-like oxidants require the presence of transition metals such as iron or copper.[28] In our experiments, the various buffers and solutions and the water were routinely passed through columns of Chelex-100 resin to remove contaminating metals. This is important, since 100 mM phosphate buffer can contain more than 5 μM iron present as a contaminant. Since ferric iron is insoluble at neutral pH, most researchers have

TABLE 2. Microsomal Oxidation of ·OH Scavengers

Substrate	Product	Rate of Product Formation[a]	
		−Azide	+Azide[b]
KMB	Ethylene	0.52 ± 0.13	1.36 ± 0.09
DMSO	Formaldehyde	0.55 ± 0.20	2.50 ± 0.35
t-Butanol	Formaldehyde	0.60 ± 0.10	2.15 ± 0.25
	Acetone	0.35 ± 0.10	1.75 ± 0.20
Benzoate	$^{14}CO_2$	0.25 ± 0.03	1.35 ± 0.20

[a]Rates are expressed as nanomoles of product per minute per milligram microsomal protein.
[b]Final concentration of azide was 1.0 mM.

TABLE 3. Effect of EDTA and Ferric-EDTA on Microsomal Oxidation of ·OH Scavengers and Ethanol

Scavenger Substrate	Rate of Product Formation (nmol/min/mg Protein)		
	Control	200 μM +EDTA	50 μM +Ferric-EDTA
10 mM KMB	0.7	2.0	25.9
33 mM DMSO	0.2	2.3	8.6
33 mM *t*-Butanol	0.4	2.8	10.1
50 mM Ethanol	5.8	13.3	23.6

added chelated iron to their systems. Probably the most popular iron chelate used is ferric-EDTA. Results in TABLE 3 show that the addition of ferric-EDTA to microsomes results in an expected increase in the oxidation of ·OH scavengers.[39] The stimulation by ferric-EDTA is inhibited by catalase (plus the omission of azide) but not by superoxide dismutase. There is an increase in NADPH oxidation upon addition of ferric-EDTA to microsomes, suggesting that ferric-EDTA can serve as an electron acceptor from either the NADPH-P-450 reductase or from cytochrome P-450. Results in TABLE 3 show that EDTA itself can also stimulate the oxidation of the ·OH scavengers, which probably reflects the chelation by EDTA of iron present in the isolated microsomes.

Desferrioxamine (Desferal) is a potent iron chelating agent used clinically to remove iron from the liver. This chelator has been used to prevent the iron-dependent production of ·OH by a variety of systems.[28] Desferrioxamine proved to be a potent inhibitor of the microsomal oxidation of ·OH scavengers (TABLE 4). However, desferrioxamine did not inhibit the oxidation of aminopyrine, a typical drug cytochrome P-450-dependent mixed-function oxidase activity of the microsomes. The latter results suggest that the inhibition of oxidation of ·OH scavengers by desferrioxamine is not due to interference with the microsomal electron transfer chain or to nonspecific effects on the microsomes. The inhibition by desferrioxamine implicates a role for nonheme iron in the microsomes in the catalysis of ·OH generation, but not in the oxidation of aminopyrine, thereby dissociating the pathway of microsomal drug metabolism from that of the ·OH generation.[40] This disassociation suggests that either cytochrome P-450 does not play a significant role in the generation of ·OH or that different forms (isozymes) of cytochrome P-450 participate in the two reactions. If cytochrome P-450 is not involved in the generation of ·OH, other sources of microsomal iron would most likely play a role in the production of ·OH. Whether this iron is adventitious iron or microsomal iron available from, *e.g.*, heme oxygenase

TABLE 4. Effect of Desferrioxamine on Microsomal Oxidation of ·OH Scavengers and Aminopyrine

Substrate	Rate of Oxidation (nmol/min/mg Protein)	Percent Inhibition by Desferrioxamine	
		3.3 μM	100 μM
Aminopyrine	8.5 ± 0.9	0	4
DMSO	3.5 ± 0.4	58	89
t-Butanol	1.5 ± 0.1	86	90
Benzoate	1.4 ± 0.1	66	92
KMB	2.6 ± 0.6	60	80

activity is not known. In view of the chelex treatment of buffers, contaminating iron in the solutions is minimal.

Ethanol Oxidation

The ability of isolated rat liver microsomes to oxidize ethanol has been well characterized by Lieber and co-workers.[24,25] Results in TABLE 3 show that in the absence of any additions, ethanol is oxidized by an order of magnitude higher rate than that found for the ·OH scavengers. When EDTA or ferric-EDTA is added to the microsomes, the rate of ethanol oxidation increases, analogous to results with the ·OH scavengers. The control rate of ethanol oxidation was slightly sensitive to inhibition by low concentrations of ·OH scavengers, whereas the increased rate produced by the addition of EDTA or ferric-EDTA was very sensitive to inhibition by ·OH scavengers. These results appear to suggest that in the absence of EDTA or ferric-EDTA, ethanol is oxidized primarily by a ·OH-independent pathway. In the presence of EDTA or ferric-EDTA, the role of ·OH becomes more significant in contributing towards the overall oxidation of ethanol by microsomes. Indeed, desferrioxamine had little or no effect on the control rate of oxidation of ethanol or 2-butanol (TABLE 5) at concentrations that nearly completely inhibited the microsomal generation of ·OH (TABLE 4). However, desferrioxamine did inhibit the microsomal oxidation of alcohols in the presence of EDTA (TABLE 5). Actually, desferrioxamine appeared to completely block the stimulation of alcohol oxidation produced by EDTA, confirming a partial role for ·OH in the oxidation of alcohols when (iron) EDTA is present.

Experiments with Reconstituted Mixed-Function Oxidase Systems

The above experiments suggested that ·OH-like species could be generated during microsomal electron transfer by an iron-catalyzed reaction. To evaluate the locus of ·OH generation, experiments were conducted with NADPH-P-450 reductase and cytochrome P-450 purified from phenobarbital-treated rats.[41,42] Initial experiments compared the rate of oxidation of ethanol and 3 typical ·OH scavengers as catalyzed by the complete reconstituted hepatic mixed-function oxidase system to the rate that occurs in the absence of cytochrome P-450 (reductase-dependent system). TABLE 6 shows that in the presence of NADPH and reductase (plus phospholipid), there was oxidation of KMB, DMSO, and *t*-butyl alcohol. This oxidation was time dependent, required the presence of NADPH, scavengers, and reductase (phospholipid enhanced

TABLE 5. Effect of Desferrioxamine on Microsomal Oxidation of Alcohols

Alcohol	Concentration of Desferrioxamine (mM)	Rate of Alcohol Oxidation −EDTA (nmol/min/mg Microsomal Protein)	Rate of Alcohol Oxidation +EDTA[b] (nmol/min/mg Microsomal Protein)
Ethanol[a]	0	7.6 ± 0.4	16.7 ± 2.4
	0.25	6.3 ± 0.8 (− 17%)	8.3 ± 0.7 (− 50%)
2-Butanol[a]	0	8.5 ± 1.1	13.8 ± 1.3
	0.25	7.6 ± 1.0 (− 11%)	8.9 ± 0.8 (− 36%)

[a]Concentrations of ethanol and 2-butanol were 50 and 33 mM, respectively.
[b]Added to a final concentration of 0.1 mM.

TABLE 6. Oxidation of ·OH Scavengers and Ethanol by Purified Components of the Liver Microsomal Mixed-Function Oxidase System[a]

Substrate	Rate of Oxidation (nmol Product/min)	
	Reductase	Reductase + Cytochrome P-450-PB
KMB	0.8	0.8
DMSO	1.9	2.0
t-Butanol	1.8	1.7
Ethanol	3.2	8.5

[a]The reaction system contained 100 μg of phospholipid + 10,000 units of reductase in the absence and presence of 1 nmol of cytochrome P-450.

the rate but was not absolutely necessary), and was linearly dependent on units of reductase added. The addition of cytochrome P-450-PB to the reaction system had no effect on the oxidation of the ·OH scavengers. This is in contrast to results with aminopyrine, a typical mixed-function oxidase substrate, whose oxidation to formaldehyde was completely dependent on the presence of P-450.

Results with ethanol as the substrate were more complex. Similar to the ·OH scavengers, ethanol could be oxidized to acetaldehyde by the reductase-dependent system itself. However, in contrast to the ·OH scavengers, the addition of P-450-PB resulted in a threefold increase in the rate of ethanol oxidation (TABLE 6). Titration experiments indicated that whereas the oxidation of the ·OH scavengers was independent of P-450 over a wide range of P-450/reductase ratios, the oxidation of ethanol was dependent on the concentration both of reductase as well as of P-450-PB. The results with ethanol, where the addition of cytochrome P-450 increased the rate of ethanol oxidation over the rate found in the presence of reductase alone, suggested that ethanol oxidation might be occurring via two independent pathways. Hence, unique differences exist between the oxidation of ethanol (which appears to involve both a reductase-dependent and a cytochrome P-450-dependent component), typical ·OH scavengers (reductase-dependent only), and classical drug substrates (cytochrome P-450-dependent).

Because the reductase alone was shown to catalyze the oxidation of ·OH scavengers, the reductase-dependent pathway of ethanol oxidation probably reflects the interaction of ethanol with ·OH generated from the reductase during NADPH-dependent electron transfer. When cytochrome P-450 was present in the system, there was no enhancement of the reductase-dependent oxidation of the ·OH scavengers, suggesting that the cytochrome P-450-dependent pathway of ethanol oxidation may not involve oxygen radicals. The effect of competitive ·OH scavengers on the rate of ethanol oxidation catalyzed by the reductase and by the complete reconstituted system was, therefore, evaluated. TABLE 7 shows that ethanol oxidation by the reductase-dependent pathway was inhibited by Me_2SO and benzoate. In a similar manner, ethylene production from KTBA was also inhibited by competing ·OH scavengers, including ethanol. Thus, ethanol inhibits the oxidation of an ·OH scavenger, whereas ·OH scavengers block the oxidation of ethanol. Clearly, a role for ·OH in the reductase-dependent pathway of ethanol oxidation can be discerned. By contrast, when cytochrome P-450 was added to the incubation mixture (note the increase in rate of ethanol oxidation), the ability of Me_2SO and benzoate to act as inhibitors of ethanol oxidation was attenuated (TABLE 7). In fact, most of the decrease in ethanol oxidation produced by Me_2SO and benzoate in the complete system appears to be due to inhibition of the reductase-dependent activity, with little or no effect on the increased

TABLE 7. Effect of Competing ·OH Scavengers on the Oxidation of KMB and Ethanol[a]

Reaction System	Addition	Activity (nmol/min)	Effect of Addition (%)
a) Reductase + KMB	—	0.30	—
	30 mM DMSO	0.08	− 73
	30 mM Benzoate	0.14	− 53
	50 mM Ethanol	0.10	− 67
b) Reductase + Ethanol	—	2.70	—
	30 mM DMSO	0.90	− 66
	30 mM Benzoate	1.50	− 46
c) P-450 + Reductase + Ethanol	—	6.50	—
	30 mM DMSO	5.40	− 17
	30 mM Benzoate	5.20	− 20
d) Increase of Ethanol	—	3.80	—
Oxidation by P-450	30 mM DMSO	4.50	+ 18
(c minus b)	30 mM Benzoate	3.70	− 3

[a]Substrate concentrations were either 53 mM (ethanol) or 10 mM (KMB). The reaction system contained 5,000 units reductase plus 100 μg phospholipid for all experiments. In experiment c, 0.5 nmol of cytochrome P-450 was also present.

rate of ethanol oxidation produced by the addition of P-450. Other experiments indicated that the P-450-catalyzed increment in ethanol oxidation was insensitive to inhibition by catalase or superoxide dismutase. Thus, it appears that the reductase is responsible for the generation of ·OH and, hence, ·OH-dependent oxidation of ethanol. Cytochrome P-450 does not catalyze the production of ·OH but does catalyze the oxidation of ethanol by an oxygen-radical independent process.

Since reduced iron is necessary for the oxidative decomposition of hydrogen peroxide resulting in hydroxyl radical production, it can be anticipated that iron played a role in the generation of ·OH during NADPH oxidation by the reductase. To determine that iron catalyzed the production of ·OH, the different properties of two iron chelators, namely EDTA and desferrioxamine, were exploited. The former is known to form an iron chelate that potentiates the oxidation of ·OH scavengers, while the latter inhibits the iron-catalyzed production of ·OH in microsomal systems. Therefore, a stimulation by EDTA with a concomitant inhibition by desferrioxamine of substrate oxidation would be consistent with a role for iron in generating ·OH in these systems. TABLE 8 shows that the reductase-dependent oxidation of ethanol, KTBA, Me_2SO, and *t*-butanol was stimulated 50, 257, 155, and 166%, respectively, by 50 μM EDTA, whereas desferrioxamine inhibited the same reactions by approximately 60% in parallel experiments. In view of the above, the addition of iron would be expected to increase the generation of ·OH and, subsequently, augment the oxidation of ·OH scavengers and ethanol. The addition of 10 μM iron-EDTA resulted in a large increase in the oxidation of all four substrates (TABLE 8).

In contrast to the strong inhibition by desferrioxamine of ethanol oxidation by the reductase system, the increase in ethanol oxidation produced by the addition of P-450-PB was insensitive to desferrioxamine. The effect of desferrioxamine on ethanol oxidation was 0, −2, and −13% in three separate experiments.

Taken as a whole, the above results suggest that an active oxygen species with the characteristics of a ·OH can be generated by NADPH-cytochrome P-450 reductase, in the absence of cytochrome P-450. It is recognized that, because we used the phenobarbital-inducible isozyme of cytochrome P-450 in these studies, some caution should be exercised in interpreting these results. For instance, one could envision that the ·OH scavengers might serve as more suitable substrates for other isozymes of cytochrome P-450. Another possibility that cannot be excluded is that, when the cytochrome P-450 is added to the reductase, a shift in the locus of ·OH production occurs, *i.e.*, the cytochrome P-450, by mediating an efficient electron transfer, would serve to reduce the probability of autoxidation of reductase. Autoxidation of oxy-cytochrome P-450 would then result in the production of oxyradicals. Further studies will be needed to clarify these possibilities.

Similar to the typical ·OH scavengers, ethanol could be oxidized by the reductase alone. This pathway of ethanol oxidation was sensitive to inhibition by competitive ·OH scavengers and by desferrioxamine, but was stimulated either by EDTA or by iron. These results suggest that the reductase-dependent pathway of ethanol oxidation can be attributed to the interaction of ethanol with ·OH generated by the reductase via an iron-catalyzed Haber-Weiss reaction. The overall significance of ·OH as a mechanism for microsomal oxidation of ethanol would be highly dependent on the concentrations of iron, and as shown below, the nature of the chelating agent. From results described above with microsomes and reconstituted systems, ·OH would probably represent a minor contribution towards microsomal ethanol oxidation ($< 20\%$) under normal conditions. The major pathway of microsomal ethanol oxidation involves an oxygenated cytochrome P-450 intermediate that is independent of ·OH and does not support the oxidation of ·OH scavengers.

TABLE 8. Effect of Iron Chelators and Iron on the NADPH-Cytochrome P-450 Reductase-Dependent Oxidation of Ethanol and ·OH Scavengers[a]

Substrate	Addition	Activity (nmol Product/min)	Effect of Addition (%)
Ethanol	None	1.8	—
	Desferrioxamine	0.7	− 61
	EDTA	2.7	+ 50
	Fe-EDTA	11.4	+ 530
KMB	None	0.14	—
	Desferrioxamine	0.06	− 57
	EDTA	0.36	+ 257
	Fe-EDTA	3.70	+ 2,543
Me_2SO	None	0.58	—
	Desferrioxamine	0.23	− 60
	EDTA	0.90	+ 155
	Fe-EDTA	5.30	+ 814
t-Butanol	None	0.27	—
	Desferrioxamine	0.10	− 63
	EDTA	0.45	+ 166
	Fe-EDTA	4.30	+ 1,493

[a]Substrate concentrations were ethanol, 53 mM; KMB, 10 mM; Me_2SO, 30 mM; *t*-butanol, 32 mM. Final concentrations of desferrioxamine, EDTA, and Fe-EDTA were 330, 50, and 10 μM, respectively. Reactions were carried out in the presence of 2,500 units of the reductase and 50 μg of dilauroyl phosphatidyl choline. Results are from two or three experiments.

Role of Iron Chelates in ·OH Generation

The above results have emphasized the importance of iron in catalyzing ·OH generation by microsomes and reductase. As mentioned in the Introduction, iron appears to play a critical role in the generation of ·OH or ·OH-like species in most biological systems studied to date. Studies were undertaken to evaluate the ability of different iron chelates (ferric-redox state) to catalyze the generation of ·OH by microsomes and reductase. Results were compared to experiments utilizing two other model ·OH-generating systems, the xanthine oxidase and the ascorbate reaction systems.[43]

TABLE 9 represents a summary of experiments designed to evaluate the relative capacity of various iron chelates to mediate the oxidation of KMB and ethanol by NADPH-cytochrome P-450 reductase. The addition of unchelated iron (probably iron phosphate in the 100 mM phosphate buffer) had little effect on the oxidation of KMB and ethanol. Similarly, the presence of iron chelated by adenine nucleotides or by citrate only increased the generation of ·OH two- or threefold. By contrast, when iron was present as either the EDTA- or DTPA-chelates, striking stimulations of KMB and ethanol oxidation were observed. As expected, ferric desferrioxamine did not promote ·OH production.

The poor ability of ferric-adenine nucleotides, ferric citrate, or unchelated ferric iron to catalyze the generation of ·OH by reductase could reflect the poor ability of these particular chelates to be reduced by the system, *i.e.*, to promote the flow of electrons from NADPH. To test this possibility, the relative capacity of the iron chelates to promote the oxidation of NADPH by NADPH-cytochrome P-450 reductase was studied. TABLE 10 shows that unchelated iron, iron citrate, and iron-adenine nucleotides had no effect on the rate of NADPH oxidation. However, consistent with the results of TABLE 9, ferric-EDTA or ferric-DTPA markedly stimulated the oxidation of NADPH. Thus, the ferric chelates that promote ·OH production by the reductase are the chelates that serve as effective electron acceptors from the reductase.

Essentially, similar results are obtained with intact rat liver microsomes, the xanthine oxidase, and the ascorbate model ·OH-generating systems.[43] These results indicate that while iron appears to play an important role in the generation of ·OH, the effect of iron seems to be dependent on the nature of the iron chelate. For example, ferric-EDTA stimulates ·OH production in all the systems tested, whereas ferric desferrioxamine is inhibitory. On the other hand, several physiological chelates such as citrate, ADP, and ATP were not very effective in promoting the generation of ·OH. It

TABLE 9. Effect of Iron Chelates on the Oxidation of KMB and Ethanol by NADPH-Cytochrome P-450 Reductase[a]

Addition	Percent Stimulation by Addition	
	KMB	Ethanol
Fe^{3+} Ammonium Sulfate	218	215
Fe^{3+} EDTA	2,991	1,588
Fe^{3+} DTPA	1,664	941
Fe^{3+} Citrate	227	212
Fe^{3+} ADP	227	174
Fe^{3+} ATP	236	176

[a]Concentration of ferric was 25 μM. Rates of oxidation in the absence of added ferric chelator were 2.2 and 30 nmol/30 min/unit of reductase activity for KMB and ethanol, respectively.

TABLE 10. Effect of Iron Chelates on NADPH Oxidation by Cytochrome P-450 Reductase

Addition	NADPH Oxidation (nmol min^{-1})	Effect (%)
Control	1.2 ± 0.1	
Fe^{3+} Ammonium Sulfate	1.2 ± 0.1	0
Fe^{3+} EDTA	10.2 ± 0.6	+ 750
Fe^{3+} DTPA	13.3 ± 2.5	+ 1,000
Fe^{3+} Citrate	1.7 ± 0.2	+ 42
Fe^{3+} Desferrioxamine	1.1 ± 0.1	− 8
Fe^{3+} AMP	1.4 ± 0.2	+ 17
Fe^{3+} ADP	1.4 ± 0.2	+ 17
Fe^{3+} ATP	1.4 ± 0.3	+ 17

will be important to study the interaction of the above, and other physiologically relevant chelating agents, including iron-containing proteins such as ferritin, with oxygen radical-generating systems, in order to understand the role of iron in promoting lipid peroxidation, the production of ·OH, and, consequently, the oxidation of ethanol via ·OH. In this regard, it is of interest that desferrioxamine has been shown to produce about 15 to 20% inhibition of the *in-vivo* rate of ethanol oxidation by rats.[44] Further studies will be needed to ascertain whether these effects of desferrioxamine reflect chelation of iron and, therefore, prevention of ·OH production. The results suggest caution when EDTA is included in buffer or reaction systems, since ferric-EDTA is strikingly effective as compared to other ferric chelates in catalyzing ·OH generation by a variety of biological, enzymatic, and chemical systems.

Increased ·OH Generation by Microsomes after Chronic Ethanol Consumption

The microsomal ethanol oxidizing system increases in activity after chronic ethanol consumption.[24,25] This increase is due to the induction of an alcohol-preferring isozyme of cytochrome P-450.[45,46] Experiments were carried out to determine (a) the role of ·OH in ethanol oxidation by microsomes from ethanol-fed rats; (b) if the rate of ·OH generation by the microsomes is altered after chronic ethanol consumption; and (c) if the increase in the rate of ethanol oxidation by these induced microsomes correlates with an increase in the rate of generation of ·OH.[47]

The oxidation of two typical ·OH scavengers, KMB and DMSO, by microsomes was increased after chronic consumption of ethanol (TABLE 11) as compared to rates found with microsomes from pair-fed controls. There was little or no production of formaldehyde or ethylene in zero-time controls (acid added before microsomes) or in the absence of microsomes or the NADPH-generating system with both kinds of microsomal preparations. The presence of 1 mM azide resulted in an increase in the oxidation of KMB and DMSO by both microsomal preparations, indicating that H_2O_2 was the precursor of the oxidant responsible for these oxidations. In the absence or presence of azide, the rate of ·OH generation was greater with microsomes from the ethanol-fed rats. The oxidation of KMB was inhibited by DMSO, ethanol, and other ·OH scavengers; the oxidation of DMSO was inhibited by KMB, ethanol, and other ·OH scavengers. The inhibition of product formation by either catalase (absence of azide) or competitive ·OH scavengers links the oxidation of KMB and DMSO by both microsomal preparations to the generation of ·OH or ·OH-like species.

TABLE 11. Effect of Chronic Ethanol Consumption on the Microsomal Oxidation of DMSO and KMB

Substrate	Rate of Product Formation (nmol/min/mg Microsomal Protein)			
	− Azide		+ Azide	
	Pair Fed	Ethanol Fed	Pair Fed	Ethanol Fed
Me_2SO (5)	0.70 ± 0.15	1.58 ± 0.28[a]	3.02 ± 0.57	8.07 ± 1.46[a]
KMB (7)	0.90 ± 0.18	1.78 ± 0.37[a]	2.19 ± 0.27	3.82 ± 0.62[b]

[a] $p < 0.02$
[b] $p < .001$.

Results in TABLE 12 show that the microsomal oxidation of ethanol was doubled after chronic ethanol treatment. Desferrioxamine, at concentrations that block almost completely the generation of ·OH by the microsomes, did not significantly inhibit the oxidation of ethanol (< 15%). Thus, ethanol oxidation was occurring primarily by a desferrioxamine-insensitive, *i.e.*, ·OH-independent pathway. Moreover, the increase in MEOS activity by microsomes from the ethanol-fed rats was not blocked by desferrioxamine, thus disassociating this increase from ·OH. Other experiments showed that in the presence of iron, and especially iron-EDTA, the oxidation of ethanol was increased by a ·OH-dependent process, and this ·OH-dependent oxidation of ethanol was twofold higher in microsomes from ethanol-fed rats (TABLE 12).

In view of their reactivity and toxicity, there is considerable interest in the production and the properties of reactive oxygen species such as the superoxide anion radical, H_2O_2, singlet oxygen, and the hydroxyl radical. With regard to alcohol, there is renewed interest in ethanol-mediated lipid peroxidation, a complex series of events that involve, in general, oxygen radicals and appear to require catalysis by transition-state metals such as iron. Recent studies have shown that the rate of production of superoxide radical,[48] H_2O_2,[49,50] and ·OH[47] is greater in microsomes from chronically ethanol-fed rats than from controls. There are also reports that iron overload may occur under certain conditions in alcoholic liver disease[51,52] and that the cellular content of glutathione, an important component of the cellular antioxidant defensive mechanism, is lowered after ethanol treatment.[53–55] It is intriguing to speculate that one consequence of the above series produced by ethanol could be increased oxidative stress in the liver. On the other hand, ethanol is an effective ·OH and ·OR scavenging agent. Therefore, ethanol should remove potent oxidants such as ·OH and RO· (and, by analogy, perhaps lipoxyl radicals, LO·). The ability of ethanol to promote or to protect against oxygen radical-mediated events is complex and will require further study.

From a metabolic point of view, the oxidation of ethanol by ·OH is probably of

TABLE 12. NADPH-Dependent Oxidation of Ethanol by Microsomes from Chronic Ethanol-Fed Rats and Pair-Fed Controls

Reaction Condition	Rate of Acetaldehyde Production (nmoles/min/mg Microsomal Protein) Pair-Fed Control	Chronic Ethanol
Control	7.6 ± 2.1	15.0 ± 2.1
Desferrioxamine	6.5 ± 1.1	13.3 ± 1.7
EDTA	16.7 ± 2.4	28.5 ± 3.7
EDTA + Desferrioxamine	8.3 ± 0.7	14.7 ± 2.0
·OH-Dependent	8.4 ± 2.1	13.8 ± 1.9

minimal significance. However, from a toxicological point of view, these interactions may have important implications and perhaps play a role or contribute to the hepatotoxicity associated with ethanol consumption.

ACKNOWLEDGMENTS

I greatly appreciate the assistance of the numerous collaborators and students who were involved in various aspects of this research. Special acknowledgment is given to Drs. Gerald Cohen, Elisa Dicker, Graciela Krikun, Shelley Klein, Gary Winston, and Dennis Feierman. I thank Ms. Roslyn C. King for typing the manuscript.

REFERENCES

1. Anbar, M. & P. Neta. 1967. J. Appl. Radiat. Isotop. **18:** 493–523.
2. Dorfman, L. M. & G. E. Adams. 1973. NSRRS, National Bureau of Standards. Washington, D.C. 46.
3. Cohen, G. 1977. *In* Alcohol and Aldehyde Metabolizing Systems. R. G. Thurman, H. Drott, J. R. Williamson & B. Chance, Eds. 411. Academic Press. New York, N.Y.
4. Beauchamp, C. & I. Fridovich. 1970. J. Biol. Chem. **245:** 4641–4646.
5. Fong, K. L., P. B. McCay, J. L. Poyer, B. B. Keele & H. Misra. 1973. J. Biol. Chem. **248:** 7792–7797.
6. Heikkila, R. E., H. Barden & G. Cohen. 1974. J. Pharmacol. Exp. Ther. **190:** 501–506.
7. Cohen, G. & R. E. Heikkila. 1974. J. Biol. Chem. **249:** 2447–2452.
8. Cohen, G., R. E. Heikkila, B. Allis, F. Cabbat, D. Dembiec, D. MacNamee, K. Mytilineou & B. Winston. 1976. J. Pharmacol. Exp. Ther. **199:** 336–352.
9. Sippel, H. W. 1973. Acta Chem. Scand. **27:** 541–550.
10. Haber, F. & J. Weiss. 1934. Proc. R. Soc. London, Ser. A **147:** 332–351.
11. Fridovich, I. 1974. Adv. Enzymol. **41:** 35–97.
12. Fridovich, I. 1978. Science **201:** 875–880.
13. Aust, S. D., D. L. Roerig & T. C. Pederson. 1972. Biochem. Biophys. Res. Commun. **47:** 1133–1137.
14. Prough, R. A. & B. S. S. Masters. 1973. Ann. N.Y. Acad. Sci. **212:** 89–93.
15. Dybing, E., S. D. Nelson, J. R. Mitchell, H. A. Sasame & J. R. Gillette. 1976. Mol. Pharmacol. **12:** 911–920.
16. Strobel, H. W. & M. J. Coon. 1971. J. Biol. Chem. **246:** 7826–7829.
17. Michelson, A. M., J. M. McCord & I. Fridovich. 1977. Superoxide and Superoxide Dismutase. Academic Press. New York, NY.
18. Bannister, J. H. & J. V. Bannister. 1980. Biological and clinical aspects of superoxide dismutase. *In* Developments in Biochemistry. Vol. 11B. Elsevier/North Holland. Amsterdam.
19. Cohen, G. & R. A. Greenwald. 1982. Oxy Radicals and Their Scavenger Systems. Elsevier. New York, NY.
20. Halliwell, B. 1978. FEBS Lett. **92:** 321–326.
21. McCord, J. M. & E. D. Day. 1978. FEBS Lett. **86:** 139–142.
22. Walling, C. 1975. Acc. Chem. Res. **8:** 125–131.
23. Ilan, Y. & G. Czapski. 1977. Biochim. Biophys. Acta **498:** 386–394.
24. Lieber, C. S. & L. M. DeCarli. 1970. J. Biol. Chem. **245:** 2505–2512.
25. Lieber, C. S. & L. M. DeCarli. 1972. J. Pharmacol. Exp. Ther. **181:** 279–287.
26. Lagercrantz, C. & S. Forshult. 1969. Acta Chem. Scand. **23:** 811–817.
27. Klein, S. M., G. Cohen & A. I. Cederbaum. 1981. Biochemistry **20:** 6006–6012.
28. Greenwald, R. A. 1985. Handbook of Methods for Oxygen Radical Research. CRC Press. Orlando, FL.
29. Mathews, R. W. & D. F. Sangster. 1965. J. Phys. Chem. **69:** 1930–1944.
30. Winston, G. W. & A. I. Cederbaum. 1982. Biochemistry **21:** 4265–4270.

31. WINSTON, G. W., W. HARVEY, L. BERL & A. I. CEDERBAUM. 1983. Biochem. J. **216:** 415–421.
32. RALEY, J. H., F. F. RUST & W. E. VAUGHAN. 1948. J. Am. Chem. Soc. **70:** 85–95.
33. CEDERBAUM, A. I., A. QUERESHI & G. COHEN. 1983. Biochem. Pharmacol. **32:** 3517–3524.
34. BAKER, R. C., S. M. SORENSON & R. A. DEITRICH. 1982. Alcoholism; Clin. Exp. Res. **6:** 247–254.
35. CEDERBAUM, A. I. & G. COHEN. 1984. Methods Enzymol. **105:** 516–522.
36. CEDERBAUM, A. I. & G. COHEN. 1985. *In* Handbook of Methods for Oxygen Radical Research. R. A. Greenwald, Ed. 81–87. CRC Press. Orlando, FL.
37. COHEN, G. & A. I. CEDERBAUM. 1980. Arch. Biochem. Biophys. **199:** 438–447.
38. COHEN, G. & A. I. CEDERBAUM. 1979. Science **204:** 66–68.
39. CEDERBAUM, A. I., E. DICKER & G. COHEN. 1980. Biochemistry **19:** 3698–3704.
40. CEDERBAUM, A. I. & E. DICKER. 1983. Biochem. J. **210:** 107–113.
41. WINSTON, G. W. & A. I. CEDERBAUM. 1983. J. Biol. Chem. **258:**1508–1513.
42. WINSTON, G. W. & A. I. CEDERBAUM. 1983. J. Biol. Chem. **258:** 1514–1519.
43. WINSTON, G. W., D. E. FEIERMAN & A. I. CEDERBAUM. 1984. Arch. Biochem. Biophys. **232:** 378–390.
44. SINACEUR, J., C. RIBIERE, C. ABU-MURAD, J. NORDMANN & R. NORDMANN. 1983. Biochem. Pharmacol. **32:** 2371–2373.
45. OHNISHI, K. & C. S. LIEBER. 1977. J. Biol. Chem. **252:** 7124–7131.
46. KOOP, D. R., E. T. MORGAN, G. E. TARR & M. J. COON. 1982. J. Biol. Chem. **257:** 8472–8480.
47. KLEIN, S. M., G. COHEN, C. S. LIEBER & A. I. CEDERBAUM. 1983. Arch. Biochem. Biophys. **223:** 425–433.
48. BOVERIS, A., C. G. FRAGA, A. I. VARSAVSKY & O. R. KOCH. 1983. Arch. Biochem. Biophys. **227:** 534–541.
49. LIEBER, C. S. & L. M. DECARLI. 1970. Science **170:** 78–79.
50. THURMAN, R. G. 1973. Mol. Pharmacol. **9:** 670–675.
51. CHAPMAN, R. W., M. Y. MORGAN, R. BELL & S. SHERLOCK. 1983. Gastroenterology **84:** 143–148.
52. PRIETO, J., M. BARRY & S. SHERLOCK. 1975. Gastroenterology **68:** 525–533.
53. VIDELA, L. A., V. FERNANDEZ, G. UGARTE & A. VALENZUELA. 1980. FEBS Lett. **111:** 6–10.
54. MACDONALD, C. M., J. DOW & M. R. MOORE. 1977. Biochem. Pharmacol. **26:** 1529–1531.
55. SHAW, S., E. JAYATILLEKE, W. A. ROSS, E. R. GORDON & C. S. LIEBER. 1981. J. Lab. Clin. Med. **98:** 417–424.

DISCUSSION OF THE PAPER

M. J. COON (*University of Michigan, Ann Arbor, MI*): I think that we are in accord that P-450 is a leaky enzyme; it tends to reduce oxygen to superoxide, peroxide, and hydroxyl radicals. On the other hand, one must distinguish between these free hydroxyl radicals in solution and the activated oxygen generated on the surface of the P-450 at the iron atom in the heme ring. The hydroxyl radical is tightly bound to iron and does not exchange with the hydroxyl radicals in solution.

We would all like to know the relative contributions of P-450 and other mechanisms of alcohol oxidation in the intact organism. Do you have any idea to what extent this iron-stimulated Haber-Weiss reaction contributes to alcohol oxidation?

CEDERBAUM: About two years ago Sinaceur and Nordman performed an *in-vivo* experiment injecting desferrioxamine into rats to study the rates of ethanol elimination. A modest decrease in ethanol oxidation is reported to have resulted from desferrioxamine administration, an effect attributed to inhibition of hydroxyl radical formation.

However, experiments *in vivo* are complicated by effects on absorption. Perhaps desferrioxamine alters ADH, shuttle mechanisms, or other pathways. Until we truly understand the availability of iron in the liver cell, it is hazardous to ascribe a significant metabolic contribution to the hydroxyl radical. In microsomes, only small amounts of ethanol are oxidized by a ·OH-dependent pathway.

COON: Have you looked at tissues other than the liver? The alcohol-P-450 is at a high level compared to other nonhepatic tissues such as the nasal membranes and the kidney.

CEDERBAUM: In the lung and kidney hydroxyl radicals are produced by the microsomes, and their formation is stimulated by iron. However, the rates are much lower than in the liver, a finding that probably reflects the lower content of P-450 reductase in those tissues.

We agree that the oxidant produced by P-450 (the equivalent of ·OH bound to the heme) does not exchange with free ·OH generated in solution. We have employed stereoisomers of 2-butanol to evaluate free and "bound" ·OH. No stereochemical discrimination for plus or minus 2-butanol was found when ·OH was generated in solution. Some stereochemical preference was found for the P-450-dependent pathway, depending on the isozymes of P-450 present.

The Functional Implications of Acetaldehyde Binding to Cell Constituents[a]

MICHAEL F. SORRELL AND DEAN J. TUMA

Liver Study Unit
Veterans Administration Medical Center
and
Departments of Internal Medicine and Biochemistry
University of Nebraska Medical Center
Omaha, Nebraska 68105

INTRODUCTION

Chronic ethanol consumption is a major cause of liver disease. Although the underlying pathogenic mechanisms are unclear, present evidence indicates that ethanol and/or its metabolites are directly injurious to the liver. However, genetic, environmental, and nutritional factors may also modulate the hepatotoxicity of ethanol.[1-3]

Since ethanol is primarily metabolized in the liver, which is a major target organ of ethanol-induced toxicity, many of the functional and structural alterations in the liver produced by ethanol consumption have been attributed to the products of ethanol oxidation.[1-4] In this regard, formation of acetaldehyde, the first metabolite of ethanol, and the altered redox state (increased NADH/NAD ratios in the cytosolic and mitochondrial compartments) have been implicated in many of the ethanol-induced alterations of hepatic structure and function.[1-5] Recently, we have formulated a hypothesis, postulating that acetaldehyde via its covalent binding to hepatic proteins may be a critical event leading to liver injury.[6,7] Such a hypothesis is consistent with the evidence implicating ethanol as a direct hepatotoxin and, in addition, takes into account the possible role of metabolic, nutritional, and environmental factors in the pathogenesis of alcoholic liver injury. In this report, we will briefly describe some of the characteristics of acetaldehyde binding to proteins, but will mostly focus on potential consequences of such binding.

COVALENT BINDING OF ACETALDEHYDE TO PROTEINS

Because of the electrophilic nature of the carbonyl carbon, acetaldehyde is a very reactive compound and is able to react with a variety of nucleophilic groups.[8] Since proteins possess many nucleophilic groups, it is not surprising that acetaldehyde readily reacts with proteins. Indeed, numerous studies have demonstrated this reactivity of acetaldehyde with proteins under physiological conditions (pH 7.4 and 37° C). Acetaldehyde has been shown to covalently bind to albumin,[9,10] plasma proteins,[11] erythrocyte membrane proteins,[12] hepatic microsomal proteins,[13] enzymes,[14] and hemoglobin.[15] Although acetaldehyde can also bind to lipids[16,17] and nucleic acids,[18] the

[a]Supported by National Institute on Alcohol Abuse and Alcoholism Grant AA04961 and by the Veterans Administration.

reaction products are mainly unstable and reversible and are formed to a lesser extent when compared to proteins.[13,16,17,19]

We have recently characterized the covalent binding of acetaldehyde to proteins and have shown that both stable and unstable adducts form as a result of this binding. In this study,[9] unstable adducts were shown to be mainly Schiff bases, resulting from the reaction of the carbonyl carbon of acetaldehyde with the ϵ-amino group of lysine. Unstable adducts were characterized by their ability to readily dissociate; however, they could be stabilized and, therefore, made detectable by reducing the Schiff bases to stable secondary amines with sodium borohydride. During short-term incubation (up to 3 hr), unstable adducts comprised the majority (75–85%) of initial reaction products. The remaining adducts were stable and relatively irreversible, since they were unaffected by treatment with acid or base, exhaustive dialysis, or gel filtration.

Additional studies investigated the effects of both long-term incubation and physiological reducing agents on the binding of acetaldehyde to proteins.[10] The results describing the binding of acetaldehyde (200 μM) to bovine serum albumin over a 10-day time course in the presence and absence of the physiological reducing agent, ascorbic acid, are shown in FIGURE 1. In the absence of ascorbate, maximum formation of total acetaldehyde-albumin adducts (stable plus unstable) occurred after 24 hr of incubation. However, the percentage of total adducts that were stable increased steadily throughout the reaction period, ranging from about 35% at the earlier time periods to over 70% after 240 hr. The presence of ascorbate caused a continuous increase in the formation of total acetaldehyde-albumin adducts. In addition, an increasingly greater percentage of the total adducts were of the stable type so that, after 72 hr, over 90% of the adducts formed were stable. These results suggest that the majority of initial reaction products of acetaldehyde with proteins are unstable adducts, but with time these unstable adducts can be stabilized either via chemical rearrangement or by spontaneous reduction. Ascorbate likely accelerated this stabilization process, resulting in a net overall increase in the formation of total adducts composed almost entirely of stable adducts. The ability of ascorbate to act as a reducing agent and, thereby, convert the unsaturated, unstable Schiff bases to stable products is the most likely mechanism by which ascorbate increases and stabilizes the formation of acetaldehyde-protein adducts; however, other explanations cannot be excluded at this time.

Based upon our previous results discussed above[9,10,14,19] and the available information in the literature concerning the reaction of amino groups with carbonyl groups,[8,20,21] the following reaction scheme that describes the formation of stable acetaldehyde-protein adducts via Schiff-base intermediates is presented in FIGURE 2. The initial reaction products are likely to be Schiff bases resulting from the reaction of the carbonyl group of acetaldehyde with an ϵ-amino group of lysine. This unstable adduct has numerous possible fates. Schiff bases can dissociate to reform acetaldehyde and protein or undergo a simple exchange reaction with another amino group. However, Schiff bases can be stabilized by additions across the double bond either by reduction or by addition of a strong nucleophile such as a thiol group. Reduction of the double bond would result in the formation of *N*-ethyllysyl and *N*-diethyllysyl residues in the protein. Since ethanol oxidation in the liver increases the availability of reducing equivalents and since a physiological reducing agent, like ascorbate, has been shown to reduce Schiff bases, we consider the stabilization of Schiff bases by reduction as a likely possibility to explain stable acetaldehyde-adduct formation in the liver during ethanol oxidation. Although other reaction schemes cannot be discounted at this time, recent results in our laboratory have indicated that lysine is the major amino acid participating in the binding of acetaldehyde to proteins. This conclusion is based upon the results of the analysis by HPLC ion-exchange chromatography of the acetaldehyde-amino acid adducts formed during the reaction of acetaldehyde with proteins. Similar products were obtained when acetaldehyde reacted with albumin, polylysine, or t-boc-lysin (unpublished data). Thus, lysine residues appear to participate in both unstable and stable acetaldehyde-protein adduct formation.

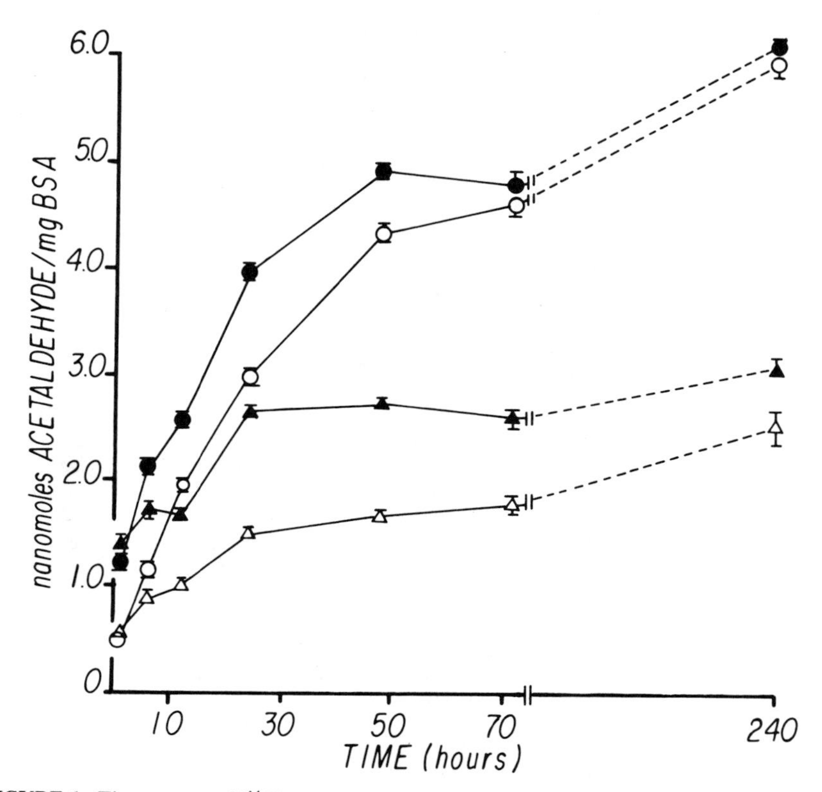

FIGURE 1. Time course of [^{14}C]acetaldehyde-adduct formation with bovine serum albumin in the presence and absence of 5 mM ascorbate. Acetaldehyde (200 μM) was incubated with albumin (6 mg/ml) in 0.1 M phosphate buffer (pH 7.4) at 37° C. Individual tubes at each time point were collected, chilled, and assayed for total and stable acetaldehyde-albumin adducts.[10] Total adducts plus ascorbate (●); stable adducts plus ascorbate (○); total adducts minus ascorbate (▲); stable adducts minus ascorbate (△). Results are the mean values (± S.E.) of four determinations. (From Tuma *et al.*[10] Reprinted by permission from *Archives of Biochemistry and Biophysics.*)

BINDING OF ACETALDEHYDE TO HEPATIC PROTEINS DURING ETHANOL OXIDATION

Recent studies from our laboratory have shown that acetaldehyde generated during ethanol oxidation does covalently bind to hepatic proteins.[19,22] In these studies, ethanol oxidation in a cell-free system (liver homogenates) and in liver slices resulted

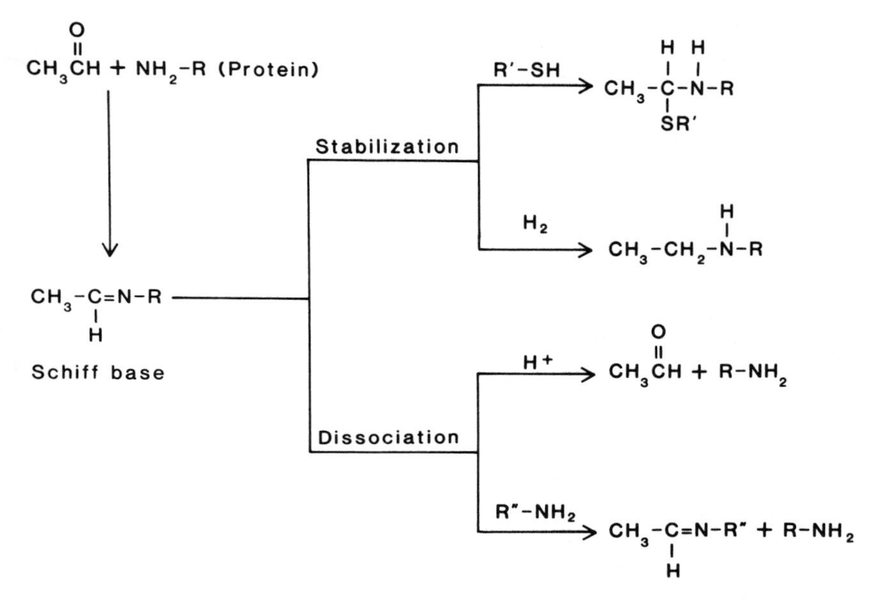

FIGURE 2. Stable acetaldehyde-protein adduct formation via Schiff-base intermediates.

in the formation of acetaldehyde, which, subsequently, reacted with hepatic proteins to form both unstable and stable acetaldehyde-protein adducts. Furthermore, evidence was obtained to suggest that unstable adducts were converted to stable adducts during incubation. Consideration of these results and the chemistry of acetaldehyde-protein adduct formation has allowed us to propose the reaction scheme depicted in FIGURE 3, which describes the formation of acetaldehyde-protein adducts in the liver during ethanol oxidation.[6,7] As indicated, ethanol oxidation via alcohol dehydrogenase results in the formation of acetaldehyde, which initially interacts with free amino groups of proteins (mainly the ϵ-amino group of lysine) to form unstable and reversible Schiff-base adducts. However, ethanol oxidation also generates excess reducing equivalents (increased NADH), which would then be available to promote the reduction of Schiff bases to form stable adducts.

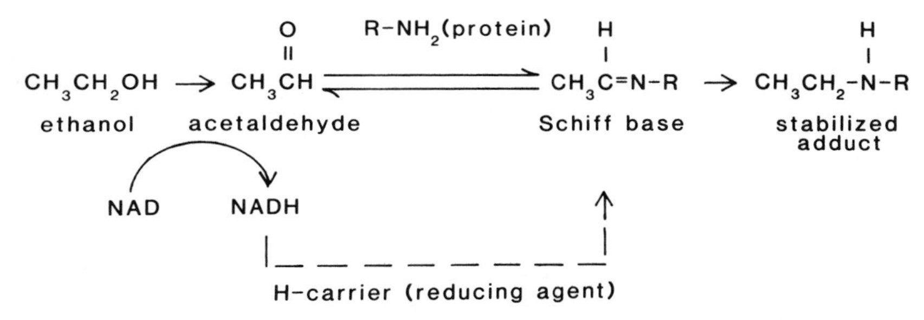

FIGURE 3. Formation of acetaldehyde-protein adducts during ethanol oxidation in the liver.

Although the formation of acetaldehyde-protein adducts in the liver during ethanol oxidation appears to be established, information concerning the extent and nature of adduct formation as well as the potential consequences to the liver is essential in determining whether adduct formation plays a role in altered hepatic structure and function. These aspects of acetaldehyde-adduct formation will be considered in the following sections of this report.

CONDITIONS MODIFYING THE BINDING OF ACETALDEHYDE IN THE LIVER

Many conditions in the liver could affect the extent of acetaldehyde binding to hepatic proteins. The most obvious would be the elevation of acetaldehyde levels during ethanol oxidation. In this regard, chronic ethanol consumption has been shown to elevate acetaldehyde levels by increasing its formation as well as by decreasing its oxidation.[23] Furthermore, the altered redox state (elevated NADH/NAD ratio) in the liver associated with ethanol oxidation could also promote acetaldehyde-adduct formation. Thus, ethanol oxidation would not only produce acetaldehyde for reaction with proteins but would also contribute reducing equivalents to further increase and stabilize adduct formation. Although the specific reducing agent that stabilizes Schiff-base adducts in the liver is unknown, ascorbic acid has been shown to increase and stabilize acetaldehyde-protein adducts.[10] This observation has added importance, since the altered redox state has been reported to be markedly exaggerated in the centrilobular region of the liver where alcoholic liver injury starts and predominates.[24]

The hepatic levels of water-soluble nucleophiles could also be important in the regulation of acetaldehyde-protein binding. Such compounds could trap acetaldehyde, and thus prevent its reaction with hepatic proteins. Our previous studies[9,19,22] demonstrated that the thiol compounds, glutathione and cysteine, were especially effective in decreasing the binding of acetaldehyde to proteins, whereas compounds with amino groups, such as lysine, were much less effective. These results would suggest that the levels of thiol compounds may be important in regulating the formation of acetaldehyde-protein adducts. Therefore, conditions such as fasting,[25] intake of drugs such as acetaminophen,[26] and particularly acute and chronic ethanol intake,[27–29] all of which lower hepatic thiol compounds, could promote the formation of acetaldehyde-protein adducts in the liver.

In summary, it appears that chronic ethanol consumption creates a milieu in the liver that favors the formation and stabilization of acetaldehyde-protein adducts. The elevated acetaldehyde levels, the decrease in thiol groups, and the exaggerated redox changes in the centrilobular region caused by chronic ethanol intake could significantly contribute to increasing the formation of stable and irreversible acetaldehyde-protein adducts. These considerations could explain, in part, the site (centrilobular region) and the slow development of alcoholic liver injury as well as its constant potentiation by continued abuse of ethanol. Furthermore, those conditions that modify adduct formation could be influenced by genetic, environmental, and nutrititional factors, and thus, could explain the role of these factors in the pathogenesis of alcoholic liver injury.

EFFECTS OF ACETALDEHYDE ADDUCTS ON THE BIOLOGICAL FUNCTION OF PROTEINS

Although the ability of acetaldehyde to bind to proteins and the formation of acetaldehyde-protein adducts in the liver during ethanol oxidation have been established, the functional consequences of such interactions have not been thoroughly investigated. Correlation of acetaldehyde-adduct formation with impaired function of a specific protein is an essential step in evaluating the role of acetaldehyde adducts in the injurious effects caused by alcohol abuse.

Since acetaldehyde can bind to proteins and form both unstable and stable adducts, functional consequences caused by both types of adducts will be considered. Extensive formation of Schiff bases (unstable adducts) could potentially modify the function of proteins. Such alterations could include the displacement of pyridoxal phosphate from its binding site on proteins[30] and interference with the activity of certain enzymes, especially those forming Schiff base-enzyme complexes as intermediates in their catalytic activity.[31] However, since the acetaldehyde-protein Schiff bases are unstable and readily reversible, the effects would be transient, and the possibility that they would lead to long-lasting dysfunction is doubtful. Furthermore, since Schiff bases readily dissociate, it would take very high concentrations of acetaldehyde to produce extensive Schiff-base formation. This factor, in view of the low concentrations of acetaldehyde in the body during ethanol oxidation,[23,32] would seem to argue against an important role of Schiff bases in alcoholic liver injury. It would appear from these considerations that Schiff-base adducts must be stabilized before any significant alterations of biological activity of proteins would be manifested and be relevant to the *in-vivo* situation.

Few studies in the literature have addressed the functional consequences of acetaldeyde adducts. Karp *et al.*[33] reported a marked inhibition of both monoacetyldiaminodiphenyl sulfone, a deputy ligand for bilirubin, and diazepam binding to stable adducts of acetaldehyde-albumin when compard to native albumin. Although they did not show direct evidence for the presence of acetaldehyde-albumin adducts in alcoholics, they suggested that their presence is likely and further implied that changes in the binding ability of albumin adducts may explain altered drug binding in alcoholics. Tsuboi *et al.*[34] showed that the formation of stable adducts of acetaldehyde with hemoglobin resulted in changes in erythrocyte-oxygen affinity. These same authors reported changes in erythrocyte morphology that could be attributed to adduct formation, especially since acetaldehyde has been shown to cross-link erythrocyte membrane proteins (spectrin and actin).[12] Dramatic loss of the biological activity of the enkephalins has also been reported when these peptides formed stable adducts with acetaldehyde.[35] Studies in our laboratory have also suggested that stable acetaldehyde-protein adducts in the liver could have potentially long-term detrimental effects. The inhibition of hepatic protein secretion induced by acute and chronic ethanol administration[5] could be a consequence of acetaldehyde binding to intracellular proteins, essential to the secretory process. This secretory defect was shown to be a consequence of ethanol metabolism and was potentiated when acetaldehyde metabolism was blocked,[5,36,37] strongly suggesting that acetaldehyde mediates this defect. In addition, the inhibitory effect of acetaldehyde on hepatic protein secretion was shown to be irreversible and dose responsive,[38] indicating that this effect could be related to the formation of stable and irreversible acetaldehyde-protein adducts. It has also been suggested by Barry and McGivan[39] that the binding of acetaldehyde to hepatic plasma membrane proteins may result in the activation of the complement system.

Our laboratory has recently undertaken studies to systematically explore the effects of acetaldehyde-protein adducts on the biological function of specific proteins. In our initial report,[14] we investigated whether the binding of acetaldehyde to selected enzymes altered their catalytic activity. In these studies, [^{14}C]acetaldehyde at various concentrations was incubated with yeast alcohol dehydrogenase, glucose-6-phosphate dehydrogenase and lactate dehydrogenase, each at 37° C and pH 7.4. Comparable degrees of acetaldehyde-adduct formation occurred among the three enzymes tested. Analysis of the binding data revealed that adduct formation was somewhat dependent on the number of lysyl residues per protein molecule, however, not completely, indicating that other factors such as the particular microenvironment of each amino group of lysine could also be a factor in determining its relative reactivity toward acetaldehyde. The effects of acetaldehyde pretreatment on the catalytic activity of these three enzymes are shown in FIGURE 4. Glucose-6-phosphate dehydrogenase activity was significantly inhibited by reaction with acetaldehyde, whereas the activities of lactate dehyrogenase and alcohol dehydrogenase were essentially unaffected. Further analyses of the data indicated that there was a direct correlation in the inhibition of glucose-6-phosphate dehydrogenase and the formation of total stabilized acetaldehyde adducts. On the other hand, despite comparable binding to acetaldehyde, the activities of the other two enzymes tested were essentially unaltered. The most likely explanation for these differential effects of acetaldehyde binding on the activity of these enzymes is that glucose-6-phosphate dehydrogenase has an essential lysine residue necessary for catalytic activity, whereas the other two do not. When acetaldehyde was directly added during the assay for glucose-6-phosphate dehydrogenase activity and the enzyme was not pretreated with acetaldehyde, minimal effects on

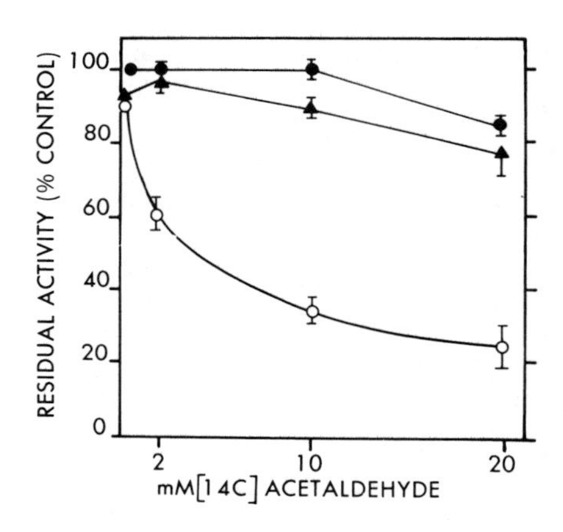

FIGURE 4. Effect of acetaldehyde-adduct formation on enzymatic activity. Alcohol dehydrogenase (▲, n = 3), lactate dehydrogenase (●, n = 3), or glucose-6-phosphate dehydrogenase (○, n = 5), each at 0.6 mg per ml was reacted for 1 hr with various concentrations of acetaldehyde. After reduction by sodium borohydride, aliquots were diluted and assayed for enzymatic activity as described.[14] Residual enzymatic activities (per cent of the unreacted control) were expressed as means ± S.E. (From Mauch *et al.*[14] Reprinted by permission from the American Association for the Study of Liver Diseases.)

activity were noted, indicating that acetaldehyde itself or rapidly forming unstable Schiff-base adducts apparently were not factors in inhibiting activity. From these results, we concluded that stable modification of enzyme lysyl residues by acetaldehyde can impair the function of certain enzymes, especially those containing an essential lysine residue at their active site.[14]

The experiments just described showed that adduct formation between glucose-6-phosphate dehydrogenase and acetaldehyde inhibited enzyme activity. These data were obtained by incubating the enzyme for short time periods (1 hr) with high concentrations of acetaldehyde in order to generate sufficient adduct formation to inhibit enzyme activity (FIG. 4). However, extended incubation of the enzyme with low concentrations of acetaldehyde (0.2 mM) for up to three days resulted in adduct formation that inhibited catalytic activity. The level of adduct formation and the corresponding degree of enzyme inhibition were similar to those observed for higher concentrations of acetaldehyde incubated for shorter periods.[14] These results indicate that when certain enzymes are exposed to low levels of acetaldehyde for extended periods of time, adduct formation sufficient to cause enzyme inhibition can occur. Such conditions could result from chronic ethanol consumption, in which hepatic proteins are repeatedly subjected to acetaldehyde, formed in stoichiometric amounts during ethanol oxidation.

These studies demonstrated that enzymes, with essential lysine residues, appear to be especially susceptible to inhibition of their catalytic activity as a result of acetaldehyde-adduct formation. This would imply that acetaldehyde is reacting with the specific lysine residue at the catalytic site of the enzyme and in this way decreasing enzyme activity. To explore this possibility, we used the enzyme, RNase, which has 10 lysyl residues including one at its active site.[40] The reason RNase was chosen as a model was that previous studies by Jentoft *et al.*[40] showed that phosphate, an active-site ligand, specifically protected lysine-41 at the catalytic site from reductive methylation by formaldehyde. Other lysyl residues were still able to react. Therefore, the effects of acetaldehyde-adduct formation with RNase were studied in a phosphate-buffered system and a phosphate-free system (Hepes buffer).[14] The results of this experiment are shown in FIGURE 5. The time course of acetaldehyde-adduct formation was similar to both buffer systems (FIG. 5B), but the effects on enzymatic activity markedly differed (FIG. 5A). Although binding in the two systems was similar, phosphate prevented the inhibition of RNase activity. These results would be expected if the catalytically essential lysine was selectively protected from reaction with acetaldehyde, and if binding to that particular lysine was required for inhibition of enzyme activity. Later studies, using [^{13}C]-NMR analysis of acetaldehyde-RNase adducts confirmed that the protective action of phosphate was due to decreased binding to the active-site lysyl residue.[41] Therefore, it appears that stabilized adducts formed between acetaldehyde and catalytically essential lysyl residues can result in marked alterations of enzymatic activity. This conclusion would suggest that the biological activity of proteins, which have essential lysine residues, would be especially susceptible to being functionally altered by acetaldehyde-adduct formation.

In companion studies, described in another article in this volume,[42] we demonstrated that acetaldehyde-adduct formation decreased the ability of tubulin to assemble into microtubules. In addition, certain lysine residues in the α-chain of the tubulin dimer had a selective and enhanced reactivity toward acetaldehyde. This selectivity of binding to the α-chain occurred for stable adducts but not for unstable ones. Furthermore, impaired tubulin assembly correlated with stable but not unstable adduct formation. These findings further substantiate our previous conclusion that binding of acetaldehyde to essential lysine residues on protein can alter their biological function. Furthermore, they indicate that certain lysine residues on proteins can have an enhanced reactivity toward acetaldehyde, suggesting that certain proteins in the liver (or other organs) could be preferential targets of acetaldehyde binding and subsequently could have their function selectively impaired.

There are other examples in the literature that also show that the covalent binding of other agents to lysine residues of certain proteins alters the properties and functions of the target protein. The most widely studied has been the nonenzymatic binding of glucose to a variety of proteins. As was the case for acetaldehyde, initial binding appears to involve formation of a Schiff base with an amino group (ϵ-amino groups of lysine or terminal α-amino groups); in this case the Schiff base can rearrange (Amandori rearrangement) and form a relatively stable product.[43,44] Various alterations in the biological function of nonenzymatic glycosylated proteins have been reported, and it has been suggested that extensive formation of these glycosylated products could be the basis for many of the complications occurring in long-term diabetes.[45] In a recent review, Brownlee *et al.*[45] reported numerous examples of the biological alterations caused by nonenzymatic glycosylation of macromolecules. These include: inactivation of enzymes, inhibition of regulatory molecule binding, cross-linking of proteins, decreased susceptibility to proteolysis, altered endocytosis, and

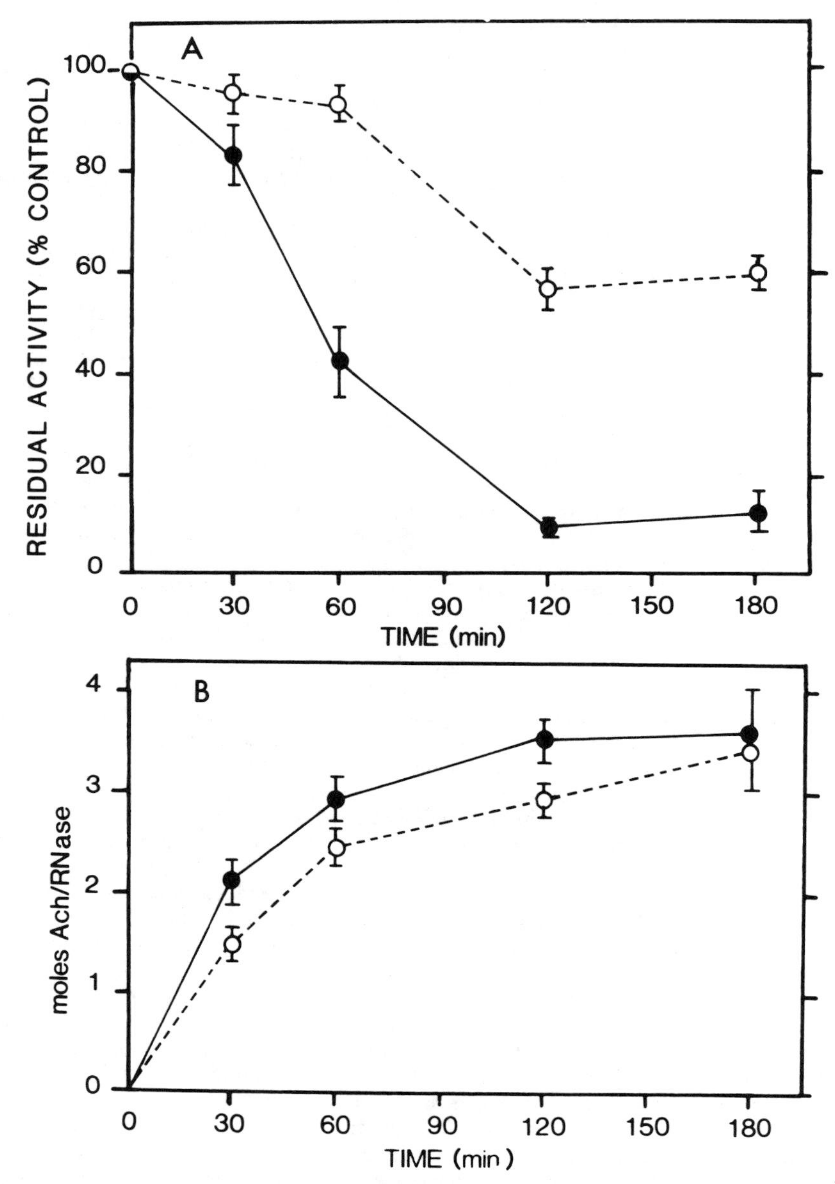

FIGURE 5. Effect of phosphate on acetaldehyde Ach/RNase-adduct formation and enzyme inhibition. RNase A (0.6 mg/ml) was reacted with 20 mM Ach at 37° C (pH 7.4) for 3 hr in either 0.2 M sodium phosphate buffer (○, n = 5) or in 0.2 M Hepes buffer (●, n = 7) in the presence of 20 mM sodium cyanoborohydride, a reducing agent that stabilizes Schiff bases. At the indicated time points, aliquots of each reaction mixture were assayed for Ach-adduct formation and for enzymatic activity.[14] Residual activity (**A**) and Ach-adduct formation (**B**) were expressed as means ± S.E. (From Mauch *et al.*[14] Reprinted by permission from the American Assoication for the Study of Liver Diseases.)

increased immunogenicity. Other aldehydes such as glyceraldehyde[46] and benzaldehyde[47] also react with lysine groups of hemoglobin by a similar mechanism, resulting in altered oxygen-affinity of hemoglobin. Lysine residues of proteins are also susceptible to nonenzymatic acetylation by agents such as acetyl CoA.[48] Covalent steroid-protein adducts, also involving lysine groups, have been suggested as a contributing factor in the pathological effects produced by elevated levels of certain endogenous steroids.[49] Recent studies by Tsai *et al.*[50] showed that modification of lysine residues of horse liver alcohol dehydrogenase by acetylation, glycosylation, or methylation, interestingly, increased the oxidative activity of this enzyme. All of the above studies point out that modification of certain lysine residues of certain proteins can markedly alter their biological function and further imply that this process can potentially initiate pathological processes in the body.

In summary, the covalent binding of acetaldehyde can impair the biological activity or function of certain proteins, especially those containing essential lysine residues. Stable acetaldehyde adducts appear to be more deleterious than unstable adducts in modifying the properties and functions of proteins. Proteins also appear to contain various lysine residues that possess relatively different degrees of reactivity toward acetaldehyde. Furthermore, certain lysine residues seem to be especially reactive with acetaldehyde, indicating a type of selective or preferential binding of acetaldehyde to certain key lysine residues in the polypeptide chain. If these reactive lysine residues are intimately involved in the biological activity of a protein, a low degree of adduct formation could potentially have serious effects on the function of these proteins. Therefore, it is possible that stable binding of acetaldehyde could occur with certain preferential target proteins in the liver (or other organs) and that a relatively low level of such binding could potentially be damaging and could play a role in alcoholic liver injury and other manifestations of alcohol toxicity.

REFERENCES

1. LIEBER C. S. 1984. Alcohol and the liver: 1984 update. Hepatology **4:** 1243–1260.
2. ORREGO, H., Y. ISRAEL & L. M. BLENDIS. 1981. Alcoholic liver disease: information in search of knowledge. Hepatology **1:** 267–283.
3. MEZEY, E. 1985. Metabolic effects of alcohol. Fed. Proc. **44:** 134–138.
4. SORRELL, M. F. & D. J. TUMA. 1979. Effects of alcohol on hepatic metabolism: selected aspects. Clin. Sci. **57:** 481–489.
5. TUMA, D. J. & M. F. SORRELL. 1984. Effect of ethanol on hepatic secretory proteins. *In* Recent Developments in Alcoholism. M. Galanter, Ed. Vol. 2: 159–180. Plenum Press. New York, NY.
6. TUMA, D. J. & M. F. SORRELL. 1985. Covalent binding of acetaldehyde to hepatic proteins: role in alcoholic liver injury. *In* Aldehyde Adducts in Alcoholism. M. A. Collins, Ed. Prog. Clin. Biol. Res. Vol. 183: 3–17. Alan R. Liss, Inc. New York, NY.
7. SORRELL, M. F. & D. J. TUMA. 1985. Hypothesis: Alcoholic liver injury and the covalent binding of acetaldehyde. Alcoholism: Clin. Exp. Res. **9:** 306–309.
8. O'DONNELL, J. P. 1982. The reaction of amines with carbonyls: its significance in the nonenzymatic metabolism of xenobiotics. Drug Metab. Rev. **13:** 123–159.
9. DONOHUE, T. M., D. J. TUMA & M. F. SORRELL. 1983. Acetaldehyde adducts with proteins: binding of [^{14}C]acetaldehyde to serum albumin. Arch. Biochem. Biophys. **220:** 239–246.
10. TUMA, D. J., T. M. DONOHUE, V. A. MEDINA & M. F. SORRELL. 1984. Enhancement of acetaldehyde-protein adduct formation by L-ascorbate. Arch. Biochem. Biophys. **234:** 377–381.
11. LUMENG, L., R. MINTER & T. K. LI. 1982. Distribution of stable acetaldehyde adducts in blood under physiological conditions. Fed. Proc. **41:** 765.
12. GAINES, K. C., J. M. SALHANY, D. J. TUMA & M. F. SORRELL. 1977. Reactions of acetaldehyde with human erythrocytes membrane proteins. FEBS Lett. **75:** 115–119.

13. NOMURA, F. & C. S. LIEBER. 1981. Binding of acetaldehyde to rat liver microsomes: enhancement after chronic alcohol consumption. Biochem. Biophys. Res. Commun. **100:** 131–137.
14. MAUCH, T. J., T. M. DONOHUE, R. K. ZETTERMAN, M. F. SORRELL & D. J. TUMA. 1986. Covalent binding of acetaldehyde selectively inhibits the catalytic activity of lysine-dependent enzymes. Hepatology **6:** 263–269.
15. STEVENS, V. J., W. J. FANTL, C. B. NEWMAN, R. V. SIMS, A. CERAMI & C. M. PETERSON. 1981. Acetaldehyde adducts with hemoglobin. J. Clin. Invest. **67:** 361–369.
16. KENNY, W. C. 1982. Acetaldehyde adducts of phospholipids. Alcoholism: Clin Exp. Res. **6:** 412–416.
17. KENNY, W. C. 1984. Formation of Schiff base adduct between acetaldehyde and rat liver microsomal phosphatidylethanolamine. Alcoholism: Clin. Exp. Res. **8:** 551–555.
18. RISTOW, H. & G. OBE. 1978. Acetaldehyde induces cross-links in DNA and causes sister chromatid exchanges in human cells. Mutat. Res. **58:** 115–119.
19. MEDINA, V. A., T. M. DONOHUE, M. F. SORRELL & D. J. TUMA. 1985. Covalent binding of acetaldehyde to hepatic proteins during ethanol oxidation. J. Lab. Clin. Med. **105:** 5–10.
20. BAILEY, A. J. 1968. Intermediate labile intermolecular cross-links in collagen fibres. Biochim. Biophys. Acta **160:** 447–453.
21. OAKES, T. R. & G. W. STACY. 1972. Reactions of thiols with Schiff bases in nonaqueous solvents. Addition equilibria, cleavage, and reduction. J. Am. Chem. Soc. **94:** 1594–1600.
22. DONOHUE, T. M., D. J. TUMA & M. F. SORRELL. 1983. Binding of metabolically derived acetaldehyde to hepatic proteins in vitro. Lab. Invest. **49:** 226–229.
23. NUUTINEN, H., K. O. LINDROS & M. SALASPURO. 1983. Determinants of blood acetaldehyde levels during ethanol oxidation in chronic alcoholics. Alcoholism: Clin. Exp. Res. **7:** 163–168.
24. JAUHONEN, P., E. BARAONA, H. MIYAKAWA & C. S. LIEBER. 1982. Mechanism for selective perivenular hepatotoxicity of ethanol. Alcoholism: Clin. Exp. Res. **6:** 350–357.
25. PESSAYRE, D., A. DOLDER, J. Y. ARTIGOU, J. C. WANDSCHEER, V. DESCATOIRE, C. DEGOTT & J. P. BENHAMOU. 1979. Effect of fasting on metabolite-mediated hepatotoxicity in the rat. Gastroenterology **77:** 264–271.
26. GILLETTE, J. R. 1981. An integrated approach to the study of chemically reactive metabolites of acetaminophen. Arch. Intern. Med. **141:** 375–379.
27. MACDONALD, C. M., J. DOW & M. R. MOORE. 1977. A possible protective role for sulphydryl compounds in acute alcoholic liver injury. Biochem. Pharmacol. **26:** 1529–1531.
28. VINA, J., J. M. ESTRELA, C. GUERRI & F. J. ROMERO. 1980. Effect of ethanol on glutathione concentration in isolated hepatocytes. Biochem. J. **188:** 549–552.
29. FERNANDEZ, V. & L. A. VIDELA. 1981. Effect of acute and chronic ethanol ingestion on the content of reduced glutathione of various tissues of the rat. Experientia **37:** 392–393.
30. LUMENG, L. 1978. The role of acetaldehyde in mediating the deleterious effect of ethanol on pyridoxal 5′ phosphate metabolism. J. Clin. Invest. **62:** 286–293.
31. GRAZI, E., H. MELOCHI, G. MARTINEZ, W. A. WOOD & B. L. HORECKER. 1963. Evidence for Schiff base formation in enzymatic aldol condensations. Biochem. Biophys. Res. Commun. **10:** 4–10.
32. K. O. LINDROS. 1982. Human blood acetaldehyde levels: with improved methods, a clearer picture emerges. Alcoholism: Clin. Exp. Res. **6:** 70–75.
33. KARP, W. B., M. KINSLEY, S. B. SUBRAMANYAM & A. F. ROBERTSON. 1985. Binding properties of glycosylated albumin and acetaldehyde albumin. Alcoholism: Clin. Exp. Res. **9:** 429–432.
34. TSUBOI, K. K., D. J. THOMPSON, E. M. RUSH & H. C. SCHWARTZ. 1981. Acetaldehyde-dependent changes in hemoglobin and oxygen affinity of human erythrocytes. Hemoglobin **5:** 241–250.
35. SUMMERS, M. C. 1985. Structural and biological studies of the acetaldehyde adducts of enkephalins and related peptides: a short review. *In* Aldehyde Adducts in Alcoholism. M. A. Collins, Ed. Prog. Clin. Biol. Res. Vol. 183: 39–49. Alan R. Liss, Inc. New York, NY.
36. VOLENTINE, G. D., D. J. TUMA & M. F. SORRELL. 1984. Acute effects of ethanol on hepatic glycoprotein secretion in the rat in vivo. Gastroenterology **86:** 225–229.

37. VOLENTINE, G. D., D. K. KORTJE, D. J. TUMA & M. F. SORRELL. 1985. Role of acetaldehyde in the ethanol-induced inhibition of hepatic glycoprotein secretion in vivo. Hepatology **5:** 1056.
38. SORRELL, M. F., D. J. TUMA & A. J. BARAK. 1977. Evidence that acetaldehyde irreversibly impairs glycoprotein metabolism in liver slices. Gastroenterology **73:** 1138–1141.
39. BARRY, R. E. & J. D. MCGIVAN. 1985. Acetaldehyde alone may initiate hepatocellular damage in acute alcoholic liver disease. Gut **26:** 1065–1069.
40 JENTOFT, J. E., T. A. GERKEN, N. JENTOFT & D. G. DEARBORN. 1981. [^{13}C] methylated ribonuclease A. ^{13}C NMR studies of the interaction of lysine 41 with active site ligands. J. Biol. Chem. **256:** 231–236.
41. MAUCH, T. J., T. M. DONOHUE, R. K. ZETTERMAN, M. F. SORRELL & D. J. TUMA. 1985. Covalent binding of acetaldehyde to lysine-dependent enzymes can inhibit catalytic activity. Hepatology **5:** 1056.
42. TUMA, D. J., R. B. JENNETT & M. F. SORRELL. 1987. The interaction of acetaldehyde with tubulin. Ann. N.Y. Acad. Sci. This volume.
43. MONNIER, V. M. & A. CERAMI. 1983. Nonenzymatic glycosylation and browning of proteins in vivo. *In* The Mailliard Reaction in Foods and Nutrition. G. R. Waller & M. S. Feather, Eds. American Chemical Society Symposium Series No. 215: 431–439. The Amerian Chemical Society. Washington, D.C.
44. HIGGINS, P. J. & H. F. BUNN. 1981. Kinetic analysis of the nonenzymatic glycosylation of hemoglobin. J. Biol. Chem. **256:** 5204–5208.
45. BROWNLEE, M., H. VLASSARA & A. CERAMI. 1984. Nonenzymatic glycosylation and the pathogenesis of diabetic complications. Ann. Intern. Med. **101:** 527–537.
46. ACHARYA, A. S., L. G. SUSSMAN & J. M. MANNING. 1983. Schiff base adducts of glyceraldehyde with hemoglobin. J. Biol. Chem. **258:** 2296–2302.
47. BEDDELL, C. R., P. J. GOODFORD, G. KNEEN, R. D. WHITE, S. WILKINSON & R. WOOTTON. 1984. Substituted benzaldehydes designed to increase the oxygen affinity of human hemoglobin and inhibit the sickling of sickle erythrocytes. Br. J. Pharmacol. **82:** 397–407.
48. GARBUTT, G. J. & E. C. ABRAHAM. 1981. Non-enzymatic acetylation of human hemoglobins. Biochim. Biophys. Acta **670:** 190–194.
49. BUCALA, R., J. FISHMAN & A. CERAMI. 1982. Formation of covalent adducts between cortisol and 16 α-hydroxyestrone and protein: Possible role in the pathogenesis of cortisol toxicity and systemic lupus erythematosus. Proc. Natl. Acad. Sci. USA **79:** 3320–3324.
50. TSAI, C. S., D. J. SENIOR, J. H. WHITE & J. L. PITT. 1985. Functional and structural responses of liver alcohol dehydrogenase to lysine modifications. Bioorg. Chem. **13:** 47–56.

DISCUSSION OF THE PAPER

H. WEINER (*Purdue University, West Lafayette, IN*): The K_ms for the aldehyde are the same in liver homogenates and with the pure enzyme. This implies that binding of acetaldehyde only occurs for the short time of the assay. Is it possible that the acetaldehyde really doesn't bind, or that it does not irreversibly bind to proteins?

SORRELL: The data show that it can irreversibly bind. The question is, how much of the ethanol is metabolized to acetaldehyde, and how much acetaldehyde is bound. Our data in liver slices suggest that it may be 1–2%. Most of the acetaldehyde is not irreversibly bound, but forms a Schiff base.

E. MEZEY (*Johns Hopkins School of Medicine, Baltimore, MD*): Please comment on the concentrations of acetaldehyde that bind and form adducts.

SORRELL: Binding is more dramatic at high concentrations of acetaldehyde. It does occur at lower concentrations, and at much lower concentrations when using microtubules. The longer you expose the protein to acetaldehyde, the lower the concentration needed.

J. SENIOR (*Merion, PA*): In severe alcoholic liver disease the serum transaminases do not increase as much as they do in viral hepatitis. Is there adduct formation with the transaminases?

SORRELL: That's possible, but we have not investigated this problem.

T.-K. LI (*Indiana University School of Medicine, Indianapolis, IN*): The lower transaminase activities in cirrhosis may be due to the lack of saturation of the transaminases with pyridoxal phosphate. You can increase the activity by adding pyridoxal phosphate to sera. We have shown that acetaldehyde competes with pyridoxal phosphate binding to transaminases and other enzymes.

Y. ISRAEL (*University of Toronto, Toronto, Canada*): Have you been able to stabilize your adducts with molecules that have SH groups? If that reaction actually occurs, two proteins can cross-link by this mechanism.

SORRELL: We are studying the problem.

ISRAEL: We have now shown that these adducts can become neoantigens and that one can produce immunoglobulin monoclonal antibodies against the acetaldehyde portion of the molecule. Such antibodies might attack cells.

SORRELL: Your finding may reopen the concept of an immunologic basis of alcoholic liver diseaes.

C.S. LIEBER (*Veterans Administration Medical Center, Bronx, NY*): What happens after chronic alcohol feeding in terms of the adduct formation with specific enzymes? A few years ago we found a striking increase in adducts after chronic alcohol feeding. Have you studied this situation?

SORRELL: We have not.

In-Vivo and *In-Vitro* Inhibition of the Low K_m Aldehyde Dehydrogenase by Diethylmaleate and Phorone

ELISA DICKER AND ARTHUR I. CEDERBAUM

Department of Biochemistry
Mount Sinai School of Medicine
One Gustave L. Levy Place
New York, New York 10029

INTRODUCTION

Mitochondria isolated from rat liver[1] contain low and high K_m aldehyde dehydrogenases. The oxidation of acetaldehyde occurs primarily by the low K_m mitochondrial aldehyde dehydrogenase.[1,2] Formaldehyde also can be metabolized by the low K_m aldehyde dehydrogenase and by a specific-cytosolic-glutathione-dependent formaldehyde dehydrogenase.[3,4] In experiments conducted to evaluate the contribution of each of these enzyme systems to the oxidation of formaldehyde by intact rat hepatocytes, it was found that the oxidation of formaldehyde was inhibited 30–50% by cyanamide, a potent inhibitor of the low K_m aldehyde dehydrogenase, or by competitive substrates such as acetaldehyde and crotonaldehyde. Mitochondrial oxidation of formaldehyde was also inhibited by cyanamide or acetaldehyde, whereas the glutathione-dependent formaldehyde dehydrogenase was not affected.[5,6] These experiments suggested that both the mitochondrial and the cytosolic formaldehyde dehydrogenases may be contributing equally toward the overall metabolism of formaldehyde.[5,6] To characterize further these enzyme systems, attempts were made to inhibit the glutathione-dependent formaldehyde dehydrogenase system by removing the glutathione cofactor of the enzyme with the glutathione-depleting agents diethylmaleate (DEM) and phorone. Results from these studies are presented in this report.

METHODS

Hepatocytes were prepared from fed 250–300 gm male Sprague Dawley rats as described previously.[7] Oxidation of formaldehyde (final concentration of 0.2 and 1.0 mM and specific activity of 0.083 μCuries per μmol) to $^{14}CO_2$ and formate by isolated hepatocytes was assayed as previously published.[7] Mitochondrial and cytosolic fractions were prepared by standard methods of differential centrifugation.[7]

In-vitro experiments were performed by the addition of DMSO, DEM, or phorone to mitochondria for 10 min before starting the reactions with acetaldehyde or formaldehyde and by incubating hepatocytes with DEM for 60 min before starting the reaction with formaldehyde. *In-vivo* experiments were performed by IP injection of DEM or phorone 60 min before preparation of liver fractions.

TABLE 1. Effect of *In-Vivo* Treatment and *In-Vitro* Addition of Diethylmaleate and Phorone on the Oxidation of Formaldehyde and Acetaldehyde by Rat Liver Mitochondria[a]

		Oxidation (nmol/min/mg Mitochondrial Protein)			
		Formaldehyde	Effect %	Acetaldehyde	Effect %
Treatment					
Corn oil		15.59 ± 1.69		25.14 ± 1.04	
DEM		3.00 ± 1.13	−81[b]	2.78 ± 0.24	−89[b]
Phorone		9.93 ± 0.95	−36[c]	17.59 ± 1.80	−30[c]
Addition (mM)					
DMSO	0.5	9.86 ± 1.11		19.15 ± 0.84	
DEM	0.5	6.10 ± 0.66	−38[b]	10.09 ± 1.20	−47[b]
DEM	1.0	3.02 ± 0.66	−69[b]	4.97 ± 2.33	−74[b]
Phorone	0.5	6.05 ± 0.98	−39[b]	9.44 ± 2.64	−51[c]

[a] *In-vivo* treatment was performed by IP injections of either corn oil, DEM, or phorone into rats. The oxidation of 1.0 mM formaldehyde or 0.2 mM acetaldehyde was assayed in mitochondria prepared from the livers of the treated rats. The oxidation of formaldehyde and acetaldehyde was assayed in mitochondria from rat liver controls in the presence of the indicated concentrations of DMSO (control), DEM, or phorone. Results are from 3 to 4 experiments each in duplicate.

[b] $p < 0.001$.

[c] $p < 0.005$.

TABLE 2. Effect of DEM on the Oxidation of Formaldehyde by Isolated Hepatocytes[a]

	Product Formation from Formaldehyde (nmol/min/mg Cell Protein)					
Addition	CO_2	Effect %	Formate	Effect %	Formate + CO_2	Effect %
			0.2 mM Formaldehyde			
None	0.67 ± 0.02		0.25 ± 0.04		0.91 ± 0.05	
Cyanamide	0.53 ± 0.05	−21[b]	0.09 ± 0.04	−65[b]	0.65 ± 0.08	−28[b]
DEM	0.63 ± 0.07	−3	0.16 ± 0.03	−25	0.79 ± 0.07	−8
DEM, washed	0.58 ± 0.09	−11	0.19 ± 0.03	−11	0.77 ± 0.09	−10
			1.0 mM Formaldehyde			
None	0.95 ± 0.06		1.52 ± 0.08		2.47 ± 0.13	
Cyanamide	0.39 ± 0.03	−59[c]	0.32 ± 0.05	−79[c]	0.72 ± 0.06	−71[c]
DEM	0.58 ± 0.03	−38[d]	0.63 ± 0.12	−59[c]	1.21 ± 0.14	−51[c]
DEM, washed	0.72 ± 0.17	−24	0.94 ± 0.21	−38[d]	1.66 ± 0.25	−33[d]

[a] Hepatocytes were incubated with 1 mM DEM for 60 min before starting reactions with formaldehyde. Washed hepatocytes were resuspended in fresh buffer after the incubation period. When present, cyanamide was used at a concentration of 0.1 mM. Results are from 3 to 4 experiments.

[b] $P < 0.05$.

[c] $P < 0.005$.

[d] $P < 0.02$.

RESULTS AND DISCUSSION

Metabolism of formaldehyde proceeds primarily by two different pathways; the low K_m mitochondrial aldehyde dehydrogenase, and the specific-glutathione-dependent cytosolic formaldehyde dehydrogenase. When the first pathway was blocked by cyanamide or acetaldehyde there was an approximately 50% inhibition of formaldehyde oxidation.[5,6] The second pathway has no known inhibitors, but the glutathione cofactor can be depleted by treatment with diethylmaleate or phorone. IP injections of these compounds reduced the total glutathione content of rat livers by about 90% and mitochondrial content by about 40%. Although the mitochondrial low K_m enzyme does not require glutathione, the oxidation of acetaldehyde and formaldehyde by mitochondria decreased by more than 80% (TABLE 1). *In-vitro* addition of DEM or phorone to isolated mitochondria produced identical inhibition. This inhibition was similar to that produced by cyanamide.[5,6] Cyanamide, DEM, and phorone appeared to act at the same enzyme site, since under conditions of reduced rates of acetaldehyde or formaldehyde oxidation produced by cyanamide, no further inhibition was observed with DEM or phorone.

The activity of the high K_m aldehyde dehydrogenase was not inhibited by either the *in-vivo* or the *in-vitro* treatments. As a consequence of removal of the glutathione cofactor, the activity of the cytosolic formaldehyde dehydrogenase was also decreased; the activity was restored by the addition of glutathione, indicating that these effects of DEM are due to the loss of GSH and not to direct effects on the enzyme.[7]

In isolated hepatocytes, the rates of oxidation of acetaldehyde and high (but not low) concentrations of formaldehyde were decreased in the presence of DEM (TABLE 2). Rates were restored to almost control values when the DEM was removed by washing.[7] Formaldehyde oxidation that is insensitive to DEM or cyanamide occurs via the GSH-formaldehyde dehydrogenase pathway, which has a higher affinity for formaldehyde than mitochondrial aldehyde dehydrogenase. In the presence of DEM, there was an accumulation of acetaldehyde during the oxidation of ethanol by isolated hepatocytes, which probably reflects impairment of the low K_m enzyme (results not shown). The results presented in this report suggest that DEM and phorone are effective *in-vivo* and *in-vitro* inhibitors of the low K_m aldehyde dehydrogenase in addition to acting as glutathione-depleting agents. Therefore, the effects of these agents should be interpreted with caution, especially in studies involving metabolism of aldehydes such as formaldehyde and in evaluating alcohol-induced lipid peroxidation.

REFERENCES

1. SIEW, C., R. A. DEITRICH & U. G. ERWIN. 1976. Arch. Biochem. Biophys. **176:** 638–649.
2. TOTTMAR, O., H. PETTERSSON & K. H. KIESSLING. 1973. Biochem. J. **135:** 577–586.
3. GOODMAN, J. J. & T. R. TEPHLY. 1971. Biochim. Biophys. Acta **252:** 489–505.
4. UOTILA, L. & M. KOIVUSALO. 1974. J. Biol. Chem. **249:** 7653–7663.
5. DICKER, E. & A. I. CEDERBAUM. 1984. Arch. Biochem. Biophys. **232:** 179–188.
6. DICKER, E. & A. I. CEDERBAUM. 1984. Arch. Biochem. Biophys. **234:** 187–196.
7. DICKER, E. & A. I. CEDERBAUM. 1985. Biochim. Biophys. Acta **843:** 107–113.

Developmental Profile of Alcohol and Aldehyde Dehydrogenases in Long-Sleep and Short-Sleep Mice[a]

T. N. SMOLEN, A. SMOLEN, AND J. VAN DE KAMP

Institute for Behavioral Genetics
University of Colorado
Boulder, Colorado 80309

We have been studying the developmental pattern of alcohol sensitivity in Long-Sleep (LS) and Short-Sleep (SS) mice. Young mice are less sensitive to acute and chronic ethanol treatment than are adults. Furthermore, young LS and SS mice differ in their sleep time response at 15 days of age, and this difference increases with age.[1] We believe these differences between young and adult mice, and young LS and SS mice are due to differences in CNS sensitivity to ethanol, but differences in metabolism could also account for this relative tolerance. Ethanol is oxidized by alcohol dehydrogenase (ADH), and the metabolite, acetaldehyde, by aldehyde dehydrogenase (AlDH). These metabolizing enzymes were measured in LS and SS mice of both sexes at 3, 6, 12, 16, 20, 24, 28, 32, and 55–65 (adult) days of age. ADH (16 mM ethanol) was measured in cytosol. High- (10 mM acetaldehyde) and low- (0.05 mM acetaldehyde) K_m forms of AlDH were measured in cytosol, mitochondria, and microsomes. The subcellular fractions were isolated by differential centrifugation.[2] There were no sex differences in ADH and AlDH activities at any age, thus, the values in TABLES 1 and 2 are the average of both sexes.

The pattern of ADH activity was the same for both LS and SS mice: the activity rose linearly from day 3, peaked at day 28, then declined 20% to adult levels. Although ADH activity for LS mice increases dramatically between 16 and 20 days of age, their blood ethanol elimination rates are not different at these two ages, nor are they different from SS mice.[1]

The AlDH data confirm earlier studies that showed the majority of high-K_m AlDH activity is in the cytosol of mouse liver.[2-4] Cytosolic high-K_m AlDH activity remained low until day 24, and then rose linearly to adult levels. The low-K_m form of the enzyme remained low at all ages.

In mitochondria, the high-K_m AlDH activity pattern was very different for LS and SS mice, which may reflect a developmental difference. The low-K_m mitochondrial AlDH pattern is nearly identical for LS and SS mice: low activity was observed at 3 days of age and rose to adult levels by 16 days of age. The developmental profiles of the high- and low-K_m AlDH in cytosol and mitochondria are very different, suggesting that these are different enzymes.[4]

Microsomal high-K_m AlDH activity peaked at 12 days of age, and declined thereafter to low adult levels. The pattern was similar for LS and SS mice.

[a]Supported in part by National Institute on Alcohol Abuse and Alcoholism Grants AA 06487 and AA 06527 to T. N. Smolen, and National Institute of Neurological and Communicative Disorders and Stroke Grant NS 20748 and National Institute of Child Health and Human Development Grant HD 21709 to A. Smolen.

TABLE 1. Developmental Profile of Alcohol and High-K_m Aldehyde Dehydrogenases in LS and SS Mouse Liver[a]

Age (Days)	Line	ADH	AlDH Cytosol	AlDH Mitochondria	AlDH Microsomes
3	LS	1.58 ± 0.28	1.18 ± 0.30	3.86 ± 0.35	1.96 ± 0.32
	SS	2.54 ± 0.17	1.64 ± 0.32	1.80 ± 0.52	0.75 ± 0.20
6	LS	2.11 ± 0.30	1.27 ± 0.11	4.22 ± 0.38	1.11 ± 0.33
	SS	2.34 ± 0.25	1.57 ± 0.15	2.52 ± 0.30	0.69 ± 0.21
12	LS	2.96 ± 0.39	1.99 ± 0.23	4.64 ± 0.92	0.98 ± 0.36
	SS	4.27 ± 0.36	3.05 ± 0.36	4.66 ± 1.07	1.72 ± 0.48
16	LS	3.67 ± 0.50	2.97 ± 0.17	6.61 ± 0.70	0.50 ± 0.22
	SS	5.91 ± 0.28	3.29 ± 0.38	4.38 ± 0.84	0.25 ± 0.10
20	LS	10.05 ± 0.68	5.25 ± 0.70	5.58 ± 1.31	0.40 ± 0.14
	SS	9.63 ± 0.28	4.87 ± 0.75	4.17 ± 1.06	0.22 ± 0.12
24	LS	10.23 ± 0.33	4.87 ± 0.86	0.87 ± 0.34	0.30 ± 0.21
	SS	9.35 ± 0.51	9.91 ± 1.69	5.90 ± 0.71	0.12 ± 0.09
28	LS	10.61 ± 1.00	10.94 ± 1.09	5.02 ± 0.69	0.70 ± 0.24
	SS	11.52 ± 0.57	12.12 ± 1.46	3.98 ± 0.79	0.30 ± 0.13
32	LS	10.11 ± 0.41	13.90 ± 1.00	2.08 ± 0.57	0.03 ± 0.03
	SS	11.80 ± 0.63	14.72 ± 0.80	5.27 ± 1.03	0.26 ± 0.16
60	LS	8.47 ± 0.42	17.96 ± 1.30	5.69 ± 1.32	0.86 ± 0.42
	SS	7.84 ± 0.47	18.82 ± 0.91	6.87 ± 0.82	0.72 ± 0.31

[a]Enzyme activities were measured in subcellular fractions prepared by differential centrifugation.[2] Tabled values are the mean ± S.E.M. of 9 to 19 mice per group. Enzyme specific activities are expressed as nmol NADH formed per min per mg protein at 37° C and pH 7.4 with 1.0 mM NAD as cofactor. Blanks contained no substrate. ADH activity was measured with 10 mM acetaldehyde as substrate. Cytosolic AlDH assays included 1.0 mM pyrazole.

TABLE 2. Developmental Profile for Low-K_m Acetaldehyde Dehydrogenase Activity in LS and SS Mouse Liver[a]

Age (Days)	Line	Cytosol	Mitochondria
3	LS	0.41 ± 0.09	1.65 ± 0.19
	SS	0.54 ± 0.09	1.73 ± 0.25
6	LS	0.24 ± 0.05	1.71 ± 0.15
	SS	0.63 ± 0.15	1.74 ± 0.18
12	LS	0.40 ± 0.06	3.27 ± 0.36
	SS	0.40 ± 0.09	3.29 ± 0.49
16	LS	0.51 ± 0.09	4.19 ± 0.33
	SS	0.72 ± 0.14	4.60 ± 0.69
20	LS	0.84 ± 0.12	3.85 ± 0.58
	SS	0.84 ± 0.13	4.30 ± 0.53
24	LS	0.86 ± 0.24	3.41 ± 0.48
	SS	0.99 ± 0.14	4.91 ± 0.43
28	LS	0.77 ± 0.13	4.48 ± 0.63
	SS	2.58 ± 0.42	4.03 ± 0.43
32	LS	2.95 ± 0.59	4.20 ± 0.58
	SS	2.40 ± 0.38	5.06 ± 0.50
60	LS	1.72 ± 0.19	4.36 ± 0.47
	SS	1.78 ± 0.17	4.33 ± 0.32

[a]Enzyme activities were measured in subcellular fractions prepared by differential centrifugation.[2] Tabled values are the mean ± S.E.M. of 9 to 19 mice per group. Enzyme specific activities are expressed as nmol NADH formed per min per mg protein at 37° C and pH 7.4 with 1.0 mM NAD as cofactor. Blanks contained no substrate. AlDH activity was measured with 0.05 mM acetaldehyde as substrate. Cytosolic AlDH assays included 1.0 mM pyrazole.

In conclusion, although there are slight differences between the LS and SS lines in the developmental profiles of both ADH and A1DH (SS mice increased toward adult activity levels approximately 4 days sooner than LS mice), this does not appear to correlate with any of the behavioral or physiological differences we have observed between LS and SS mice at these ages. Furthermore, the young mice have achieved adult enzyme activity at a time when they are more tolerant to ethanol than adults. Therefore, metabolic factors do not appear to play a role in the relative insensitivity of young mice to ethanol.

REFERENCES

1. SMOLEN, T. N., A. SMOLEN & J. VAN DE KAMP. 1985. Pharmacologist **9:** 281.
2. SMOLEN, A., T. N. SMOLEN & A. C. COLLINS. 1982. Comp. Biochem. Physiol. **73B:** 815–822.
3. SMOLEN, A., A. L. WAYMAN, T. N. SMOLEN, D. R. PETERSEN & A. C. COLLINS. 1981. Comp. Biochem. Physiol. **69C:** 199–204.
4. LITTLE, R. G., II & D. R. PETERSEN. 1983. Comp. Biochem. Physiol. **74C:** 271–279.

Alcohol and Acetaldehyde Metabolism:

Concluding Remarks

ESTEBAN MEZEY

Johns Hopkins School of Medicine
Baltimore, Maryland 21205

The genetic and acquired differences in ethanol and acetaldehyde metabolism described in this section may predispose to alcoholism and alcohol-associated tissue damage.

Great strides have been made in the identification and characterization of multiple molecular forms of human alcohol dehydrogenase, which differ widely in kinetic properties. The distribution of the molecular forms of alcohol dehydrogenase varies in different populations and most likely accounts for individual differences in ethanol elimination. Also important to remember is that alcohol dehydrogenase has a broad substrate range with multiple exogenous and endogenous substrates. The specificity and affinity of the multiple molecular forms of alcohol dehydrogenase for various substrates is largely unknown but most likely very variable. Hence, the effect of ethanol ingestion on the metabolism of endogenous substrates of alcohol dehydrogenase may differ depending on the enzymatic makeup of the individual. The effects of other influences, such as nutritional and hormonal, on alcohol dehydrogenase activity and ethanol metabolism, which are well documented in animals, remain to be defined in man.

Microsomal oxidation of ethanol has now conclusively been demonstrated to be catalyzed by a unique cytochrome P-450 isoenzyme. The per-cent contribution of microsomal oxidation of ethanol metabolism is small except at high ethanol concentrations. The importance of this pathway relates to the interaction between ethanol and the metabolism of endogenous compounds, drugs, and carcinogens. Acute ethanol ingestion inhibits, while chronic ethanol ingestion induces, metabolism of these compounds. The implications of these observations are enormous as regards organ injury.

All present evidence indicates that liver mitochondrial aldehyde dehydrogenase is principally responsible for acetaldehyde oxidation. Decrease in the activity of this enzyme or inactive enzyme as found in many orientals results in marked accumulation of acetaldehyde and flushing on the ingestion of ethanol. The accumulation of acetaldehyde may have toxic consequences, but interestingly enough individuals with the inactive enzyme may have a lower risk of alcoholism. The cytoplasmic aldehyde dehydrogenase, which is more sensitive to disulfiram inhibition and decreases after chronic ethanol consumption, apparently also contributes to acetaldehyde oxidation. We have been particularly interested in the role of erythrocyte aldehyde dehydrogenase in oxidizing circulating acetaldehyde released from the liver and hence decreasing the level of acetaldehyde reaching extrahepatic organs. The erythrocyte aldehyde dehydrogenase has a K_m of 49 μM for acetaldehyde, and its activity increases after acute ethanol ingestion but decreases after chronic ethanol consumption.

The varied effects of ethanol as regards oxygen-derived radicals and their possible consequences is very intriguing. Chronic ethanol ingestion results in increased production of superoxide ion and hydroxyl radicals, which leads to increased lipid peroxidation. On the other hand, ethanol is a scavenger of hydroxyl radicals, which causes its metabolism to acetaldehyde. It would be important to know whether or not the presence of ethanol suppresses enhanced lipid peroxidation observed after chronic ethanol feeding.

Finally, acetaldehyde is a very reactive metabolite, and its binding to cell components has been postulated to be a major mechanism of tissue injury. The binding of acetaldehyde to lysyl groups of the catalytic site of enzymes resulting in decreased enzyme activity indicates a specific mechanism for metabolic alterations in the cell. Studies to correlate changes in enzyme activity produced by acetaldehyde binding with metabolic and physiological consequences may provide new insights into mechanisms of tissue injury in alcoholism.

General Anesthetic and Specific Effects of Ethanol on Acetylcholine Receptors[a]

KEITH W. MILLER, LEONARD L. FIRESTONE, AND STUART A. FORMAN

Departments of Anaesthesia and Pharmacology
Massachusetts General Hospital and Harvard Medical School
Boston, Massachusetts 02114

Ethanol is effective at much higher concentrations than almost any other drug. It is sufficiently lipid-soluble to gain access to all parts of the body, and consequently it exerts a wide variety of actions at many different loci. More recently it has become clear that ethanol may exert a number of different actions even at a single well-defined target.[1] The question of interest here is how many molecular mechanisms underlie ethanol's actions in producing different effects on a single target protein. We shall approach this question within the framework of current thinking about the mechanism of general anesthesia, and we shall use as our paradigm the acetylcholine receptor from *Torpedo*. Of all ethanol's loci of action on excitable membranes it is the best characterized at the molecular level.

Theories of general anesthesia fall into two overall classes; those that suppose anesthetics to act indirectly on proteins via perturbations of their surrounding lipids and those that postulate a direct anesthetic-protein interaction. Although controversy exists over which type of interaction underlines general anesthesia itself, it is clear that both types of action do occur and must be considered.[2,3]

LIPID THEORIES OF ANESTHETIC ACTION

General anesthetics are a structurally diverse group of molecules that all share the property of being lipid soluble. Indeed the correlation between anesthetic potency and solubility in lipid bilayers is extremely good—for some two dozen general anesthetics for which data is currently available the concentration in lipid bilayers during anesthesia falls in a narrow range around 25–50 mM.[4]

How does the anesthetic, once in the lipid bilayer, perturb protein function? Many attempts have been made to address this problem. Their relative success has been reviewed recently.[2–7] All the theories share the feature that anesthetics perturb the lipid bilayer's structure in some way and this perturbation is then transmitted to those membrane proteins whose function is altered. A general statement of all lipid-perturbation hypotheses[4] can be written as:

$$E^{50} = C^{50} \cdot P_L \cdot T_P \qquad \textbf{(1)}$$

[a]Supported by National Institute of General Medical Sciences Training Grant GM-07592 and Center Grant GM-15904 to the Harvard Anaesthesia Research Center and by the Department of Anesthesia, Massachusetts General Hospital.

where E^{50} means that a half-maximal change has occurred in some functional assay, such as a group of animals being anesthetized or an excitable membrane being inhibited, and when C^{50} is the concentration of the anesthetic in the lipid bilayer. P_L is a perturbation of the lipid bilayer produced by unit concentration of anesthetic in the lipid bilayer, and T_P describes the transmission of that lipid perturbation to the function of a membrane protein.

When the last two terms in Equation 1 are ignored it reduces to the lipid-solubility hypothesis. The lipid-perturbation hypotheses assume that the last term is constant and ascribe various different meanings to P_L, such as lipid fluidity or disorder, membrane expansion, lateral phase separation in a lipid bilayer, and so on. Since C^{50} is usually constant, it follows that if the hypothesis is correct P_L will usually be constant too; the size of the perturbation will be directly proportional to the concentration of anesthetic in the lipid bilayer (this is what is referred to as a colligative property). The value of T_P should be independent of the anesthetic, but it should depend on the protein and its lipid environment in order to explain why some membrane proteins are more sensitive than others to anesthetics.

Specific Lipid Perturbations

Although a general anestheticlike interaction must obey the above rules, it is interesting to consider how lipid perturbation mechanisms of the type in Equation 1 might also lead to quite selective actions of anesthetics. This can happen if C^{50} or P_L depends on the nature of the anesthetic. In this case E^{50} is no longer a constant independent of the anesthetic, but varies. In pharmacological terms anesthetics differ in their efficacy. This might occur for two reasons. First, the anesthetics may be distributed within the bilayer in a nonuniform way so that the average concentration in the membrane does not accurately represent the anesthetic's microdistribution. For example, a polar or amphipathic anesthetic would be concentrated at the bilayer interfaces relative to the middle of the bilayer. Second, the nature of the membrane perturbation may not be colligative. For example, it might vary with the shape of the anesthetic.

The most highly studied example of such effects is provided by the polycyclic alcohol, cholesterol. The hydroxyl group of cholesterol is located in the interfacial region of the bilayer and is not randomly distributed across the bilayer.[8] In addition cholesterol's shape determines its ability to order phospholipid bilayers; chemical modification generally leads to a decrease in this ordering ability.[9]

Less information is available about the more pharmacologically interesting alcohols. Comparison of the free energy for transferring hydrocarbons[10] and alcohols[11] from water to lipid bilayers suggests that the alcohols in the bilayer retain a hydrogen bond for a good proportion of the time. This conclusion is supported by nuclear magnetic resonance (NMR) studies of benzyl alcohol.[12] Thus we may expect ethanol to be preferentially located at the acyl-chain–polar-region interface of the lipid bilayer, with smaller quantities penetrating into the nonpolar interior. This property is probably shared by longer-chain alcohols.[4] Ethanol will only penetrate a short distance into a lipid bilayer, separating the acyl chains at the ester-bond level of the phospholipids. Since free space cannot exist in the bilayer and the entropic cost of allowing water to contact the acyl chains is prohibitively high, the acyl chains will flex to fill this space. In contrast, hexadecanol will separate neighboring phospholipids without perturbing the acyl chains, because it matches them in length.

An expression of this differential perturbation due to shape factors (in this case length) is provided by the action of alcohols in depressing the gel to liquid crystalline phase transitions of phospholipids. Medium chain-length 1-alkanols depress this phase-transition temperature in a colligative manner; longer 1-alkanols raise this temperature, while shorter 1-alkanols lower it at low concentrations and raise it at high concentrations.[13,14] Ethanol's action was elucidated by X-ray-diffraction studies that showed that as the phase transition temperature began to rise the two lipid leaflets interdigitated and the bilayer thickness decreased from 4.2 nm to 3.0 nm.

Furthermore, a given protein might be sensitive to a number of different changes in its surrounding lipid bilayer (see below).

Specific Lipid-Protein Coupling in Axonal Membranes

The action of general anesthetics on axonal conduction has been studied in great detail by Haydon and his colleagues (for review, see REFS. 15, 16). At first sight the effects of different classes of anesthetics seem to be specific. They found, for example, that steady-state inactivation curves[15] (in Hodgkin-Huxley terminology, h_∞ verses voltage) were shifted to the left by hydrocarbons but not at all by alcohols. This could be explained if the hydrocarbons, but not the alcohols, increased membrane thickness. An inactivation-voltage sensor in the membrane would then experience a decreased-voltage gradient only with hydrocarbons. This interpretation was consistent with capacitance measurements that suggested that, indeed, only hydrocarbons changed the axonal membrane's thickness. Thus in terms of Equation 1 P_L is the bilayer thickness and T_P is the relationship of the voltage sensor to the voltage gradient across the bilayer. The picture that emerges from these studies is one in which specific effects can often be explained consistently within the framework of the lipid theories, provided that the complexity of anesthetic-lipid interactions is taken into account.

ANESTHETIC-PROTEIN INTERACTIONS

General anesthetics are known to interact with proteins.[2] This is not surprising, because about half the amino acids of proteins have hydrophobic side chains. In membrane proteins these amino acids may be in the lipid bilayer (see above), but in regions of the protein outside the membrane, or in cytoplasmic proteins, they will be internalized to avoid contact with water. Sometimes hydrophobic residues may be folded around an often rigid cofactor (for example, nicotinamide or flavin nucleotides). Alternatively proteins may aggregate to form oligomers with opposed hydrophobic faces.[17]

Two types of anesthetic-protein interactions can be distinguished. In the first there is great specificity, and only a few anesthetics take part. In the second there is less specificity, and the site may appear to mimic that of general anesthesia.

Specific Interactions

The best example of the first type is provided by myoglobin. This protein may be crystallized and its structure studied by X ray diffraction.[18,19] Xenon at a partial pressure of 2.5 atm binds near the heme in an empty pocket surrounded by nonpolar amino-acid side chains. Difluorodichloromethane, which is slightly larger than xenon, can only bind in the same pocket by pushing some of these side chains aside. Further increase in the anesthetic's size prevents binding.

Nonspecific Interactions

Typical of the much less selective type of anesthetic-protein interaction are the luciferases of bacteria and fireflies.[20] Recent work on the firefly enzyme shows that inhibition is competitive with a cofactor, luciferin. The anesthetic potency of a wide range of volatile anesthetics and alcohols, including ethanol, correlate with their inhibitory action on the enzyme.[20] Ethanol caused half inhibition at 600 mM, some three times its general anesthetic concentration. Inhibition was thought to require two ethanol molecules at the site. This was so for alcohols up to hexanol, but octanol and decanol required only one.

The luciferases provide as good a model of the site of general anesthesia as any protein studied to date. Unfortunately they have not been crystallized, so their structure is unknown. However, cofactors often bind in hydrophobic clefts that may be hinged at their closed end. This might explain how the site on the luciferases accommodates a wider variety of anesthetics than most proteins.

MOLECULAR THEORIES OF ETHANOL'S ACTION

Hypotheses of ethanol's action have centered around perturbation of lipid bilayers in biomembranes. The evidence was reviewed recently[7] and need not be summarized here. A deeper understanding requires the choice of a model system that will enable molecular-level questions to be answered. Such a model should be an excitable membrane protein, sensitive to ethanol and available in high purity and yield. Currently only one such is available, the acetylcholine receptor from the electroplaques of *Torpedo*. The number of subunits and their amino-acid sequence in this nicotinic receptor resemble those in the equivalent receptors at the mammalian neuromuscular junction.[21] The expectation that their pharmacology might be similar has been fulfilled (see below), and therefore we may use the *Torpedo* receptor with some confidence that our findings will be of general applicability.

The nicotinic acetylcholine receptor has a molecular weight of 280,000 daltons and consists of five subunits; two are identical, and the others are homologous with them.[21-23] Theoretical analysis suggests that all these subunits have a similar arrangement in the lipid bilayer with five alpha-helices per subunit passing through the lipid bilayer. Two of these helices are polar, or charged, on one face and nonpolar on the other and may be involved in lining the ion channel passing through the middle of the protein. The other three helices are nonpolar, making an outer annulus between the channel helices and the bilayer.[24] The homology with mammalian receptor is such that when mammalian subunits are substituted for their *Torpedo* counterparts the hybrids function normally.[25] The *Torpedo* receptor's lipid composition is typical of plasma membranes with some 17% of the phospholipid carrying a net negative charge and 46% of the total lipid being cholesterol.[23] About one hundred and fifty lipid molecules surround the protein. Cholesterol is more concentrated next to the protein than in the rest of the lipid bilayer (for reviews, see Refs. 22, 23). Experiments in which the native lipids are totally replaced by synthetic ones show that the presence of cholesterol is essential for recovery of activity.[26]

Acetylcholine receptors contain binding sites for agonists (*e.g.* acetylcholine) and a cation-specific transmembrane channel. To a good approximation the receptor may be thought of as existing in either an active (A) or desensitized (D) state (see Equation 2). Two agonist molecules (here represented as L for simplicity) bind to the active state (A) with low affinity leading to channel opening (AL^o) within a few hundred microseconds. If agonist persists in the vicinity of the receptor, another conformational change leads to the desensitized state (D), which has a high affinity for agonists and does not conduct ions.[27] The horizontal reactions in Equation 2 all occur rapidly (milliseconds), while the vertical reactions occur slowly (seconds to minutes).

$$\begin{array}{ccccc} L + A & \xrightleftharpoons{K_{LOW}} & AL & \underset{b}{\overset{a}{\rightleftharpoons}} & AL^{o} \\ k_4 \uparrow\downarrow k_3 & & & & k_2 \uparrow\downarrow k_1 \\ L + D & \underset{K_{HI}}{\rightleftharpoons} & DL & & \end{array} \quad (2)$$

Actions of Ethanol on the Acetylcholine Receptor's Channel

Ethanol, propanol, butanol, and pentanol decrease the rate of decay of postsynaptic miniature end-plate currents (mepc's).[28] Noise analysis[28,29] suggests that ethanol increases the channel lifetime. However, other measurements suggest that a change in agonist affinity is the explanation.[30]

In contrast, heptanol and longer alcohols, like other general anesthetics, inhibit agonist-induced postsynaptic depolarization and increase the rate of mepc decay (for review, see REFS. 28, 31). Bradley *et al.*[29] studied agonist concentration-response relationships using electrophysiological techniques—not a very reliable procedure. Ethanol (up to 0.5 M) and propanol (70 mM) increased the amplitude of the maximum response and shifted the concentration-response curves to the left, suggesting that the affinity of acetylcholine for active receptors increased. Hexanol and octanol shifted acetylcholine concentration-response curves to the right and decreased the maximum response.

The above studies of ethanol's effects on AChR suggest several models of action, but none of the studies is complete enough to establish a mechanism. Gage *et al.*[28] suggested that short-chain alcohols increase the dielectric constant of the lipid bilayer, which decreases the transmembrane electric field sensed by the dipole responsible for channel closure (this decreases rate constant b in Equation 2, leading to a decrease in the overall equilibrium constant for channel opening). A contrasting model invokes direct ethanol-protein interactions.[29] All alcohols are assumed to bind to one saturable hydrophobic site in the channel lumen, which is only accessible to the open-channel state. Long-chain alcohols block the channel completely. Short-chain alcohols only partially occlude the channel, stabilizing the open state (longer channel lifetime but lower conductivity and higher affinity). A combination of ethanol and hexanol acted nonindependently, consistent with competition for a common site.

Ethanol and Receptor Desensitization

Biochemical studies have shown that ethanol shares with the longer alcohols and general anesthetics the ability to increase the fraction of desensitized receptors (D). Their relative potency in this is related to their lipid solubility.[32–34] Thus while the short-chain alcohols, including ethanol, act uniquely on the channel, their action as desensitization agents is held in common with the general anesthetics. Indeed, this action is pressure-reversible,[35] and the ability of ethanol and some general anesthetics to desensitize the receptor correlate with their membrane/buffer partition coefficients (FIG. 1).[36]

EXPERIMENTAL METHODS

Methods used here were described previously.[35–37] Briefly, all experiments were carried out at 4° C on acetylcholine receptor-rich membranes isolated by differential and sucrose-density gradient centrifugation. [^{3}H]-Acetylcholine binding was measured in the presence of diisopropylfluorophosphate using filtration on glass-fiber filters with correction for nondisplaceable binding. Channel properties were measured as the $^{86}Rb^+$ efflux elicited from preloaded membrane vesicles by exposure to agonist for 10 sec. The effects of exposure to all tested alcohols under the conditions shown were completely reversible.

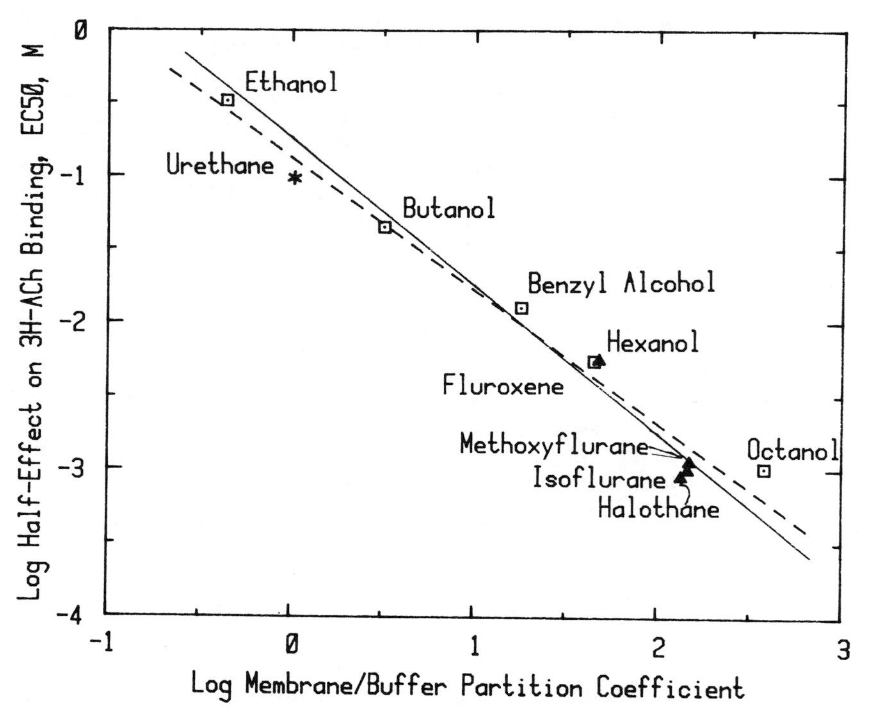

FIGURE 1. Comparison of desensitizing effect of anesthetics (EC_{50}) on cholinergic binding with membrane-buffer partition coefficient. Correlation of the EC_{50} with partition coefficient in lipid bilayers. *Squares,* alcohols; *triangles,* volatile anesthetics; *star,* urethane. The *solid line* has been fitted with a slope of -1.0 as required by the lipid hypothesis. The *dashed line* is from an unconstrained least-squares fit and has a slope of -0.90 ± 0.054 (s.d.) which does not differ significantly from -1.0 ($p > 0.1$). (From L. L. Firestone *et al.*[37] Reprinted by permission from *Anesthesiology.*)

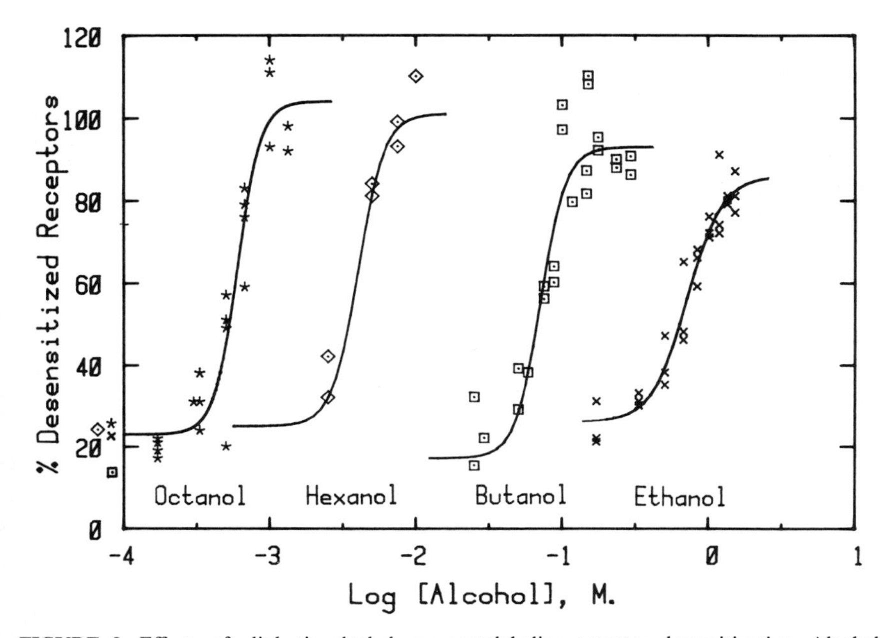

FIGURE 2. Effects of aliphatic alcohols on acetylcholine receptor desensitization. Alcohol effects on desensitization were determined at 4° C by filtration assay, designed to measure AChRs driven to a conformation with high affinity for [^{3}H]-ACh (acetylcholine). Aliquots of membrane suspensions (final concentration = 25 nM sites) were preincubated in capped glass vials (30 min, 4° C) with dilutions of *Torpedo* Ringer's solution saturated with alcohol. [^{3}H]-ACh (50 nM final concentration) was then added to the suspensions for 5 sec, vortexed, and the mixture filtered (Whatman GF/F glass fiber). The total [^{3}H]-ACh concentration in the suspension and the free [^{3}H]-ACh concentration in the filtrate were determined by scintillation counting. The total bound concentration was the difference between them. Nondisplaceably bound [^{3}H]-ACh was $\leq$5% of the bound, as determined in separate controls using the irreversible nicotinic antagonist, alpha-Bungarotoxin. The starting concentration of receptor sites (B_{max}) was determined by incubating an aliquot of membrane to equilibrium (60 min, 4° C) with saturating [^{3}H]-ACh. The AChRs present in the high-affinity state (R_{HI}) was calculated as: ($[B_{5sec}/B_{max}] \times 100$)%.

[^{3}H]-ACh itself failed to induce a significant proportion of low-to-high affinity-state transitions. Alcohol concentrations were monitored by gas chromatography. *Data points* are single determinations, and the *line* the best fit by a nonlinear least-squares fitting routine. Half-effect (R_{HI}^{50}) concentrations are listed in TABLE 1.

RESULTS

Desensitization

Since the desensitized receptor (D) has a higher affinity for acetylcholine than the resting receptor (A), and the interconversion between these states is slow, its incidence can be estimated by brief exposure of receptors to a concentration of [^{3}H] acetylcholine sufficiently low to bind only receptors in the D state. Preincubation with alcohols increases the proportion of desensitized receptors from about 20% to 95%. The effect shows sigmoid dependence on concentration, and the relative potency of the alcohols increases with chain length (FIG. 2).

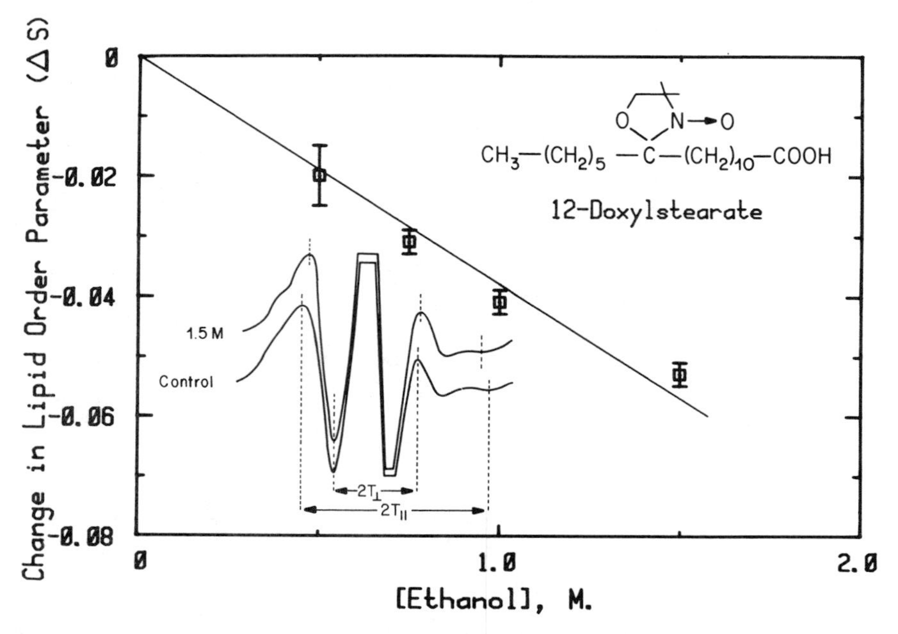

FIGURE 3. Alcohol effects on lipid order of acetylcholine receptor-rich membranes. Native receptor-rich membranes from *Torpedo* were spin-labeled for electron spin resonance (ESR) spectroscopy by depositing a thin film of a spin-labeled lipid fatty acid (12-doxylstearate [12-DS], *upper right*) from methanolic stock solution and gently shaking with membranes overnight at 4° C. Final probe concentrations were ≤1 mole % in membrane lipids. Dilutions of *Torpedo* Ringer's saturated with alcohols were mixed with spin-labeled membranes. Samples were immediately transferred to thin-walled glass capillary tubes and flame-sealed on ice. Spectroscopy was performed at 4.0° C on a Varian E109E ESR spectrometer, operating at 9.2 GHz, with microwave and magnetic fields set to minimize signal distortion (microwave power = 10 mW, magnetic field strength = 3,250 Gauss, filter time constant = 1 sec, scan time = 8 min, and magnetic field modulation amplitude = 1.0 Gauss). The temperature in the ESR sample cavity was maintained at the set point ±0.1° C by a stream of thermostated N_2 passing through a dewar insert containing the sample. Spectral splittings (*lower left:* $2T_\parallel$ and $2T_\perp$) were used to calculate the lipid-order parameter, S, by the method of Hubbell and McConnell (REF. 38):

$$S = (T_\parallel - T_\perp)/\{T_{zz} - 0.5(T_{xx} + T_{yy})\}$$

where T_{xx}, T_{yy}, and T_{zz} are single crystal measurements, and $T_\parallel$ and $T_\perp$ are the outer and inner (spectral) hyperfine splitting parameters. To correct for solvent polarity, S was multiplied by a/a′ (= $[T_{xx} + T_{yy} + T_{zz}]/[T_\parallel + 2T_\perp]$).

The data are from a single representative experiment with triplicate determinations at each concentration of ethanol. They were fitted by linear least squares, and, since the y-intercept was not significantly different than zero, the *line* was constrained to pass through the origin. The results of this analysis are shown in TABLE 1. The control order parameter was 0.628 ± 0.012 (mean ± s.d.; n = 3).

Lipid Disorder

All the alcohols disordered the *Torpedo* membrane. 12-DS reported a linear decrease in order parameter with increasing ethanol concentration (FIG. 3). Again the potency increased with chain length (TABLE 1). At the highest concentrations studied the disordering shifted the high- and low-field peaks so far towards the center of the spectra that the underlying immobilized spectral component was clearly revealed.

Channel Effects

Using vesicles without spare receptors we found that all the alcohols except methanol reduce the $^{86}Rb^+$ flux elicited by saturating concentrations of agonist, F_A(max), in a concentration-dependent manner, but to varying degrees, depending on chain length (see FIG. 4). Butanol, hexanol, and octanol induce total blockade of ion flux at high concentrations, with IC_{50}s shown in TABLE 1. Propanol at the highest concentrations tested reduces F_A(max) by about 75%, and ethanol only causes a 15% reduction of F_A(max) at concentrations up to 3 Molar. Methanol does not block ion flux, but causes a small increase in the maximum flux response to carbachol.

TABLE 1. The Change in Order Parameter, ΔS, Associated with Desensitization, R_{HI}^{50}, and Inhibition of the Channel, IC_{50}

Alkanol	$-\Delta S^{12}/M$	R_{HI}^{50} (mM)	$-\Delta S$ at R_{HI}^{50}	IC_{50} (mM)	$-\Delta S$ at IC_{50}
Ethanol	0.038	706	0.027	>2,700	>0.103
Butanol	0.351	71	0.025	22	0.008
Hexanol	5.49	6.3	0.035	1.6	0.009
Octanol	51.8	0.60	0.031	0.02	0.001
Range	1,363×	1,177×	1.3×	>135,000×	>100×

We further investigated alcohol effects by measuring carbachol concentration-response curves in the presence of drugs. Octanol at 5 and 10 μM reduced F_A(max) by 19 and 31% respectively. However, in each case the concentration of agonist eliciting a half-maximal response, C_{50}, remained unchanged at 113 ± 4 μM.

Because ethanol reduces F_A(max) by only 15%, we were able to study its effects on C_{50} over a wide concentration range. FIGURE 5 shows a series of carbachol concentration-response curves from a single acetylcholine receptor vesicle preparation in the presence of ethanol. The concentration-response curves reveal a twofold reduction of C_{50}, SC_{50}, (TABLE 1) with 270 mM ethanol and a leftward shift of more than two decades with 2.7 M ethanol. The inset shows that the leftward-shift effect of ethanol is not saturable.

Alkanol interactions were also tested by studying the effects of alcohol mixtures on carbachol concentration-response curves. The combined effects of 10 μM octanol and 1 M ethanol on the carbachol concentration-response curves show that octanol has no effect on the leftward shift induced by 1 M ethanol. However, ethanol reverses octanol's inhibitory effect on F_A(max). Thus, ethanol's action is independent of octanol's, but octanol's action is not independent of ethanol's.

The interpretation of these observations is not simple. A possibility is that two processes underlie the concentration-response curve. Thus, at saturating agonist

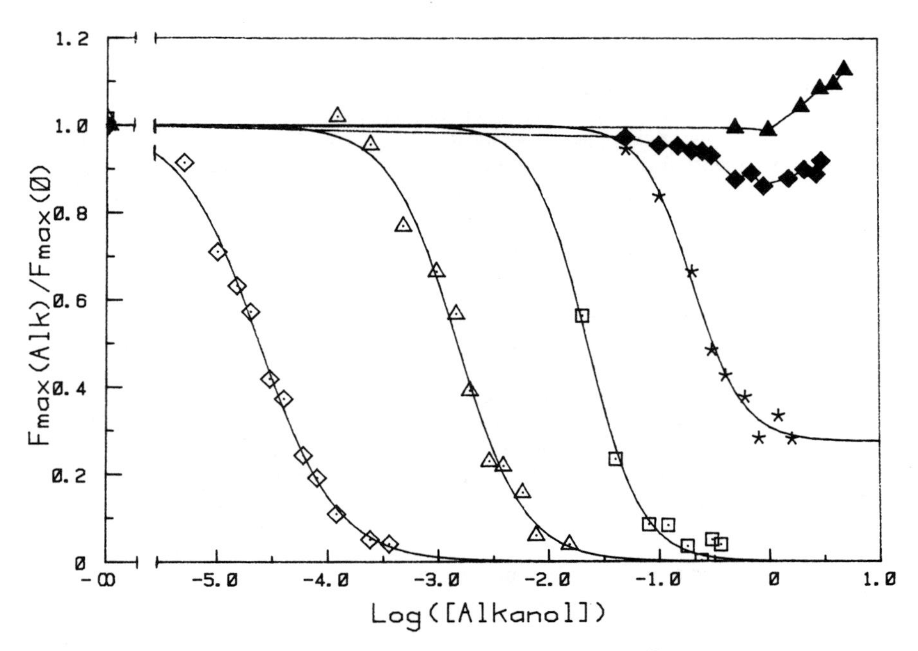

FIGURE 4. Effects of n-alkanols on agonist-induced flux. 10 sec $^{86}Rb^+$ efflux from *Torpedo* vesicles at 4° C was measured as follows. Vesicles were incubated overnight with 100–200 uCi/ml $^{86}RbCl$ and enough alpha-Bungarotoxin to block 80% of [^3H]-ACh sites. These vesicles were passed through a gel filtration column (Sephadex G-50) to remove extravesicular $^{86}Rb^+$. Efflux was initiated when an aliquot of vesicles was added to buffer containing 5 mM carbamylcholine (Carb) and alkanol where appropriate. After 10 sec the vesicle suspension was poured through a filter (Whatman GF/F) in a vacuum manifold. Filtrates were counted in scintillation fluid. Efflux measurements were corrected for passive leak (measured in the absence of agonist). IC_{50} from nonlinear least-squares fits are given in TABLE 1.

concentrations flux may be terminated by fast desensitization, whereas at low agonist concentration it may also be terminated by dissociation of acetylcholine. This experimental uncertainty can be resolved by carrying out the assay on a millisecond time scale so that initial flux rate can be resolved.

DISCUSSION

Several Actions on One Protein

Clearly alcohols exert a number of different functional effects on the acetylcholine receptor. At the channel, homologous alcohols seem not to have a common action—long-, but not short-chain alcohols are inhibitors; while short-, but not long-chain alcohols shift agonist concentration-response curves to the left. On the other hand, all the alcohols enhance acetylcholine-receptor desensitization.

Their chain-length dependence further distinguishes these effects (FIG. 6). Inhibition of the channel stands out as having the steepest chain-length dependence, while the leftward shift of flux concentration-response curves, desensitization, and lipid disordering all share similar chain-length dependencies.

Desensitization is Associated with Lipid Perturbations

The concentration causing half-maximal desensitization, EC_{50}, correlates perfectly with the concentration causing a 0.01 unit change in order parameter ($r = 0.9993$; slope $= 1.04 \pm 0.028$, not significantly different from the expected value of one). The percentage change in order parameter at ethanol's EC_{50} is 4.4.

This desensitization would reduce the safety factor for synaptic conduction, because occupation of desensitized receptors by agonist does not open channels. It is only slowly reversed on removal of alcohols.[32]

Membrane Permeability is Associated with Lipid Perturbation

The leftward shift of flux concentration-response curves was observed from methanol to butanol. A twofold shift was associated with a 1.6% change in order parameter, and the effect was not saturable even at ten times these concentrations. The degree of leftward shift was independent of the presence of octanol, suggesting an independent mode of action for these two drugs (although this needs confirmation on a

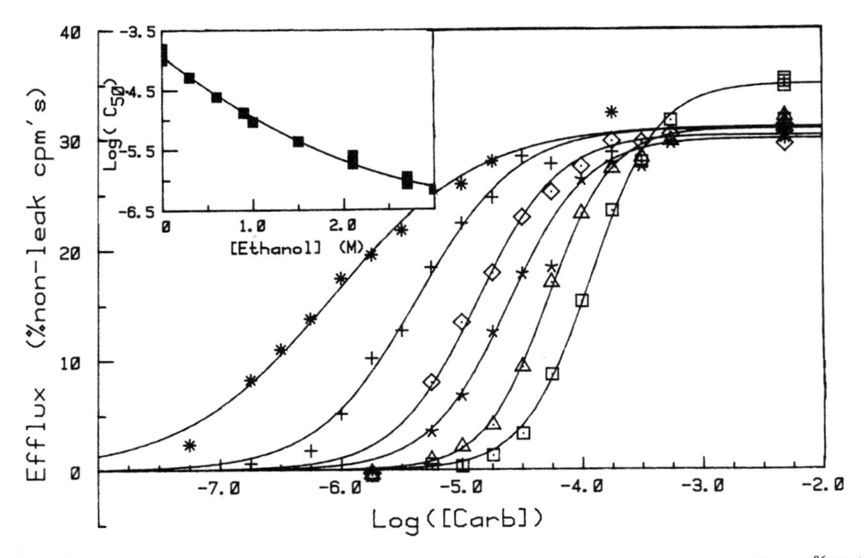

FIGURE 5. Ethanol's effect on carbamylcholine concentration-response curves. 10 sec $^{86}Rb^{+}$ efflux was measured as described in the legend to FIGURE 4. All measurements were performed on a single batch of vesicles. Carb concentration was varied and ethanol concentration fixed for each set of experiments. C_{50} is the Carb concentration eliciting half-maximal efflux, and was derived from nonlinear least-squares fits of the data. *Inset:* Log(C_{50}) is plotted against ethanol concentration for 12 experiments.

faster time scale, see above). The only fact preventing assigning a nonspecific mechanism to this effect is that the longer-chain alcohols do not exhibit it. There are two possible explanations. First, the leftward shift could result from a noncolligative action of the short-chain alcohols analogous to their ability to cause interdigitation, as discussed above. Second, it is possible that these alcohols do exert a colligative effect, but the ability of the ones with longer chains to cause the leftward shift are masked by inhibition of the channel at lower concentrations. This view is supported by FIGURE 6, which shows that the predicted twofold-shift concentration SC_{50} of octanol is 0.75 mM, 38-fold above its IC_{50}.

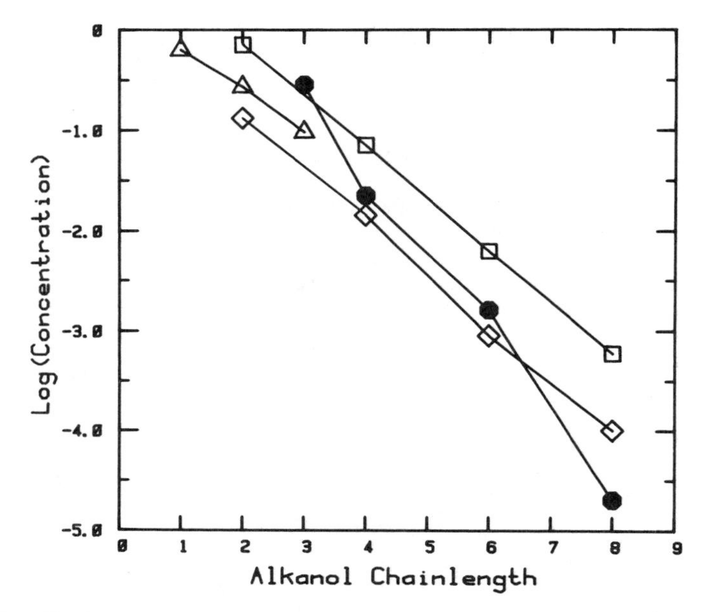

FIGURE 6. The dependence of various effects on acetylcholine receptors on alcohol chain length. The concentration of alcohols that cause a 1/2% change in order parameter (*diamonds*), 50% increase in the number of desensitized receptors (R_{HI}^{50}, *squares*), 50% inhibition of maximum flux (F_A(max), *solid circles*) and a twofold leftward shift in carbachol concentration-response curves (SC_{50}, *triangles*) are plotted against the number of carbon atoms in each alcohol.

In terms of Equation 2 the leftward shift in concentration-response curves could be caused by a change in K_{LOW} or in the rate constants a or b. (Since the rate of ligand binding is diffusion-limited and invariant, these possibilities may be distinguished by measuring the dissociation rate of the ligand. Such measurements have not been undertaken). Both possibilities have been advocated by different workers. Gage *et al.*[31] suggested that b decreases because ethanol alters the membrane's dielectric constant (see above). Our results demonstrating that ethanol increases flux at low agonist concentrations in the absence of a membrane voltage without increasing maximum flux are inconsistent with this model. Quastel and colleagues[30] posited a decrease in the off-rate of acetylcholine, as did Nelson and Sachs.[39]

Inhibition of the Channel; a Specific Action?

The steep dependence of IC_{50} on chain length results from octanol blocking channels at concentrations well below those causing other effects, while propanol does so at concentrations that cause more than a twofold shift in the agonist concentration-response curve (FIG. 6). Methanol does not inhibit, and ethanol does so only mildly (FIG. 4), even at concentrations that cause significant changes in order parameter. Thus, bulk lipid perturbation is unlikely to account for channel inhibition, even though this idea was pursued previously.[28]

Alternatively an allosteric site might be involved. Such a site might be located at the lipid-protein interface, perhaps displacing some vital lipid from this region, or in the lumen of the channel. The lack of action of the short-chain alcohols could then arise either because they bind but fail to exert an effect (Peper's model[29] above falls into this general category) or because they do not bind. The former model would be consistent with our observation that ethanol reduces the inhibition caused by octanol. If this is the case then ethanol's action should be both saturable and surmountable.

An interesting further possibility is suggested by the requirement for cholesterol when reconstituting the channel's activity into lipid bilayers.[26] It is possible that the alcohols, with the exception of those with the shortest chains, can compete with the endogenous alcohol, cholesterol, displacing it from a specific site in the lipid-protein interface. Lacking some specific structural feature(s) of cholesterol, these alcohols might be unable to support channel activity.

CONCLUSIONS

Alcohols act on the acetylcholine receptor by both nonspecific and specific mechanisms to produce a number of effects. Ethanol, however, lacks the specific action that results in channel inhibition. It acts nonspecifically to produce desensitization, a property it shares with the general anesthetics. It probably also acts nonspecifically, either colligatively or not, to increase the apparent affinity for channel activation. It is probable that these two actions are mediated by different lipid perturbations (P_L, see Equation 1). Our work does not establish a mechanism; the association with lipid disordering is no more than that.

These experiments were carried out at 4° C, a temperature at which the *Torpedo* is very active. The animal is not in plentiful enough supply to determine anesthetic potency with it, but at 3° C, 333 mM ethanol is required to anesthetize frogs[40] and at 10° C, 343 mM for tadpoles. At 300 mM ethanol the apparent affinity for channel activation has doubled and the percentage of desensitized receptors has increased from 15 to 40%. Thus, although this receptor was chosen purely as a model for elucidating molecular mechanisms, the findings are not without physiological relevance.

ACKNOWLEDGMENTS

The authors wish to thank Drs. Gergely and Seidel, Boston Biomedical Research Institute, for the use of an electron spin resonance spectrometer, and Patricia Streicher for technical assistance.

REFERENCES

1. FORMAN, S. A., A. S. VERKMAN, J. A. DIX & A. K. SOLOMON. 1985. n-Alkanols and halothane inhibit red cell anion transport and increase band 3 conformational change rate. Biochemistry **24:** 4859–4866.
2. MILLER, K. W. 1985. The nature of the site of general anesthesia. Int. Rev. Neurobiol. **27:** 1–57.
3. ROTH, S. H. 1979. Physical mechanisms of anesthesia. Annu. Rev. Pharmacol. Toxicol. **19:** 159–178.
4. JANOFF, A. S. & K. W. MILLER. 1982. A critical assessment of the lipid theories of general anaesthetic action. *In* Biological Membranes. D. Chapman, Ed. Vol. 4: 417–476. Academic Press. London.
5. FRANKS, N. P. & W. R. LIEB. 1982. Molecular mechanisms of general anaesthesia. Nature **300:** 487–493.
6. DLUZEWSKI, A. R., M. J. HALSEY & A. C. SIMMONDS. 1983. Membrane interactions with general and local anesthetics: a review of molecular hypotheses of anaesthesia. Mol. Aspects Med. **6:** 459–573.
7. GOLDSTEIN, D. B. 1984. The effects of drugs on membrane fluidity. Annu. Rev. Pharmacol. Toxicol. **24:** 43–64.
8. WORCESTER, D. L. 1975. Neutron diffraction studies of biological membranes and membrane components. *In* Neutron Scattering for the Analysis of Biological Structures. B. P. Schoenborn, Ed. Vol. 27: 37–57. Brookhaven National Laboratory. Upton, NY.
9. PRESTI, F. T. 1985. The role of cholesterol in regulating membrane fluidity. *In* Membrane Fluidity in Biology. R. Aloia & J. M. Boggs, Eds. Vol. 4: 97–146. Academic Press. New York, NY.
10. MILLER, K. W., L. HAMMOND & E. G. PORTER. 1977. The solubility of hydrocarbon gases in lipid bilayers. Chem. Phys. Lipids **20:** 229–241.
11. KATZ, Y. & J. M. DIAMOND. 1974. Thermodynamic constants for nonelectrolyte partition between dimyristoyl lecithin and water. J. Membr. Biol. **17:** 101–120.
12. COLLEY, C. M., S. M. METCALFE, B. TURNER, A. S. V. BURGEN & J. C. METCALFE. 1971. The binding of benzyl alcohol to erythrocyte membranes. Biochim. Biophys. Acta **233:** 720–729.
13. JAIN, M. K. & N. M. WU. 1977. Effect of small molecules on the dipalmitoyl lecithin liposomal bilayers: III. Phase transition in lipid bilayers. J. Membr. Biol. **34:** 157–201.
14. ROWE, E. S. 1983. Lipid chain length and temperature dependence of ethanol-phosphatidylcholine interactions. Biochemistry **22:** 3299–3305.
15. URBAN, B. W. 1985. Modifications of excitable membranes by volatile and gaseous anesthetics. *In* Effects of Anesthesia. B. G. Covino, H. A. Fozzard, K. Rehder & G. Strichartz, Eds. 13–28. American Physiological Society. Bethesda, MD.
16. HAYDON, J., J. R. ELLIOTT, B. M. HENDRY & B. W. URBAN. 1986. The action of nonionic anesthetic substances on voltage-gated ion conductances in squid giant axons. *In* Molecular and Cellular Mechanisms of Anesthetics. S. H. Roth & K. W. Miller, Eds. 267–277. Plenum Publishing Company. New York, NY.
17. WILLIAMS, R. J. P. 1979. The conformational properties of proteins in solution. Biol. Rev. **54:** 389–437.
18. SETTLE, W. 1973. Function of the myoglobin molecule as influenced by anesthetic molecules. *In* A Guide to Molecular Pharmacology and Toxicology. R. M. Featherstone, Ed. Part II: 477–493. Marcel Dekker, Inc. New York, N.Y.
19. TILTON, R. F., Jr., I. D. KUNTZ, Jr. & G. A. PETSKO. 1984. Cavities in proteins: structure of a metmyoglobin-xenon complex solved to 1.9 Å. Biochemistry **23:** 2849–2857.
20. FRANKS, N. P. & W. R. LIEB. 1984. Do general anaesthetics act by competitive binding to specific receptors? Nature **310:** 599–601.
21. NODA, M., H. TAKAHASHI, T. TANABE, M. TOYOSATO, S. KIKYOTANI, Y. FURUTANI, T. HIROSE, H. TAKASHIMA, S. INAYAMA, T. MIYATA & S. NUMA. 1983. Structural homology of Torpedo californica acetylcholine receptor subunits. Nature **302:** 528–532.
22. CONTI-TRONCONI, B. M. & M. A. RAFTERY. 1982. The nicotinic cholinergic receptor: correlation of molecular structure with functional properties. Annu. Rev. Biochem. **51:** 491–530.

23. CHANGEUX, J. -P., A. DEVILLERS-THIERY & P. CHEMOUILLI. 1984. Acetylcholine receptor: an allosteric protein. Science **225:** 1335–1345.
24. FAIRCLOUGH, R. H., J. FINER-MORE, R. A. LOVE, D. KRISTOFFERSON, P. J. DESMEULES & R. M. STROUD. 1983. Subunit organization and structure of an acetylcholine receptor. Cold Spring Harbor Symposium on Quant. Biol. **48:** 9–20.
25. SACKMAN, B., C. METHFESSEL, M. MISHINA, T. TAKAHASHI, T. TAKAI, M. KURASAKI, K. FUKUDA & S. NUMA. 1985. Role of acetylcholine receptor subunits in gating of the channel. Nature **318:** 538–543.
26. FONG, T. M. & M. G. MCNAMEE. 1986. Correlation between acetylcholine receptor function and structural properties of membranes. Biochemistry **25:** 830–840.
27. NEUBIG, R. R., N. D. BOYD & J. B. COHEN. 1982. Conformations of *Torpedo* acetylcholine receptor associated with ion transport and desensitization. Biochemistry **21:** 3460–3467.
28. GAGE, P. W. & O. P. HAMILL. 1981. Effects of anesthetics on ion channels in synapses. Int. Rev. Physiol. **25:** 1–45.
29. BRADLEY, R. J., R. STERZ & K. PEPER. 1984. The effects of alcohols and diols at the nicotinic acetylcholine receptor of the neuromuscular junction. Brain Res. **295:** 101–112.
30. MCLARNON, J. G., P. PENNEFATHER & D. M. J. QUASTEL. 1986. Mechanism of nicotinic channel blockade by anesthetics. *In* Molecular and Cellular Mechanisms of Anesthetics. S. H. Roth & K. W. Miller, Eds. 155–164. Plenum Publishing Company. New York, NY.
31. GAGE, P. W., D. MCKINNON & B. ROBERTSON. 1986. The influence of anesthetics on postsynaptic ion channels. *In* Molecular and Cellular Mechanisms of Anesthetics. S. H. Roth & K. W. Miller, Eds. 139–153. Plenum Publishing Company. New York, NY.
32. BOYD, N. D. & J. COHEN. 1984. Desensitization of membrane-bound *Torpedo* acetylcholine receptor by amine noncompetitive antagonists and aliphatic alcohols: studies of [^{3}H]-acetylcholine binding and $^{22}Na^+$ ion fluxes. Biochemistry **23:** 4023–4033.
33. YOUNG, A. P. & D. S. SIGMAN. 1981. Allosteric effect of volatile anesthetics on the membrane-bound acetylcholine receptor. Mol. Pharmacol. **20:** 498–505.
34. EL-FAKAHANY, E. F., E. R. MILLER, M. A. ABBASSY, A. T. ELDEFRAWI & M. E. ELDEFRAWI. 1983. Alcohol modulation of drug binding to the channel sites of the nicotinic acetylcholine receptor. J. Pharmacol. Exp. Ther. **224:** 289–296.
35. BRASWELL, L. M., K. W. MILLER & J. F. SAUTER. 1984. Pressure reversal of the action of octanol on postsynaptic membranes from *Torpedo.* Br. J. Pharmacol. **83:** 305–311.
36. MILLER, K. W., L. M. BRASWELL, L. L. FIRESTONE, B. A. DODSON & S. A. FORMAN. 1986. General anesthetics act both specifically and nonspecifically on acetylcholine receptors. *In* Molecular and Cellular Mechanisms of Anesthetics. S. H. Roth & K. W. Miller, Eds. 125–137. Plenum Medical Book Company. New York, NY.
37. FIRESTONE, L. L., J.-F. SAUTER, L. M. BRASWELL & K. W. MILLER. 1986. The actions of general anesthetics on acetylcholine receptor-rich membranes from Torpedo. Anesthesiology **64:** 694–702.
38. HUBBELL, W. L. & H. M. MCCONNELL. 1971. Molecular motion in spin-labeled phospholipids and membranes. J. Amer. Chem. Soc. **93:** 314–326.
39. NELSON, D. J. & F. SACHS. 1981. Ethanol decreases dissociation of agonist from nicotinic channels. Biophys. J. **33:** 121a.
40. MEYER, H. H. 1901. Zur theorie der alkoholnarkose. Arch. Exp. Pathol. Pharmakol. **46:** 338–346.

DISCUSSION OF THE PAPER

B. CHANCE (*University City Science Center, Philadelphia, PA*): These are highly dissociated interactions; the concentrations of receptors must be in the nM region, and these are. Are there any of these that approach stoichiometric binding? It seems that these are general inhibitors and in a special class of highly dissociated ones.

MILLER: We don't really have any idea about the stoichiometry of the alcohol action. The Hill coefficients for inhibition are close to one. One cannot determine it directly by binding measurements using radioactive alcohol because of the nonspecific binding in the lipid. There are local anesthetic sites on this receptor that have a stoichiometry of roughly one per protein. Thus, there is a precedent for a clear stoichiometry.

D. MCCARTHY (*University of New Mexico School of Medicine, Albuquerque, NM*): Are there cholinesterase binding sites, and are they in the membrane? Why do your kinetics not reflect some disparity in binding characteristics? One would expect an enzyme binding site and a membrane site.

MILLER: There is actually very little esterase left by the time we have purified the membrane, but it's inhibited with DFP.

H. ROTTENBERG (*Hahnemann University School of Medicine, Philadelphia, PA*): You were careful not to interpret the close association between the fluidity change and desensitization. However, isn't it a very simple explanation to say that the alcohol or the drug has to be in the membrane to cause both effects? Therefore, they are both related to the partition coefficient, and it really takes you back to where you started from.

MILLER: The point I'm making is that in this membrane, because of the purity, we should be able to establish a mechanistic link. This we have not yet established, but the correlation is very strong.

A. P. THOMAS (*Hahnemann University School of Medicine, Philadelphia, PA*): Is there any relationship between the activity of the channel, as opposed to the activity of the binding of acetylcholine? You've looked almost exclusively at acetylcholine binding, although you've studied channel activities in terms of rubidium efflux. Is there any possibility that the channel still binds the acetylcholine in one circumstance, but doesn't transport the ion, and in the other circumstance, that it transports the ion, but the binding of acetylcholine is inhibited?

MILLER: We have not seen competitive inhibition of acetylcholine binding. There is an inhibitory site for acetylcholine on the channel that has very low affinity. It is the third acetylcholine site, and we do see interactions between that site and octanol. But we haven't characterized them completely.

THOMAS: Is it possible that the channel can be inhibited even when acetylcholine is bound to what is, theoretically, not a desensitized receptor?

MILLER: Yes, it is. The action of octanol is not on the channel and is not a desensitizing action. It acts at about 20 μM, whereas it causes desensitization at a few hundred μM.

CHANCE: I was interested in your pressure effect. How many kilobars did you use?

MILLER: 300 atmospheres pulled back the binding curve perturbed by one mM octanol. Anesthesia can be reversed in a group of animals, or the ED_{50} can be shifted a factor of two by about a hundred atmospheres. Considering what happens to bulk liquids, this is quite sensitive.

I. DIAMOND (*Ernest Gallo Clinic and Research Center, San Francisco, CA*): Do specific toxins, such as bungarotoxin, block receptor-induced flux?

MILLER: Yes, bungarotoxin blocks the acetylcholine site, preventing activation and occupation.

DIAMOND: You don't see any flux change or any alcohol effect?

MILLER: We have done experiments to ask whether ethanol opens the channel by itself, independent of the acetylcholine site. If you completely titrate off all the acetylcholine sites with bungarotoxin and then add ethanol, you see a leak of ions through the membrane, which is not blocked by anything.

DIAMOND: Does that leak change with the subsequent application of acetylcholine?

MILLER: No. The bungarotoxin effect is irreversible.

DIAMOND: Does the receptor undergo a change in K_m with desensitization?

MILLER: As the receptor is occupied, its affinity increases with time, until it ends up in a high-affinity, desensitized state. The change in K_m is of several orders of magnitude.

J. M. LITTLETON (*King's College, London, England*): You have shown quite nicely that some of the alcohols can cause changes in the flux of sodium through the channel of the nicotinic acetylcholine receptor. Do they have any other effects on channel blocking agents like, for example, histrionicotoxin? Have you looked at effects on histrionicotoxin or curare to see whether alcohol may actually displace them from the channel?

MILLER: Drug cocktails are not our strong point! We have not looked at the channel-blocking actions of histrionicotoxin or curare. However, in the last few years people have realized that while acetylcholine opens its own channel at low concentrations, at high concentrations it will block that channel, apparently by a specific action. We have looked at the effect of some drugs on acetylcholine's channel-blocking action. The local anesthetics in our hands appear to compete for acetylcholine at that site. Octanol interacts with it but probably doesn't compete. Ethanol appears to have little or no interaction.

QUESTION: Some years ago in electrophysiological experiments on the neuromuscular junction, it was shown that high concentrations of ethanol prolong the decay of the end-plate potential. If I understood what you were saying, it was that ethanol increased the desensitization of the acetylcholine receptor, which might decrease the current. Could you discuss this difference?

MILLER: There have been a number of electrophysiological studies. Ethanol does increase desensitization slowly, but it causes the leftward shift in the dose-response curve that we demonstrated very rapidly. We have seen it as early as 34 milliseconds after mixing. So I do not think that the effects we studied on the channel are related to desensitization.

The advantage of biochemical studies is that one can change drug concentration very rapidly. They do not have the time resolution to look at single channels, however. The physiologists on the other hand look at desensitized systems; in other words, they dribble in their acetylcholine and ethanol and sit there and watch channels open. So the biochemical and physiological approaches are not terribly comparable. What we really need from the physiologists is single-channel studies with ethanol, but very little has been done so far. There is one abstract published, to be exact. The interpretation of Sachs's group, who published that abstract, was that the noise analysis results of Gage were not being interpreted correctly. Sachs's abstract suggests that the agonist dissociates more slowly; in other words, its affinity was higher for the acetylcholine receptor in the presence of ethanol. That explanation is similar to ours. Gage, who was the first person to look at this, argued that the channel closing rate slowed down. However, both Peper's and Quastel's groups, who carried out similar work, conclude that ethanol's action is an affinity effect and not a channel lifetime effect.

Lipid Polymorphism

C.P.S. TILCOCK AND P.R. CULLIS

Department of Biochemistry
The University of British Columbia
Vancouver, British Columbia V6T 1W5, Canada

INTRODUCTION

While it is well established that the vast majority of the combined lipids within any given biological membrane form bilayer (lamellar) structures, it is clear that the bilayer structure of membranes is not immutable. For example, the phenomenon of membrane fusion, whether intercellular (as in fertilization, myogenesis, or the formation of bone polykaryocytes), intracellular (as in the lysosomal degradation of phagocytic vacuoles or the release of vesicles from the Golgi), or occurring at the level of the plasma membrane (during such processes as the release of secretory products, endocytosis, or cell division), requires the transient, controlled destabilization of bilayer structure at the fusion site.

Within this context it is of interest that many of the lipids of biological membranes, both singly and also in mixtures with other lipids, can adopt nonlamellar (*e.g.*, hexagonal H_{II}, inverted micellar, etc.) structures in response to physiologically relevant variables such as pH, ionic strength, or the distributed presence of divalent cations and proteins. In addition, exogenous lipophilic agents such as local anesthetics and (of particular relevance to this symposium) short- and long-chain alcohols can also influence lipid polymorphism. These structures may be of relevance to membrane contact and fusion as well as the packing properties of lipids in bilayers.

In this article we provide an overview of the polymorphic phase behavior of lipids.

NUCLEAR MAGNETIC RESONANCE DETERMINATION OF LIPID PHASE STRUCTURE

Individual phospholipids can adopt a variety of phases upon hydration including lamellar (bilayer), hexagonal H_I and H_{II}, micellar, inverted micellar, or cubic, dependent upon factors such as the nature of the lipid headgroup, the unsaturation or degree of side-branching of the lipid acyl chains, water content, temperature, pH, ionic strength, or the presence of divalent cations, other lipids, or polypeptides and proteins.[1] The essential structural features of the familiar lamellar (bilayer) and hexagonal H_{II} phases are shown in FIGURE 1 together with representative ^{31}P NMR spectra and also freeze-fracture electron micrographs. The H_{II} phase is comprised of hexagonally packed lipid cylinders in which the lipids are oriented with their headgroups towards central aqueous channels (~20 Å diameter).

While X ray and neutron diffraction are the definitive techniques for the determination of lipid phase structure, both ^{31}P NMR and ^{2}H NMR may be used to conveniently monitor lipid phase behavior. A detailed discussion of any of these techniques lies beyond the scope of this article. However, see REFERENCE 1 and the many review articles cited therein for a discussion of methodology. Each technique possesses advantages and disadvantages. For example, NMR does not tell of structure directly, but rather provides information concerning the motional properties of the ensemble, which may then be correlated with structure. NMR is therefore used in an extrapolative manner, based upon direct structural determination by X ray or neutron diffraction techniques. As examined by ^{31}P NMR, all hydrated phosphodiester lipids in large (> 400-nm diameter) bilayer structures exhibit an asymmetric line shape with a low-field shoulder and high-field peak separated by 40 to 50 ppm. The actual

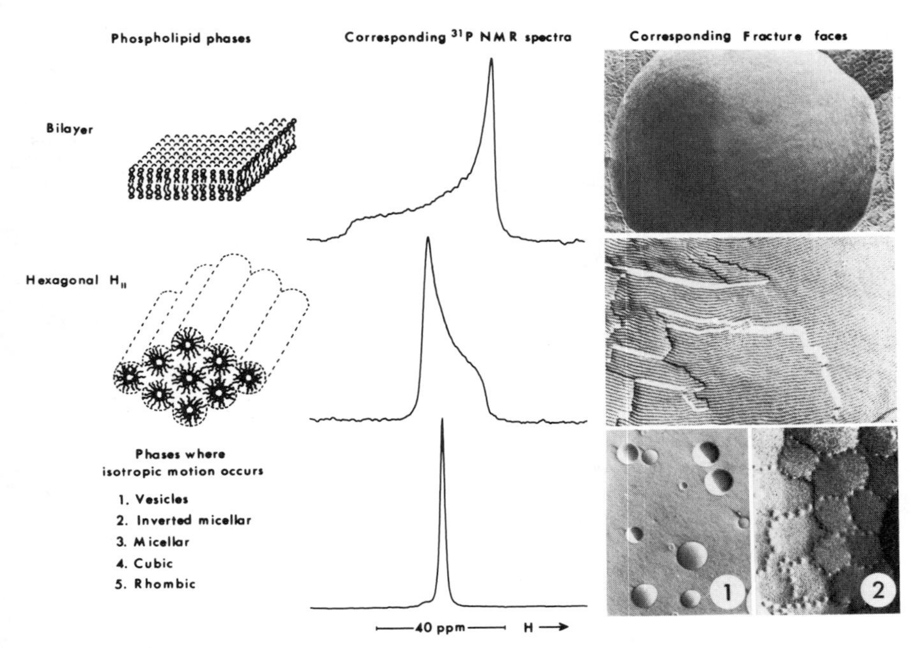

FIGURE 1. ^{31}P NMR and freeze-fracture characteristics of phospholipids in various phases. The bilayer ^{31}P NMR spectrum was obtained from aqueous dispersions of egg yolk phosphatidylcholine, and the hexagonal H_{II} phase spectrum from phosphatidylethanolamine (prepared from soybean phosphatidylcholine). The ^{31}P NMR spectrum representing isotropic motion was obtained from a mixture of 70 mol% soya phosphatidylethanolamine and 30% egg yolk phosphatidylcholine after heating to 90° C for 15 min. All preparations were hydrated in 10 mM tris-acetic acid (pH 7.0) containing 100 mM NaCl, and the ^{31}P NMR spectra were recorded at 30° C in the presence of proton decoupling. The freeze-fracture micrographs represent typical fracture faces obtained from bilayer and H_{II}-phase systems as well as structures giving rise to isotropic motional averaging. The bilayer configuration (total erythrocyte lipids) gives rise to a smooth fracture face, whereas the hexagonal H_{II} configuration is characterized by ridges displaying a periodicity of 6 to 15 mm. Common conformations that give rise to isotropic motion are represented in the bottom micrograph: (1) bilayer vesicles (~100 nm diameter) of egg phosphatidylcholine and (2), structures containing lipidic particles (egg phosphatidylethanolamine containing 20 mol% egg phosphatidylserine at pH 4).

separation is dependent on the lipid species, temperature, and other factors. This line shape is characteristic of axially averaged motion on the NMR time scale, due to rotation of the phospholipid about its long axis.[2] The resonance position of the high-field peak of an unoriented system (*e.g.*, a lipid vesicle dispersion) corresponds to bilayers lying parallel to the magnetic field, whereas the resonance position of the shoulder corresponds to an orientation of bilayers perpendicular to the field. The asymmetry of the line shape reflects the greater probability of finding bilayers in unoriented systems lying parallel rather than perpendicular to the field.

In the hexagonal H_{II} phase, lipids experience additional motional averaging due to translational diffusion of lipid molecules around the walls of the lipid cylinders on the NMR timescale. This results in a decrease in the width of the spectrum by a factor of two and also a reversal in the sign of the chemical shift anisotropy, which is due to the fact that in the hexagonal H_{II} phase, the probability of finding a lipid cylinder parallel to the field (in which all the lipid directors are normal to the field) is less than finding a lipid cylinder normal to the field. Alternatively, in phases where motion is isotropic (*i.e.*, the lipid samples all possible orientations with respect to the applied field on the NMR timescale), ^{31}P NMR gives narrow, symmetric resonances. This occurs for lipids in structures such as small bilayer vesicles (diameter < 200 nm), micelles, inverted micelles, or phases such as cubic or rhombic.

Corresponding freeze-fracture micrographs are also shown in FIGURE 1. Lamellar phases give rise to extended smooth fracture faces, whereas hexagonal H_{II} phases exhibit rippled fracture planes. FIGURE 1 also illustrates the freeze-fracture replicas obtained from small unilamellar vesicles and from lipidic particles.[3]

^{2}H NMR may also be used to determine lipid phase structure, as illustrated in FIGURE 2 for dioleoyl phosphatidylethanolamine (PE) labelled at the C_{11} position of both acyl chains. Each deuteron gives rise to a doublet whose separation in an unoriented system depends upon the average orientation of the deuterium nucleus with respect to the applied field.[4] As the lipid undergoes a lamellar to hexagonal H_{II} transition, the additional motional averaging due to lateral diffusion of the lipid around the lipid cylinders causes a reduction in the (quadrupolar) splitting by a factor of two (or more). Particular advantages of ^{2}H NMR compared to ^{31}P NMR are that ^{2}H NMR allows quantification of order in the acyl chains and that it is possible to monitor the phase behavior of a single ^{2}H-labelled lipid in a mixed system.

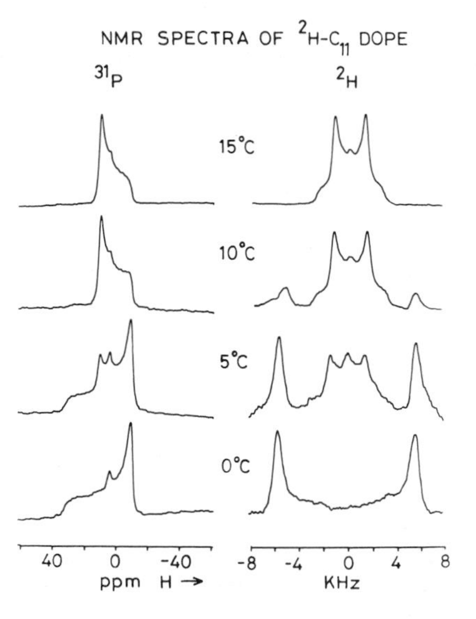

FIGURE 2. ^{31}P and ^{2}H NMR spectra as a function of temperature of fully hydrated dioleoyl phosphatidylethanolamine (DOPE), which is ^{2}H-labelled at the C_{11} position of the acyl chains ($[C_{11}\text{-}^{2}H_2]$DOPE). The ^{31}P NMR spectra were obtained at 81.0 MH_2 in the presence of proton decoupling, whereas the ^{2}H NMR spectra were obtained at 30.4 MHz.

TABLE 1. Lamellar to Hexagonal H_{II} Transition Temperatures for Various Synthetic and Naturally Derived Phosphatidylethanolamines

Species	Phase	Conditions
Diacyl species		
20:0/20:0	H_{II}	96° C
18:0/18:0	H_{II}	100° C
16:0/16:0	H_{II}	109–123° C
14:0/14:0	H_{II}	85° C
$16{:}0/18{:}1_c$	H_{II}	75° C
$18{:}1_c/16{:}0$	H_{II}	70° C
$18{:}1_c/18{:}1_c$	H_{II}	60° C
$18{:}1_c/18{:}1_c$	H_{II}	10° C
18:2/18:2	H_{II}	−15° C
18:3/18:3	H_{II}	−15° C
20:4/20:4	H_{II}	−30° C
22:6/22:6	H_{II}	−30° C
Egg	H_{II}	25–35° C
	L	pH 8.5, pressure
From egg PC—	H_{II}	40–45° C
E. coli	H_{II}	55–60° C
Human erythrocyte	H_{II}	8° C
Porcine erythrocyte	L + H_{II}	20–40° C, 10–90% water
Rat liver e.r.	H_{II}	7° C
Rabbit s.r.	H_{II}	0° C
Soya bean	H_{II}	−10° C
Rat mitochondrial	H_{II}	10° C
Dialkyl species		
18:1/18:1	H_{II}	80° C
16:0/16:0	H_{II}	86° C
14:0/14:0	H_{II}	93° C, excess water
		78° C, salt NaCl
12:0/12:0	H_{II}	100° C, excess water
		70° C, low water
Effect of acyl chain linkage		
Vinyl ether	H_{II}	30° C
Alkyl ether	H_{II}	53° C
Acyl ether	H_{II}	68° C

THE POLYMORPHIC PHASE BEHAVIOR OF INDIVIDUAL LIPID SPECIES

The lamellar to hexagonal H_{II} transition temperature (T_{bh}) for a variety of phosphatidylethanolamines (PE) of both synthetic and natural origin are shown in TABLE 1. Several general conclusions may be drawn. First, many species of naturally occurring PE preferentially adopt a hexagonal H_{II} phase at physiological temperatures. It is evident that for di-unsaturated species, T_{bh} increases with increasing saturation or the presence of trans-unsaturated acyl chains. For di-saturated dialkyl (ether linkage) or diacyl (ester linkage) species, T_{bh} decreases with increasing chain length. Decreasing the water content or increasing the ionic strength results in a

TABLE 2. Polymorphic Phase Preferences of Liquid Crystalline Unsaturated Lipids

	Phase Preferences	
Lipid	Physiological Conditions[a]	Other Conditions
Phosphatidylcholine	L	H_{II} low hydration and high temp
Sphingomyelin	L	
Phosphatidylethanolamine	H_{II}	L, pH 8.5 low temp
Phosphatidylserine	L	H_{II}, pH 3.5
Phosphatidylglycerol	L	H_{II}, high temp, high salt conc.
Phosphatidylinositol	L	
Cardiolipin	L	H_{II}, divalent cations, pH 3, high salt
Phosphatidic acid	L	H_{II}, divalent cations, pH 3.5, high salt
Monoglucosyldiglyceride	H_{II}	
Diglucosyldiglyceride	L	
Monogalactosyldiglyceride	H_{II}	
Digalactosyldiglyceride	L	
Cerebroside	L	
Cerebroside sulfate	L	
Ganglioside	M	
Lysophosphatidylcholine	M	
Cholesterol		Induces H_{II} phase in mixed lipid systems
Unsaturated fatty acids		Induce H_{II} phase

[a] L = lamellar, H_{II} = hexagonal, M = micellar.

decrease in T_{bh}. Alkaline pH stabilizes PE in a lamellar phase, whereas acidic conditions favor hexagonal H_{II} and cubic phases.

Many other species of lipid can also adopt nonlamellar phases under various conditions. The polymorphic phase preferences of many of the major classes of phospholipid found in biological membranes is shown in TABLE 2. It is clear that many lipids, under the appropriate conditions, in isolation adopt the hexagonal H_{II} phase. For example, low pH induces the hexagonal H_{II} phase for unsaturated phosphatidylserine (PS),[5] or the addition of calcium converts cardiolipin from a lamellar to a hexagonal H_{II} phase-preferring species.[6]

POLYMORPHISM IN MIXED LIPID SYSTEMS

While of great intrinsic interest, studies upon individual lipids are far removed from the complexity of a biological membrane. It is therefore of interest to examine the phase behavior of lipids in mixtures in order to determine how the lamellar/nonlamellar phase preferences of such systems may be modulated by physiologically relevant variables.

Phosphatidylcholine (PC) can stabilize unsaturated PEs in a lamellar phase in a manner that is dependent upon various factors including the acyl chain unsaturation, temperature, and also the molar ratio of PC to PE.[7,8] In general, increased unsaturation, high temperatures, and the presence of less than 20–25 mole percent PC are all factors that favor destabilization of lamellar structure and formation of hexagonal H_{II}, cubic, or inverted micellar phases. This effect of PC is quite general in that any lipid that adopts a lamellar phase in isolation will stabilize a lamellar phase in mixtures with PE. This is illustrated in FIGURE 3 for mixtures of PE with 15–30 mole percent of various acidic phospholipids. Also shown is the ability of calcium to induce lamellar to hexagonal H_{II} transitions in those systems. For PE/PS mixtures, dependent upon the acyl chain unsaturation, calcium induces a lateral phase separation of the PS component into an anhydrous lamellar Ca^{2+}/PS complex, leaving the PE free to revert to the hexagonal H_{II} phase it preferentially adopts in isolation.[9] For PE/PG mixtures, calcium reduces the ability of PG to stabilize a lamellar phase without inducing a lateral phase separation, both lipids in the mixture participating in the lamellar-H_{II} transition.[10] The mechanism in the case of PE/PI mixtures is more equivocal, but there may be partial phase separation.[11] Since cardiolipin alone, in the presence of calcium, adopts a hexagonal H_{II} phase, addition of calcium to PE/CL results in a lamellar-H_{II} transition for both species.

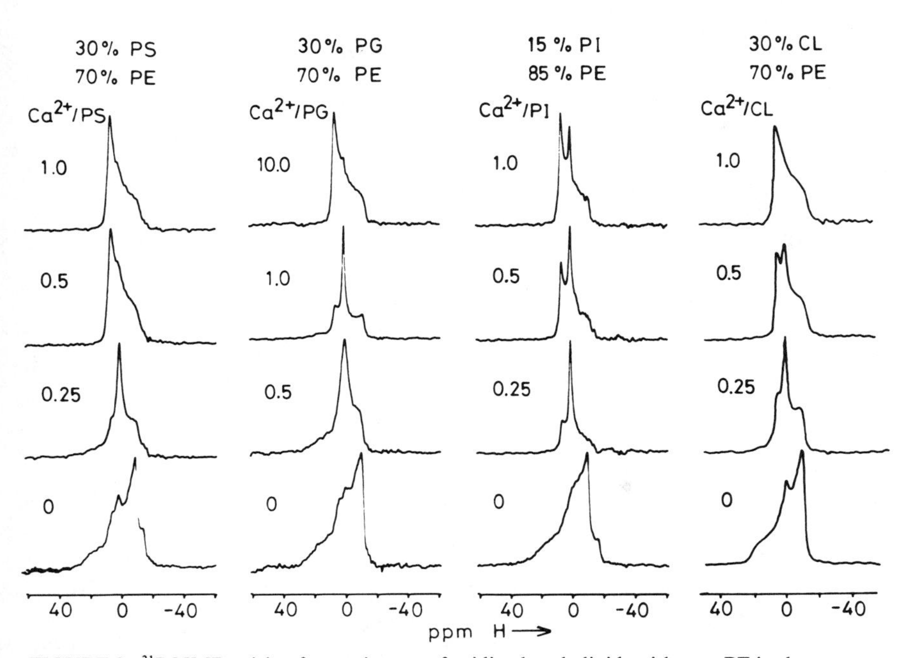

FIGURE 3. ^{31}P NMR arising from mixtures of acidic phospholipids with soya PE in the presence of various molar ratios of calcium.

EFFECTS OF CHOLESTEROL

In addition to its known ability to inhibit the formation of gel-state lipid and to decrease the permeability of lamellar systems, cholesterol is also able to destabilize lamellar structure and promote hexagonal H_{II} structure in a variety of unsaturated lipid mixtures. This is illustrated for a dioleoyl PE/dioleoyl PC (4:1) mixture in FIGURE 4. It can be seen that 20 mole percent cholesterol completely induces hexagonal H_{II} structure and that even as little as 2 mole percent can perturb the phase behavior. This figure also shows that ^2H NMR spectra from [C_{11}-$^2H_2$]-DOPC/DOPE (1:4) and DOPC/[C_{11}-$^2H_2$]-DOPE (1:4) mixtures are very similar, indicating that the DOPE and DOPC partition equally amongst the lamellar, hexagonal H_{II}, and isotropic phases, *i.e.*, on the NMR time scale, cholesterol does not exhibit a preferential association with either lipid component. The presence of cholesterol in PE/PS systems is also known to modify the response of such systems to divalent cations. First, the lamellar phase is destabilized at lower molar ratios of calcium to PS than in the absence of cholesterol. Second, magnesium can also induce hexagonal H_{II} phase structure, an effect not observed in the absence of cholesterol, and third, cholesterol inhibits the ability of divalent cations to induce lateral phase separations.[9,12] The bilayer destabilizing effects of cholesterol are quite general in that various sterols such

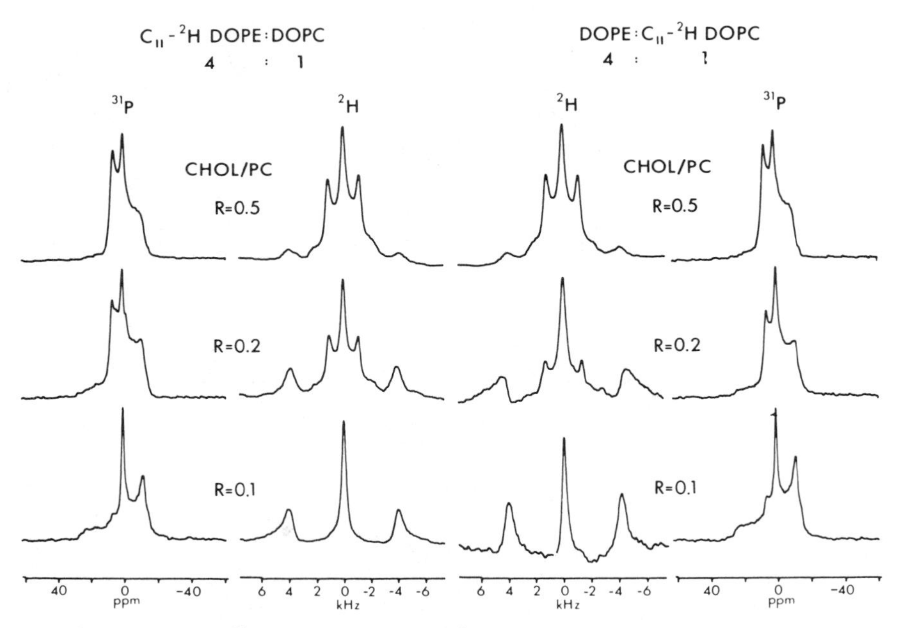

FIGURE 4. 81.0-MHz ^{31}P NMR and 30.7-MHz ^2H NMR spectra at 30° C arising from aqueous dispersions of mixtures of DOPE, DOPC, and cholesterol (CHOL) at a DOPE/DOPC molar ratio of 4:1 where either the DOPE is ^2H-labelled at the C_{11} position ([C_{11}-$^2H_2$]DOPE) or the DOPC is ^2H-labelled at the C^{11} position ([C_{11}-$^2H_2$]DOPC). The ratio R refers to the molar ratio of cholesterol to DOPC.

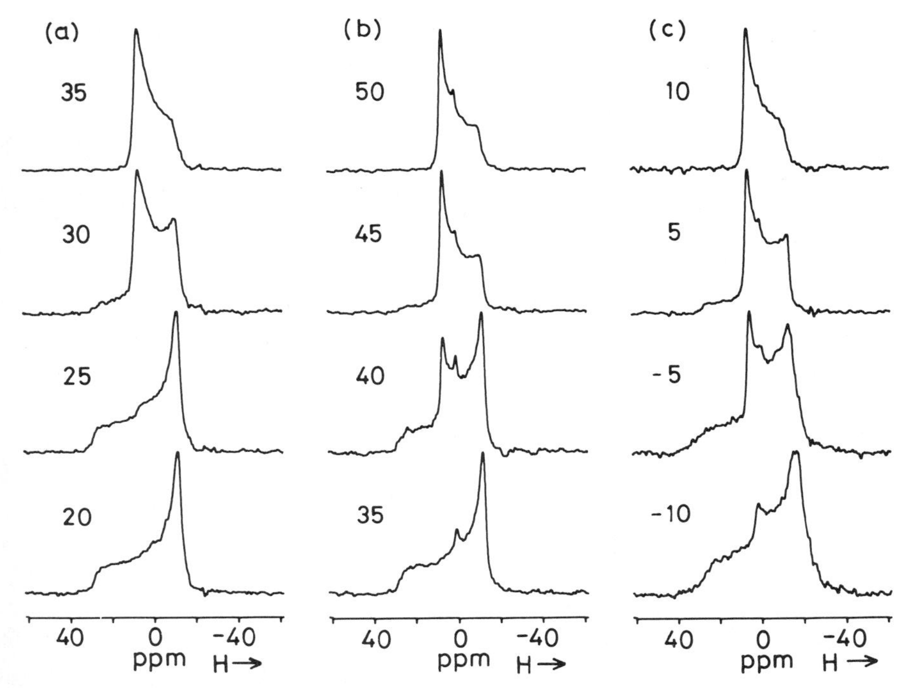

FIGURE 5. 81.0-MHz ^{31}P NMR of egg PE at the indicated temperatures, (**a**) in the absence of alcohols, (**b**) in the presence of ethanol (ethanol to lipid molar ratio = 4.5), and (**c**) in the presence of decanol (decanol to lipid molar ratio = 0.45).

as ergosterol, coprostanol, epicoprostanol, stigmasterol, and androstanol can all induce the same effects (Cullis and Tilcock, unpublished observations).

EFFECTS OF ANESTHETICS AND ALCOHOLS

The local anesthetics, dibucaine and chlorpromazine, can induce hexagonal H_{II} phase structure in mixtures with unsaturated cardiolipin or phosphatidic acid.[13,14] Alternatively, in mixtures with unsaturated PE, chlorpromazine, dibucaine, tetracaine, and procaine all stabilize lamellar structure.[15] The amount of anesthetic required to induce these effects depends upon the lipid composition, since the presence of acidic phospholipids can markedly increase the partition coefficient of positively charged anesthetics.

The effect of ethanol and decanol on the phase behavior of egg PE is shown in FIGURE 5. In the absence of alcohol, egg PE exhibits a T_{bh} of approximately 30° C. Addition of ethanol to give an ethanol/phospholipid molar ratio of 4.5 caused an increase in T_{bh} to about 42° C, whereas addition of decanol to a molar ratio of 0.45 resulted in a decrease in T_{bh} to 5° C. The effect of various alcohols upon the T_{bh} of egg PE is shown in FIGURE 6. Shorter chain alcohols ($C \leq 4$) may be considered to stabilize

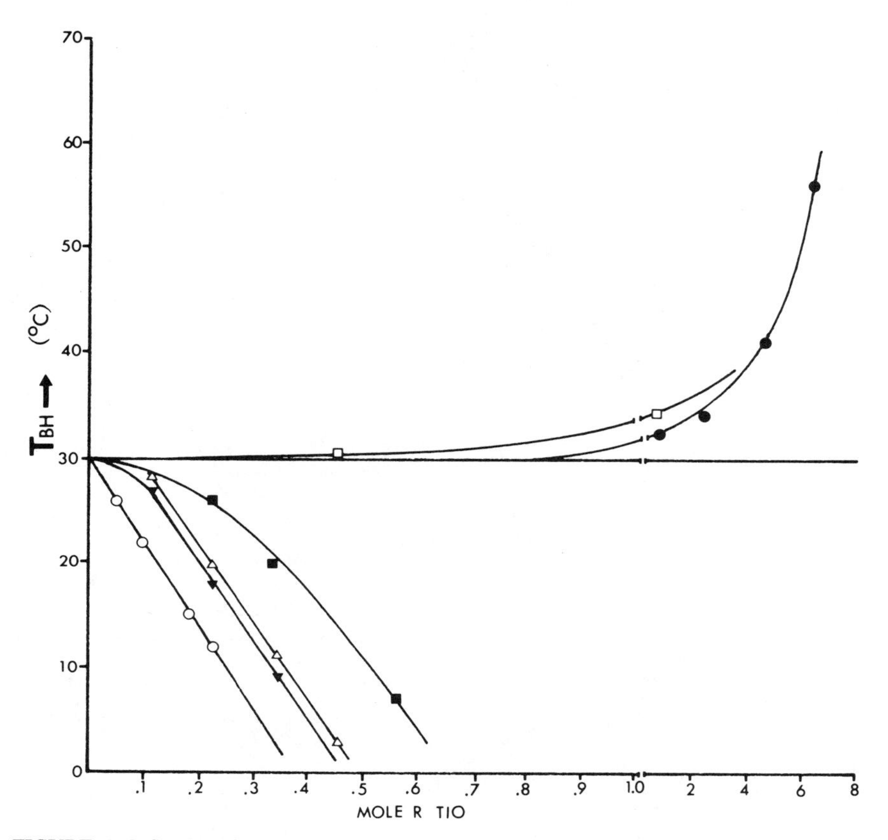

FIGURE 6. Influence of varying amounts of normal alcohols on the bilayer to hexagonal H_{II} phase transition temperature (T_{bh}) of egg PE. T_{bh} is estimated as that temperature where 50% of the lipid is in a lamellar organization and 50% is in the hexagonal H_{II} phase: ●-ethanol; □-butanol; ■-hexanol; ▼-octanol; △-decanol; O-lauryl alcohol.

lamellar structure in that they cause an increase in T_{bh}, whereas longer chain alcohols ($C > 6$) promote formation of the hexagonal H_{II} phase at lower temperatures.

LIPIDIC PARTICLES

Mixtures of lipids such as PC, which preferentially adopts a lamellar phase in isolation, and PE, which adopts the hexagonal H_{II} phase, often exhibit narrow isotropic ^{31}P NMR spectra that are associated with the appearance of lipidic particles as visualized by freeze-fracture techniques.[16] Lipidic particles are thought to represent intrabilayer inverted micellar structures, as illustrated in FIGURE 7, which have been suggested to arise as intermediates in the lamellar-H_{II} transition. Such lipidic particles have been postulated to occur during membrane fusion, as illustrated in FIGURE 8.

THE MOLECULAR BASIS OF LIPID POLYMORPHISM

To a first approximation, the phase behavior of lipids may be rationalized in terms of their dynamic molecular shape, as illustrated in FIGURE 9. It may be considered that

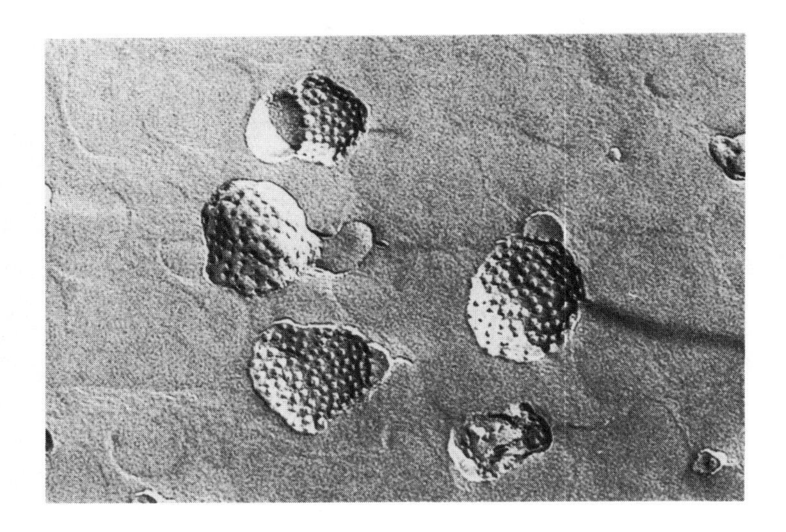

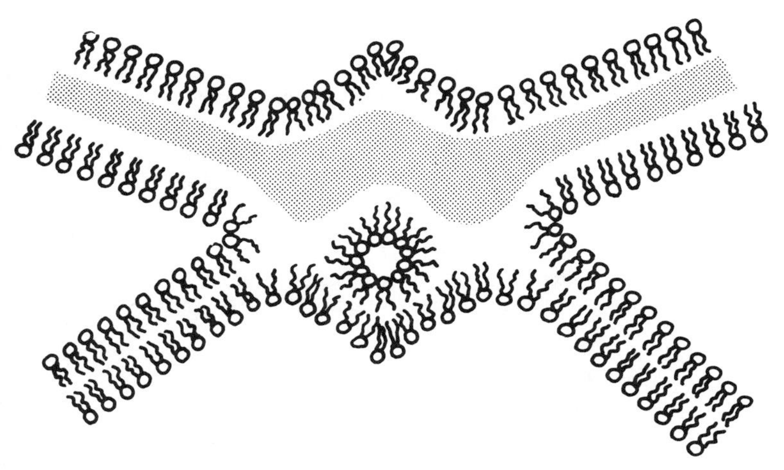

FIGURE 7. Freeze-fracture micrograph of lipidic particles induced by calcium in a cardiolipin/soya PE (1:4) mixture. A model of the lipidic particle as an inverted micelle is depicted *below* the micrograph. The *shaded area* represents the fracture region.

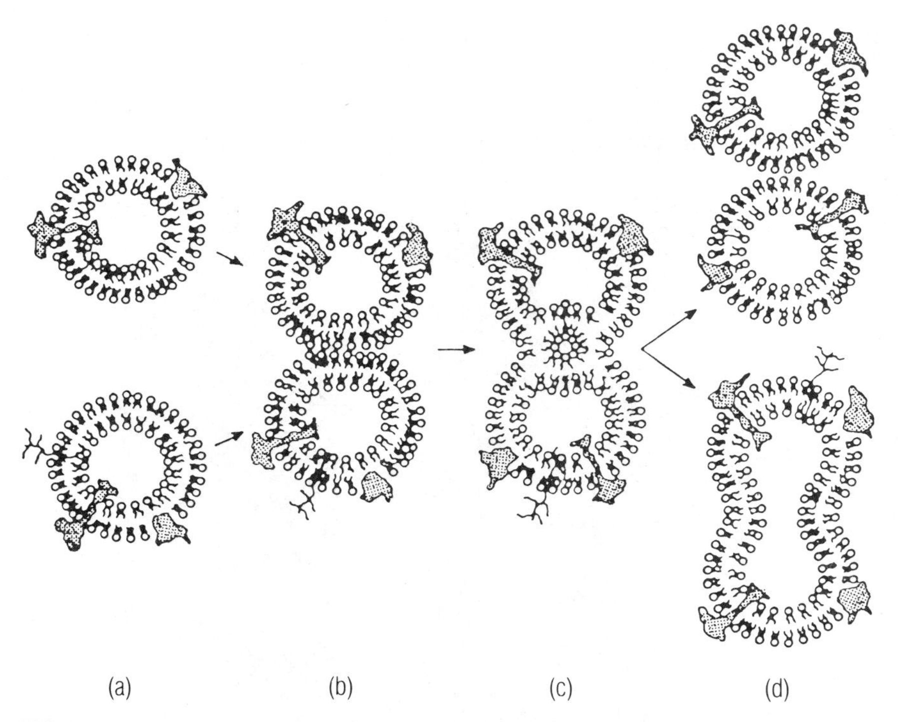

FIGURE 8. Proposed mechanism of membrane fusion proceeding via an inverted cylinder or inverted micellar intermediate.

lipids such as lysophosphatidylcholines possess a dynamic cross section that is inverted conical, the area subtended by the headgroup at the membrane/water interface being greater than that subtended by the acyl chains. On purely geometric grounds one might expect that such lipids would pack into micellar or hexagonal H_I aggregates so as to minimize hydrocarbon/water contacts. Lipids such as phosphatidylcholine or phosphatidylserine at neutral pH may be considered to possess a cylindrical dynamic cross section and would thus pack most readily into lamellar assemblies. Alternatively, lipids such as PE or PS at pH 3 (where the charge on the headgroup would be suppressed), may be considered to possess a conical dynamic cross section with the area subtended by the headgroup being less than that of the acyl chains. Such lipids would most readily adopt inverted structures such as inverted micellar or hexagonal H_{II}.

This simple rationale provides a qualitatively reasonable explanation for lipid phase behavior. In the case of PEs, increasing temperature leads to increased entropic splay of the acyl chain region, *i.e.* the swept hydrophobic volume increases relative to the headgroup, an effect that would favor a lamellar-H_{II} transition. Similarly for PS or phosphatidic acid, low pH results in protonation of the headgroup and an effective decrease in the effective headgroup area.

One prediction of this shape hypothesis is that mixtures of lipids that adopt the hexagonal H_{II} phase in isolation, and those that adopt micelles in isolation, should form lamellar structures. This has been demonstrated to occur for mixtures of egg PE and various detergents,[17] as illustrated in FIGURE 10.

It is important to recognize that the dynamic shape is a consequence of many interrelated factors such as the size and motion of the lipid headgroup and acyl chains, the extent of headgroup hydration and charge, temperature, counterion binding, hydrogen bonding associations, and various other factors. Also the dynamic shape is an

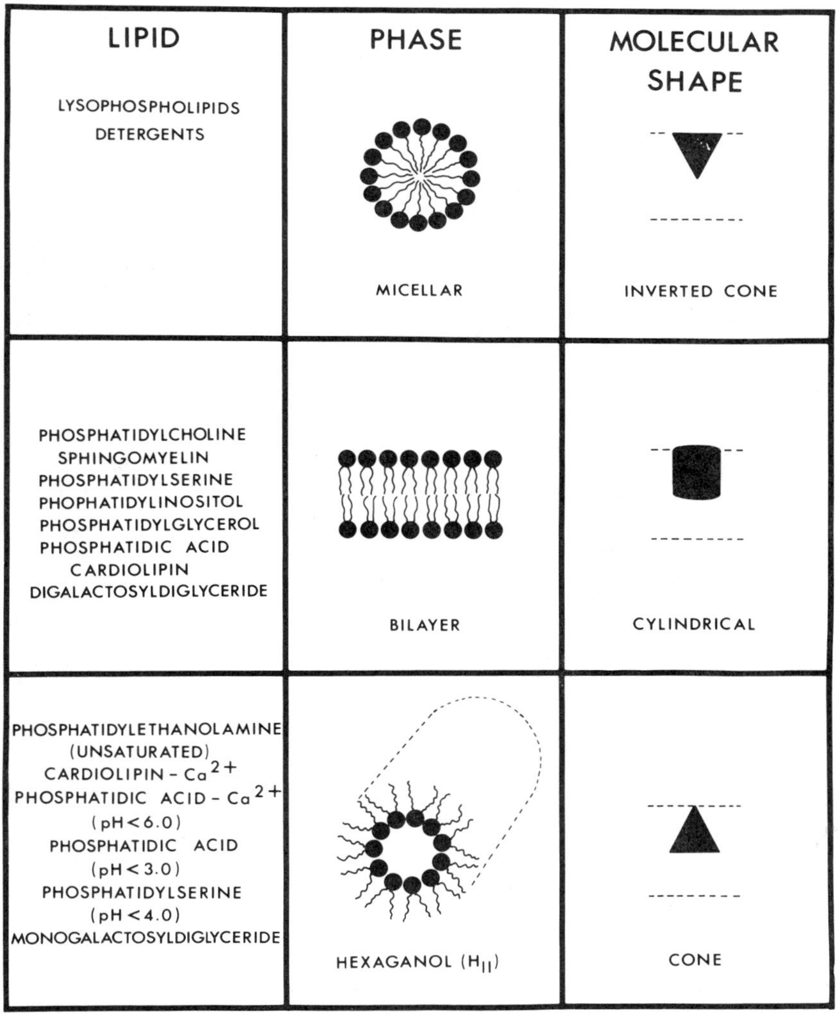

LIPID	PHASE	MOLECULAR SHAPE
LYSOPHOSPHOLIPIDS DETERGENTS	MICELLAR	INVERTED CONE
PHOSPHATIDYLCHOLINE SPHINGOMYELIN PHOSPHATIDYLSERINE PHOPHATIDYLINOSITOL PHOSPHATIDYLGLYCEROL PHOSPHATIDIC ACID CARDIOLIPIN DIGALACTOSYLDIGLYCERIDE	BILAYER	CYLINDRICAL
PHOSPHATIDYLETHANOLAMINE (UNSATURATED) CARDIOLIPIN - Ca^{2+} PHOSPHATIDIC ACID - Ca^{2+} ($pH<6.0$) PHOSPHATIDIC ACID ($pH<3.0$) PHOSPHATIDYLSERINE ($pH<4.0$) MONOGALACTOSYLDIGLYCERIDE	HEXAGANOL (H_{II})	CONE

FIGURE 9. Polymorphic phases and corresponding dynamic molecular shapes of component lipids.

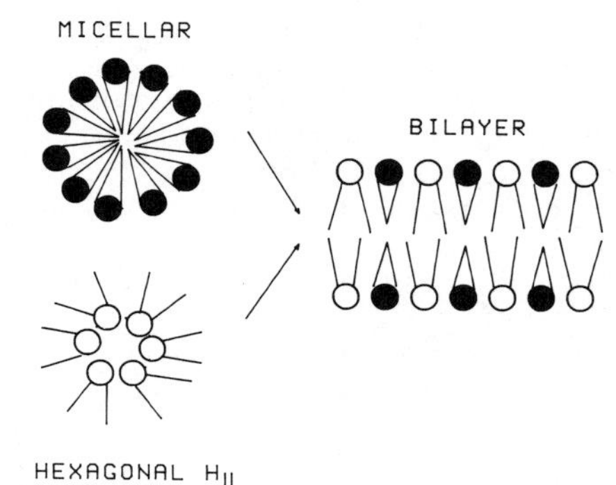

FIGURE 10. A net bilayer structure arising from mixtures of cone-shaped (H_{II} phase) lipids and inverted cone (micellar) lipids due to shape complementarity effects.

ensemble property, not an intrinsic property of a given lipid, and is thus modulated by interaction with surrounding lipids. This is illustrated in FIGURE 11, which shows the effect of dioleoyl PE on the quadrupole splitting (ΔQ) of deuterium-labelled C_{11}-dioleoyl PC. With increasing PE content in the mixture, ΔQ increases, indicating increased hydrocarbon chain order, at least at the C_{11} position. This effect may be rationalized by considering that the acyl chains of the "conical" PE molecules

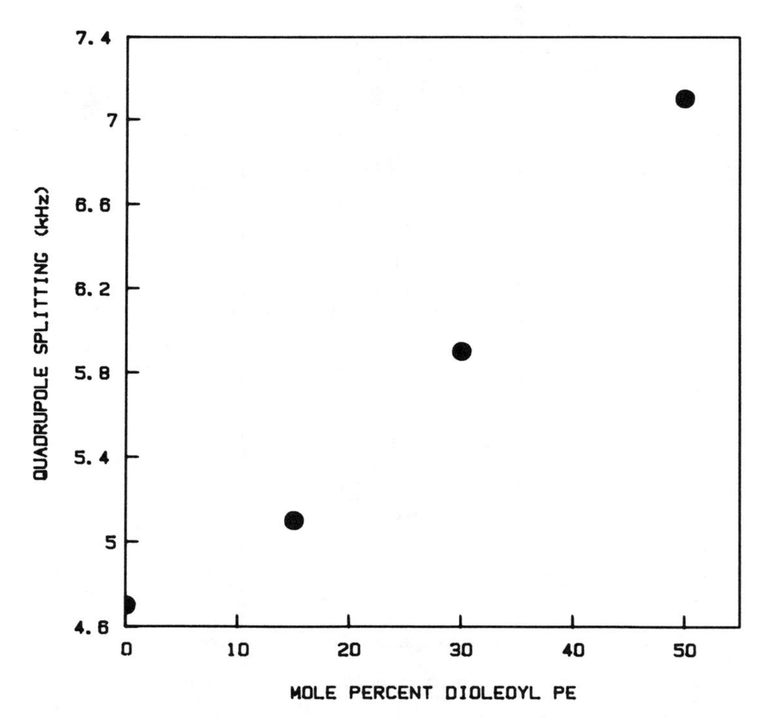

FIGURE 11. Effect of dioleoyl PE on the quadrupole splitting of $^2H_2$-C_{11}-dioleoyl PC at 30° C.

compress the surrounding PC acyl chains, in order to minimize water penetration into the hydrocarbon region that would otherwise be exposed.

SUMMARY

The phase behavior of phospholipids may be monitored using ^{31}P or ^{2}H NMR techniques, which provide information concerning the motional properties of the lipid ensemble, which may then be correlated with structure.

The lamellar/nonlamellar phase preferences of many lipids, either synthetic or naturally derived, may be controlled by factors such as variation in temperature, hydration, or of greater physiological relevance, pH, ionic strength, the presence of divalent cations such as calcium, or the presence of lipid soluble agents such as anesthetics and alcohols. The ability of short-chain alcohols to stabilize a bilayer structure for egg PE may be rationalized in terms of the packing of lipids whose dynamic shapes are complementary, as illustrated in FIGURE 11. On this basis, short-chain alcohols would partition preferentially at the membrane/water interface and would thereby stabilize a lamellar structure. Larger-chain alcohols may partition deeper into the hydrophobic acyl chain region in order to minimize hydrocarbon/water contact and so may perturb the acyl chain packing, increasing the effective swept volume of the chains and so promoting hexagonal H_{II} phase formation.

REFERENCES

1. CULLIS, P. R., M. J. HOPE, B., DE KRUIJFF, A. J. VERKLEIJ & C. P. S. TILCOCK. 1985. *In* Phospholipids and Cellular Regulation. J. F. Kuo, Ed. Vol. 1:2–59. CRC Press. Boca Raton, FL.
2. SEELIG, J. 1978. Biochim. Biophys. Acta **515:** 104.
3. VERKLEIJ, A. J. 1984. Biochim. Biophys. Acta **779:** 43.
4. DAVIS, J. H. 1983. Biochim. Biophys. Acta **737:** 117.
5. HOPE, M. J. & P. R. CULLIS. 1980. Biochim. Biophys. Res. Commun. **92:** 846.
6. SEDDON, J. M., R. D. KAYE & D. MARSH. 1983. Biochim. Biophys. Acta **734:** 347.
7. CULLIS, P. R., P. W. M. VAN DIJCK, B. DE KRUIJFF & J. DE GIER. 1978. Biochim. Biophys. Acta **513:** 21.
8. TILCOCK, C. P. S., M. B. BALLY, S. B. FARREN & P. R. CULLIS. 1982. Biochemistry **21:** 4596.
9. TILCOCK, C. P. S., M. B. BALLY, S. B. FARREN, P. R. CULLIS & S. M. GOUNER. 1984. Biochemistry **23:** 2696.
10. FARREN, S. B. & P. R. CULLIS. 1980. Biochim. Biophys. Res. Commun. **97:** 182.
11. NAYAR, R., S. L. SCHMID, M. J. HOPE & P. R. CULLIS. 1982. Biochim. Biophys. Acta **688:** 169.
12. BALLY, M. B., C. P. S., TILCOCK, M. J. HOPE & P. R. CULLIS. 1983. Can. J. Biochem. Cell Biol. **61:** 346.
13. CULLIS, P. R. & A. J. VERKLEIJ. 1979. Biochim. Biophys. Acta **522:** 546.
14. VERKLEIJ, A. J., R. DE MAAGD, J. LEUNISSEN-BIJVELT & B. DE KRUIJFF. 1982. Biochim. Biophys. Acta **684:** 225.
15. HORNBY, A. P. & P. R. CULLIS. 1981. Biochim. Biophys. Acta **647:** 285.
16. DE KRUIJFF, B., A. J. VERKLEIJ, C. J. A. VAN ECHTELD, W. J. GERRITSEN, C. MOMBERS, P. NOORDAM & J. DE GIER. 1979. Biochim. Biophys. Acta **555:** 200.
17. MADDEN, T. D. & P. R. CULLIS. 1982. Biochim. Biophys. Acta **684:** 149.

DISCUSSION OF THE PAPER

B. CHANCE (*University City Science Center, Philadelphia, PA*): There is a transition in mitochondrial oxidative phosphorylation below 15 degrees. At about 12 degrees they refuse to accept adenosine diphosphate, for example. Have you tried to correlate structure and function?

CULLIS: We have tried to do that in the case of membrane fusion. The ability of lipids to adopt these different structures, particularly different phase structures such as the hexagonal phase, is mainly related in terms of function to processes such as fusion where you have membrane contact.

L. L. M. VAN DEENEN (*State University of Utrecht, Utrecht, The Netherlands*): Why are the effects of ethanol and decanol on PE of opposite nature? You did not give an explanation for this. Could these differences be explained by differences in the shapes of these molecules? The opposite effects of ethanol and decanol may be tied in with your explanation for lipid polymorphism.

CULLIS: These differences are quite compatible with the shape explanation. The ability of ethanol to reside just inside the membrane/water interface would be consistent with its having approximately an inverted cone shape. This would stabilize the bilayer, whereas the octanol would partition further into the membrane, increasing the hydrocarbon chain area thereby promoting hexagonal H_{II} phase organization.

D. B. GOLDSTEIN (*Stanford University School of Medicine, Stanford, CA*): Your diagram of membrane fusion (FIG. 8) shows that the phospholipids from the external membrane monolayer form the hexagonal H_{II} phase. PE is usually located in the inner monolayer. Does that bother you and do you have an explanation for this?

CULLIS: If one is worried about fusion in an intracellular context, then it doesn't bother me, because the PE is on the right side of the membrane. With regard to cell-cell fusion, if one had a process that put two cells in very close apposition, that would entail removing water from that region. Such dehydration promotes nonbilayer structure.

GOLDSTEIN: So you wouldn't need PE for that?

CULLIS: No, you wouldn't. The presence of cholesterol could be quite sufficient.

Ethanol-Induced Adaptation in Biological Membranes[a]

DORA B. GOLDSTEIN

Stanford University School of Medicine
Stanford, California 94305

ALCOHOLISM AS AN ADAPTATION

That a drug should set in motion physiological processes to counteract its effect is a familiar notion. Drug tolerance and physical dependence fit the general rubric of adaptive changes, in that they are responses that offset the original effects of a drug and they reveal a state in which the organism does not function comfortably in the absence of the accustomed agent. This was noted 40 years ago by Himmelsbach[1] with respect to opiates. His observation that withdrawal reactions are generally opposite in nature to the primary drug effects (for example, hyperexcitability after withdrawal of a sedative drug) clearly suggested an adaptive mechanism. This is not a situation unique to addiction, since many other drugs, especially hormones, elicit a dependent state by homeostatic mechanisms. The organism becomes dependent on the exogenous drug. Adaptive changes can be considered beneficial in general, although the failure of recovery to keep pace with drug elimination after withdrawal leaves the organism in a temporary unbalanced state that may be severe. This paper deals primarily with adaptive mechanisms in cell membranes, a topic included in several recent reviews.[2–6]

Time Courses of Adaptations

To the extent that alcoholism develops over decades of drinking, it cannot be adaptive in nature. Evolution would not have maintained an adaptive mechanism that acts so slowly. We know of adaptive responses that happen almost instantaneously, those that take days, and seasonal changes that vary over months. Anything slower cannot be dealt with in this framework. To test whether some aspects of alcoholism might develop quickly enough to fit the adaptation hypothesis we developed a model in mice, consisting of a brief, intense period of ethanol administration by inhalation. Ours is not a model of alcoholism but a method for testing the homeostat hypothesis in its application to a few typical effects of chronic ethanol intake. Tolerance arises to ethanol-induced ataxia (measured by ability to balance on a dowel)[7] and physical dependence is manifested by convulsions elicited by handling after withdrawal.[8] Both tolerance[9] and physical dependence,[8] as we measure them, are dose-related and can develop during a single session of intoxication, lasting a few hours. Our model will pick up effects that are too evanescent for some other methods, and it ignores slow changes. The phenomena that arise over years in alcoholic people must be much more complex than those we observe in the laboratory, but they probably include the relatively simple responses that we study.

[a]Supported by United States Public Health Service Grant AA 01066 and by the Alcoholic Beverage Medical Research Foundation.

In nature, animals adapt to a single stimulus, such as a change in temperature, by several mechanisms simultaneously, using fast neuronal responses first, then more efficient biochemical ones. Several such mechanisms may be activated at once, to come into play sequentially. We also expect that quickly arising responses will die away quickly as they are replaced by succeeding events or when the stimulus is removed, whereas slowly developing adaptations will decay with a long time course. Analogously, multiple responses may occur when the body adapts to a drug in the internal milieu. Some discrepancies in the literature about the time courses of recovery after exposure to ethanol[10,11] may reflect the different innate time courses of the physiological processes chosen for study in different laboratories.

Studies of adaptive changes cannot help us to identify the site of action of a drug. The deranged function is not necessarily restored at its primary site. For example, a change in blood pressure, sensed at some specific receptors, may set in motion an array of responses in different organs and tissues. Observing changes in the vascular smooth muscle would not tell whether the original disruption began there, rather than in the heart or kidney.

Membrane-Bound Enzymes

Enzymes that reside in membranes are often inhibited by ethanol *in vitro,* then become insensitive to it after chronic administration. They may develop greater intrinsic activity after chronic ethanol treatment, suggesting both tolerance and physical dependence. The Na^+-K^+-ATPase follows this pattern.[3] Similarly, John *et al.*[12] observed an increase in phospholipase A_2 and in phospholipid base exchange enzymes after chronic ethanol treatment, in contrast to inhibition of these enzymes by ethanol *in vitro.* Various observations of receptor up- or down-regulation are also interpreted in terms of adaptive changes, although these have often been difficult to reproduce under different conditions of ethanol administration. We do not know how gene expression is regulated to produce these changes after exposure to ethanol.

Sometimes the apparent adaptation takes the form of an altered affinity of a ligand for an enzyme or receptor, a process that implies a change in protein conformation rather than protein synthesis. Since these proteins are embedded in cell membranes, this brings us to lipid bilayers as a possible site of the adaptive response.

MEMBRANE BILAYERS

Ethanol-Induced Disorder

The notion that ethanol may exert its acute effects in membranes derives directly from the Meyer-Overton hypothesis of anesthetic action.[13] Ethanol is one of the drugs whose (weak) potency can be predicted from its (low) solubility in lipids. We may then surmise that adaptations to the acute effect may also occur in membranes, although this need not follow, as mentioned above. Much of the research in my laboratory over the past several years has explored these possibilites. My colleagues Jane Chin, Robbe Lyon, Edward Gallaher, Linda Parsons, Janet McComb, Barry Perlman, and Christopher Daniels have shared in this work.

We use two methods to estimate membrane order, electron paramagnetic resonance (EPR) and fluorescence polarization. They are similar but sometimes give qualitatively different results, because the different probes report from regions of the membrane with different physical properties and responses to drugs. Both techniques, like all other methods of measuring the fluidity of membranes, show that ethanol disorders the bilayer.[2] That is, the drug causes increased molecular motion, such as faster rotation of phospholipids about their long axes and increased wobbling motion of the acyl chains of the phospholipids. Ethanol and other anesthetic agents allow exogenous probes (spin labels or fluorescent dyes) to move more rapidly or with greater amplitude than in the absence of drugs. Ethanol reduces the order parameter of phospholipid model membranes, synaptosomal plasma membranes, and erythrocyte membranes spin-labeled with doxylstearic acid probes.[14,15] Ethanol also reduces the steady state fluorescence anisotropy of biomembranes labeled with diphenylhexatriene.[16] The effects are linearly concentration-related and they vary directly with the lipid solubility of different alkanols,[17] suggesting that ethanol's effects are proportional to its concentration in the hydrophobic phase, rather than its binding to a receptor. Binding should give a logarithmic concentration relation.

Nevertheless, the effects are always small. This requires that we attempt to correlate the membrane disordering with some measure of the drug's actions *in vitro*, as an indication whether the tiny changes are important. Indeed, ethanol does affect membranes from ethanol-sensitive mice more strongly than membranes from mice that are resistant to intoxication, *e.g.* in comparisons of genetic variants[18] or in normal vs. tolerant animals.[19] This indicates that the observed disorder of the bulk lipids accurately reflects some important action of ethanol, as yet undiscovered. We can definitely feel the earthquake but we may be remote from its epicenter.

Adaptation to Membrane Disorder

Now we are in a position to predict what kind of an adaptation might occur after chronic exposure of a biological system to ethanol. The too-fluid membrane should pull itself together and adjust its chemical composition to compensate for the continuous presence of the drug. The membrane lipids of poikilothermic animals are known to do so in response to changes in ambient temperature.[20]

Indeed, there are many examples of increased membrane order after chronic administration of ethanol. But the effect is not universal; it depends on the tissue, on the method of examining the membranes, and perhaps on the conditions of ethanol administration. The order parameter of mouse synaptosomal plasma membranes from ethanol-treated mice was higher than that of controls[21] when the spin label was 12-doxylstearic acid but not when it was 5-doxylstearic acid.[19,21] Excess order could be seen in fluorescence polarization experiments with diphenylhexatriene as probe.[22] Thus the deeper probes (diphenylhexatriene and 12-doxylstearic acid) reveal an increase in order not seen near the membrane surface (with 5-doxylstearic acid).

Increased membrane order after chronic ethanol treatment has been reported by other investigators. The order of synaptic membranes of ethanol-treated mice was found to be increased, whether it was measured by fluorescence polarization with diphenylhexatriene as probe[23] or even by EPR with 5-doxylstearic acid,[24] although there was no change in the order of the extracted membrane lipids.[23] Hepatic microsomal membranes of chronically ethanol-treated rats, spin-labeled with 5-doxylstearic acid, are more ordered than controls,[25] but brain microsomal membranes are not.[26] Nor are hepatic microsomal membranes labeled in deeper parts of the bilayer with 12-doxylstearic acid or its phospholipid analog,[27] or hepatic mitochondrial membranes whether spin-labeled[28] or studied by fluorescence polarization.[29]

Furthermore, hepatic plasma membranes are unique in that they become more fluid during chronic ethanol treatment of the animals[30–31] or in culture in the presence of ethanol.[32] In cultured hepatoma cells,[32] ethanol metabolism seems to be responsible for the increase in fluidity, since no such change occurred in a fibroblast line unable to metabolize ethanol or in experiments where hepatoma cells were treated with methylpyrazole to block ethanol metabolism.

Bacterial membranes, like those of mammals, are disordered by ethanol *in vitro.* When *Escherichia coli* is grown in the presence of ethanol, the intact membranes become stiffer, just as mammalian membranes often do, but the extracted membrane lipids become more fluid. The stiffness of whole membranes is explained by their higher-than-normal ratio of protein to phospholipid, and the fluidity of the lipids is apparently due to increased vaccenic acid (18:1) in proportion to palmitic (16:0).[33] In the eukaryotic micro-organism *Tetrahymena,* growth in ethanol was followed by increased order of microsomal lipids but decreased order in the lipids of mitochondria and plasma membranes.[34]

An intriguing observation is that the erythrocyte membranes of recently drinking alcoholic subjects are more ordered than controls.[35] It has not yet been shown whether the membranes of these individuals differ from controls in the absence of ethanol consumption. Such a finding might suggest a genetic difference predating alcoholism.

Tolerance in Membranes

Whether or not the membranes of ethanol-treated animals are more ordered than normal, they usually are resistant to ethanol-induced disordering *in vitro.* In our inhalation model, tolerance develops in the intact mice over a few days of ethanol exposure, and it decays rapidly within about 30 hr after withdrawal.[7] The magnitude of the tolerance is about twofold. That is, the tolerant mice can balance on a dowel at a brain ethanol concentration nearly twice that of controls. The magnitude of the membrane tolerance *in vitro* depends on the probe used[21] and is usually greater than that seen in the whole animals. For example with 12-doxylstearic acid as probe, the tolerant membranes require 200 mM more ethanol than those of controls for a given degree of disorder, although they have experienced blood levels of only 50 mM during chronic intoxication. The decreased sensitivity of the membranes to ethanol *in vitro* is reversible at about the same rate as the recovery from tolerance in the whole animals.[36]

This tolerance in synaptic membranes of mice or rats after chronic ethanol administration has been observed by others.[23,24,37] Furthermore, tolerance arises in several other types of membranes, even including the hepatic plasma membranes that become more fluid after chronic ethanol treatment.[30] Hepatic microsomal[25,27] and mitochondrial[24,28] membranes and brain microsomal membranes[26] become tolerant to ethanol after treatment of the animals, as do mouse erythrocytes.[19] A rare exception is the lack of tolerance in *Tetrahymena* membranes after growth in the presence of ethanol, despite increased order of some of its membranes.[34] The increased order of red cell membranes of alcoholics, noted above,[35] is accompanied by tolerance to the disordering effect of ethanol *in vitro.*

Rottenberg and co-workers[24,38] have demonstrated a decreased solubility of ethanol (and other compounds) in membranes from ethanol-treated rats, and they suggest that the membrane tolerance is simply the result of lower concentrations of the drug within the membranes of the ethanol-treated animals. This might indeed explain the tolerant membranes. Decreased partition coefficients could not be the cause of the increased membrane order, but perhaps (conversely) the solubility of ethanol is reduced when the membranes are rigid.

Chemical Basis for Increased Order

The experiments described above demonstrate that membrane tolerance and increased order after chronic ethanol administration are frequent but not universal findings. The data on accompanying changes in the chemical composition of membranes are even less consistent. The search for altered lipids centers on cholesterol and on saturated acyl chains of phospholipids because increased amounts of these components would be expected to stiffen membranes and to render them relatively resistant to ethanol-induced disorder. Indeed, in some situations, the membrane cholesterol/phospholipid ratio does increase during ethanol treatment, for example in mouse synaptosomal membranes[39,40] after treatment of mice with ethanol by diet and also in hepatic plasma membranes of neonatal mice that had received ethanol *in utero*.[41] Increased cholesterol/phospholipid ratios were also seen in hepatic mitochondrial and microsomal membranes of monkeys that had been treated with ethanol for a year.[42]

But cholesterol does not always increase after chronic ethanol treatment. In our inhalation experiments where tolerance and physical dependence develop in mice and where the membranes show corresponding changes,[21] and in similar experiments by La Droitte *et al.*,[43] there was no change in the cholesterol/phospholipid ratio of the brain synaptosomal plasma membranes. Nor was there any change in the cholesterol/phospholipid ratio of human[44] or rat[43] red cells or of cultured HeLa cells[45] or hepatoma cells[32] after chronic ethanol exposure. Others have even found decreased cholesterol/phospholipid ratios in hepatic plasma membranes[30,31] or synaptosomal membranes[23] from ethanol-treated animals.

Increased saturation of the acyl chains of phospholipids is the mechanism for homeoviscous adaptation to temperature in *E. coli*[46] and in fish.[20] Membranes from ethanol-treated animals often show an apparently analogous increased saturation of acyl chains,[26,34,42–44,47] but this is not a universal finding.[32,48] Sometimes the membrane fatty acids become more unsaturated, contrary to predictions.[41,45,49–51] Several authors, including Smith *et al.*,[52] have commented on the possible confounding effects of dietary and nutritional factors on these different responses to ethanol.

The ratio of saturated to unsaturated fatty acids in phospholipids is controlled in part by the activity of hepatic microsomal acyl-CoA desaturase. This enzyme activity is decreased after chronic ethanol treatment[53,54] but not by ethanol *in vitro*.[54] The desaturase is also decreased by chronic ethanol treatment in *Tetrahymena*.[34] In *E. coli*, exposure to ethanol paradoxically produces increased proportions of unsaturated fatty acids in the membrane; this is mediated by preferential inhibition of saturated fatty acid synthesis, rather than by a change in desaturase activity.[55] As mentioned above, adaptations may proceed by various mechanisms simultaneously. If the primary objective of the adaptation is to counteract disorder, the organism may use cholesterol, fatty acid chains, and other mechanisms simultaneously, and one or another may be detectable to the investigator after different periods of ethanol exposure.

The proportions of different phospholipid head groups may change during ethanol administration, as may the amounts of exposed carbohydrate. These must be important, but we do not know their relation to membrane order. Thus we cannot connect such changes with an adaptive response to membrane order.

A possible mechanism for changes in membrane lipids (when they do occur) is suggested by our experiments on passive transfer of lipids into erythrocyte membranes *in vitro*. Exchange of cholesterol with various donors including high- and low-density serum lipoproteins is accelerated by the presence of ethanol,[56] and ethanol preferentially increases the rate of transfer of palmitic (relative to oleic) acid into erythrocyte membranes.[56a] Alling *et al.*[57] suggest that ethanol, by disordering membranes, may make them more responsive than normal to changes in plasma lipids.

CONCLUSIONS

Although the homeostat hypothesis remains attractive, testing it has revealed many inconsistencies; it emerges tattered. One may still find a body of observations that fit the general outlines of the idea, but there are many exceptions. Perhaps these will be clues to the actual mechanism of responses to ethanol.

One of the most difficult problems is the small magnitude of the supposed stimulus for the adaptation, the disordering of membranes by ethanol. Although the disordering effect is statistically significant at 10 or 20 mM ethanol,[14] it is less than a 1% change in the order parameter. Not only is the change in bulk lipid small in absolute terms, but it is small in relation to changes caused by warming. At a lethal concentration of ethanol the disorder matches that of only 1 degree of warming, a change that animals undergo daily in their circadian rhythms. Clearly ethanol does not exactly mimic warming.

Nevertheless, the fine correspondence between sensitivity of membranes and that of the animals from which they were taken suggests that we are on the right track. The disorder in bulk lipids, monitored by spin labels and fluorescent probes, may reflect a corresponding but more disruptive change elsewhere, in regions where the probes do not penetrate well or in specific locations that are averaged out in studies of the whole membrane. The possibility that some membrane proteins lack cholesterol in their boundary lipids, though not yet well established, suggests that there may be ethanol-sensitive proteins whose immediate environment favors entry of ethanol. This is where we should seek the epicenter of the disruption we now feel only distantly.

REFERENCES

1. Himmelsbach, C. K. 1943. Can the euphoric, analgetic and physical dependence effects of drugs be separated? Fed. Proc. **2:** 201–203.
2. Goldstein, D. B. 1984. The effects of drugs on membrane fluidity. Annu. Rev. Pharmacol. Toxicol. **24:** 43–64.
3. Hunt, W. A. 1985. Alcohol and Biological Membranes. The Guilford Press. New York, NY.
4. Littleton, J. M. 1983. Tolerance and physical dependence on alcohol at the level of synaptic membranes: A review. J. R. Soc. Med. **76:** 593–601.
5. Michaelis, E. K. & M. L. Michaelis. 1983. Physico-chemical interactions between alcohol and biological membranes. Res. Adv. Alc. Drug Prob. **7:** 127–173.
6. Taraschi, T. F. & E. Rubin. 1985 Effects of ethanol on the chemical and structural properties of biologic membranes. Lab. Invest. **52:** 120–131.
7. Goldstein, D. B. & R. Zaechelein. 1983. Time course of functional tolerance produced in mice by inhalation of ethanol. J. Pharmacol. Exp. Ther. **227:** 150–153.
8. Goldstein, D. B. 1972. Relationship of alcohol dose to intensity of withdrawal signs in mice. J. Pharmacol Exp. Ther. **180:** 203–215.
9. Gallaher, E. J., L. M. Parsons & D. B. Goldstein. 1982. The rapid onset of tolerance to ataxic effects of ethanol in mice. Psychopharmacology **78:** 67–70.
10. Goldstein, D. B. 1974. Rates of onset and decay of alcohol physical dependence in mice. J. Pharmacol. Exp. Ther. **190:** 377–383.
11. Branchey, M., G. Rauscher & B. Kissin. 1971. Modifications in the response to alcohol following the establishment of physical dependence. Psychopharmacologia **22:** 314–322.
12. John, G. R., J. M. Littleton & P. T. Nhamburo. 1985. Increased activity of Ca^{++}-

dependent enzymes of membrane lipid metabolism in synaptosomal preparations from ethanol-dependent rats. J. Neurochem. **44:** 1235–1241.
13. MEYER, K. H. 1937. Contributions to the theory of narcosis. Trans. Faraday Soc. **33:** 1062–1068.
14. CHIN, J. H. & D. B. GOLDSTEIN. 1977. Effects of low concentrations of ethanol on the fluidity of spin-labeled erythrocyte and brain membranes. Mol. Pharmacol. **13:** 435–441.
15. CHIN, J. H. & D. B. GOLDSTEIN. 1981. Membrane-disordering action of ethanol. Variation with membrane cholesterol content and depth of the spin label probe. Mol. Pharmacol. **19:** 425–431.
16. PERLMAN, B. J. & D. B. GOLDSTEIN. 1984. Genetic influences on the central nervous system depressant and membrane-disordering actions of ethanol and sodium valproate. Mol. Pharmacol. **26:** 547–552.
17. LYON, R. C., J. A., MCCOMB, J. SCHREURS & D. B. GOLDSTEIN. 1981. A relationship between alcohol intoxication and the disordering of brain membranes by a series of short-chain alcohols. J. Pharmacol. Exp. Ther. **218:** 669–675.
18. GOLDSTEIN, D. B., J. H. CHIN & R. C. LYON. 1982. Ethanol disordering of spin-labeled mouse brain membranes: Correlation with genetically determined ethanol sensitivity of mice. Proc. Natl. Acad. Sci. USA **79:** 4231–4233.
19. CHIN, J. H. & D. B. GOLDSTEIN. 1977. Drug tolerance in biomembranes. A spin label study of the effects of ethanol. Science **196:** 684–685.
20. JOHNSTON, P. V. & B. I. ROOTS. 1964. Brain lipid fatty acids and temperature acclimation. Comp. Biochem. Physiol. **11:** 303–309.
21. LYON, R. C. & D. B. GOLDSTEIN. 1983. Changes in synaptic membrane order associated with chronic ethanol treatment in mice. Mol. Pharmacol. **23:** 86–91.
22. PERLMAN, B. J. & D. B. GOLDSTEIN. 1983. Ethanol and sodium valproate disordering of membranes from chronic ethanol treated mice. Fed. Proc. **42:** 2123.
23. HARRIS, R. A., D. M. BAXTER, M. A. MITCHELL & R. J. HITZEMANN. 1984. Physical properties and lipid composition of brain membranes from ethanol tolerant-dependent mice. Mol. Pharmacol. **25:** 401–409.
24. ROTTENBERG, H., A. WARING & E. RUBIN. 1981. Tolerance and cross-tolerance in chronic alcoholics: Reduced membrane binding of ethanol and other drugs. Science **213:** 583–585.
25. PONNAPPA, B. C., A. J. WARING, J. B. HOEK, H. ROTTENBERG & E. RUBIN. 1982. Chronic ethanol ingestion increases calcium uptake and resistance to molecular disordering by ethanol in liver microsomes. J. Biol. Chem. **257:** 10141–10146.
26. ALOIA, R. C., J. PAXTON, J. S. DAVIAU, O. VAN GELB, W. MLEKUSCH, W. TRUPPE, J. A. MEYER & F. S. BRAUER. 1985. Effect of chronic alcohol consumption on rat brain microsome lipid composition, membrane fluidity and Na^+-K^+-ATPase activity. Life Sci. **36:** 1003–1017.
27. TARASCHI, T. F., A. WU & E. RUBIN. 1985. Phospholipid spin probes measure the effects of ethanol on the molecular order of liver microsomes. Biochemistry **24:** 7096–7101.
28. WARING, A. J., H. ROTTENBERG, T. OHNISHI & E. RUBIN. 1981. Membranes and phospholipids of liver mitochondria from chronic alcoholic rats are resistant to membrane disordering by alcohol. Proc. Natl. Acad. Sci. USA **78:** 2582–2586.
29. GORDON, E. R., J. ROCHMAN, M. ARAI & C. S. LIEBER. 1982. Lack of correlation between hepatic mitochondrial membrane structure and functions in ethanol-fed rats. Science **216:** 1319–1321.
30. SCHULLER, A., J. MOSCAT, E. DIEZ, C. FERNANDEZ-CHECA, F. G. GAVILANES & A. M. MUNICIO. 1984. The fluidity of plasma membranes from ethanol-treated rat liver. Mol. Cell. Biochem. **64:** 89–95.
31. YAMADA, S. & C. S. LIEBER. 1984. Decrease in microviscosity and cholesterol content of rat liver plasma membranes after chronic ethanol feeding. J. Clin. Invest. **74:** 2285–2289.
32. POLOKOFF, M. A., T. J. SIMON, R. A HARRIS, F. R. SIMON & M. IWAHASHI. 1985. Chronic ethanol increases liver plasma membrane fluidity. Biochemistry **24:** 3114–3120.
33. DOMBEK, K. M. & L. O. INGRAM. 1984. Effects of ethanol on the *Escherichia coli* plasma membrane. J. Bacteriol. **157:** 233–239.

34. GOTO, M., Y. BANNO, S. UMEKI, Y. KAMEYAMA & Y. NOZAWA. 1983. Effects of chronic ethanol exposure on composition and metabolism of *Tetrahymena* membrane lipids. Biochim. Biophys. Acta **751:** 286–297.
35. BEAUGE, F., H. STIBLER & S. BORG. 1985. Abnormal fluidity and surface carbohydrate content of the erythrocyte membrane in alcoholic patients. Alcoholism (NY) **9:** 322–326.
36. LYON, R. C. & D. B. GOLDSTEIN. 1983. Restoration of synaptic membrane order following chronic ethanol treatment in mice. Biophys. J. **41:** 198a.
37. JOHNSON, D. A., N. M. LEE, R. COOKE & H. H. LOH. 1979. Ethanol-induced fluidization of brain lipid bilayers: Required presence of cholesterol in membranes for the expression of tolerance. Mol. Pharmacol. **15:** 739–746.
38. KELLY-MURPHY, S., A. J. WARING, H. ROTTENBERG & E. RUBIN. 1984. Effects of chronic ethanol consumption of the partition of lipophilic compounds into erythrocyte membranes. Lab. Invest. **50:** 174–183.
39. CHIN, J. H., L. M. PARSONS & D. B. GOLDSTEIN. 1978. Increased cholesterol content of erythrocyte and brain membranes in ethanol-tolerant mice. Biochim. Biophys. Acta **513:** 358–363.
40. SMITH, T. L. & M. J. GERHART. 1982. Alterations in brain lipid composition of mice made physically dependent to ethanol. Life Sci. **31:** 1419–1425.
41. ROVINSKI, B. & E. A. HOSEIN. 1983. Adaptive changes in lipid composition of rat liver plasma membrane during postnatal development following maternal ethanol ingestion. Biochim. Biophys. Acta **735:** 407–417.
42. CUNNINGHAM, C. C., R. E. BOTTENUS, P. I. SPACH & L. L. RUDEL. 1983. Ethanol-related changes in liver microsomes and mitochondria from the monkey, *Macaca fascicularis*. Alcoholism (NY) **7:** 424–430.
43. LA DROITTE, P., Y. LAMBOEUF & G. DE SAINT-BLANQUAT. 1984. Lipid composition of the synaptosome and erythrocyte membranes during chronic ethanol-treatment and withdrawal in the rat. Biochem. Pharmacol. **33:** 615–624.
44. LA DROITTE, P., Y. LAMBOEUF, G. DE SAINT-BLANQUAT & J.-P. BEZAURY. 1985. Sensitivity of individual erythrocyte membrane phospholipids to changes in fatty acid composition in chronic alcoholic patients. Alcoholism (NY) **9:** 135–137.
45. KEEGAN, C., P. A. WILCE, E. RUCZKAL-PIETRZAK & B. C. SHANLEY. 1983. Effect of ethanol on cholesterol and phospholipid composition of HeLa cells. Biochem. Biophys. Res. Commun. **114:** 985–990.
46. SINENSKY, M. 1974. Homeoviscous adaptation—a homeostatic process that regulates the viscosity of membrane lipids in *Escherichia coli*. Proc. Natl. Acad. Sci. USA **71:** 522–525.
47. LITTLETON, J. M. & G. JOHN. 1977. Synaptosomal membrane lipids of mice during continuous exposure to ethanol. J. Pharm. Pharmacol. **29:** 579–580.
48. SCHILLING, R. J. & R. C. REITZ. 1980. A mechanism for ethanol-induced damage to liver mitochondrial structure and function. Biochim. Biophys. Acta **603:** 266–277.
49. SUN, G. Y. & A. Y. SUN. 1979. Effect of chronic ethanol administration on phospholipid acyl groups of synaptic plasma membrane fraction isolated from guinea pig brain. Res. Commun. Chem. Path. Pharmacol. **24:** 405–408.
50. MORRISSON, M., P. A. WILCE & B. C. SHANLEY. 1984. Influence of ethanol on fatty acid composition of phospholipids in cultured neurons. Biochem. Biophys. Res. Commun. **122:** 516–521.
51. INGRAM, L. O., K. D. LEY & E. M. HOFFMANN. 1978. Drug-induced changes in lipid composition of *E. coli* and of mammalian cells in culture: Ethanol, pentobarbital, and chlorpromazine. Life Sci. **22:** 489–494.
52. SMITH, T. L., A. E. VICKERS, K. BRENDEL & M. J. GERHART. 1982. Effects of ethanol diets on cholesterol content and phospholipid acyl composition of rat hepatocytes. Lipids **17:** 124–128.
53. UMEKI, S., H. SHIOJIRI & Y. NOZAWA. 1984. Chronic ethanol administration decreases fatty acyl-CoA desaturase activities in rat liver microsomes. FEBS Lett. **169:** 274–278.
54. WANG, D. L. & R. C. REITZ. 1983. Ethanol ingestion and polyunsaturated fatty acids: Effects on the acyl-CoA desaturases. Alcoholism (NY) **7:** 220–226.

55. BUTTKE, T. M. & L. O. INGRAM. 1980. Ethanol-induced changes in lipid composition of *Escherichia coli:* Inhibition of saturated fatty acid synthesis *in vitro.* Arch. Biochem. Biophys. **203:** 565–571.
56. DANIELS, C. K. & D. B. GOLDSTEIN. 1982. Movement of free cholesterol from lipoproteins or lipid vesicles into erythrocytes. Acceleration by ethanol *in vitro.* Mol. Pharmacol. **21:** 694–700.
56a. CHIN, J. H. & D. B. GOLDSTEIN. 1985. Differential effects of ethanol on uptake of saturated and unsaturated fatty acids into red cell membranes *in vitro.* Fed. Proc. **44:** 1239.
57. ALLING, C., S. LILJEQUIST & J. ENGEL. 1982. The effect of chronic ethanol administration on lipids and fatty acids in subcellular fractions of rat brain. Med. Biol. **60:** 149–154.

DISCUSSION OF THE PAPER

B. CHANCE (*University City Science Center, Philadelphia, PA*): Is fluorescence anisotropy a more sensitive measure of ethanol perturbation of membrane structure than electron spin resonance?

GOLDSTEIN: I don't know how to equate EPR order parameters to fluorescence anisotropy. The total range that they can undergo is different, and the interpretation is different. I work pretty much empirically, and over the range of ethanol concentrations that we use it's easier to find a certain percent change in fluorescence anisotropy than in order parameters. This depends on the spin label, because the deeper you go into the membrane, the more sensitive the EPR measures get.

J. M. LITTLETON (*King's College, London, England*): Dr. Goldstein, you mentioned that physical dependence could occur within days but you didn't mention that tolerance could occur within hours.

GOLDSTEIN: You are right about that. Physical dependence can also occur within hours.

LITTLETON: I wonder whether your changes in the physical parameters of lipids parallel the rate of development of tolerance or physical dependence?

GOLDSTEIN: Physical dependence, like tolerance, as both of us have shown, can develop in hours. Mice given a single injection of ethanol show convulsions when picked up by the tail just as they show tolerance in your experiments and mine. But I haven't measured the rate of onset of tolerance and physical dependence in the membranes. I have measured the rate of its disappearance. It goes down within a day just as both tolerance and physical dependence do after a three-day exposure to ethanol.

Partition of Ethanol and Other Amphiphilic Compounds Modulated by Chronic Alcoholism[a]

HAGAI ROTTENBERG

Pathology Department
Hahnemann University School of Medicine
Philadelphia, Pennsylvania 19102

INTRODUCTION

Chronic alcoholism is associated with the development of tolerance to and dependence on ethanol, which in turn leads to a severe withdrawal syndrome when ethanol is withdrawn abruptly.[1,2] Although the mechanism of the acute effects of ethanol on behavior is not fully understood, there is little doubt that it is closely related to the pharmacological effects elicited by general anesthetics.[3] It is clear that the site of the acute behavioral effects of ethanol is located in the plasma membrane of the central nervous system. However, alcohol distributes in all organs of the body and partitions into all cell membranes, as well as the membranes of cell organelles. While the presence of alcohol in these sites may not be responsible for the acute behavioral effects, it may, nevertheless, contribute to the long-term effects observed in chronic alcoholism. Indeed, pathological manifestations of chronic alcoholism are observed in various other organs, such as the liver, heart, and skeletal muscle. These may be related to the effect of alcohol on the function of the membranes of these organs.[4] In addition, and unlike other drugs, alcohol is a caloric-rich metabolite of intermediate metabolism; its consumption in quantities that lead to chronic alcoholism has profound effects on intermediate metabolism in the liver.[5]

Recent studies with animal models have indicated that in parallel with the development of tolerance and dependence there are changes in the structure and function of various membranes.[6-12] One of the most compelling observations is that membranes isolated from animals that were fed or that inhaled ethanol over an extended period of time become resistant to the *in-vitro* effects of ethanol on membrane function and its physical properties.[8-12]

These observations are compatible with the suggested existence of a biological adaptation process in which the organism modulates its membrane composition to compensate for the effects of the drugs.[13] It has been reported that the resistance to the fluidizing effect of ethanol in plasma membranes of chronic alcoholic mice or rats is associated with increased cholesterol content.[12,14,15] Other changes in membrane composition were reported, but no clear pattern was observed.[7,11,16-18] We have previously shown that liver mitochondria from ethanol-fed rats are resistant to the uncoupling effect of ethanol on membrane enzymes[10] and also to the fluidizing effect of ethanol[11] This resistance is associated with a major change in the fatty acid composition of the phospholipid cardiolipin.[11]

[a]Supported by United States Public Health Service Grants AA03842 and AA05662.

CROSS-TOLERANCE TO MEMBRANE DISORDERING IN CHRONIC ALCOHOLISM

The resistance to membrane disordering in mitochondrial membrane from ethanol-fed rats is not limited to the effect of ethanol.[9] Other membrane perturbing reagents, such as general anesthetics, are also considerably less effective in membranes from ethanol-fed rats. FIGURE 1 shows the effect of ethanol and halothane on mitochondrial membrane structure in ethanol-fed rats and their control. The lower panel shows the effect on the order parameter, S, which is calculated from the electron paramagnetic resonance (EPR) signal of membrane-embedded 5-doxylstearic acid. Both alcohol and

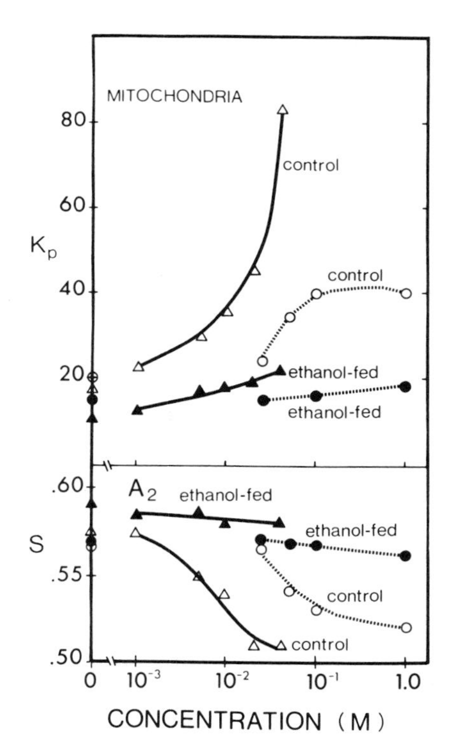

FIGURE 1. Effect of ethanol (●, O) and halothane (▲, △) on the 5N10 partition coefficient K_p (*upper panel*) and on the 5-doxylstearate order parameter S (*lower panel*) in liver mitochondrial membranes from ethanol-fed rats and their pair-fed controls. EPR spectra were obtained at 35° ± 0.2° C with a Varian E-109 spectrophotometer. The nitroxide-decane derivative 5N10 was added to membrane suspensions (10 mg of protein per milliliter) to give an apparent concentration of 0.1 mM. Partition coefficients were calculated by estimating the amount of bound and free probes from the EPR spectra. Sufficient 5-doxylstearic acid was added to the membrane suspension to obtain a probe-to-lipid ratio of 1:200. Rats were given ethanol (14g per kilogram of body weight per day) for 35 days; pair-fed controls received isocaloric carbohydrate instead of ethanol. For EPR measurements the membranes were suspended in 0.25 M sucrose, 20 mM tris-Cl (pH 7.4), 2 μM rotenone, and 10 μM ferrocyanide. At least three pairs were tested for each of the effects shown. The results are typical examples of these experiments.[19]

halothane are much less effective in lowering the order parameter in membranes from ethanol-fed rats. Notice that the concentration of ethanol required to produce disordering is much higher than that of halothane, because the membrane water partition coefficient of halothane is considerably higher. The upper panel shows the effect of ethanol and halothane on the membrane partition of the spin-labeled decane, 5N10. Increasing membrane disorder increases the partition of the probe into the membrane. With this probe also, it is observed that membrane from ethanol-fed rats is resistant to the disordering effects of both ethanol and halothane. Notice also that the partition of the decane spin probe in membranes from ethanol-fed rats is lower even in the absence of added ethanol.

The cross-tolerance to the disordering effect of halothane in membranes from ethanol-fed rats and the difference in the partition of the decane spin probes is not specific to mitochrondrial membranes. It was observed in all membrane preparations we tested. FIGURE 2 shows experiments similar to those of FIGURE 1 but with synaptosomes instead of liver mitochondria.[19] The results are nearly identical. FIGURES 3 and 4 show similar experiments with plasma membranes prepared from red blood cells (RBCs).[20] Again, tolerance to ethanol, cross-tolerance to halothane, and lower partition coefficient of the decane spin probes was observed in RBC membranes prepared from ethanol-fed rats. We believe that the generality of these observations and their strong correlation provide a clue to the molecular mechanism of alcohol

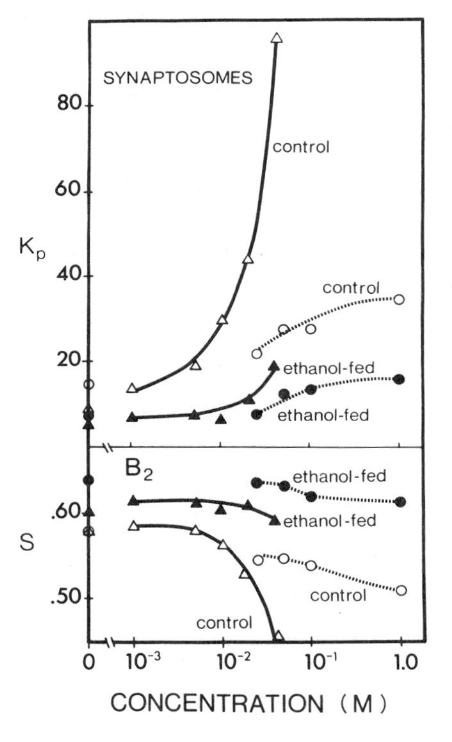

FIGURE 2. Effect of ethanol and halothane on 5N10 partition coefficient K_p (*upper panel*) and on the 5-doxylstearate order parameter S (*lower panel*) in brain synaptosomes from ethanol-fed rats and their pair-fed controls. Conditions and symbols as in FIGURE 1.[19]

tolerance in these membranes. We hypothesized that during adaptation to ethanol feeding, changes in membrane composition lead to lowering of the membrane water partition coefficients of amphiphilic molecules. This would then lead to tolerance to the effects of alcohol on both membrane structure and function, since these effects depend on alcohol concentration in the membrane and not on the medium. Similarly, tolerance to the effect of halothane or any anesthetic is also expected for the same reason. The first indication for the plausibility of this hypothesis was the fact that there was a significant lowering of the partition of the decane spin probe in all membrane preparations that exhibit tolerance to ethanol and halothane. It was, however,

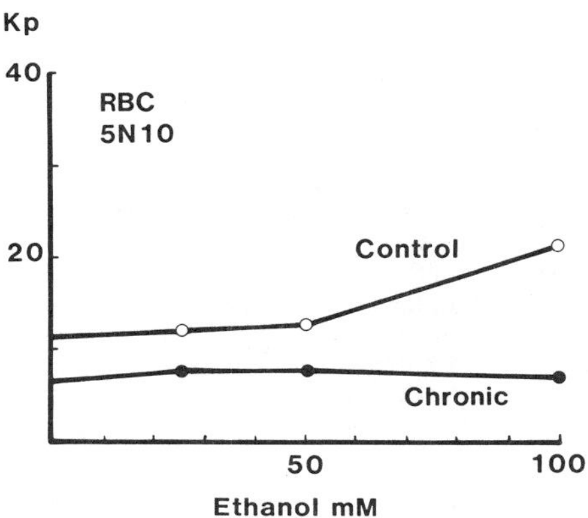

FIGURE 3. Effect of ethanol on the partition coefficient of 5N10 in red blood cell membranes (Waring, Rottenberg, and Rubin, unpublished observation).

necessary to test this hypothesis by a direct determination of the partition coefficients of alcohol and anesthetics.

PARTITION COEFFICIENTS OF ETHANOL AND ANESTHETICS IN MEMBRANES FROM ETHANOL-FED RATS

Partition coefficients can be measured directly from the distribution of radiolabeled compounds between the membrane and the suspending medium. This is a relatively simple procedure for compounds of relatively high partition coefficients, such as most common anesthetics, but it becomes technically difficult in the case of ethanol because of its relatively low partition coefficient. However, in mitochondria, which are more fluid and have a higher dialectric constant than most biological

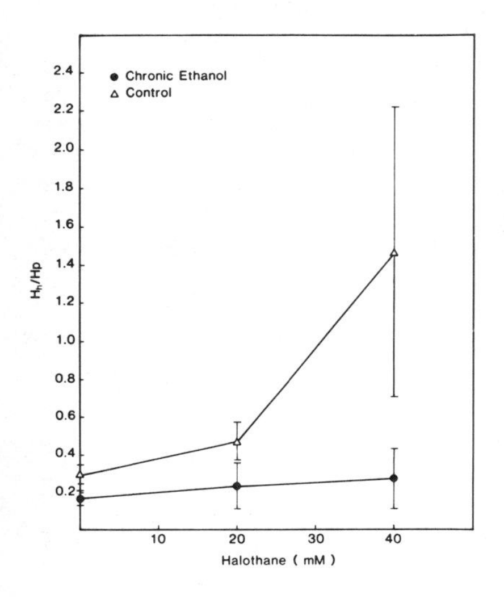

FIGURE 4. Effect of halothane on the partition parameter of 5N10 in erythrocyte membranes from control and ethanol-fed animals. H_h corresponds to the probe in the membrane phase and H_p to that in the aqueous solution. Points represent the means ± standard deviation of four pairs of rats.[20]

TABLE 1. Partition Coefficients of Ethanol and Anesthetics in Liver Mitochondria[a]

	Ethanol	Halothane	Phenobarbital
Control	3.60	28.6	33.0
Chronic ethanol	1.17	21.4	22.1
Control/ethanol[b]	4.24	1.35	1.54

[a]Liver mitochondrial membranes from ethanol-fed and control rats. The coefficients were determined by incubating the membranes with a ^{14}C-labeled compound and ^3H_2O. The $^{14}C/^3H$ ratio in the supernatant and the pellet was determined after sedimentation of the membranes. Partition coefficients were calculated on the assumption that the amount of ^{14}C-labeled compound in the pellet in excess of the amount dissolved in the pellet water is dissolved in the membrane lipids. Values are means ± standard deviations. All the experimental values are significantly different from the corresponding control values at $P < 0.01$ (paired t-test).[19]
[b]Average pair ratio.

membranes, these measurements are possible. TABLE 1 shows the partition coefficients of alcohol, halothane, and phenobarbital in liver mitochondria from ethanol-fed rats and their controls.[19] In all cases, partition coefficients are lower in membrane from ethanol-fed, confirming our prediction. TABLE 2 shows similar experiments with synaptosomes.[19] In these membranes, the partition coefficient of ethanol is lower and the method is at its limit of sensitivity. Nevertheless, a clear lowering of the partition was observed for both ethanol and halothane. In plasma membrane from RBC, ethanol partition is even lower and could not be reliably estimated by this method. However, halothane partition could be measured and these measurements (FIG. 5) confirmed the lowering of the partition coefficient of halothane in this system as well. The lower partition coefficient of halothane was also demonstrated by another method. A membrane-embedded dye, diphenylhexatriene (DPH), is quenched by halothane. The extent of quenching depends on the membrane concentration of halothane and, hence, on its partition coefficient. FIGURE 6 shows that halothane is considerably less effective as a quencher of membrane-embedded DPH in RBC membranes from ethanol-fed rats. This difference is apparently entirely due to the difference in partition coefficient. FIGURE 7 shows that when the quenching is related to *membrane* concentration (which is calculated on the basis of the results of FIG. 5), there is no significant difference in the extent of quenching between the two preparations. To verify that ethanol partition is also reduced in these membranes, we developed yet another method to investigate the partition coefficient. It is well known[21] that anesthetics and other compounds protect red blood cells from hypotonic hemolysis. The protection depends on the partition coefficient of the agent, since it arises from the increased surface area of the cell by the incorporated drug. FIGURE 8 shows the protection by ethanol and isobutanol of red blood cells from ethanol-fed rats and controls.[22] Both ethanol and isobutanol are much

TABLE 2. Partition Coefficients of Ethanol and Anesthetics in Brain Synaptosomes[a]

	Ethanol	Halothane
Control	1.00	27.5
Chronic ethanol	0.33	21.7
Control/ethanol[b]	3.07	1.38

[a]Brain synaptosomal membranes from ethanol-fed and control rats. Conditions and procedures as in TABLE 1.[19]
[b]Average pair ratio.

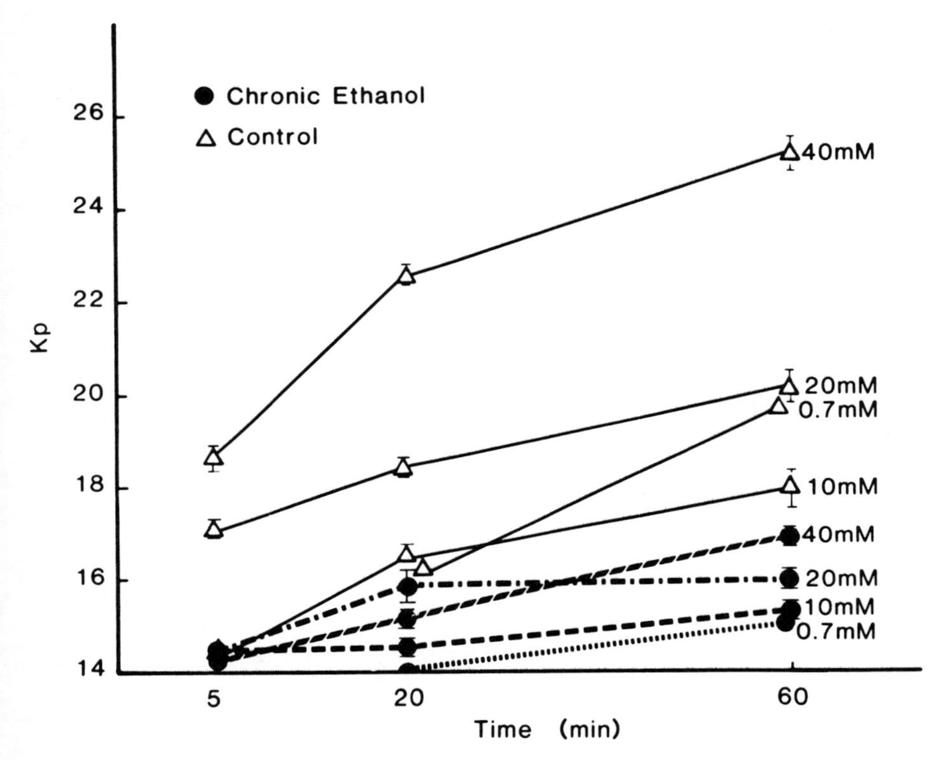

FIGURE 5. Concentration of radioactive halothane in the erythrocyte membrane as a function of time. Ratio of the membrane concentration to external concentration, K_p, at each time interval, was determined by incubating the membranes with ^{14}C-labeled halothane and ^3H_2O. $^{14}C/^3H$ ratios in the supernatant and the pellet were estimated after sedimentation of the membranes.[20]

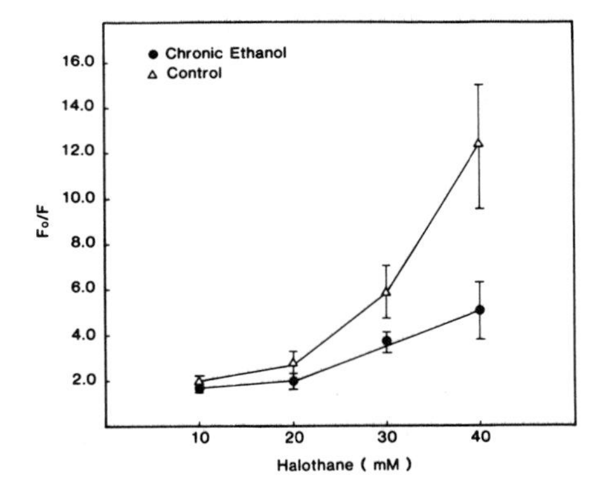

FIGURE 6. Stern-Volmer plots for the steady-state quenching of DPH fluorescence by halothane in erythrocyte membranes. Measurements were taken 30 minutes after addition of the quenching agent to the dye-membrane samples. F_o represents initial fluorescence and F the value after 30 minutes. Each data point represents the mean ± standard deviation from five pairs of rats.[20]

less effective in protecting blood cells from ethanol-fed animals, indicating lower partition coefficients. High concentrations of ethanol in membrane from ethanol-fed rats enhance the hemolysis, since the dilution of the salt solution by the ethanol is more effective in increasing hemolysis than the membrane protection afforded by the incorporated ethanol. In subsequent studies, the method was slightly modified to account for the dilution effect. The decreased partition coefficient of several alcohols and halothane was clearly demonstrated by this method (FIG. 9).

CHOLESTEROL AND THE DEVELOPMENT AND DISAPPEARANCE OF MEMBRANE TOLERANCE TO ETHANOL

It was reported that resistance to the disordering effect of ethanol in membrane from alcohol-tolerant rats and mice is associated with increased membrane cholesterol content.[12,14,15] Since it has been shown that cholesterol reduces the partition of various amphiphilics into phospholipid vesicles,[23] this may explain the observation that the partition coefficients are lowered in ethanol-fed rats. We, therefore, embarked on a study of the relationship between the development (and disappearance) of membrane

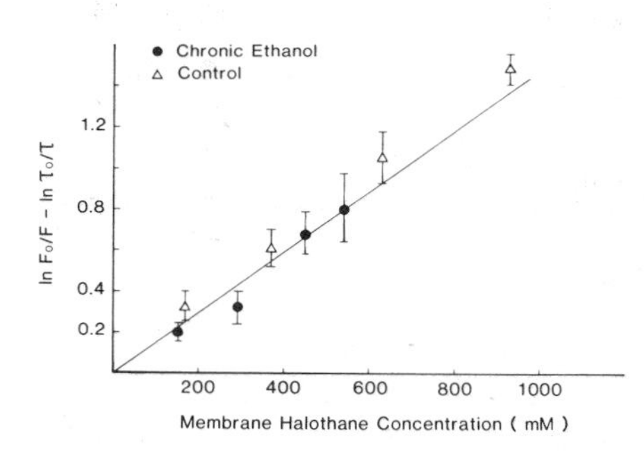

FIGURE 7. Fluorescence steady-state measurements plotted according to a modified Stern-Volmer equation as a function of membrane halothane concentration.[20]

alcohol tolerance and membrane and serum cholesterol. FIGURE 10 (lower panel) shows the time course of the reduction in alcohol partition coefficient, as indicated by the alcohol protection of RBC from hypotonic hemolysis. As ethanol feeding progresses, the protection is reduced, reaching a stable lower value after approximately 20 days. In the initial 4–5 days of feeding the rats do not consume large amounts of ethanol, so that the period on full dose (14 g/kg/day) is only about 2 weeks prior to the establishment of maximal effect. When the alcohol was withdrawn from the diet, the protection from hemolysis returned to a normal value within 24–48 hours. After 4 days without ethanol feeding, the rats were fed ethanol again, and the protection was gradually lowered to its previous low level after 14 days. This experiment indicates that while the development of the membrane changes is slow (14 days), the disappearance after alcohol withdrawal is very fast (24–48 h). It is interesting to note that the time course of the disappearance of the membrane perturbation parallels the time course of the withdrawal syndrome. The level of plasma cholesterol was determined in parallel (FIG. 10, upper panel). It was observed that in ethanol-fed rats there was a significant increase in serum cholesterol. The cholesterol level increases in parallel with the loss of protection from hypotonic hemolysis during ethanol feeding and decreased abruptly upon withdrawal. There was, however, no change in the ratio of free to esterified

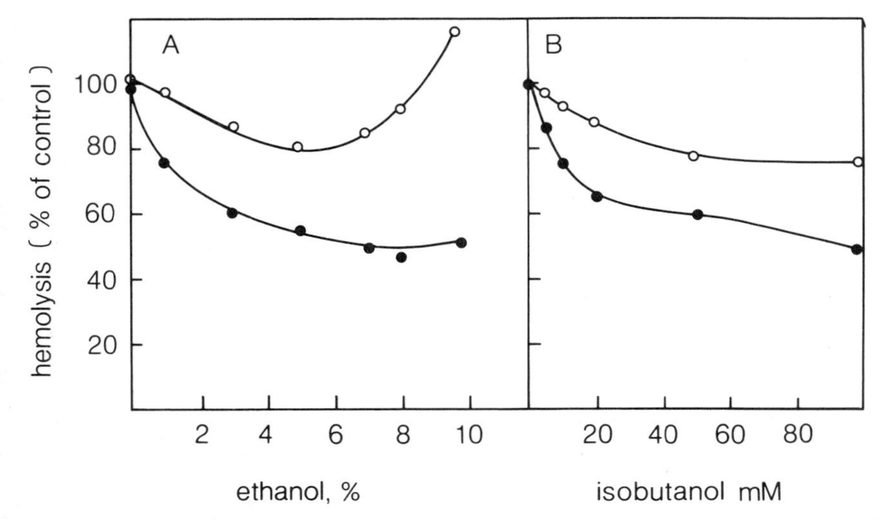

FIGURE 8. Protection by alcohols of red blood cells from hypotonic hemolysis in an ethanol-fed rat and its pair-fed control. (**A**) The effect of ethanol (1–10%) on the relative extent of hemolysis in ethanol-fed (O) and control (●). Procedure as described in Materials and Methods in REFERENCE 22. Ethanol feeding: 35 days. Hemolysis without drugs (taken as 100%) was 32% of total number of cells. (**B**) The effect of isobutanol (5–100 mM). Other conditions and symbols as in (**A**).[22]

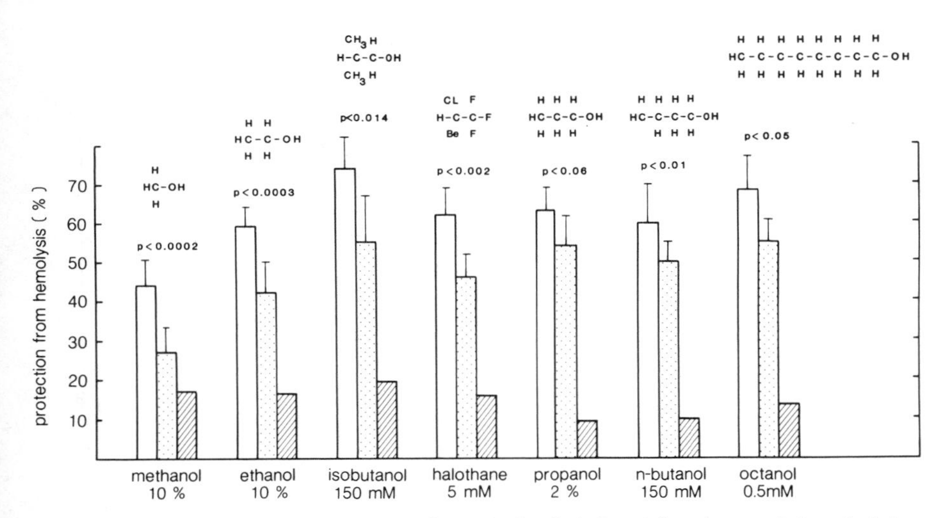

FIGURE 9. The protection by alcohols and anesthetics (halothane) from hypotonic hemolysis in ethanol-fed rats and their controls. The figure shows the effect of methanol (10%), ethanol (10%), isobutanol (150 mM), propanol (2%), n-butanol (150 mM), and octanol (0.5 mM). The concentrations of the various compounds were selected to give approximately the same protection as ethanol in controls (60%). The *open bars* show the percentage protection in controls, the *stippled bars* the percentage protection in ethanol-fed, and the *hatched bars* the averaged pair difference. There were eight pairs in the ethanol and methanol experiments and five pairs in the others. The standard deviation is shown above each average and the probability (paired t-test) is indicated as well. Assay as described in Materials and Methods in REFERENCE 22. Ethanol feeding: 35 days.[22]

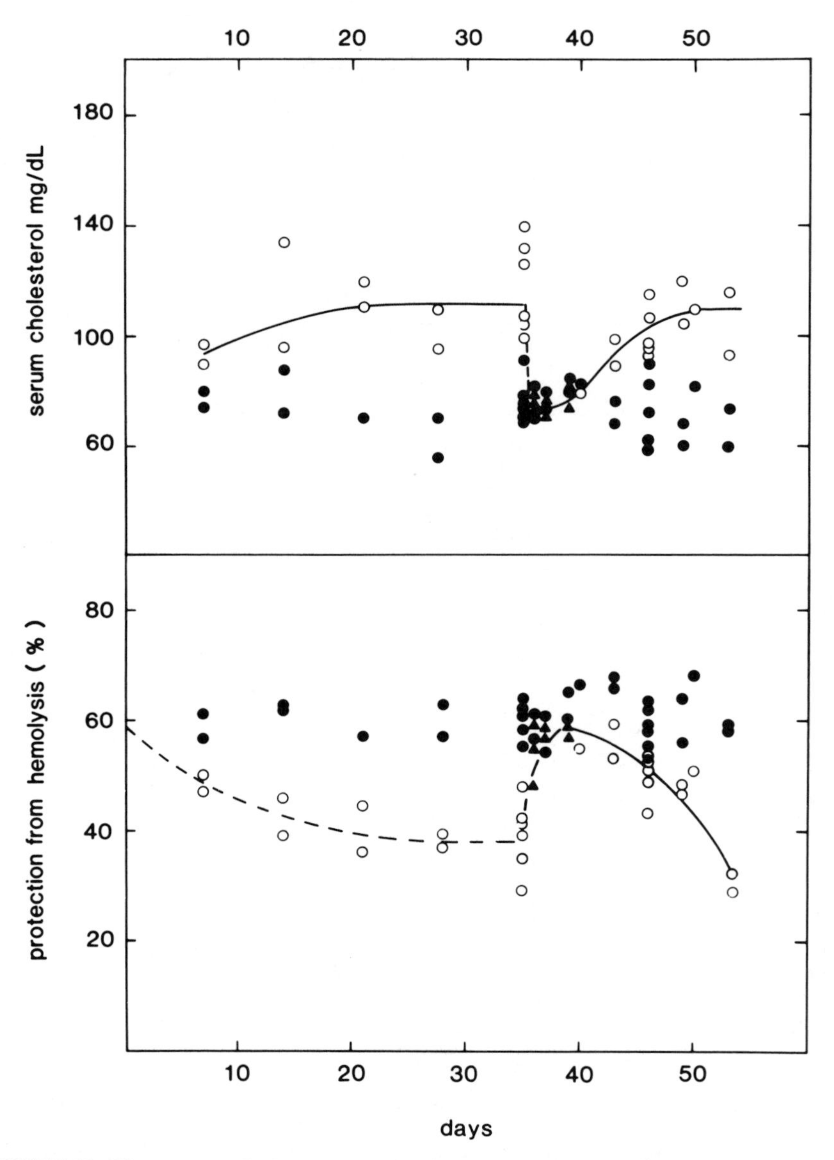

FIGURE 10. Time course of ethanol protection from hypotonic hemolysis (*bottom*) and total plasma cholesterol (*top*) in ethanol feeding, its withdrawal, and re-feeding. Ethanol concentration was 10%. ●, control; O, ethanol-fed during ethanol feeding; ▲, ethanol-fed after withdrawal (ethanol was fed for up to 35 days, then withdrawn for up to 4 days and re-fed for up to 14 days).[22]

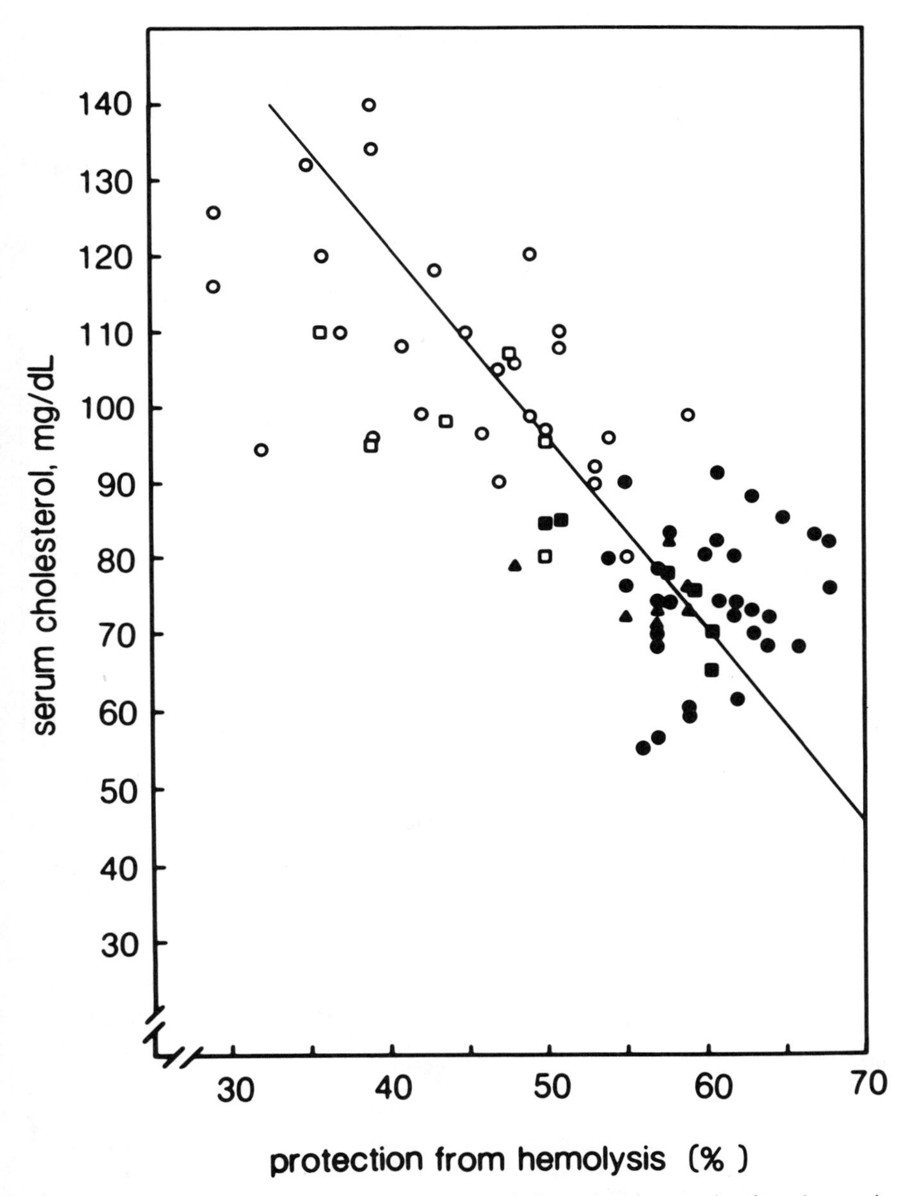

FIGURE 11. The correlation between total serum cholesterol and protection from hypotonic hemolysis by 10% alcohol. Data are from FIGURE 10 and also include the six pairs of fat-rich diet. ●, controls; O, ethanol-fed; ▲, ethanol withdrawn; ■, controls of fat-rich diet; and □, fat-rich diet.[22]

cholesterol in the plasma. The strong correlation between alcohol protection and plasma cholesterol levels raises the question whether the increased cholesterol level is not a necessary and sufficient cause for the reduction in alcohol protection from hemolysis. To test this hypothesis we fed rats a diet rich in saturated fat (coconut oil) and cholesterol in order to raise their blood cholesterol. We obtained a modest, but significant increase in plasma cholesterol (89 ± 11 versus 76 ± 8) in parallel with a modest, but significant decrease in alcohol protection from hemolysis, suggesting that the increased level of blood cholesterol is sufficient to induce the membrane changes associated with alcohol tolerance. FIGURE 11 shows the correlation between plasma cholesterol and protection by ethanol from hypotonic hemolysis. Clearly, both in ethanol feeding and cholesterol feeding the increased cholesterol is associated with a decreased protection. It was initially suspected that high blood cholesterol leads to high

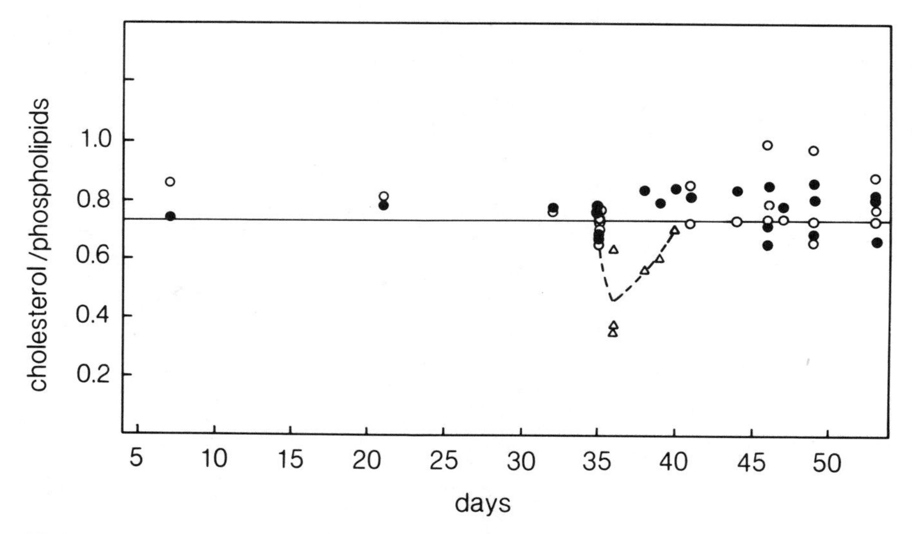

FIGURE 12. Membrane cholesterol/phospholipid ratio as a function of ethanol feeding and its withdrawal. Membranes were prepared from pairs shown in FIGURE 10. ●, controls; O, ethanol-fed; Δ, withdrawn from ethanol.[22]

membrane cholesterol, which in turn leads to lower partition coefficient. This explanation, however, was found to be wrong. FIGURE 12 shows the cholesterol/phospholipid ratio in membrane from control and ethanol-fed taken from the experiment of FIGURE 10. There is a slight, insignificant, increase in this ratio, which cannot fully explain the large reduction in the partition coefficients. One might hypothesize that during alcohol feeding and its associated elevated serum cholesterol there is a change in membrane lipid composition that is intended to *lower* the partition of *cholesterol* into the membrane as a protection from elevated cholesterol levels. Accordingly, the resistance to ethanol and other membrane perturbing agents is a side effect of the resistance of these membranes to the incorporation of cholesterol. If this hypothesis can be verified, we will have to radically modify our view of the meaning and implications of the widely observed tolerance to ethanol in membranes from ethanol-fed animals.

REFERENCES

1. MENDELSON, J. H. 1971. *In* The Biology of Alcoholism. B. Kissin & H. Begleiter, Eds. Plenum Press. New York, NY.
2. GOLDSTEIN, D. B. & N. POL. 1971. Science **172:** 288.
3. HERBERT, B. 1979. *In* Biochemistry and Pharmacology of Ethanol. E. Majchrowicz & E. P. Noble, Eds. Vol 2: 511. Plenum Press. New York, NY.
4. RUBIN, E. & H. ROTTENBERG. 1982. Fed. Proc. **41:** 2465.
5. REITZ, R. C. 1979. *In* Biochemistry and Pharmacology of Ethanol. E. Majchrowicz & E. P. Noble, Eds. Vol. 1: 353. Plenum Press. New York, NY.
6. LITTLETON, J. M. & G. JOHN. 1977. J. Pharm. Pharmacol. **29:** 579.
7. FRENCH, S. W., T. J. IHRIG & R. S. MORIN. 1970. Q. J. Stud. Alcohol **31:** 801.
8. CHIN, J. H. & D. B. GOLDSTEIN. 1977. Science **196:** 684.
9. CURRAN, M. & P. SEEMAN. 1977. Science **197:** 910.
10. ROTTENBERG, H., D. E. ROBERTSON & R. RUBIN. 1980. Lab. Invest. **42:** 318.
11. WARING, A., H. ROTTENBERG, T. OHNISHI & E. RUBIN. 1981. Proc. Natl. Acad. Sci. USA **78:** 2582.
12. GOLDSTEIN, D. B. 1984. Annu. Rev. Pharmacol. Toxicol. **24:** 43.
13. HILL, M. W. & A. D. BANGHAM. 1975. Adv. Exp. Med. Biol. **59:** 1.
14. CHIN, J. H., L. M. PARSON & D. B. GOLDSTEIN. 1978. Biochim. Biophys. Acta **513:** 358.
15. JOHNSON, D. A., N. M. LEE, R. COOKE & H. H. LOH. 1979. Mol. Pharmacol. **15:** 739.
16. SUN, G. Y. & A. Y. SUN. 1958. Alcoholism **9:** 164.
17. SMITH, T. L., A. E. VICKERS, K. BRENDEL & M. J. GERHARD. 1982. Lipids **17:** 124.
18. WING, D. R., D. J. HARVEY, J. HUGHES, P. G. DUNBARD, K. A. MCPHERSON & W. D. M. PATON. 1982. Biochem. Pharmacol. **31:** 3431.
19. ROTTENBERG, H., A. WARING & E. RUBIN. 1981. Science **213:** 583.
20. KELLY-MURPHY, S., A. WARING, H. ROTTENBERG & E. RUBIN. 1984. Lab. Invest. **50:** 178.
21. SEEMAN, P. 1972. Pharmacol. Rev. **28:** 583.
22. ROTTENBERG, H. 1986. Biochim. Biophys. Acta **855:** 211.
23. KORTEN, K., T. J. SOMMER & K. W. MILLER. 1980. Biochim. Biophys. Acta **599:** 271.

DISCUSSION OF THE PAPER

L. L. M. VAN DEENEN (*State University of Utrecht, Utrecht, The Netherlands*): I find your results very interesting. I am not surprised to see that your increase in cholesterol level of the serum lipoproteins was not reflected in the erythrocyte membrane. Only under extreme conditions (as Dr. Goldstein showed) where you have this enormous increase of cholesterol, is this reflected. With respect to these experiments about ordering and disordering in the membrane being related to cholesterol content, I was wondering whether it is feasible to modify erythrocyte cholesterol content after you have caused adaptation to the ethanol? You can remove in a subtle manner some of the cholesterol by incubating them with liposomes, which have a lower cholesterol content, and then determine if there is any effect on ordering of the membrane by doing this.

ROTTENBERG: Yes, one nice thing about this idea is that it's relatively easy to check as opposed to some other maybe more interesting ideas. Right now we are doing experiments similar to the one that you just described. Unfortunately, I don't yet have the answer. We separate the red blood cells from the serum of untreated animals and incubate the serum with red cells from the ethanol-fed animals and vice versa. If our idea is right, we expect to see changes in the cholesterol/phospholipid distribution that would support this view.

B. CHANCE (*University City Science Center, Philadelphia, PA*): Dr. Rottenberg, you didn't close the circle where you showed the striking effect of chronic alcohol treatment on the respiration rate, yet you ended up saying the membranes weren't changed. Do you want to close that loop?

ROTTENBERG: Yes. You are referring to the mitochondria experiment. There are other changes in the mitochondria such as the content of some of the oxidative enzymes and probably ATPase. What I was saying is that as far as the sensitivity of the enzyme to ethanol goes, it appears that it's solely a result of reduced partition coefficient of ethanol into this membrane. The sensitivity of the enzyme to ethanol is probably not changed, rather there's less ethanol in these membranes.

D. MCCARTHY (*University of New Mexico School of Medicine, Albuquerque, NM*): Can you tell us anything about the changes in the aqueous phase in the membranes during these adaptations? Your hypothesis is that ethanol is being added to the membrane, but if one looks at Dr. Cullis' model and other models one might wonder if in fact ethanol does not progressively remove water from some of the aqueous phase in the membrane and quite markedly change the hydrogen bonding medium for phospholipid groups.

ROTTENBERG: I really don't have any direct evidence for or against this idea, which has been extensively discussed in relation to the action of anesthetics in general. I want to emphasize that as far as the measurement of the partition coefficient is concerned, changes in the water content of the membrane will not have any effect on these numbers.

D. B. GOLDSTEIN (*Stanford University School of Medicine, Stanford, CA*): I would like to ask you about your experiments on the antihemolytic effect. What you've essentially shown is that tolerance develops to the antihemolytic effect. You have not directly shown a change in the partition coefficient, and you assume, I think, a little too easily, that it is a change in the partition coefficient that explains this. I agree with you that this may involve a change in the lipids, but it could be protein cytoskeleton, and quite a lot of other things could have changed.

ROTTENBERG: As I said, it's certainly not a direct measurement, and I won't try to convince you otherwise. However, I think that mine is the best interpretation so far, namely, that ethanol protects membranes, particularly red blood cell membranes, from hemolysis, because it incorporates into the membrane and expands the membrane surface. This interpretation may be only qualitatively correct, since I did not calculate the partition coefficient. I would say qualitatively there is a difference in a partition coefficient. So that's the extent that I'm willing to interpret my data as change of partition based on this experiment. If someone can dispute that the protection of red blood cell membranes from hemolysis is not related to partition into the lipid and expansion of the membrane, then of course I'll have to rethink my interpretation. But at this point I think it's a safe interpretation.

Effects of Ethanol on Membrane Order: Fluorescence Studies[a]

R. ADRON HARRIS, RITA BURNETT,
SUSAN MCQUILKIN, ANNE MCCLARD,
AND FRANCIS R. SIMON

Denver Veterans Administration Medical Center
Alcohol and Hepatobiliary Research Centers
and
Departments of Pharmacology and Medicine
University of Colorado School of Medicine
Denver, Colorado 80262

INTRODUCTION

The close correlation of membrane solubility with intoxicating potency for a variety of alcohols and other intoxicant-anesthetics indicates a membrane site of action for alcohol and related drugs (*e.g.*, REF. 1). One theory is that alcohol alters the functional properties of cell membranes because it alters their physical properties. A detailed discussion of this concept is beyond the scope of this paper but is the topic of several recent reviews.[2–5] This paper will discuss the use of fluorescent probes in the study of effects of ethanol on membrane physical properties.

Ethanol is often said to alter membrane "fluidity." Fluidity does not have a precise definition and is usually taken to mean the relative motional freedom of membrane constituents, particularly lipids. In a more rigorous sense, fluidity consists of rotational diffusion, which is related to microviscosity and hindered anisotropic rotations, which are related to lipid order.[4,6] One focus of this paper is how fluorescence measurements can be used to distinguish between effects of alcohol on these two properties—microviscosity and lipid order.

FLUORESCENCE TECHNIQUES

In the last 15 years fluorescence polarization has been widely used to analyze membrane structure; the significance of the approach was strengthened by the advent of time-resolved techniques.[6,7] The basic principle of measuring fluorescence polarization is that the sample is excited with vertically polarized light, and emission is measured through polarizers both parallel and perpendicular to the excitation polarizer.[6] For measurements of the rate and/or range of membrane motion, a polar fluorophore such as diphenylhexatriene (DPH) is partitioned into the lipid matrix of the membrane. DPH is a particularly ideal probe in that it has a high extinction coefficient but does not fluoresce in aqueous solutions.[8] For "steady state" measure-

[a]Supported by the Veterans Administration and by United States Public Health Service Grants AA06399 and AA03527 to the Alcohol Research Center, and AM15851 and AM34914 to the Hepatobiliary Research Center.

ments, emission is measured under continuous illumination; emission is measured parallel ($\|$) and perpendicular ($\perp$) to the polarized exciting light. Historically, the data have been more frequently reported as polarization (P) where

$$P = \frac{I_{\|} - I_{\perp}}{I_{\|} + I_{\perp}}$$

However, from a mathematical and theoretical viewpoint,[6] the preferable expression is in terms of anisotropy (r) where

$$r = \frac{I_{\|} - I_{\perp}}{I_{\|} + 2I_{\perp}}$$

Both P and r are related to the ratio $I_{\|}/I_{\perp}$. The limiting anisotropy (r_o) refers to the anisotropy in the absence of depolarizing processes such as rotational diffusion or energy transfer. This can be determined experimentally by immobilizing the probe in a rigid matrix, such as propylene glycol at $-70°$ C. In an isotropic solution, *i.e.,* an homogeneous oil, the range of r_o values must be $-0.2 \leq r_o \leq 0.4$ (P_o ranges from $-0.33 \leq P_o \leq 0.50$).

Until recently, fluorescence polarization (anistropy) data were interpreted in terms of microviscosity using the Perrin equation.[9]

$$(1/r - 1/3) = (1/r_o - 1/3)(1 + 3\,\tau/\bar{\rho})$$

where τ is fluorescence lifetime and $\bar{\rho}$ is the rotational relaxation time. Alternatively, the Perrin equation may be expressed as

$$r = r_o/(1 + \tau/T_c)$$

where T_c is rotational correlation time. Under some conditions, viscosity (η) can be determined from

$$\eta = T_c RT/V$$

where T is absolute temperature, V is the volume of the rotating unit, and R is the gas constant. When τ is constant and homogenous, η can be calculated directly from the steady state anisotropy. τ can be measured by pulse lifetime or phase-modulation measurements. The relative merits of both approaches have been discussed by Lackowicz.[6] In our experience,[10,11] drugs do not appreciably affect τ. Thus, one could calculate T_c or η directly from the steady state anisotropy as long as the fluoropore, *e.g.,* DPH had the same freedom (isotropy) of depolarizing rotations as in a homogeneous reference oil solution. A variety of data indicate this is *not* the case. For example, the article by van Blitterswijk *et al.*[12] illustrates the problem. DPH is incorporated into a membrane, and anisotropy is measured as a function of time after a brief pulse of polarized light. Initially, the emitting dipoles of DPH are parallel to the direction of polarization and r approaches r_o. With time, DPH rotates and the anisotropy decreases. In an isotropic oil, the anisotropy falls to zero. However, in real membranes, r reaches a limiting value r_∞ because the motion of the probe is hindered. The steady state anisotropy r_s is then composed of two components: r_k, the fast decaying component related to rotational diffusion, and r_∞, the slowly decaying component related to membrane order (the S parameter).

$$r_s = r_\infty + r_k$$

The ratio r_∞/r_0 has been shown on theoretical grounds[13-15] to be related to the square of the order parameter(s). Thus,

$$r/r_o = S^2, 0 \le S \le 1$$

S is the same orientational order parameter measured in nuclear magnetic resonance (NMR) and electron spin resonance (ESR). The mathematical similarities between S_{DPH} and S_{NMR} have been elegantly described by Lipari and Szabo.[15] The question of whether or not it will be possible to develop similar relationships for other fluorescence probes remains to be answered. To some degree this issue is dealt with in a recent article by Hare.[16]

DPH remains the most widely used probe partly because its behavior in the membrane is understood better than other probes and partly because it distributes throughout the hydrophobic core of lipid bilayers. However, probes with other properties and localization are also useful. Trimethylammonium-DPH (TMA-DPH) has spectral properties identical to DPH but is anchored at the surface of the membrane because of the positively charged group.[17] An anthroyloxy group can be attached to any carbon of a fatty acid,[18,19] resulting in fluorescent "depth" probes, analogous to the widely used electron paramagnetic resonance (EPR) probes.[20] Pyrene derivatives are useful because their fluorescence emission spectra are dependent on the mobility of the probes in the membrane and are thus an accurate measure of microviscosity (not order).[21] In this paper, we discuss the use of these techniques in the study of membrane actions of ethanol.

ACUTE EXPOSURE TO ETHANOL *IN VITRO*

Studies with DPH indicate that exposure to a sufficiently large concentration of ethanol will decrease the steady state polarization of the probe in any cell membrane or lipid bilayer, indicating a membrane fluidizing action of ethanol. However, different membranes vary in their sensitivity to ethanol. Brain synaptic plasma membranes (SPM) are more sensitive than brain myelin,[22] and synthetic membranes formed from pure phospholipids are remarkably resistant to effects of ethanol on polarization of DPH.[23] These observations led us to ask which membrane components might increase the sensitivity of phospholipid vesicles to ethanol. We found that addition of hydrophobic proteins (*e.g.*, gramicidin), of cholesterol, or of lipids differing in phospholipid headgroup or acyl unsaturation did not markedly alter the sensitivity of dimyristoyl phosphatidylcholine (DMPC) to ethanol. However, inclusion of gangliosides, galactolipids found in plasma membranes—particularly SPMs, markedly enhanced the membrane fluidizing actions of ethanol and other anesthetics (FIG. 1).[23,24] The sensitivity of membranes (with or without gangliosides) to ethanol is not related to their intrinsic (basal) fluidity (FIG. 1). We have suggested that gangliosides act by increasing the partitioning of alcohol into the membrane,[23] but this remains to be studied in detail.

The decrease in steady state polarization of DPH produced by ethanol raises the question of whether ethanol decreases the order or viscosity of the membrane. We answered this question by measuring the effects of acute, *in vitro* exposure to ethanol on the steady state polarization, fluorescence lifetime, and rotation rate (estimated from differential polarized phase measurements) of DPH in mouse brain SPM (TABLE 1). Ethanol affects only lipid order (limiting anisotropy) and does not alter the kinetic component of the depolarization. Together with our observation that ethanol does not alter the fluorescence spectrum of pyrene derivatives incorporated into membranes (unpublished), these results indicate that ethanol (at least at concentrations below 500 mM) does not alter the microviscosity of membrane lipids.

There is now no doubt that pharmacologically relevant concentrations of ethanol disorder cell membranes, but the relevance of these changes to membrane function remains controversial. The main problem is that the changes in order are miniscule, and similar changes produced by increasing body temperature (by 1° or 2° C) do not result in anesthesia (*e.g.*, REF. 25). However, heat may be a much less selective perturbation than ethanol. Also, the measured changes are an average of all membrane komains, and there may be larger, localized changes. One approach to the relevance of membrane perturbations has been to ask whether animals that differ in genetic

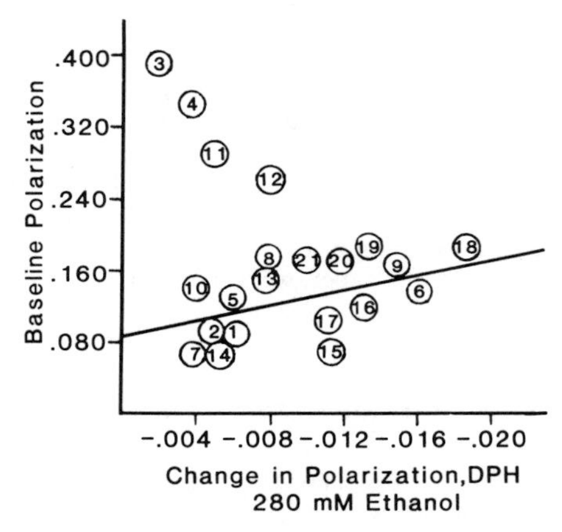

FIGURE 1. Comparison of baseline polarization with ethanol-induced change in polarization for different phospholipid vesicles. The ordinate presents the baseline polarization of DPH, and the abscissa shows the change in polarization produced by *in-vitro* exposure to 280 mM ethanol. The composition of the vesicles was: (**1**) Dioleoylphosphatidylcholine (DOPC); (**2**) DOPC + brain gangliosides (BG); (**3**) DipalmitoylPC (DPPC); (**4**) DPPC + BG; (**5**)DPPC at 47° C; (**6**) DPPC + BG at 47° C; (**7**) DimyristoylPC (DMPC) at 47° C; (**8**) DMPC + cholesterol (33 mole%) at 47° C; (**9**) DMPC + cholesterol + BG at 47° C; (**10**) DMPC; (**11**) DMPC + cholesterol; (**12**) DMPC + cholesterol + BG; (**13**) DMPC + sphingomyelin; (**14**) DistearoylPC (DSPC) at 65° C; (**15**) DSPC + BG at 47° C; (**16**) DPPC/DOPC 3:2 + BG; (**17**) DPPC/DOPC 2:3 + BG; (**18**) DMPC + asialoganglioside; (**19**) DMPC + disialoganglioside; (**20**) DMPC + monosialoganglioside; (**21**) DMPC + sphingosine. All galactolipids were tested at a concentration of 10 mole%. Temperature was 30° C unless noted otherwise. See Harris *et al.*[23] for experimental details.

sensitivity to the behavioral effects of ethanol also differ in the membrane actions of ethanol. A study with an EPR probe (5-doxylstearate) of the upper regions of the acyl chains indicates that there is indeed a genetic correlation between behavioral and membrane sensitivity in Long-Sleep (LS)/Short-Sleep (SS) and heterogeneous stock (HS) mice.[26] However, we (TABLE 2) and others[27] found that the LS and SS mice do not differ in ethanol sensitivity as judged by DPH. This discrepancy could be due to the fact that DPH is located much deeper in the membrane than 5-doxylstearate. To test this hypothesis, we studied the effects of ethanol on TMA-DPH, a probe of the upper regions of the acyl chains, in brain SPM from LS and SS mice. In contrast to DPH,

TABLE 1. *In-Vitro* Effects on Ethanol on the Hindered and Unhindered Rotations of DPH in Brain Synaptic Plasma Membranes[a]

Ethanol Concentration	Change in Anisotropy ($\times 10^3$)		
	r_s	r_∞	r_k
70 mM	-1.3 ± 0.2[b]	-2.9 ± 0.7[b]	0.1 ± 0.05
280 mM	-4.8 ± 0.2[c]	-6.2 ± 1.2[b]	-0.1 ± 0.06

[a]Values represent the change in anisotropy produced by *in-vitro* exposure to given concentrations of ethanol. r_s is the steady state anisotropy, r_∞ the limiting anisotropy, and r_k the kinetic component of the anisotropy. r_s was measured directly; r_∞ and r_k were calculated from the fluorescence lifetime and the differential phase shift as described in the text. All measurements were obtained with an SLM4800 spectrofluorimeter. SPM were prepared from DBA/2 mice, and DPH was incorporated as described previously.[33] Values are mean ± S.E.M., n = 6.

[b]Significant effect of ethanol, $p < 0.05$.

[c]Significant effect of ethanol, $p < 0.01$.

TMA-DPH polarization was more sensitive to ethanol in membranes from LS than SS mice (TABLE 2). Thus, the genetic difference between these mice appears to involve the membrane surface rather than the core.

CHRONIC EXPOSURE TO ETHANOL *IN VIVO* AND *IN VITRO*

Studies of the acute disordering actions of ethanol on membranes raised the question of whether chronic exposure might produce membrane adaptation. It was suggested that a sort of homeoviscous adaptation could result in a more rigid membrane that would be resistant to ethanol, and these changes could be the basis of tolerance and dependence.[28] An EPR study first demonstrated that brain membranes from ethanol-tolerant mice are resistant to the disordering action of ethanol.[29] DPH was soon used to demonstrate the ethanol resistance of vesicles formed from extracted brain membrane lipids from rodents treated chronically with ethanol.[30–32] Subsequent studies showed that chronic ethanol exposure reduces the ability of ethanol to alter the polarization of DPH in plasma membranes from brain,[33,34] liver,[35] and cultured hepatoma cells.[36] Despite the ubiquity of this membrane tolerance, there is no consensus about changes in lipid composition that could produce the resistance. Neither cholesterol nor gangliosides appear responsible.[30,33,37]

In addition to membrane tolerance, the homeoviscous adaptation hypothesis requires that chronic ethanol exposure produce a membrane that is more rigid in the

TABLE 2. Effects of Ethanol on Steady State Polarization of TMA-DPH and DPH Incorporated into SPM from Long-Sleep (LS) and Short-Sleep (SS) Mice[a]

Mouse Line	Decrease in Polarization ($\times 10^3$)	
	TMA-DPH	DPH
LS	8.8 ± 0.8	10.9 ± 0.9
SS	5.5 ± 0.5[b]	11.0 ± 1.0

[a]Values represent the decrease in fluorescence polarization produced by *in-vitro* exposure to 280 mM ethanol. Measurements were made as described previously.[33] Values are mean ± S.E.M., n = 18.

[b]Significant genetic difference, $p < 0.005$.

absence of ethanol. Studies have noted such an increase in rigidity of intact *Escherichia coli* plasma membranes,[38] intact brain SPM,[33,37,39] and extracted lipids from brain membranes.[30] However, in the case of liver plasma membranes the homeoviscous adaptation clearly does not occur following chronic ethanol exposure. Three groups independently reported more fluid membranes following ethanol treatment.[35,36,40] We have extended these findings to demonstrate that chronic ethanol consumption decreases the order, not viscosity, of liver plasma membranes from rats. In addition, this change is restricted to the sinusoidal rather than the canalicular region of the cell (TABLE 3). We also found that chronic ethanol ingestion reduced the order of basolateral membranes from ileum and jejunum but did not alter the order of brush border membranes from these tissues (unpublished). These observations raise the question of why chronic alcohol treatment disorders plasma membranes from liver and gut, yet has either a rigidifying action or no effect on brain membranes (and muscle sarcoplasmic reticulum[41]). We suggested that the disordering action requires metabolism of ethanol and will be observed only in tissues that contain alcohol dehydrogenase (ADH).[36] This is supported by the observation that inhibitors of ADH prevent the disordering action of chronic ethanol on cultured hepatoma cells. Also, Chinese hamster ovary cells, which lack ADH, do not respond to chronic ethanol exposure.[36]

TABLE 3. Effects of Chronic Ethanol on Steady State Polarization of DPH in Plasma Membranes from Brain, Liver, and Cultured Cells[a]

Membrane	Polarization ($\times 10^3$)	
	Control	Chronic Ethanol
Mouse Brain SPM	269 ± 1	281 ± 2[b]
Rat Liver sPM	203 ± 2	171 ± 2[c]
Rat Liver cPM	283 ± 7	293 ± 4
H35 PM	205 ± 6	182 ± 6[c]
H35 PM—4-MP	204 ± 4	203 ± 1
CHO-K1 PM	240 ± 3	238 ± 6

[a]Mouse brain synaptic plasma membranes (SPM) were obtained from mice fed an ethanol-containing diet or an equicaloric diet for 7 days;[33] liver sinusoidal (sPM) and canalicular (cPM) plasma membranes were prepared from rats fed ethanol or a control diet for 4 weeks. Values were similar for lab chow controls and pair-fed controls. Plasma membranes (PM) were prepared from hepatoma (H35) cells grown with 80 mM ethanol for 3 weeks or with 4-methylpyrazole (4-MP) (0.2 mM) and 10 mM ethanol for 3 weeks. CHO cells were grown with 80 mM ethanol for 3 weeks. These experiments are described in detail elsewhere.[36] Values are mean ± S.E.M., n = 6 – 15.

[b]Significant effect of chronic ethanol, $p < 0.05$.

[c]Significant effect of chronic ethanol, $p < 0.01$.

CONCLUSIONS

During the past ten years, a sophisticated technology has evolved to study membrane physical properties by use of fluorescent probes. These techniques have been applied to the problem of the mechanism of alcohol intoxication, tolerance, and dependence and have provided information that is complementary to and consistent with EPR studies. In general, these experiments have shown that acute exposure to ethanol preferentially disorders the hydrophobic core of cell membranes. One question is why the disordering action is greater deeper in the membrane. This does not appear to be consistent with the fact that ethanol is not very lipid-soluble, and the highest concentration will be at the membrane surface. However, recent NMR studies suggest that ethanol *orders* the membrane surface and disorders the core.[42] Thus, the effect of ethanol at a given membrane depth may reflect the sum of these ordering/disordering actions.

A remarkably consistent finding is that chronic ethanol treatment results in membranes that are resistant to the disordering actions of ethanol. Effects of chronic ethanol exposure on basal membrane order are far less consistent and range from small increases in order to rather large decreases in order, depending upon the type of membrane studied. We speculate that the disordering action is a consequence of alcohol metabolism, but this remains to be proved (or disproved). Although the disordering action of chronic ethanol exposure cannot be a homeoviscous adaptation, it could be a functional adaptation. For example, if acute ethanol exposure inhibited a membrane transport process (due to a direct action on the carrier protein), then disordering of the membrane could increase the activity of the carrier (*e.g.*, REF. 43), resulting in ethanol tolerance.

It is pleasing that a general consensus is evolving about the effects of ethanol on membrane order, but the key question is whether any of these effects are of physiological importance. Investigators have pointed out that the observed changes may be too small to alter membrane function. At present there is no consensus as to how large a change in order is required to influence membrane proteins or as to which membrane proteins are most sensitive to lipid perturbations (*e.g.*, REF. 44). These are important topics for future research, but at present the only information regarding the physiological importance of lipid order is the correlation of differences in ethanol sensitivity *in vitro* and *in vivo*. The observation that mice, differing in ethanol sensitivity due either to genetic differences or to tolerance development, display appropriate differences in membrane sensitivity provides strong circumstantial evidence for involvement of membrane lipids in alcohol action. It is particularly interesting that the different responses of TMA-DPH and DPH implicate the membrane surface in genetic differences in alcohol sensitivity. However, it is always possible that the genetic relationship is fortuitous; *i.e.*, some of the genes controlling ethanol sensitivity may be closely linked to genes coding for membrane order but do not directly influence membranes.

Because genetic differences in sensitivity are so important in interpreting biochemical data, we must consider the implications of the observation that the ethanol sensitivity of the membrane core (as evaluated by fluorescence polarization of DPH) does not differ between LS and SS mice. Does this mean that effects of ethanol on DPH have nothing to do with ethanol intoxication? Not necessarily; it merely means that the genes segregated in the LS/SS lines do not alter the environment sensed by DPH. It is likely that all of the genes relevant to ethanol sensitivity will not segregate in any single selective breeding study. Thus, it is possible that a different type of genetic analysis would detect a relationship between alcohol sensitivity *in vivo* and DPH polarization.

In summary, study of membrane physical properties remains a promising approach to understanding alcohol intoxication, tolerance, and dependence, but a firm link between membrane order and physiological function has yet to be established.

ACKNOWLEDGMENTS

We thank Larry Zaccaro for determining fluorescence lifetimes and dynamic depolarization and Dr. Robert Hitzemann for generating stimulating discussions.

REFERENCES

1. McCreery, M. J. & W. A. Hunt. 1978. Physico-chemical correlates of alcohol intoxication. Neuropharmacology **17:** 451–461.
2. Harris, R. A. & R. J. Hitzemann. 1981. Membrane fluidity and alcohol action. *In* Currents in Alcoholism. M. Gallanter, Ed. Vol. 8: 379–404. Grune and Stratton. New York, NY.
3. Miller, K. W. 1985. The nature of the site of general anesthesia. Int. Rev. Neurobiol. **27:** 1–61.
4. Hitzemann, R. J., R. A. Harris & H. H. Loh. 1985. Pharmacological, developmental, and physiological regulation of synaptic membrane phospholipids. *In* Phospholipids and Cellular Regulation. J. F. Kuo, Ed. Vol. 1: 97–129. CRC Press, Inc. Boca Raton, FL.
5. Michaelis, E. K. & M. L. Michaelis. 1982. Physico-chemical interactions between alcohol and biological membranes. Res. Adv. Alcohol and Drug Prob. **7:** 1–90.
6. Lakowicz, J. R. 1983. Principles of Fluorescence Spectroscopy. Plenum Press. New York, NY.
7. Weber, G. 1981. Resolution of the fluorescence lifetimes in a heterogeneous system by phase and modulation measurements. J. Phys. Chem. **85:** 749.
8. Shinitsky, M. & Y. Barenholz. 1974. Dynamics of the hydrocarbon layer in liposomes of lecithin and sphingomyelin containing dicetylphosphate. J. Biol. Chem. **249:** 2652.
9. Perrin, F. 1926. Polarization of light of fluorescence, average life of molecules in the excited state. J. Phys. Radium **7:** 390.
10. Harris, R. A. & F. Schroeder. 1982. Effects of barbiturates and ethanol on the physical properties of brain membranes. J. Pharmacol. Exp. Ther. **223:** 424.
11. Hitzemann, R. J. & R. A. Harris. 1984. Developmental changes in synaptic membrane fluidity: a comparison of 1,6-diphenyl-1,3,5-hexatriene (DPH) and 1-[4-(trimethylamino)phenyl]-6-phenyl-1,3,5-hexatriene (TMA-DPH). Dev. Brain Res. **14:** 113–120.
12. van Blitterswijk, W. J., R. P. van Hoeven & B. W. van der Meer. 1981. Lipid structural order parameters (reciprocal of fluidity) in biomembranes derived from steady-state fluorescence polarization measurements. Biochim. Biophys. Acta **644:** 323.
13. Dale, R. E., L. A. Chen & L. Brand. 1977. Rotational relaxation of the "microviscosity" probe diphenyl-hexatriene in paraffin oil and egg lecithin vesicles. J. Biol. Chem. **252:** 7500.
14. Jahnig, F. 1979. Structural order of lipids and proteins in membranes: evaluation of fluorescence anisotropy data. Proc. Natl. Acad. Sci. USA **76:** 6361.
15. Lipari, G. & A. Szato. 1980. Effect of liberational motion on fluorescence depolarization and nuclear magnetic resonance relaxation in macromolecules and membranes. Biophys. J. **30:** 489.
16. Hare, F. 1983. Simplified derivation of angular order and dynamics of rodlike fluorophores in models and membranes. Biophys. J. **42:** 205.
17. Prendergast, F. G., R. P. Haugland & P. J. Callahan. 1981. 1-[4-(Trimethylamino)phenyl]-6-phenylhexa-1,3,5-triene: Synthesis, fluorescence properties, and use as a fluorescence probe of lipid bilayers. Biochemistry **20:** 7333–7338.
18. Tilley, L., K. R. Thulborn & W. H. Sawyer. 1979. An assessment of the fluidity gradient of the lipid bilayer as determined by a set of n-(9-anthroyloxy) fatty acids (n = 2,6,9,12,16). J. Biol. Chem. **254:** 2592–2594.
19. Vincent, M. & J. Gallay. 1984. Time-resolved fluorescence anisotropy study of effect of a cis double bond on structure of lecithin and cholesterol-lecithin bilayers using n-(9-anthroyloxy) fatty acids as probes. Biochemistry **23:** 6514–6522.
20. Chin, J. H. & D. B. Goldstein. 1981. Membrane-disordering action of ethanol: variation with membrane cholesterol content and depth of the spin label probe. Mol. Pharmacol. **19:** 425–431.
21. Melnick, R. L., H. C. Haspel, M. Goldenberg, L. M. Greenbaum & S. Weinstein. 1981. Use of fluorescent probes that form intramolecular excimers to monitor structural changes in model and biological membranes. Biophys. J. **34:** 499–515.
22. Harris, R. A. & F. Schroeder. 1981. Ethanol and the physical properties of brain membranes: fluorescence studies. Mol. Pharmacol. **20:** 128–137.
23. Harris, R. A., G. Groh, D. M. Baxter & R. J. Hitzemann. 1984. Effects of gangliosides

on the physical properties and ethanol sensitivity of membrane lipids. Mol. Pharmacol. **25:** 410–417.

24. HARRIS, R. A. & G. I. GROH. 1985. Membrane disordering effects of anesthetics are enhanced by gangliosides. Anesthesiology **62:** 115–119.
25. INGRAM, L. O., V. C. CAREY & K. M. DOMBEK. 1982. On the relationship between alcohol narcosis and membrane fluidity. Substance and Alcohol Actions/Misuse **2:** 213–224.
26. GOLDSTEIN, D. B., J. H. CHIN & R. C. LYON. 1982. Ethanol disordering of spin-labeled mouse brain membranes: correlation with genetically determined ethanol sensitivity of mice. Proc. Natl. Acad. Sci. USA **79:** 4231–4233.
27. PERLMAN, B. J. & D. B. GOLDSTEIN. 1984. Genetic influences on the central nervous system depressant and membrane-disordering actions of ethanol and sodium valproate. Mol. Pharmacol. **26:** 547–552.
28. HILL, M. W. & A. D. BANGHAM. 1975. General depressant drug dependency: a biophysical hypothesis. Adv. Exp. Med. Biol. **59:** 1–9.
29. CHIN, J. H. & D. B. GOLDSTEIN. 1977. Drug tolerance in biomembranes: a spin label study of the effects of ethanol. Science **196:** 684–685.
30. JOHNSON, D. A., N. M. LEE, R. COOKE & H. H. LOH. 1979. Ethanol-induced fluidization of the brain lipid bilayers: required presence of cholesterol in membranes for the expression of tolerance. Mol. Pharmacol. **15:** 739–746.
31. JOHNSON, D. A., N. M. LEE, R. COOKE & H. H. LOH. 1980. Adaptation to ethanol-induced fluidization of brain lipid bilayers: cross-tolerance and reversibility. Mol. Pharmacol. **17:** 52–55.
32. JOHNSON, D. A., H. J. FRIEDMAN, R. COOKE & N. M. LEE. 1980. Adaptation of brain lipid bilayers to ethanol-induced fluidization. Biochem. Pharmacol. **29:** 1673–1676.
33. HARRIS, R. A., D. M. BAXTER, M. A. MITCHELL & R. J. HITZEMANN. 1984. Physical properties and lipid composition of brain membranes from ethanol tolerant-dependent mice. Mol. Pharmacol. **25:** 401–409.
34. BEAUGE, F., C. FLEURET-BALTER, J. NORDMANN & R. NORDMANN. 1984. Brain membrane sensitivity to ethanol during development of functional tolerance to ethanol in rats. Alcoholism: Clin. Exp. Res. **8:** 167.
35. SCHULLER, A., J. MOSCAT, E. DIEZ, C. FERNANDEZ-CHECA, F. G. GAVILANES & A. M. MUNICIO. 1984. The fluidity of plasma membranes from ethanol-treated rat liver. Mol. Cell Biochem. **64:** 89–95.
36. POLOKOFF, M. A., T. J. SIMON, R. A. HARRIS & F. R. SIMON. 1985. Chronic ethanol increases liver plasma membrane fluidity. Biochemistry **24:** 3114–3120.
37. LYON, R. C. & D. B. GOLDSTEIN. 1983. Changes in synaptic membrane order associated with chronic ethanol treatment in mice. Mol. Pharmacol. **23:** 86–91.
38. DOMBEC, K. M. & L. O. INGRAM. 1984. Effects of ethanol on the *Escherichia coli* plasma membrane. J. Bacteriol. **157:** 233–239.
39. HARRIS, R. A., J. C. CRABBE, JR. & J. D. MCSWIGGAN. 1984. Relationship of membrane physical properties to alcohol dependence in mice selected for genetic differences in alcohol withdrawal. Life Sci. **35:** 2601–2608.
40. YAMADA, S. & C. S. LIEBER. 1984. Decrease in microviscosity and cholesterol content of rat liver plasma membranes after chronic ethanol feeding. J. Clin. Invest. **74:** 2285–2289.
41. MRAK, R. E. 1983. Calcium transport and fluorescence polarization of 1,6-diphenyl-1,3,5-hexatriene in sarcoplasmic reticulum from normal and ethanol-tolerant rats. Exp. Neurol. **80:** 573–581.
42. HITZEMANN, R. J., C. GRAHAM-BRITTAIN, H. E. SCHUELER & G. P. KREISHMAN. 1987. Ordering/disordering effects of ethanol in dipalmitoylphosphatidylcholine liposomes. This volume.
43. MOLITORIS, B. A., A. C. ALFREY, R. A. HARRIS & F. R. SIMON. 1985. Renal apical membrane cholesterol and fluidity in the regulation of phosphate transport. Am. J. Physiol. **249:** F12–F19.
44. HARRIS, R. A. & P. BRUNO. 1985. Membrane disordering by anesthetic drugs: relationship to synaptosomal sodium and calcium fluxes. J. Neurochem. **44:** 1274–1281.

DISCUSSION OF THE PAPER

D. McCarthy (*University of New Mexico School of Medicine, Albuquerque, NM*): With regard to the changes in the sialic acid containing gangliosides, these might change the water content of the membrane very appreciably. Did you do any experiments to see whether acute removal of the sialic acid by neuraminidase or other treatments would cause a rapid return of the tolerance to the base line?

Harris: No, we did not. Those experiments are difficult because neuraminidase will only remove the sialic acid from 2- and 3-sialic gangliosides, producing GM1, which is the one that we think is important anyway. It's not that easy to selectively clip off the sugars or all the sialic acids of the membrane. Our studies with model systems with different sialolipids, though, point to the sugar groups as being important, and not the sialic acid.

J. M. Littleton (*King's College, London, England*): You mentioned the changes in liver plasma membrane. There have been several studies now that showed there are really quite marked changes in the fatty acyl chains of liver phospholipids after chronic ethanol treatment. Have you looked at those at all, because I think some of the changes that occur are in the direction of more unsaturated fatty acids. That could certainly go some way towards explaining the changes.

Harris: Yes, we have looked at that. However, they do not appear to be the basis for the changes in the animals. In the cultured cells, the basis appears to be a change in the sphingomyelin-to-phosphatidylcholine ratio. There is also some decrease in cholesterol levels in the animal but not in the cultured cells. This seems to be a little bit like the brain, where one can see a clear change in physical properties. But it's not so easy to understand how it's related to change in the composition.

B. Chance (*University City Science Center, Philadelphia, PA*): Could you discuss the changes in cholesterol *in vivo* as well as changes in cholesterol ester?

Harris: Cholesterol ester is present only in very small amounts in our plasma membrane preparations, and the difference between chronically ethanol-treated ones and controls is not very great. We don't think that cholesterol esters explain the change in plasma membrane. Of course, if one looks at the entire cell, there are marked changes in cholesterol ester. But it seems that very little of that cholesterol ester partitions into the plasma membrane.

Question: In one of your experiments you showed that adding calcium does not change the depolarization. Could you comment on how the calcium acts in the two different types of mice?

Harris: The results we showed were that if one adds calcium to the membranes, that difference in ethanol sensitivity between the LS and SS mice disappears. The membranes from both lines become less sensitive to alcohol, but the LS changes more than the SS; so they both show the same change. This indicates that calcium can alter the properties of the membrane surface. There are two things going on making both lines less sensitive to alcohol, but they are having a greater effect on one line than the other. Interestingly, one doesn't see this effect of calcium if one looks at the core of the membrane with DPH, so it seems to have something to do with the surface. We did this experiment because of observations that injection of calcium into animals or application of calcium to Purkinje cells also reduces the difference between these lines. Some years ago Barry Hoffer's group showed a marked difference in the effects of ethanol on Purkinje cell firing in these lines. The LS line is about 30 times more sensitive to ethanol than the SS line as measured by inhibition of Purkinje cell firing. Recently,

Gene Erwin and Mike Palmer showed that if one applies calcium to the Purkinje cells, both lines become equally sensitive to ethanol. So we wondered if the membranes would respond in the same way, and to our surprise they do.

CHANCE: I would like to make a concluding remark. Never before in the history of membrane studies have so many techniques been applied ranging from ESR, to probes, to temperature, to pressure, genetics and what have you. But the key question, it seems to me, remains before us—whether the very significant effect observed at very high ethanol concentrations can be extrapolated back to a physiological function in the cell.

Membrane Adaptation to Chronic Ethanol in Man[a]

ALAN C. SWANN, EDWARD L. REILLY,
AND JOHN E. OVERALL

Department of Psychiatry
University of Texas Medical School
and
Mental Sciences Institute
Houston, Texas 77025

Despite the considerable morbidity associated with chronic ethanol abuse, little is known about the clinical relevance of the membrane effects of ethanol. Ethanol increases membrane fluidity, with consistent effects on regulation of cation transport in the rat.[1] Inhibition of Na,K-ATPase and reduction of K^+ affinity by ethanol are consistent with its effects on cation transport *in vivo*[2] and with the reversal of acute behavioral effects by K^+.[3] The concentration of ethanol required for inhibition of Na,K-ATPase by ethanol is reduced by norepinephrine, apparently via apha-l receptors.[4] During chronic ethanol treatment, sensitivity of Na,K-ATPase to ethanol is reduced,[5] and regulation of the enzyme is altered in a manner consistent with decreased membrane fluidity.[6]

There is little information on effects of ethanol on Na,K-ATPase in man. An early study reported that red blood cell Na,K-ATPase was higher in alcoholics than in controls, but there were no data on sensitivity to ethanol or on clinical characteristics of the control group.[7] More recently, Na,K-ATPase activity was reported to be reduced in alcoholic patients, with reduced sensitivity to ethanol.[8]

We report here red blood cell Na,K-ATPase and its sensitivity to ethanol in 41 newly admitted alcoholic inpatients compared to 14 healthy subjects of similar age and

TABLE 1. Red Blood Cell (Na^+, K^+)-ATPase in Alcoholic Patients and Controls[a]

	Patients (n = 41)	Controls (n = 14)
Basal activity	40.7 ± 24.5	28.8 ± 17.1
With 0.1 M ethanol/0.1 mM norepinephrine	31.6 ± 19.7[b]	16.1 ± 10.4
Inhibition	0.203 ± 0.169[c]	0.393 ± 0.183

[a]Red blood cell (Na^+, K^+)-ATPase activity is in nmol/(mg prot.min) ± standard deviation. Basal activity is (Na^+, K^+)-ATPase activity without ethanol. Inhibition is the reduction in activity by ethanol plus norepinephrine as a fraction of the basal activity. Ethanol or norepinephrine alone at these concentrations had no effect. Significance of differences by 2-tailed student t test.
[b]P 0.01.
[c]P 0.001.

[a]Supported by United States Public Health Service Grants AA05785 and MH00415.

TABLE 2. Correlations Between (Na^+, K^+)-ATPase and Behavioral Measures

	Basal	Ethanol	Sensitivity
Somatic concern	0.440[a]	0.533[a]	−0.071
Excitement	0.385[b]	0.542[a]	−0.224
Anxiety	0.012	0.234	−0.495[a]
Tension	−0.143	0.057	−0.406[b]
Behavioral withdrawal	0.071	0.314	−0.352[b]

[a]P 0.01.
[b]P 0.05.

gender, and relationships between these parameters and estimated ethanol consumption, psychiatric systems and history, and medical characteristics. The subjects were otherwise healthy male alcoholics, age 37.6 ± 11.3 (SD), and controls, age 36.5 ± 9.2. The alcoholics had no primary psychiatric or substance abuse disorder other than alcoholism, no significant medical disorder requiring pharmacologic treatment, and no laboratory or physical evidence of poor nutritional state.

Table 1 shows the red blood cell Na,K-ATPase indices in alcoholic patients and controls. The major difference is that sensitivity to ethanol is reduced in the alcoholic patients. Enzymatic activity is somewhat higher in the patients, consistent with animal data[5] and an earlier report in man.[7]

There was a significant negative correlation between sensitivity of Na,K-ATPase activity to ethanol ($r = -0.46$; P 0.02). As summarized in TABLE 2, sensitivity to ethanol also correlated significantly with ratings for anxiety ($r = -0.49$) and for agitation ($r = -0.4$). Sensitivity to ethanol did not correlate with basal activity. None of the Na,K-ATPase indices correlated with age, with depression, or, after correction for correlations with clinical variables, with medical laboratory measurements.

In summary, sensitivity of red blood cell Na,K-ATPase was lower in alcoholic patients than in controls and correlated negatively with ethanol intake. These data suggest that tolerance to membrane effects of ethanol, similar to that reported in animals,[5] occurs in man.

REFERENCES

1. ISRAEL, Y., H. KALANT & A. E. LEBLANC. 1966. Biochem. J. **100:** 27–33.
2. KALANT, H., W. MONS & M. A. MCMAHON. 1966. Canad. J. Physiol. Pharmacol. **44:** 1–12.
3. ISRAEL, Y., H. KALANT & J. LAUFER. 1965. Biochem. Pharmacol. **14:** 1803–1814.
4. RANGARAJ, N. & H. KALANT. 1980. Canad. J. Physiol. Pharmacol. **58:** 1342–1346.
5. LEVENTAL, M. & B. TABAKOFF. 1980. J. Pharmacol. Exp. Ther. **212:** 315–319.
6. SWANN, A. C. 1985. J. Pharmacol. Exp. Ther. **232:** 475–479.
7. ISRAEL, Y., H. KALANT, E. LEBLANC, J. BERNSTEIN & J. SALAZAR. 1979. J. Pharmacol. Exp. Ther. **185:** 330–336.
8. STIBLER, H., F. BEAUGE' & S. BORG. 1984. Alcoholism: Clin. Exp. Res. **8:** 522–527.

Effects of Alcohol and Halothane on the Structure and Function of Sarcoplasmic Reticulum[a]

S. TSUYOSHI OHNISHI

Membrane Research Institute
University City Science Center
Philadelphia, Pennsylvania 19104

Using heavy sarcoplasmic reticulum (SR) prepared from skeletal muscle of rats (Sprague Dawley) and pigs (Pietrain; malignant hyperthermia susceptibles and nonsusceptible controls), we studied the effects of ethanol (both acute and chronic) and halothane (acute).

ACUTE ETHANOL EFFECTS

Ethanol acts as a general anesthetic, and both acute and chronic ethanol ingestions are known to cause reduction of contractility of skeletal muscle.[1–4] However, the mechanism of this negative inotropic action of ethanol is not well understood. It is known that the maximum ethanol concentration in the blood of chronic alcoholics may reach 0.5%. At such concentrations, ethanol did not reduce either the interaction between actin and myosin[5] or troponin-calcium binding.[6] Therefore, the major effect of ethanol on muscle must be on some subcellular system other than contractile (and regulatory) proteins. Since ethanol interacts with membranes like other anesthetics,[7] it is reasonable to speculate that ethanol may act on the sarcoplasmic reticulum (SR) of skeletal muscle. However, as shown in our previous papers,[8,9] ethanol at physiologically realistic concentrations did not affect the calcium transport (ATP-induced calcium uptake activity) of the SR.

In our effort to search for any possible effect of ethanol at "physiologic" concentrations, we found that the calcium-induced calcium release phenomenon is very sensitive to ethanol,[8–11] and that this phenomenon may be related to a physiological effect of ethanol on muscle. Even at concentrations less than 0.5%, ethanol markedly enhanced the calcium release from the skeletal SR.[8] It was also observed that ethanol did not release calcium from the SR by itself, but instead enhanced the calcium efflux from the SR when such a release was activated by the addition of calcium.[8,10,11] At this point, there is a remarkable contrast between ethanol and halothane (a general anesthetic commonly used in the operating room), namely, that halothane triggered calcium release by itself.[10–12] An increase of calcium efflux in the presence of ethanol may lead to a decrease of maximum calcium content of the SR.[8] Therefore, on exposure to ethanol (in both chronic and acute cases), the SR may gradually lose calcium, and this loss may eventually cause the observed reduction of contractile force.

[a]Supported in part by National Institutes of Health Grants GM 35681 and GM 33025.

The finding that halothane triggered calcium release led us to study the effect of halothane in malignant hyperthermia (MH). Using genetically MH-susceptible pigs, it was found that the SR prepared from such pigs has a higher calcium permeability than that of SR from normal pigs.[12–14] It was also observed that the permeability of MH-SR was remarkably enhanced by the addition of halothane (concentrations from 10 to 200 μM, which were well below anesthetic levels). Ethanol did not trigger calcium release in either control or MH-susceptible pig SR.

CHRONIC EFFECTS OF ETHANOL (ADAPTATION PHENOMENON)

Chronic alcoholic rats (fed for 5 weeks with a liquid alcohol diet) were prepared at the Department of Pathology, Hahnemann University School of Medicine. It was

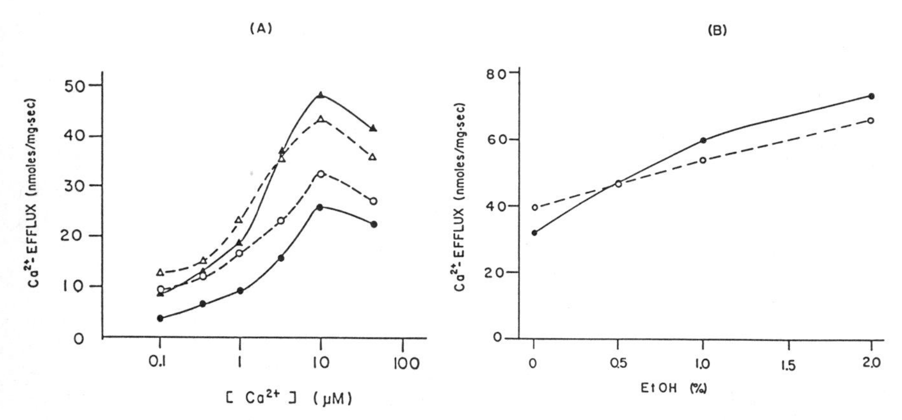

FIGURE 1. (**A**) The relationship between the passive calcium efflux and free calcium concentration. The *solid lines* represent data from normal rat SR and the *dashed lines* data from alcoholic rat SR. *Circular symbols* (●,○) indicate the measurements in the absence of ethanol, and *traingular symbols* (▲,△) represent experiments performed in the presence of 1% ethanol. (**B**) Effect of ethanol concentration on calcium efflux. The *solid line* shows control rat SR and the *dashed line* alcoholic rat SR. The free calcium concentration was 10 μM.

found that chronic alcohol ingestion did not change the activity of ATP-induced calcium uptake activity of the SR of these rats.[9] However, we found that the calcium permeability of alcoholic rat SR was higher than that of normal SR (even in a reaction medium that did not contain ethanol; FIG. 1A). We also found that calcium release from chronic rat SR was less enhanced by the acute addition of ethanol than that from normal rat SR (alcohol adaptation phenomenon; FIG. 1B).[15]

STRUCTURE-FUNCTION RELATIONSHIP

In an attempt to find a structure-function relationship for calcium movement in biological membranes, we studied the effect of ethanol on the order parameter of the

SR membrane as measured by a spin-probe technique.[9] As shown in FIGURE 2A, the order parameter of alcoholic SR is much lower than that of control rat SR (suggesting that the membrane was already more fluidized than in normal SR). However, the acute addition of ethanol did not further decrease the order parameter of alcoholic SR. In contrast, there was a greater decrease of the order parameter when the same concentration of ethanol was added to normal rat SR.

Another interesting finding was that the order parameter of MH-SR was remarkably decreased by the addition of halothane at concentrations between 10 and 200 μM, where halothane triggered the calcium release.[13] In nonsusceptible pig SR, this concentration of halothane neither released calcium nor significantly decreased the order parameter (FIG. 2B). Dantrolene, which is clinically used to suppress MH symptoms when encountered in the operating room, was found to inhibit both halothane-induced calcium release and halothane-induced decrease of the order

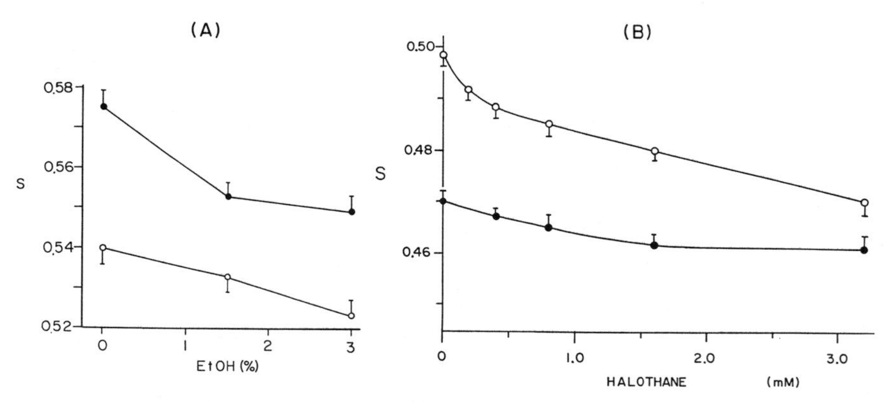

FIGURE 2. (**A**) Effect of ethanol addition on the order parameter (S) of normal rats (*filled circles*) and alcoholic rats (*open circles*) as measured by 7-doxylstearic acid probe. (**B**) Effect of halothane on the order parameter of SR prepared from normal pigs (*filled circles*) and malignant hyperthermia susceptible pigs (*open circles*) as measured by 12-doxylstearic acid probe.

parameter in the MH-SR.[13] These results suggest that changes in membrane fluidity of the SR may regulate the calcium release phenomenon, and that both ethanol and halothane influence this mechanism.

REFERENCES

1. MENDOZA, L. C., K. HELLBERG, A. RICKART, G. TILLICH & R. J. BING. 1971. Clin. Pharmacol. **11:** 165–171.
2. REGAN, T. J., G. KOROXENIDIS, C. B. MOSCOS, H. A. OLDEWURTEL, P. H. LEHAN & H. K. HELLEMS. 1966. J. Clin. Invest. **45:** 270–280.
3. LOCHER, A., A. COWLEY & A. J. BRINK. 1969. Am. Heart J. **78:** 770–780.
4. RUBIN, E. 1979. N. Engl. J. Med. **301:** 28–33.
5. ISHII, Y., S. T. OHNISHI & E. RUBIN. 1985. Biochem. Pharmacol. **34:** 203–210.
6. ISHII, Y. & S. T. OHNISHI. 1985. Biochim. Biochem. Acta **843:** 145–149.
7. SEEMAN, P. 1972. Pharmacol. Rev. **24:** 583–655.

8. OHNISHI, S. T., J. F. FLICK & E. RUBIN. 1984. Arch. Biochem. Biophys. **233:** 588–594.
9. OHNISHI, S. T., A. J. WARING, S. R. G. FANG, K. HORIUCHI & T. OHNISHI. 1985. Membr. Biochem. **6:** 49–63.
10. OHNISHI, S. T. 1979. J. Biochem. **86:** 1147–1150.
11. OHNISHI, S. T. 1981. *In* The Mechanism of Gated Calcium Transport Across Biological Membranes. S. T. Ohnishi & M. Endo, Eds. 275–293. Academic Press. New York, NY.
12. OHNISHI, S. T., S. TAYLOR & G. A. GRONERT. 1983. FEBS Lett. **161:** 103–107.
13. OHNISHI, S. T., A. WARING, S. R. G. FANG, K. HORIUCHI, J. L. FLICK, K. K. SADANAGA & T. OHNISHI. 1986. Arch. Biochem. Biophys. **247:** 294–301.
14. KIM, D., F. A. SRETER, S. T. OHNISHI, J. F. RYAN, J. ROBERTS, P. D. ALLEN, L. F. MESZAROS, B. ANTONIU & N. IKEMOTO. 1984. Biochim. Biophys. Acta **775:** 320–327.
15. S. T. OHNISHI. 1985. Membr. Biochem. **6:** 33–47.

Ordering/Disordering Effects of Ethanol in Dipalmitoylphosphatidylcholine Liposomes[a]

ROBERT J. HITZEMANN,[b] CINDY GRAHAM-BRITTAIN,[c]
HAROLD E. SCHUELER,[c] AND GEORGE P. KREISHMAN[c]

[b]Department of Psychiatry and Behavioral Sciences
State University of New York at Stoney Brook
Stoney Brook, New York 11794
and
[c]Department of Chemistry
University of Cincinnati
Cincinnati, Ohio 45221

The lipid perturbation hypothesis suggests that the ethanol-induced changes in membrane function are caused by a concentration-dependent decrease in membrane lipid order. This prediction has been confirmed in both natural and synthetic membrane systems by a variety of techniques.[1] However, there are numerous arguments against the hypothesis, which include: a) the effects of ethanol at pharmacological concentrations are small; b) over some temperature ranges the potency of ethanol decreases with increasing temperature, whereas, the partition coefficient increases;[2] and c) genetic selection studies suggest that ethanol and related hydrophilic anesthetics differ in their mechanism(s) of action from the hydrophobic anesthetics.[3]

In an attempt both to better understand the effects of ethanol on membrane order and to examine the merits of the criticisms above, we have begun a series of ^{1}H- and ^{2}H-NMR experiments that focus on ethanol-membrane interactions. Delayed Fourier Transform ^{1}H-NMR spectroscopy (DFT) is used to simultaneously monitor lipid order in three regions of the membrane matrix: the choline methyl groups, the fatty acid methylene moieties, and the terminal fatty acid methyl groups.[4] ^{2}H-NMR is used to monitor ethanol partitioning to two membrane domains that are differentiated by their pre-exchange lifetimes and are presumed to represent partitioning to the membrane surface and interior.[5] By combining ^{1}H- and ^{2}H-NMR techniques we have shown that ethanol has two diametrically opposed effects on rat synaptic plasma membranes. While ethanol bound to the interior of membrane induces disordering effects, ethanol bound to the surface induces ordering effects. The overall effect on membrane order can be either ordering or disordering depending upon both the ethanol concentration and the temperature.[6]

In the present study we used a similar approach to examine the effects of ethanol on dipalmitoylphosphatidylcholine (DPPC) liposomes. Such studies are preliminary to examining more complex artificial membranes that more closely mimic natural membranes. In FIGURE 1, DFT NMR spectra of DPPC liposomes are shown at 46° C with 0.0% and 1.0% CD_3CD_2OD. In the DFT experiment, an increase in spectral

[a]Supported by National Science Foundation Grant CHE-8102974 as partial funding for purchase of the NMR spectrometer.

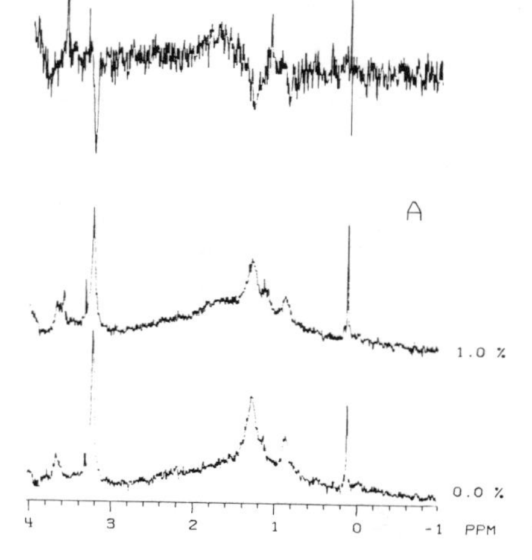

FIGURE 1. (**A**) 300 MHz DFT spectra of DPPC liposomes in D_2O/PBS buffer with 0.0% and 1.0% CD_3CD_2OD added at 46° C. (**B**) The difference spectrum of the 1.0 minus 0.0% samples.

intensity is interpreted as the result of disordering, while a decrease in spectral intensity implies an ordering effect in a specific domain of the lipid matrix. By means of an internal standard (the resonance at 0.05 ppm), a difference spectrum can be obtained (FIG. 1b). Upon the addition of ethanol, a marked *decrease* in the spectral intensity of the choline methyl resonance at 3.1 ppm is clearly evident. In the methylene spectral region (1 to 2.0 ppm), a significant *increase* in spectral intensity is

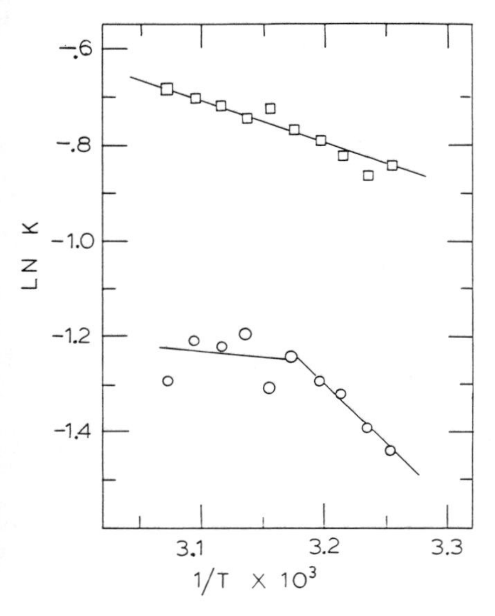

FIGURE 2. The van't Hoff plot of the ethanol partition coefficients to the interior, K_1(O), and to the surface, K_2(□), of DDPC liposomes in H_2O/PBS buffer.

observed. As in the rat SPM study, an ordering effect on the surface of the membrane and a disordering effect on the interior of the membrane are observed.

The temperature dependence of the ethanol partition coefficients to the interior (K_1) and the surface (K_2) of the liposomes as determined by ^{2}H-NMR is shown in FIGURE 2. As can be seen, K_1 is biphasic due to the melting of the DPPC bilayer, while K_2 is unaffected by the phase transition. Above the phase transition, both partition coefficients increase with increasing temperature, but the slope is greater for K_2 than K_1. Thus, with increasing temperature, the surface ordering vector increases more rapidly than the interior disordering vector. Therefore, the total membrane disordering by ethanol will be attenuated with increasing temperature.

These preliminary results on DPPC liposomes, which confirm our earlier work on neuronal membranes, clearly indicate that an understanding of the pharmacological effects of ethanol will require a careful assessment of factors that enhance or attenuate the binding of ethanol to the surface and the interior. These studies are currently in progress.

REFERENCES

1. HITZEMANN, R. J., R. A. HARRIS & H. H. LOH. 1984. *In* Physiology of Membrane Fluidity. M. Shinitsky, Ed. 109–126. CRC Press. Boca Raton, FL. Including references therein.
2. FRANKS, N. P. & W. R. LIEB. 1982. Nature **300:** 487–493. Including references therein.
3. COWAN, A. Personal communication.
4. KREISHMAN, G. P. & R. J. HITZEMANN. 1985. Brain Res. **344:** 162–166.
5. KREISHMAN, G. P., C. GRAHAM-BRITTAIN & F. J. HITZEMANN. 1985. Biochem. Biophys. Res. Commun. **130:** 301–305.
6. HITZEMANN, R. J., H. E. SCHUELER, C. GRAHAM-BRITTAIN & G. P. KREISHMAN. 1986. Biochim. Biophys. Acta **859**(1): 189–197.

PART III. EFFECTS OF ETHANOL ON BIOLOGICAL MEMBRANES: CHEMICAL STUDIES

Effects of Altered Phospholipid Molecular Species on Erythrocyte Membranes

L. L. M. VAN DEENEN, F. A. KUYPERS,
J. A. F. OP DEN KAMP, AND B. ROELOFSEN

Department of Biochemistry
State University of Utrecht
Padualaan 8
3584 CH Utrecht, The Netherlands

In an earlier volume of these *Annals,* we reported on a number of parameters that are important for lipid-lipid and lipid-protein interactions in biomembranes.[1] Special attention was given to the bewildering variation in polar headgroups and apolar side chains of membrane phospholipids. Already in 1965 relations were established between the chemical structure and the membrane function of various lipid classes. Based on model experiments with synthetic phospholipids the conclusion was reached that in natural membranes there is a strong tendency to preserve a liquid-expanded state of the lipid backbone. Several observations made in our laboratory confirmed the view that many types of cells are capable of maintaining the physical properties of the lipid core within certain limits.

At an early stage we emphasized the importance of fatty acid pairing within the phospholipid molecule as a major determinant of the physical state of the lipid bilayer. Analyses were presented that described the composition of lecithin or phosphatidylcholine (PC) in terms of molecular species. The data demonstrated that species having saturated and (poly)unsaturated fatty acids and being esterified at the 1- and 2-position of *sn*-3-phosphatidylcholine (PC) are dominant, but disaturated and di(mono)unsaturated species appeared to occur as well.[1] Significant differences were found to exist between the molecular species composition of nonhomologous membranes. Furthermore, it was demonstrated that different diets can cause striking quantitative differences in the species composition of membrane phospholipids, but at the same time these studies revealed the existence of a homeostatic mechanism, which limits the overall change in the physical properties of the lipid barrier of the membrane. Recently, it has become possible to circumvent some of the regulatory systems that control the chemical makeup of membrane lipids and to induce specific alterations in the composition of a family of phospholipid species.

During the past century the erythrocyte has served as a favourite model for membrane research. We now know, for example, in detail its phospholipid composition and the distribution of the phospholipid classes over the two leaflets of the lipid bilayer; even the rate of transbilayer movement of the phospholipids has been clarified. Limiting the discussion to one major phospholipid, we can state that PC comprises about 30% of the total phospholipid complement of the membrane of the human erythrocyte, and is distributed asymmetrically over the two layers: 75% is found in the outer leaflet and 25% in the inner leaflet. FIGURE 1 shows the major species found in this membrane, and it is noteworthy that these PC species are distributed randomly over the two compartments of the bilayer. Significant differences exist in transbilayer movement between various species with a $t\frac{1}{2}$ variation of 4–24 hr. During the 120 days' life span of the human erythrocyte, the lipid composition appears to be highly

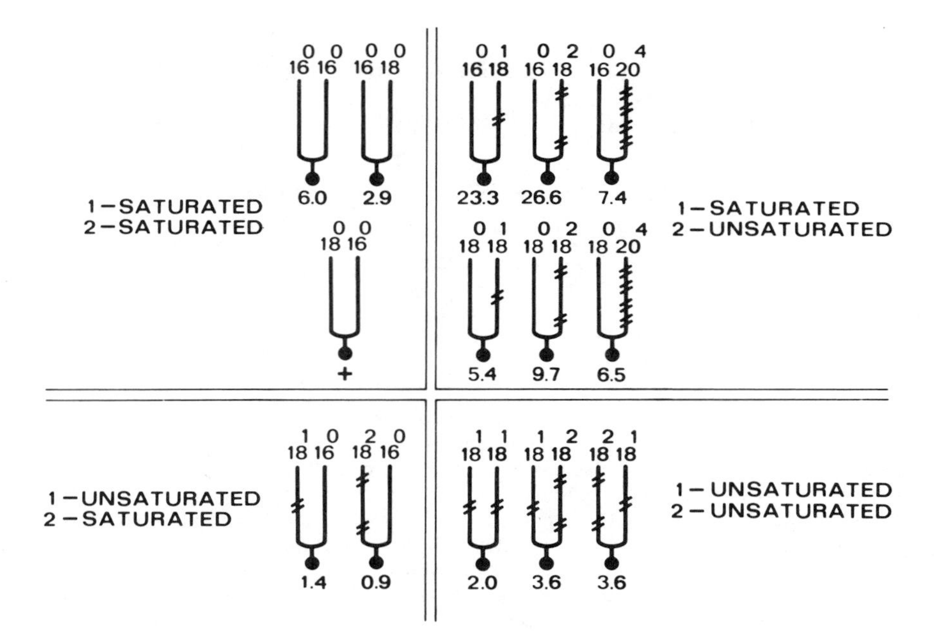

FIGURE 1. The most abundant molecular species of PC in the membrane of the human erythrocyte. The fatty acids on the *sn*1- and *sn*2-positions are denoted by the number of carbon atoms and double bonds in the chains. The relative amounts of each species is presented as % of the total PC content.

conserved, and this property serves to maintain membrane characteristics such as permeability and resistance against mechanical and osmotic stress. The differences in fatty acid composition of the erythrocyte phospholipids, which can be induced upon extreme variations in dietary fats, are less pronounced than the changes in composition of plasma lipids. Furthermore, such alterations in the membrane are subject to a given pattern, indicating the operation of a compensatory mechanism. However, the use of a protein responsible for the PC-specific transfer between membranes provides a useful tool for retailoring the composition of PC species in a controlled manner.

The PC-transfer protein has been thoroughly investigated,[2] and this protein catalyzes a one-for-one exchange of PC molecules only. If a suitable donor system is selected, native PC species can be replaced in the erythrocyte membrane by other well defined PC species (FIG. 2) without altering the original content of the various phospholipid classes and cholesterol.[3,4] Because the transbilayer movement (flip-flop) of PC in erythrocytes is a relatively slow process the substitutions in PC species composition are mainly introduced into the outer membrane leaflet. Substantial modifications can be induced and it was possible to set the limits in between which the fatty acid composition of PC can be varied without measurable effects on membrane structure and function. An increase in the content of certain species appears not to affect membrane properties, but the membrane of intact human erythrocytes appears to tolerate only limited changes in the level of other PC species (TABLE 1).

A replacement of native PC by disaturated species, such as (dipalmitoyl)-PC and (distearoyl)-PC, proceeded at a low rate, and extensive retailoring could only be achieved by repeatedly adding fresh donor vesicles.[4] The introduction of the disaturated molecules was accompanied by a gradual increase in osmotic fragility, finally resulting in hemolysis (TABLE 1). The saturated species (dipalmitoyl)-PC accounts for about 2% of the total membrane lipid. It is remarkable to note that an increase of this species to 4% has a large effect on membrane stability and finally at a relative concentration of some 6% results in lysis of human erythrocytes. On the other hand, essentially all of the PC in the outer layer of the membrane could be replaced by (1-palmitoyl,2-oleoyl)-PC and (1-palmitoyl,2-linoleoyl)-PC without altering parameters such as K^+ permeability, osmotic fragilities, and erythrocyte shape (TABLE 1). Perhaps this is not surprising inasmuch as these mixed-acid species are major membrane constituents and together with the closely related (1-stearoyl,2-oleoyl)- and (1-stearoyl,2-linoleoyl)-derivatives comprise about 65% of the total PC (FIG. 1). However, a further increase of the total degree of unsaturation of the PC fraction of the membrane appeared to modify the properties of the erythrocyte membrane considerably. Biosynthesis of PC species containing two polyunsaturated acyl chains is avoided by mammalian cells. Introduction of this type of molecule into the erythrocyte membrane, namely synthetic (dilinoleoyl)-PC, resulted in a progressive increase in osmotic fragility, and hemolysis started to occur after some 30% of the total PC had been replaced by this unnatural species, K^+ permeability was found to be increased in these cells as well. Previously, we had observed that in artificial membranes (dilinoleoyl)-PC produced leaky bilayers, indicating already that the barrier properties of this species are unsatisfactory. A native species of the erythrocyte membrane is (1-palmitoyl,2-arachidonoyl)-PC (see FIG. 1); elevation of the content of this mixed-acid species surprisingly resulted in a decreased osmotic fragility. K^+ appeared to leak out of the cells already at isotonic conditions, and due to a loss of barrier properties the osmotic resistance is shifted to the hypotonic side.[4] Experimental details about these observations and the reliability of the exchange procedure to modify the PC species composition of erythrocytes have been presented recently.[5]

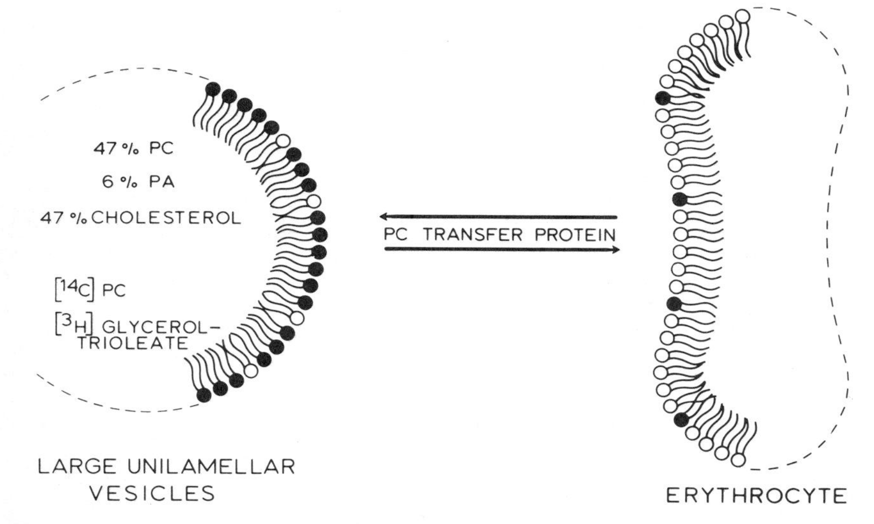

FIGURE 2. Retailoring of native erythrocyte PC by well defined PC species using the PC-specific transfer protein from bovine liver.

TABLE 1. Substitution of Phosphatidylcholine Species in Outer Monolayer of Intact Human Red Cells

Species		16:0/18:1	16:0/18:2	16:0/20:4	16:0/16:0	18:2/18:2
Initial content[a]	% of PC	23.3	26.6	7.4	6.0	—
	% of total lipid	5.1	5.9	1.6	1.3	—
Maximal substitution[a]	% of PC	80–100	80–100	~80	~53	~53
	% of total lipid	22	22	~18	~12	~12
Unsaturation index PC[a]		0.50	1.00	~1.8	~0.4	~1.5
K^+-leakage		normal	normal	increasing	normal	increasing
Osmotic fragility		normal	normal	decreasing	increasing	increasing
Hemolysis		none	none	+[b]	+[b]	+[b]
Shape		normal	normal	spheroechinocytes	echinocytes	stomatocytes
Survival *in vivo*		normal	n.d.[c]	decreasing	decreasing	n.d.[c]

[a] In *outer* monolayer.
[b] Starts at the indicated maximal extent of substitution.
[c] Not determined.

This new approach demonstrates that the structure and the combination of fatty acyl residues within the phospholipid molecule are critical in the control of permeability and mechanical strength of the membrane. The results allow limits to be set between which the degree of unsaturation of PC in the outer monolayer of human erythrocytes can be varied without disturbing the cell surface. The unsaturation index (*i.e.* the number of double bonds per fatty acyl residue in PC) should be 0.5–1.0; otherwise the alignment of the molecules within the membrane is disordered, which leads to abnormal behaviour. A loss of membrane stability results after an increase in the degree of saturation (unsaturation index below 0.5). Apparently, such species as (1-palmitoyl,2-oleoyl)-PC and (1-palmitoyl,2-linoleoyl)-PC fit most optimally into the erythrocyte membrane.

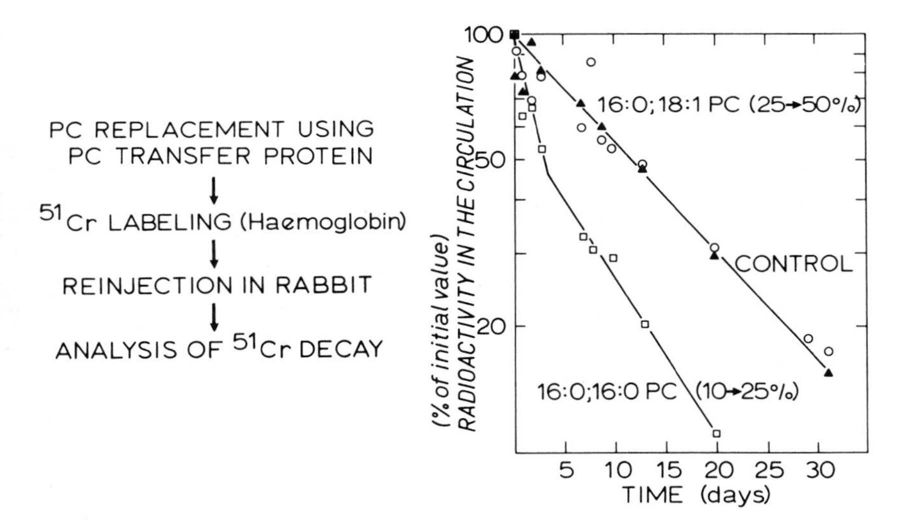

FIGURE 3. *Left side:* experimental design; *right side:* survival curves of rabbit erythrocytes based on ^{51}Cr recovery in the circulation. *Upper line* shows control cells (○) and cells in which 40% of the native PC has been replaced by (1-palmitoyl,2-oleoyl)-PC (▲). *Lower line* shows cells in which 15% of native PC has been replaced by (1,2-dipalmitoyl)-PC (□).

These conclusions were confirmed by experiments in which the survival of erythrocytes *in vivo* was measured after retailoring the PC species composition.[6] Replacement of native PC species with (1-palmitoyl,2-oleoyl)-PC up to 40% of the total PC complement had no effect on the survival time of rabbit or horse erythrocyte when reinjected into the animal. Replacement of PC in both erythrocyte species with (dipalmitoyl)-PC up to 15% appeared to produce a significant reduction in the circulation time of the modified cells (FIG. 3). At about 30% replacement by this saturated PC species the reinjected cells appeared to be cleared from circulation within 24 hr. In this context it is relevant to discuss the change in cell shape induced by the incorporation of PC species into the erythrocyte membrane.

In the first experiments on the retailoring of PC species in rat erythrocytes, it was observed that after insertion of disaturated PC molecules gross morphological changes of the cells accompanied the alterations in the intrinsic membrane properties.[3] With the progression of the incorporation of disaturated PC a formation of echinocytes and spherocytes was observed. This finding was confirmed and extended in experiments on human erythrocytes.[7] FIGURE 4 shows that echinocytes are formed when about 25% of the native PC has been replaced by (dipalmitoyl)-PC, and a further increase in the content of this saturated PC brings about the appearance of spheroechinocytes. Because replacement of PC species can only be measured as an average value of the total cell population, a precise quantitative correlation between PC replacement and changes in erythrocyte morphology is difficult to make. In accordance with observations on osmotic fragility the introduction of mixed-acid species such as (1-palmitoyl,2-oleoyl)-PC and (1-palmitoyl,2-linoleoyl)-PC did not cause any change in the discoid shape of the cells (FIG. 4). On the other hand, an increase of the content of (1-palmitoyl,2-arachidonoyl)-PC caused a formation of spheroechinocytelike erythrocytes (FIG. 5). This type of change in cell shape was also found after the introduction of the unsaturated species (dimyristoyl)-PC. Furthermore, cells with an increased level of (1-palmitoyl,2-arachidonoyl)-PC not only revealed a spheroechinocytic form but, in addition, showed typical dimples. Another surprising morphological alteration was

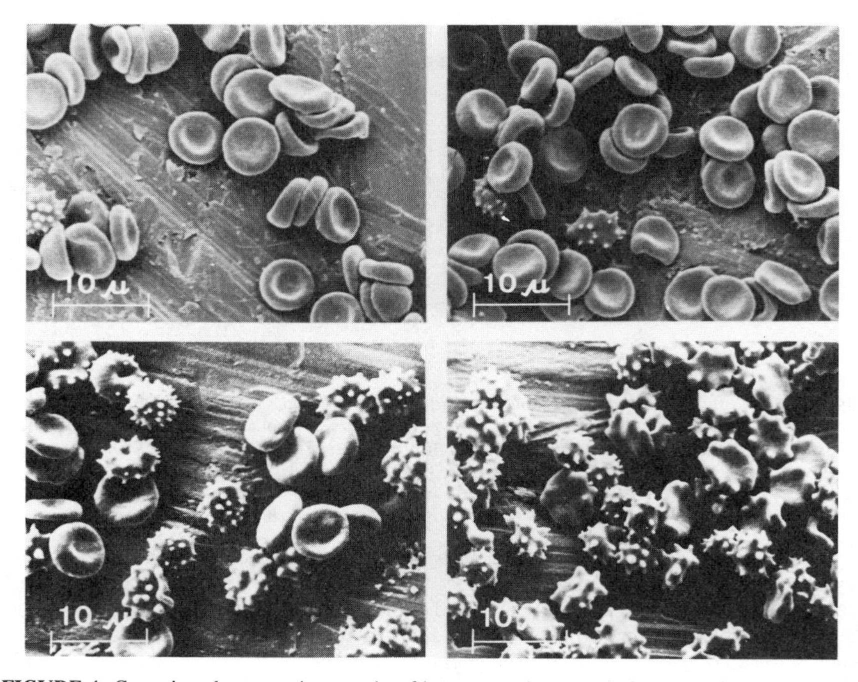

FIGURE 4. Scanning electron micrographs of human erythrocytes before and after PC replacement. *Bars:* 10μm. *Upper left:* control erythrocytes at the start of the experiments. *Upper right:* cells in which 50% of the native PC has been replaced by (1-palmitoyl,2-oleoyl)-PC during a 20-hr incubation. Similar pictures (not shown) were obtained by incubating control cells for a long period and by replacement incubations using, as donor PC, (1-palmitoyl,2-linoleoyl)-PC, egg-PC, or PC from rat liver microsomes. *Lower left:* erythrocytes in which ±30% of the PC had been replaced by (1,2-dipalmitoyl)-PC in a 15-hr incubation. *Lower right:* erythrocytes in which 40% of the PC had been replaced by (1,2-dipalmitoyl)-PC in a 20-hr incubation. Similar erythrocytes were obtained.

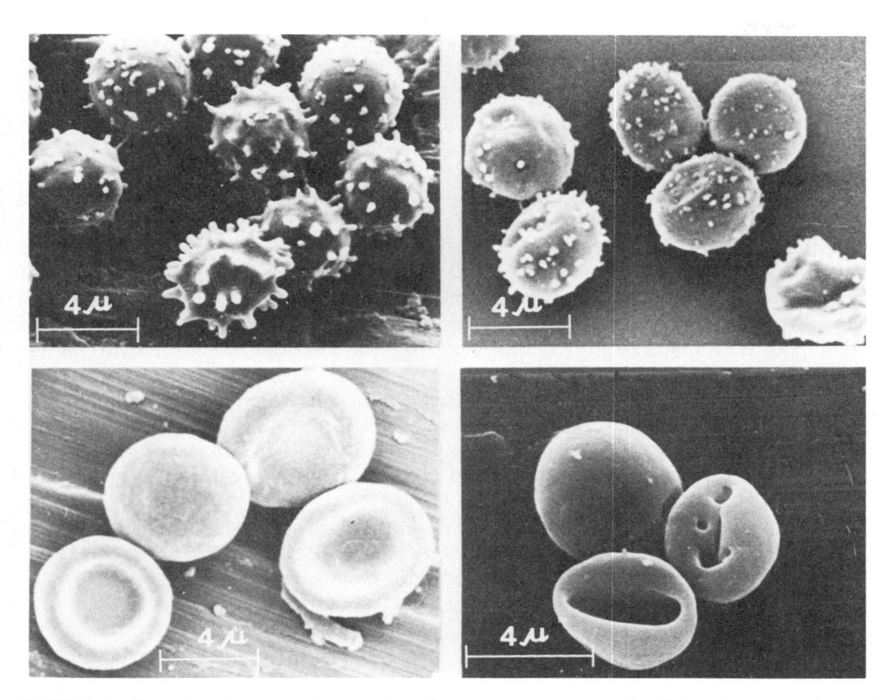

FIGURE 5. Scanning electron micrographs of human erythrocytes after PC replacement. *Bars:* 4 μm. *Upper left:* spheroechinocytes obtained by replacing 40% of the erythrocyte PC by (1,2-dimyristoyl)-PC in a 12-hr incubation. *Upper right:* erythrocytes after replacement of 20% of the native PC by (1-palmitoyl,2-arachidonoyl)-PC in a 15-hr incubation. *Lower left:* erythrocytes in which 48% of the native PC had been replaced by (1,2-dioleoyl)-PC in a 15-hr incubation. *Lower right:* erythrocytes after replacing 30% of the native PC by (1,2-dilinoleoyl)-PC in a 6-hr incubation.

observed after incorporation of (dilinoleoyl)-PC in the membrane. After replacement of 20–30% of the PC with this di-polyunsaturated species, the cells transformed into typical stomatocytes (FIG. 5); a further increase in the concentration of this species appeared to cause membrane disruption. Some tendency to adopt a stomatocytic morphology was observed after replacement of more than 40% of the PC with synthetic (dioleoyl)-PC. The changes in morphology appear to relate to changes in membrane properties (fragility and permeability) as induced by PC variation. The most dramatic alterations in erythrocyte morphology are observed when the unsaturation index becomes lower than 0.5 or higher than 1.0.

A possible explanation for the phenomenon that retailoring of PC species causes a change in cell shape is related to data on molecular cross sections of PC species presented some 20 years ago in these *Annals*.[1] Measurements on monomolecular films of a series of chemically synthesized PCs revealed that the molecular packing is governed to an appreciable extent by chain length and the number of unsaturated bonds of the fatty acid constituents of the PC species. Striking differences were found to exist in the molecular cross section of PC species with a different unsaturation index. The changes in erythrocyte shapes are proposed to result from differences in the geometry of the individual PC molecules, suggesting that replacement of one species by another may disturb the pre-existing packing of lipid molecules with the membrane.[7] This idea, presented in FIG. 6, indicates how substitution of the moderately cone-shaped native species (1-palmitoyl,2-oleoyl)- and (1-palmitoyl,2-linoleoyl)-PC by a more cylindrically shaped species such as a disaturated PC leads to the formation of echinocytes. By the introduction of more disaturated PC the area occupied by the apolar tails is decreased and a bending outwards of the outer monolayer will maintain the most effective packing of the lipids in the membrane. To appreciate the model fully, it has to be remembered that the modification is introduced primarily into the outer leaflet, and that such a unilateral alteration is not compensated for by a rearrangement of (phospho)lipid molecules between outer and inner layer. Substitution of the native mixed-acid PC species by (dilinoleoyl)-PC, which has a more pronounced cone shape because of a larger cross section at the end of the hydrophobic moiety, will lead to a bending of the outer leaflet of the membrane to the inward direction (FIG. 6). We suppose that the inner monolayer and membrane skeleton follow the behaviour of the outer layer and that in this way the cell adopts a stomatocytic morphology.[7] Presently, it is not possible to provide a simple explanation for the rather complex effects observed after replacement of native PC by (1-palmitoyl,2-arachidonoyl)-PC.

As argued above, the introduction of differently shaped lipids in an existing matrix leads to a bending of the outer membrane layer to allow a correct packing in this layer. This event forces the inner region to follow suit, resulting in a change of the overall contour of the cell envelope. Further support for this proposal was found by a

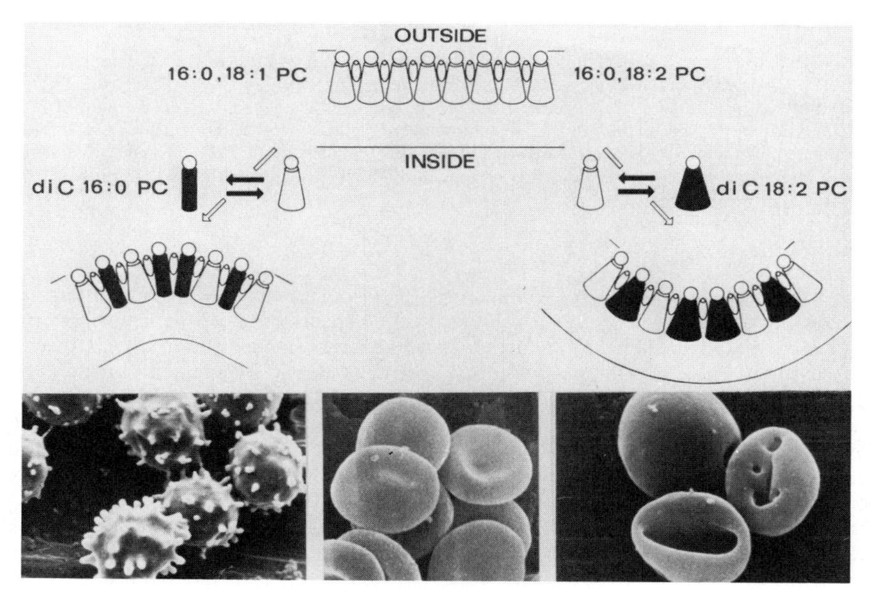

FIGURE 6. The effect of changes in lipid molecular shape on modification in the erythrocyte shape. At the *left* of the figure the replacement of native PC species with (dipalmitoyl)-PC is depicted and the resulting echinocytes are shown. Replacement by the (dilinoleoyl)-PC and the resulting stomatocytes are shown on the *right* side. The central electron micrograph shows control erythrocytes.

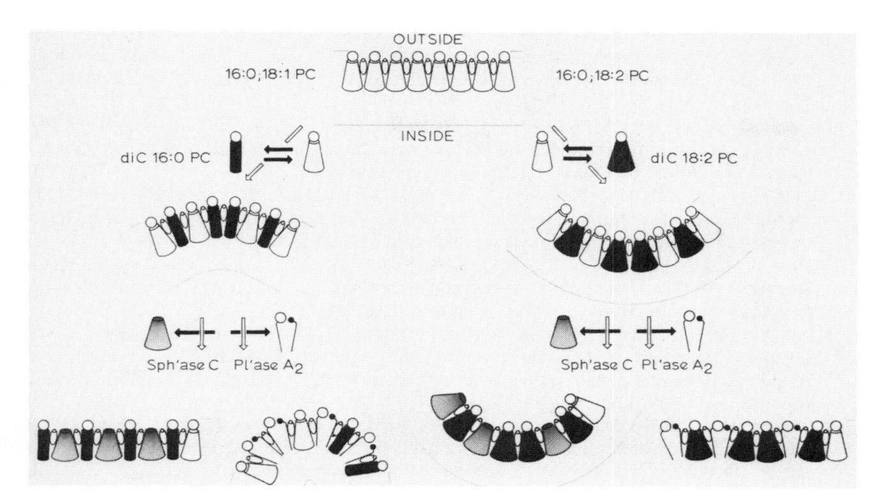

FIGURE 7. Model to explain shape changes of erythrocytes induced by changes in lipid molecular shape. PC exchange is depicted in the first step of the scheme: on the *left* side the replacement with (dipalmitoyl)-PC; on the *right* the replacement with (dioleoyl)-PC. Subsequent treatments with phospholipase A_2 (PL'ase A_2) and sphingomyelinase C (Sph'ase C) are depicted *below*. The *small figures* between the phospholipid molecules represent cholesterol. For sake of clarity they are depicted this way, although they actually occupy a cone shape.

manipulation of the PC species composition followed by treatment of the cells with phospholipases. This combination of lipid modifications should give predictable changes in erythrocyte morphology.[8] The results and their interpretation are schematically represented in FIGURE 7. The morphological changes induced by the two different types of lipid manipulations applied sequentially to the cells appeared to be additive, and indeed can be explained on the basis of the difference due to variation of molecular shape. Erythrocytes enriched with (dipalmitoyl)-PC in the outer layer tend to form echinocytes; a nonlytic hydrolysis of PC in the outer layer by phospholipase A_2 will produce free fatty acid and lyso-PC (inverted cone shape) and as a result further creation occurs. Sphingomyelinase C treatment produces ceramides (cone-shaped), and a (slow) restoration to normal morphology is attained because the geometric forms of ceramide and (dipalmitoyl)-PC molecules are complementary. On the other hand, a rapid further invagination of erythrocytes occurred in cells in which the sphingomyelinase treatment did increase the number of cone-shaped molecules present (FIG. 7).

These studies may demonstrate that the red cell continues to serve as an interesting model for biomembranes. A protein responsible for a specific transfer of PC between membranes provides a useful tool for a retailoring of the PC species composition in the membrane. This manipulation can induce striking alterations in membrane properties, and the results may help explain alterations observed in membrane phospholipids triggered by chronic ethanol consumption.

REFERENCES

1. DEENEN, L. L. M. VAN. 1966. Some structural and dynamic aspects of lipids in biological membranes. Ann. N.Y. Acad. Sci. **137:** 717–730.

2. WIRTZ, K. W. A., P. MOONEN, L. L. M. VAN DEENEN, R. RADHAKRISHNAN & H. G. KHORANA. 1980. Identification of the lipid binding site of the phosphatidylcholine exchange protein with a photo-sensitive nitrene and carbene precursor of phosphatidylcholine. Ann. N.Y. Acad. Sci. **348:** 244–255.
3. LANGE, L. G., G. VAN MEER, J. A. F. OP DEN KAMP & L. L. M. VAN DEENEN. 1980. Hemolysis of rat erythrocytes by replacement of the natural phosphatidylcholine by various phosphatidylcholines. Eur. J. Biochem. **110:** 115–121.
4. KUYPERS, F. A., B. ROELOFSEN, J. A. F. OP DEN KAMP & L. L. M. VAN DEENEN. 1984. The membrane of intact human erythrocytes tolerates only limited changes in the fatty acid composition of its phosphatidylcholine. Biochim. Biophys. Acta **769:** 337–347.
5. CHILD, P., J. J. MYHER, F. A. KUYPERS, J. A. F. OP DEN KAMP, A. KUKSIS & L. L. M. VAN DEENEN. 1985. Acyl selectivity in the transfer of molecular species of phosphatidylcholines from human erythrocytes. Biochim. Biophys. Acta **812:** 321–332.
6. KUYPERS, F. A., E. W. EASTON, R. VAN DEN HOVEN, F. WENSING, B. ROELOFSEN, J. A. F. OP DEN KAMP and L. L. M. VAN DEENEN. 1985. Survival of rabbit and horse erythrocytes *in vivo* after changing the fatty acyl composition of their phosphatidylcholine. Biochim. Biophys. Acta **819:** 170–178.
7. KUYPERS, F. A., B. ROELOFSEN, W. BERENDSEN, J. A. F. OP DEN KAMP & L. L. M. VAN DEENEN. 1984. Shape changes in human erythrocytes induced by replacement of the native phosphatidylcholine with species containing various fatty acids. J. Cell Biol. **99:** 2260-2267.
8. CHRISTIANSON, A., F. A. KUYPERS, B. ROELOFSEN, J. A. F. OP DEN KAMP & L. L. M. VAN DEENEN. 1985. Lipid molecular shape affects erythrocyte morphology: a study involving replacement of native phosphatidylcholine with different species followed by treatment of cells with sphingomyelinase C or phospholipase A_2. J. Cell Biol. **101:** 1455–1462.

DISCUSSION OF THE PAPER

QUESTION: In the sickled red cells, are there also compositional alterations in the lipids or altered molecular species, or is that not known?

VAN DEENEN: To the best of my knowledge (and I think several laboratories have investigated them) there are no differences in the lipid composition of a sickle cell compared to a normal cell. The only effect that has been observed is this increase in the flip-flop rate, which we believe is occurring in regions where no cytoskeleton is present. Because of that, one can conclude that components like phosphatidylserine can come towards the outer monolayer. This may contribute to the final stages of this disease, where you have chemical changes such as blockage of the capillaries because of the deformation of the erythrocyte shape and increased coagulation. This is one of the consequences of this altered phospholipid flip-flop.

K. W. MILLER (*Massachusetts General Hospital, Boston, MA*): Could you say something about exchanging just a few of the erythrocyte lipids with exchange proteins?

VAN DEENEN: We must be careful when we talk about percentages, because these graphs show, for instance, that 25% of the phosphatidylcholine of the outer layer has been replaced by a given species, which sounds like a high number. Phosphatidylcholine is only 1/5th of the total lipid, and we have only altered a few percent of this, which is small compared to the total lipid count.

J. M. LITTLETON (*King's College, London, England*): You've managed to go through your entire lecture without mentioning the one lipid that I'm going to talk about exclusively, and that's the inositol phospholipids. Could you mention what structural role they might have in the red cell membrane?

VAN DEENEN: We know that phosphatidylinositol (PI) is localized in the inner red cell monolayer. This is also the case in the platelets. A particular role has not been attributed (to the best of my knowledge) to PI in the erythrocyte membrane. Phosphatidylserine has been demonstrated to be active in the sodium potassium ATPase; PI may also be a component of one of the enzyme systems, but I could not really say much more.

R. A. HARRIS (*University of Colorado School of Medicine, Denver, CO*): Brain membranes contain a large amount of their PC as (dipalmitoyl)-PC (DPPC). In view of the effects of this lipid on red cells, I wonder if this is because of the small radius of curvature of axons and nerve endings in brain, or do you think there's another reason why there's so much DPPC in brain?

VAN DEENEN: There's even more DPPC in lung, where it is the most predominant species, and a component of lung surfactant. To come back to your question, I don't think that I have a good answer. I can only mention that erythrocytes from another species, the rat, also have a large percentage of (dipalmitoyl)-PC. Actually, this paper illustrates again how poor our knowledge still is, as was also emphasized by Dr. Cullis in his paper about certain aspects of rat brain lipids.

D. B. GOLDSTEIN (*Stanford University School of Medicine, Stanford, CA*): Is anything known about the distribution of different phospholipids along the plane of the membrane that would explain the biconcave disk shape of the red cell? Or is that entirely a property of the cytoskeleton?

VAN DEENEN: Yes, there is a possibility that there are domains in time depending on the time scale you are looking at. But no relation between lateral distribution of lipids and red cell shape is known.

Dihydropyridine-Sensitive Ca^{2+} Channels and Inositol Phospholipid Metabolism in Ethanol Physical Dependence[a]

M. J. HUDSPITH, C. H. BRENNAN, S. CHARLES, AND J. M. LITTLETON

Department of Pharmacology
King's College
Strand
London WC2R 2LS, England

There has been much speculation about the role of Ca^{2+} in the acute and chronic effects of ethanol on the central nervous system.[1,2] Current evidence suggests that the presence of ethanol is inhibitory to the depolarisation-induced entry of Ca^{2+} into neurones through voltage-operated channels.[3,4] This effect seems of sufficient magnitude to account for the frequently reported reduction in depolarisation-induced release of neurotransmitters by ethanol *in vitro,*[5,6] and this in turn could play a role in the central depressant effects of the drug. In this case one would expect that the development of tolerance and physical dependence would be associated with a loss of the effect of ethanol *in vitro* on depolarisation-induced Ca^{2+} entry and with an increase in Ca^{2+} entry observed in the absence of ethanol.[7] While the first alteration has been observed,[3,4] there are no reports of an increase in depolarisation-induced Ca^{2+} entry in preparations of brain obtained from ethanol-dependent animals; indeed it has been reported that Ca^{2+} entry is reduced in preparations from such animals.[3,4] Despite this tendency to reduced Ca^{2+} entry, however, many studies have shown that depolarisation-induced neurotransmitter release from these preparations is either normal or enhanced.[6,8,9]

Such observations suggest strongly that some change in the ability of Ca^{2+} to induce transmitter release has occurred in central neurones during the induction of tolerance and dependence. On the basis of experiments on the ability of depolarisation and Ca^{2+} ionophores to release dopamine from striatal preparations[6,9] and on the activity of Ca^{2+}-dependent enzymes of phospholipid metabolism[10,11] we have suggested that about a fourfold increase in sensitivity to external Ca^{2+} concentrations is necessary to explain the altered characteristics of preparations from ethanol-dependent animals. If such a change occurs in the absence of any bulk increase in Ca^{2+} flux then some changes in the sensitivity of intracellular processes to Ca^{2+} must have occurred. Current knowledge of the inositol phospholipid "secondary messenger" system indicates that both arms of this system augment the intracellular effects of Ca^{2+}.[12] Thus the production of inositol phosphates can induce the release of Ca^{2+} from intracellular stores, whereas diacylglycerols potentiate the effect of Ca^{2+} on phospholipid-dependent protein kinase.[12] Although the role of intracellular Ca^{2+} stores in stimulus: release coupling in neurones is probably not great, it has been shown that

[a]Supported in part by the Wellcome Trust. M. J. Hudspith is a Medical Research Council scholar.

phorbol esters (which mimic the effect of diacylglycerols) can potentiate neurotransmitter release.[13] An increased turnover of neuronal inositol phospholipids is therefore a potential mechanism for the increased neuronal Ca^{2+} sensitivity associated with the development of tolerance and physical dependence on ethanol.

Previous work from this and other laboratories has shown an increased turnover of membrane phospholipids in brain from ethanol-tolerant and physically dependent rats. It is not surprising therefore that evidence for increased turnover of inositol phospholipids, including those that have been basal[11] depolarisation-induced[11] and receptor-stimulated,[14] has been obtained in similar preparations. Several problems of interpretation remain before it can be asserted that this increase in inositol lipid breakdown has any functional role either in the neuronal "Ca^{2+} sensitivity" discussed earlier or in the tolerance and physical dependence which it was suggested might result from this. The two major problems involve cause and effect relationships and can be summarised as: (1) Does the increased inositol lipid turnover simply reflect the increased Ca^{2+} sensitivity in the preparations from ethanol-dependent animals or does it cause this increased sensitivity? (2) Does the increased basal and depolarisation-induced inositol lipid turnover simply reflect the greater spontaneous and depolarisation-induced release of transmitters (which then stimulate inositol lipid breakdown via postsynaptic receptors) in the preparations from ethanol-dependent animals, or does it cause this increased release?

We have attempted to answer the first question by investigating the effect of the dihydropyridine Ca^{2+} channel activators and inhibitors on inositol phospholipid breakdown in preparations from control and ethanol-dependent animals. The stimulus for these experiments were the observations that the Ca^{2+} channel activator BAYK8644 can potentiate both neurotransmitter release[15] and depolarisation-induced inositol lipid breakdown,[16] whereas the dihydropyridine Ca^{2+} channel inhibitors are unable to alter neurotransmitter release[15] but do have an inhibitory effect on inositol lipid breakdown.[16] These results indicate that under normal circumstances Ca^{2+} entry through the subclass of channels sensitive to dihydropyridines influences inositol lipid breakdown but has little effect on neurotransmitter release; an above-normal entry through these channels, however, as caused by BAYK8644, stimulates both inositol lipid breakdown[16] *and* neurotransmitter release.[15] These changes are clearly similar to those induced by chronic ethanol administration, and a further link between dihydropyridine sensitive Ca^{2+} channels and ethanol dependence is provided by the recent observation that the dihydropyridine Ca^{2+} channel inhibitors can prevent the ethanol withdrawal syndrome at doses which have no discernible effects in control animals[17] and secondly that BAYK8644 at low concentrations acts as an ethanol antagonist.[18]

The question of whether inositol lipid breakdown on depolarisation occurs presynaptically (and modifies transmitter release) or whether it is a postsynaptic consequence of neurotransmitter release is a very difficult one. We have attempted to answer this question in brain slices by comparing the requirements for Ca^{2+} of the two processes. If inositol lipid breakdown is a consequence of neurotransmitter release then it should have exactly the same ionic requirements. In fact inositol lipid breakdown occurs at Ca^{2+} concentrations which support very little transmitter release.[11] We have also examined a peripheral tissue, the isolated rat vas deferens, where it is possible to follow the functional consequences of nerve stimulation or receptor occupation directly and where it is also possible to denervate the preparation using 6-hydroxydopamine *in vitro*. The effects of depolarisation on inositol phospholipid turnover in innervated and denervated vasa have been studied in tissues from control and ethanol-dependent animals. If any part of the depolarisation-induced inositol lipid turnover is a consequence of lipid breakdown in the nerve terminals then this should be reduced in denervated preparations.

These experiments are therefore designed to investigate the mechanisms for the increased inositol phospholipid turnover found in preparations from ethanol-dependent rats and to establish whether this has functional consequences both in brain and in peripheral nerves.

MATERIALS AND METHODS

Induction of Ethanol Tolerance and Dependence

Male Sprague Dawley rats (250–300 g) were exposed to ethanol vapour in the way previously described[9] or were assigned in equal numbers to a control group housed under similar conditions except that ethanol was absent from the inspired air. Treatment duration was 6 or 7 days during which all animals had free access to food (Spiller modified 41B) and drinking water. Animals inhaling ethanol vapour showed a mild to moderate degree of ataxia and sedation during the treatment period. On withdrawal from this we have previously observed a moderate to severe physical withdrawal syndrome consisting of locomotor excitation followed by relative immobility with coarse tremor, piloerection and some spontaneous tonic-clonic convulsions. After 6–7 days animals were killed while still intoxicated by stunning and decapitation. The head was taken for dissection of the brain and the abdomen was opened for removal of the vasa deferentia.

Inositol Phospholipid Turnover in Cortical Slices

Cortical slices were prepared as described previously,[11] and preincubated in a Krebs-Ringer bicarbonate medium containing 1.3 mM Ca^{++} gassed O_2:CO_2 (95:5) at 37° C for 60 minutes with two intermediate changes of buffer. Forty microlitre aliquots of gravity-settled slices were pipetted into plastic tubes containing 0.3 μM myo[2^3H]-inositol and 10 mM Li^+ in 240 μl buffer. The tubes were gassed, capped and incubated for a further 30 minutes in a shaking water bath at 37° C. For investigations into receptor-mediated inositol phospholipid metabolism, agonists were added to the labelled slices in a volume of 10 μl. In experiments involving dihydropyridines [^{3}H] inositol-labelled slices were exposed to BAYK8644 for 10 minutes prior to elevation of external K^+; where antagonists were used they were added either 10 minutes prior to the elevation of K^+ or 10 minutes prior to the addition of BAYK8644. Normal K^+ was elevated from 6 mM by addition of KCl without an equivalent reduction of Na^+; responses to K^+ obtained in this manner have been found to be identical to those where K^+ is substituted for Na^+. After 45 minutes 0.94 ml $CHCl_3$:MeOH (1:2v/v) followed by 0.31 ml each of $CHCl_3$ and water were added. 750 μl of the upper phase were diluted to 3 ml with water and the [^{3}H]-inositol phosphates extracted by the addition of 0.5 ml of a 50% (w/v) slurry of Dowex formate. After four washes with 3 ml of 5 mM inositol the inositol phosphates were eluted with 0.5 ml of 1 M NH_4 formate/0.1 M formic acid. 0.4 ml of this eluate were counted with 4.5 ml of Ecoscint in a Packard Tri-Carb scintillation counter.

Inositol Phospholipid Turnover in Vasa Deferentia

Vasa deferentia were dissected and freed from adherent blood vessels on ice; transverse slices of 1 mm thickness were prepared with a McIlwain tissue chopper and pre-incubated as described above. Slices were then transferred to a Krebs buffer containing 0.3 μM myo[2^3H]-inositol and 10 mM Li^+ substituted for Na^+. After 2 hours individual slices were removed, washed with unlabelled buffer and placed in plastic tubes containing 240 μl of unlabelled 10 mM Li^+ buffer: for K^+-depolarisation studies up to 40 mM, K^+ was substituted for Na^+ in this buffer. In experiments involving receptor-mediated stimuli, agonists were added in a volume of 10 μl. Antagonists when used were added 30 minutes prior to the addition of agonists to the labelled slices. After 45 minutes stimulation, inositol phosphates were extracted and counted as described above.

Contractile Responses of Vasa Deferentia

Vasa deferentia were dissected and divided into three sections of approximately equal lengths. Experiments on contractile responses were carried out either on the most epididymal or most prostatic section; the middle section was discarded. The sustained response of the epididymal section of the vas forms the contractile response on which the results shown here are based. Sections of vasa were set up in an organ bath of volume 30 ml containing oxygenated Krebs solution at 37° C. Transmural stimulation was performed with platinum wire electrodes positioned on each side of the preparation. Stimuli at 50 V with a pulse width of 0.5 m sec were delivered from a Grass 544 stimulator. Contractions of the longitudinal smooth muscle were measured using a Grass FT 10G transducer coupled to a Grass polygraph recorder. In "denervation" experiments sections of vasa were incubated overnight at 4° C in 6-hydroxydopamine (6-OHDA) at a concentration of 5×10^{-4} M. Controls for this procedure were incubated overnight at 4° C in Krebs solution with no 6-OHDA present.

Materials

Myo[2–^{3}H]-inositol (16.3 Ci/mmol) in water : ethanol (9:1 v/v) was purchased from Amersham International (UK). Dowex 1×8 (100–200 mesh) Cl^- form was obtained from BIO-RAD Laboratories (UK) Ltd.

Dihydropyridines (BAYK8644, nitrendipine and nimodipine) were gifts from BAYER (UK). Animals were purchased from Charles River (UK) and ethanol (AR grade) from James Burroughs Ltd (UK). All other materials were from BDH (Poole, UK) or from Sigma (UK).

RESULTS

Effects of Dihydropyridines on Cortical Inositol Phospholipid Breakdown

Increasing K^+ concentrations from 6 mM to 40 mM consistently produced an increase in accumulation of [^{3}H] inositol phosphates in cortical slices. The accumulation was consistently greater in preparations from ethanol-dependent animals (FIGURE 1). The effect of BAYK8644 on depolarisation-induced [^{3}H] inositol phosphate accumulation was highly concentration-dependent, 5×10^{-7} M producing a potentiation whereas 5×10^{-5} M produced inhibition. 5×10^{-6} M produced virtually no effect whereas 5×10^{-8} M produced a potentiation similar to that at 5×10^{-7} M BAYK8644 (data not shown). As can be seen in FIGURE 1 the effect of BAYK8644 at 5×10^{-7} M was considerably greater in preparations from ethanol-dependent animals.

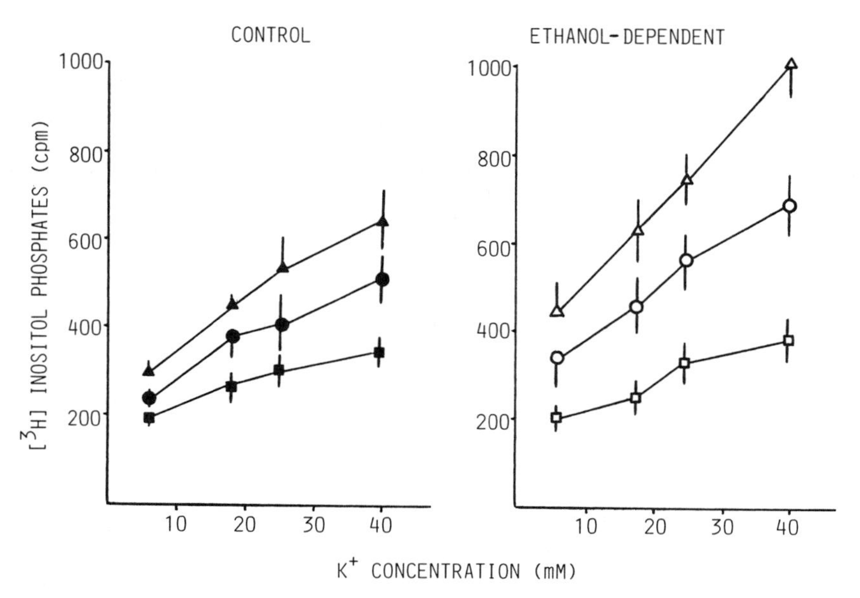

FIGURE 1. The effect of the Ca^{2+} channel activator BAYK8644 on depolarisation-induced accumulation of [^{3}H] inositol phosphates in cortical slices. Results show values from *control* animals (n = 5) on the left (filled symbols) and from *ethanol-dependent* animals (n = 5) on the right (open symbols). Error bars are S.E.M. *Circles* show the effect of increasing K^+ concentration (6 mM, 18 mM, 25 mM, 40 mM) on the accumulation of [^{3}H] inositol phosphates in the absence of BAYK8644. *Triangles* show the potentiating effect of BAYK8644, 5×10^{-7} M, on [^{3}H] inositol phosphate accumulation, whereas *squares* show the inhibitory effect of BAYK8644 at 5×10^{-5} M.

Since BAYK8644 had little effect on preparations from control animals the effects of dihydropyridine Ca^{2+} antagonists were investigated mainly in preparations from ethanol-dependent animals. In the limited numbers of experiments on control preparations performed similar trends were observed. The dihydropyridine Ca^{2+} channel inhibitor nitrendipine had little effect on the depolarisation-induced accumulation of [^{3}H] inositol phosphates at any concentration studied in preparations from ethanol-dependent rats (FIGURE 2). Similar results were obtained with nimodipine and nifedipine (data not shown). 10^{-5} M nitrendipine did however significantly inhibit the potentiation of depolarisation-induced [^{3}H] inositol phosphate accumulation produced by BAYK8644 (FIGURE 2).

Receptor-Mediated Inositol Phospholipid Breakdown

Both carbachol and norepinephrine produced a concentration-related increase in [^{3}H] inositol phosphate accumulation in cortical slices and in both cases the increase was greater in preparations from ethanol-dependent animals (FIGURE 3). It is difficult to assess the significance of these changes since basal accumulation of [^{3}H] inositol phosphates differs between the two groups (FIGURE 3). When epididymal sections (or indeed prostatic sections—data not shown) of vasa deferentia were incubated with norepinephrine there was also an accumulation of [^{3}H] inositol phosphates which was almost completely prevented by prazosin, 5×10^{-6} M. The accumulation of [^{3}H] inositol phosphates was consistently and significantly ($p < 0.02$) greater in preparations from ethanol-dependent animals (FIGURE 4).

Depolarisation-Induced Inositol Phospholipid Breakdown in Vasa

Depolarisation with 40 mM K^+ produced a marked, prazosin-insensitive, accumulation of [^{3}H] inositol phosphates in epididymal sections of vasa deferentia. This accumulation was significantly greater ($p < 0.02$) in preparations from ethanol-dependent animals (FIGURE 5). After overnight incubation at 4° C in 6-hydroxydopamine K^+-depolarisation still produced accumulation of [^{3}H] inositol phosphates, but this no longer differed in preparations from control and ethanol-dependent animals (FIGURE 5).

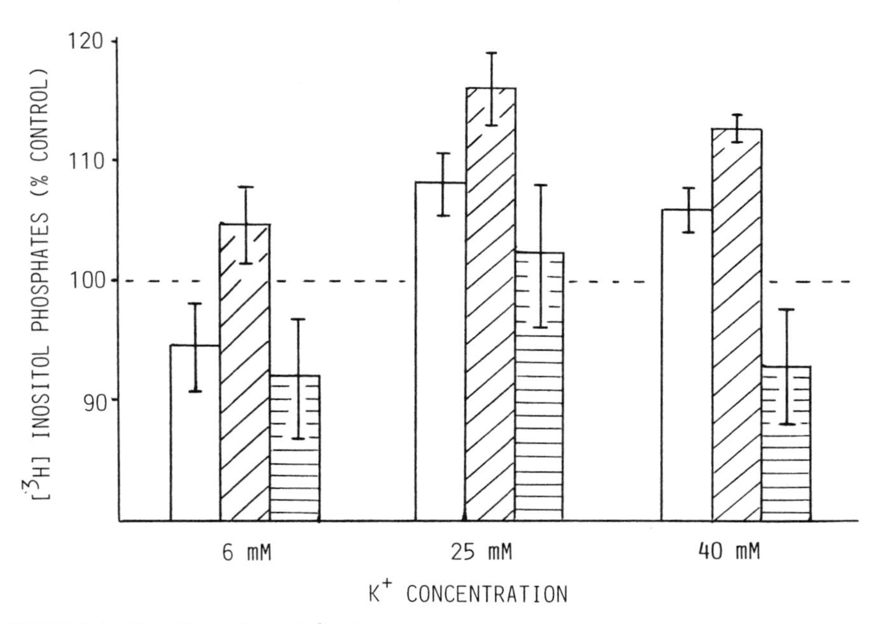

FIGURE 2. The effect of the Ca^{2+} channel inhibitor nitrendipine on the depolarisation- and BAYK8644-induced accumulation of [^{3}H] inositol phosphates in cortical slices. Results show values from ethanol-dependent animals ($n = 4$). Error bars are S.E.M. In order to reduce the effect of variation between animals the effects of BAYK8644 (5×10^{-7} M—*diagonally hatched columns*) and Nitrendipine (10^{-5} M—*open columns*) have been shown as a percentage of the control response, *i.e.* the [^{3}H] inositol phosphate accumulation in the absence of drugs, at each K^+ concentration. The only significant changes ($p < 0.05$) are the increases produced by BAYK8644 in K^+-depolarised slices. These changes were prevented by the presence of nitrendipine (10^{-5} M—*horizontally hatched columns*).

Contractile Responses of Vasa Deferentia

Norepinephrine produced a concentration-dependent, calcium-dependent, sustained contraction of epididymal sections of vasa deferentia. No contractions in the extreme prostatic sections of the longitudinal muscle of the vas were seen although these still produced a consistent inositol phospholipid response to norepinephrine. The norepinephrine-induced contractions of the epididymal sections were consistently and significantly ($p < 0.05$) greater in preparations from ethanol-dependent rats (FIGURE 6).

Electrical transmural stimulation of the whole or epididymal section of the vas deferens produced contractions with an initial phasic and sustained tonic portion. The sustained contraction was inhibited by prazosin, 5×10^{-7} M, the response to transmural stimulation was highly dependent on Ca^{2+} concentrations and was markedly greater ($p < 0.001$) in preparations from ethanol-dependent rats (FIGURE 7). After overnight incubation with 6-hydroxydopamine it was not possible to observe any response to electrical transmural stimulation at normal parameters. Such preparations

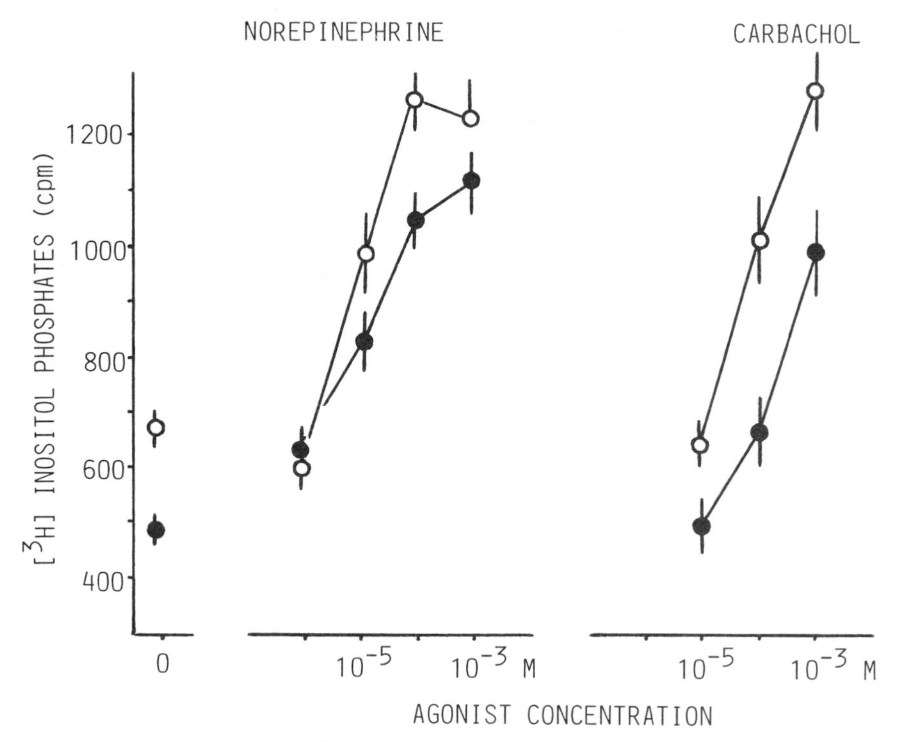

FIGURE 3. [^{3}H] inositol phosphate accumulation in rat cortical slices induced by norepinephrine and by carbachol. Values show the effect of *norepinephrine* (on left) and *carbachol* (on right) on the accumulation of [^{3}H] inositol phosphates in cortical slices from *control* rats (filled symbols) or *ethanol-dependent* rats (open symbols). Error bars show S.E.M.; $n = 5$. *Agonist concentrations* were 10^{-6} M to 10^{-3} M for norepinephrine and 10^{-5} M to 10^{-3} M for carbachol.

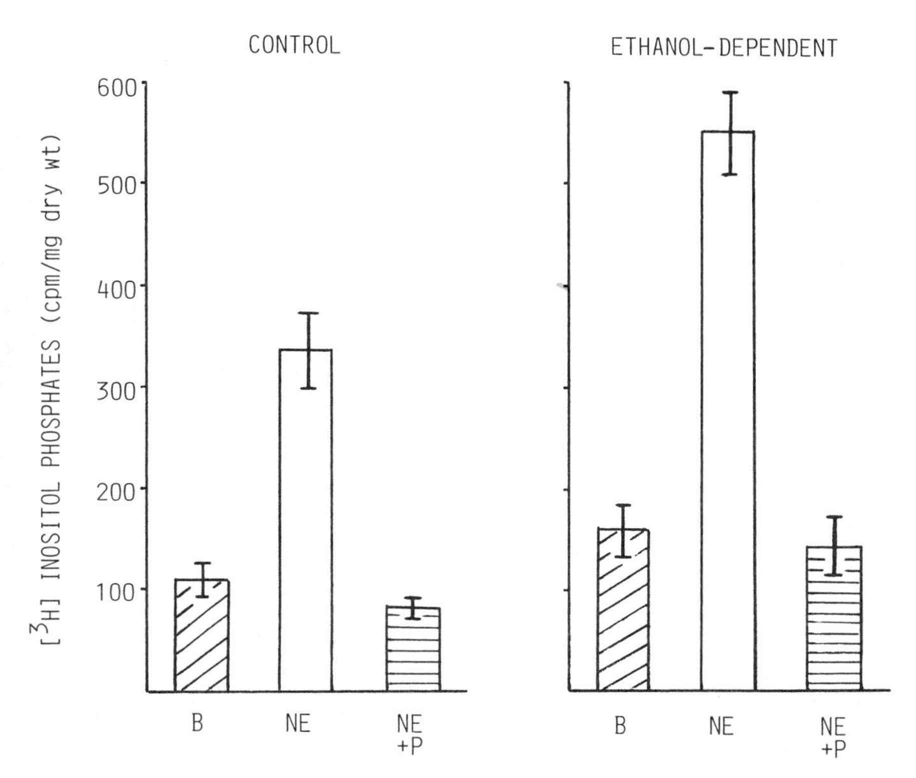

FIGURE 4. [^{3}H] inositol phosphate accumulation in rat vasa deferentia induced by norepinephrine and its inhibition by prazosin. Histograms show the basal (*diagonally hatched columns*) accumulation of [^{3}H] inositol phosphates in vasa deferentia from *control* animals (on left) and *ethanol-dependent* animals (on right). The effect of incubation with norepinephrine, 10^{-5} M, is shown in the *unshaded columns,* whereas the effect of adding prazosin, 10^{-5} M, together with norepinephrine is shown in the *horizontally hatched columns.* Error bars are S.E.M.; n = 28 determinations except in the prazosin experiments where n = 16 determinations. For further details see text.

would however respond to norepinephrine or to K^+ depolarisation with sustained contractions.

DISCUSSION

The experiments described were designed to investigate the relation between the enhanced neuronal Ca^{2+} sensitivity and inositol phospholipid metabolism which occurs in the central nervous system of rats made physically dependent on ethanol. In addition, using a peripheral tissue it was hoped to establish whether the increased inositol lipid breakdown occurred pre- and/or postsynaptically and whether it produced functional consequences in the tissue concerned. The results suggest that the changes described occur both pre- and postsynaptically, that both have important

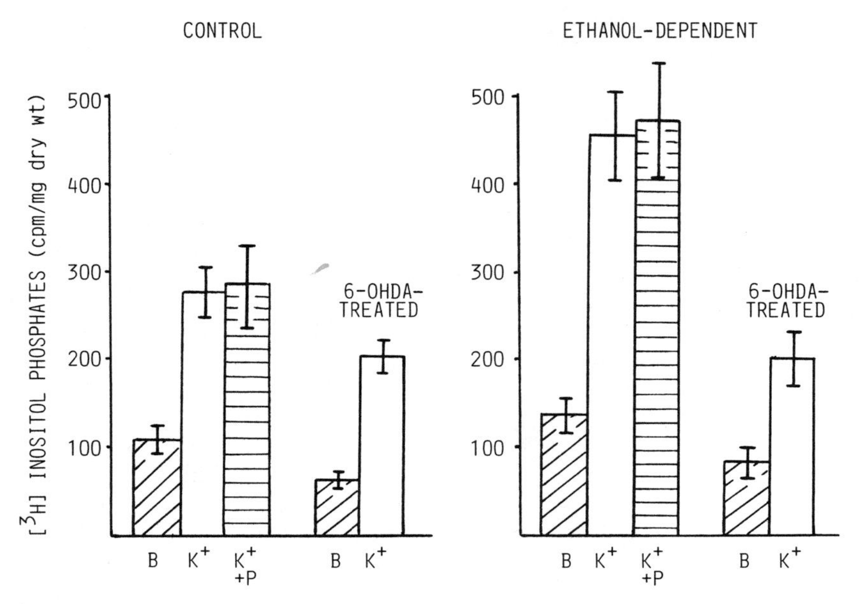

FIGURE 5. K^+-depolarisation-induced accumulation of [^{3}H] inositol phosphates in intact and 6-OHDA "denervated" vasa deferentia. Histograms show basal (*diagonally hatched columns*), 40 mM K^+-stimulated (*unshaded columns*) and 40 mM K^+ with 10^{-5} M prazosin (*horizontally hatched columns*) accumulation of [^{3}H] inositol phosphates in vasa from *control* animals (on left) and *ethanol-dependent* animals (on right). The first three columns show results obtained from fresh intact vasa; the last two columns show results obtained from vasa incubated overnight at 4° C with 6-hydroxydopamine, 5×10^{-4} M, before rewarming and investigation for [^{3}H] inositol phosphate accumulation. Error bars are S.E.M.; n = a minimum of 16 determinations. For further details see text.

consequences for neurotransmission and that they may result from alterations in dihydropyridine-sensitive Ca^{2+} channels. If this is so then the dihydropyridine "Ca^{2+} antagonists," which can reverse enhanced depolarisation-induced inositol lipid breakdown could be of great value in treating the consequences of chronic ethanol consumption.

In the experiments involving the effect of dihydropyridines on K^+-depolarisation-induced accumulation of [^{3}H] inositol phosphates in cortical slices our previous observation[11] of increased inositol lipid breakdown in preparations from ethanol-dependent animals was confirmed. The Ca^{2+} channel activator BAYK8644 potentiated this effect at low concentrations and was very much more effective in preparations from ethanol-dependent rats. The dihydropyridine Ca^{2+} channel inhibitor nitrendipine had little effect on depolarisation-induced inositol lipid breakdown but prevented the enhancement due to BAYK8644. These latter results show similarities to a previous report[16] in which BAYK8644 enhanced inositol lipid turnover and was inhibited by the dihydropyridine PN-200–110. In the previous experiments however the inhibitor was also effective against depolarisation-induced inositol lipid breakdown. It will be of interest to examine the effect of this drug on the enhanced depolarisation-induced inositol lipid turnover associated with ethanol dependence. The results obtained suggest strongly that the depolarisation-induced turnover of inositol lipids in cortical slices is coupled closely to Ca^{2+} entry through dihydropyridine-sensitive voltage-operated Ca^{2+} channels. The turnover may be enhanced in the preparations from ethanol-dependent animals because there is a functional increase in these channels. Increased [^{3}H] dihydropyridine binding to brain preparations has been reported after ethanol administration both to mice[19] and to rats.[20] This effect might modify neurotransmission either presynaptically (modulating neurotransmitter release) or postsynaptically (modulating receptor activation).

The possibility that changes in receptor-mediated inositol lipid breakdown are associated with ethanol dependence was investigated both in cortical slices and in the isolated vas deferens preparation. In cortical slices both norepinephrine and carbachol-

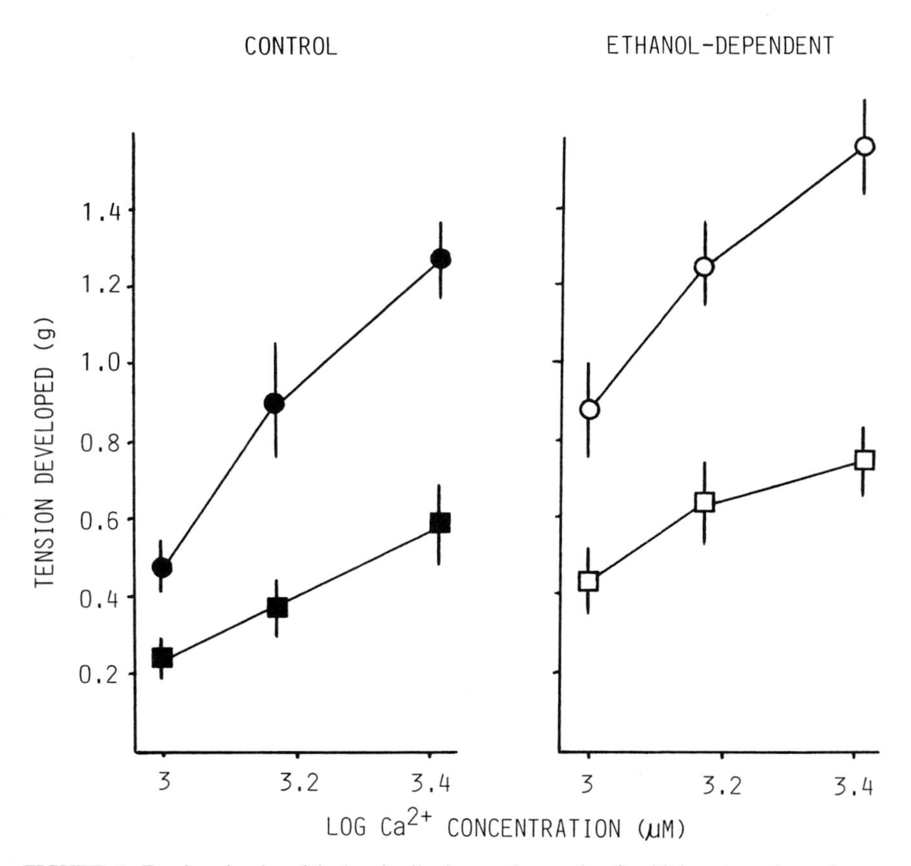

FIGURE 6. Tension developed in longitudinal smooth muscle of epididymal sections of vasa deferentia in response to norepinephrine. Values shown are the tension developed by the epididymal one third of vasa deferentia from *control* animals (filled symbols on left) and *ethanol-dependent* animals (open symbols on right) in response to norepinephrine at two concentrations. The effect of the higher concentration, 5.9×10^{-5} M, is shown by *circles*, that of the lower, 5.9×10^{-6} M, by *squares*. Experiments were performed at three Ca^{2+} concentrations, 1 mM, 1.5 mM and 2.6 mM; preparations were virtually unresponsive in the absence of added Ca^{2+}. Error bars are S.E.M.; each point is the mean of 8 experiments.

stimulated inositol lipid breakdown to a greater extent in preparations obtained from ethanol-dependent animals. In the rat vas deferens the epididymal end of the preparation responds to norepinephrine via α_1-adrenoceptors with a sustained contraction. This was shown to be sensitive to external Ca^{2+} and to be consistently greater in preparations from ethanol-dependent rats. When the effect of norepinephrine on [^{3}H] inositol phosphate production was studied in similar preparations it was found that

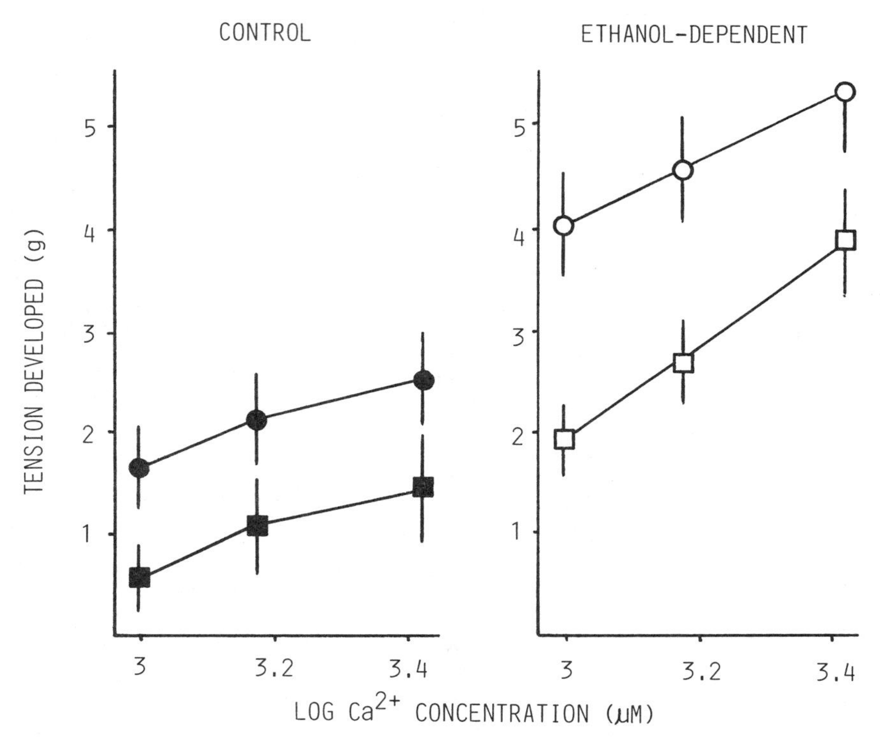

FIGURE 7. Tension developed in longitudinal smooth muscle of epididymal sections of vasa deferentia in response to electrical transmural stimulation. Values shown are the sustained tension developed by the epididymal one third of vasa deferentia from *control* animals (filled symbols on left) and *ethanol-dependent* animals (open symbols on right) in response to transmural electrical stimulation for 30 secs. Transmural stimulation was delivered from platinum electrodes at a pulse width of 0.5 sec at 50 V. Two frequencies were used; the effect of the higher, 50 Hz, is shown by *circles,* whereas that of the lower, 10 Hz, is shown by *squares.* Experiments were performed at three Ca^{2+} concentrations, 1 mM, 1.5 mM and 2.6 mM; preparations were unresponsive to transmural stimulation in the absence of added Ca^{2+}. Error bars are S.E.M.; each point is the mean of 8 experiments.

vasa from ethanol-dependent animals showed a significantly greater response than those from controls. This response was also sensitive to prazosin, a specific α_1-adrenoceptor antagonist.

These results suggest that the modification which ethanol induces in the Ca^{2+}/inositol lipid signalling system occurs in peripheral tissues as well as in the central nervous system and that it applies to postsynaptic receptor-modulated responses. Since these have recently been shown to be very sensitive to membrane depolarisation[21] the mechanism proposed previously (*i.e.* some alteration in voltage-operated Ca^{2+} channels) might also apply here. The increased tension in response to norepinephrine developed in isolated vasa from ethanol-dependent animals has been reported before[22] and argues that the enhanced inositol lipid response has functional consequences.

Most work on the inositol lipid system has related alterations in the turnover of these lipids to receptor activation. However, in the central nervous system it is clear that depolarisation and Ca^{2+} entry can cause enhanced breakdown of inositol lipids.[23] It is difficult to show whether this is a direct effect in nerve terminals or cell bodies or whether it is an indirect (*i.e.* receptor-mediated) consequence of neurotransmitters released by the Ca^{2+} entry. In our view it is probably all three. Kendall & Nahorski[24] have shown that it *can* be a consequence of transmitter release, but, under normal circumstances, the depolarisation-induced inositol lipid turnover cannot be blocked by a "cocktail" of antagonists. We have shown that the Ca^{2+} dependence of inositol lipid turnover is not identical to that of transmitter release. These findings suggest that at least a part of the enhanced inositol lipid turnover which accompanies K^+-depolarisation is a direct consequence of the depolarisation, but it is still not clear whether it occurs presynaptically or postsynaptically. Again we turned to experiments on the vas deferens to answer this point.

When the epididymal section of the rat vas deferens is stimulated electrically and transmurally the sustained contractions represent the effect of norepinephrine released from the endogenous sympathetic neurones. This response is sensitive to external Ca^{2+} and is inhibited by prazosin (*i.e.* it is an α_1-adrenoceptor response). The electrically-stimulated contractions produced in vasa from ethanol-dependent rats were much greater than those from controls. Since the difference is much greater than that to exogenous norepinephrine this suggests that increased release of norepinephrine from the sympathetic terminals occurs on transmural stimulation in the vasa from ethanol-dependent animals.

We have attempted to follow the accumulation of [^{3}H] inositol phosphates after transmural electrical stimulation of the vas deferens. There is measurable accumulation but this is variable and does not differ greatly from basal levels. We are in the process of refining the technique. Here we have demonstrated that K^+-depolarisation of the intact vas produces a large and consistent breakdown of inositol lipids and that this is greater in the preparations from ethanol-dependent animals. It has previously been reported[22] that K^+-depolarisation-induced contractures of the vas are greater in ethanol-treated animals. Prazosin, in concentrations shown to be effective against the norepinephrine-induced breakdown of inositol lipids had no effect on the depolarisation-induced breakdown, suggesting that very little of this could be a consequence of released norepinephrine acting on postsynaptic α_1-adrenoceptors. It is presumably therefore a direct consequence of depolarisation of nerve terminals or smooth muscle cells.

Incubation of vasa overnight with 6-hydroxydopamine effectively denervated the tissue, since this no longer produced any contractile response to transmural stimulation. K^+-depolarisation however still produced a contractile response and also caused inositol phospholipid breakdown. The accumulation of labelled inositol phosphates was now similar in vasa from control and ethanol-dependent animals. This argues strongly that some breakdown of inositol phospholipids occurs in nerve terminals as a result of K^+-depolarisation and that this is considerably greater in the preparations from ethanol-dependent animals. Some inositol lipid breakdown occurs in K^+-depolarised smooth muscle cells, but this appears not to differ in the two types of preparation. As in the central nervous system it would be very interesting to investigate the effect of depolarisation on receptor-stimulated accumulation of [^{3}H] inositol phosphates[21] in the vasa in relation to the contractile response produced. In the case of presynaptic inositol lipid turnover we suggest that this may play a role of the greater response of ethanol-dependent vasa to transmural stimulation by increasing neurotransmitter release from sympathetic nerve terminals.

By this combination of experiments on central and peripheral tissues we believe we are now in a position to suggest that increased Ca^{2+} sensitivity is a widespread cellular phenomenon in ethanol dependence. Our evidence shows that it occurs both presynaptically, where depolarisation-induced inositol lipid breakdown may be a consequence of Ca^{2+} flux through dihydropyridine-sensitive channels, and postsynaptically, where inositol lipid breakdown is mediated by receptor occupation. The mechanism may in fact be similar, since depolarisation has been reported to modulate receptor-stimulated inositol lipid breakdown.[21]

There seems little doubt that these large and consistent increases in inositol lipid responses in ethanol-dependent animals have marked functional effects both pre- and postsynaptically. The findings have important implications for therapy, since they suggest that drugs which interfere with the inositol lipid signalling system and/or the dihydropyridine-sensitive Ca^{2+} channels, such as lithium salts and the dihydropyridine "Ca^{2+}-antagonists," might reverse or prevent many of the chronic effects of alcohol intake including physical dependence. While the effect of lithium salts in treating alcoholism is controversial[25] this present information may suggest a more logical regime. Dihydropyridines have not yet been tested in alcoholics, but in animal studies they have recently been shown to prevent the ethanol physical withdrawal syndrome at doses which have virtually no effect on control animals.[17,18] The therapeutic potential of centrally acting Ca^{2+} antagonists in alcoholism is very exciting indeed.

REFERENCES

1. MAYER, J. M., J. M. KHANNA & H. KALANT. 1980. Eur. J. Pharmacol. **68:** 223–227.
2. ROSS, D. H., M. A. MEDINA & H. L. CARDENAS. 1974. Science **186:** 63–65.
3. HARRIS, R. A. & W. F. WOOD. 1980. J. Pharmacol. Exp. Ther. **213:** 562–568.
4. LESLIE, S. W., E. BARR, C. JUDSON & S. P. FARRAH. 1983. J. Pharmacol. Exp. Ther. **225:** 571–575.
5. CARMICHAEL, F. J. & Y. ISRAEL. 1975. J. Pharmacol. Exp. Ther. **193:** 824–834.
6. LYNCH, M. A. & J. M. LITTLETON. 1983. Nature **303:** 175–176.
7. LITTLETON, J. M. 1984. *In* Pharmacological Treatments for Alcoholism. G. Edwards & J. Littleton, Eds. 119–144. Croom Helm. London.
8. CLARK, J. W., H. KALANT & F. J. CARMICHAEL. 1977. Can. J. Physiol. Pharmacol. **55:** 758–768.
9. LYNCH, M. A., D. SAMUEL & J. M. LITTLETON. 1985. Neuropharmacology **24:** 479–485.
10. JOHN, G. R., J. M. LITTLETON & P. T. NHAMBURO. 1985. J. Neurochem. **44:** 1235–1241.
11. HUDSPITH, M. J., G. R. JOHN, P. T. NHAMBURO & J. M. LITTLETON. 1985. Alcohol **2:** 133–138.
12. BERRIDGE, M. J. & R. F. IRVINE. 1984. Nature **312:** 315–321.
13. TANAKA, C., H. FUJIWARA & Y. FUJII. 1986. FEBS Lett. **195:** 129–134.
14. TABAKOFF, B., T. SAITO, J. M. LEE & P. L. HOFFMAN. 1985. Abstract 59. Res. Society on Alcoholism. South Carolina.
15. MIDDLEMISS, D. N. & M. SPEDDING. 1984. Nature **303:** 175–176.
16. KENDALL, D. A. & S. R. NAHORSKI. 1985. Br. J. Pharmacol. **85:** 526P.
17. DOLIN, S. J., M. J. HALSEY & H. J. LITTLE. 1986. Br. J. Pharmacol. In press.
18. DOLIN, S. J., M. J. HALSEY & H. J. LITTLE. 1987. This volume.
19. ROSS, D. H. & N. MONIS. 1984. Alcohol Clin. Exp. Res. **8:** 115.
20. LUCHI, L., S. GOVONI, F. BATTAINI, G. PASINETTI & M. TRABUCHI. 1985. Brain Res. **332:** 376–379.

21. FOWLER, C. J., A.-M. O'CARROL, J. A. COURT & J. M. CANDY. 1986. J. Pharm. Pharmacol. **38:** 201–208.
22. DE MOREAS, S. & F. R. CAPAZ. 1984. J. Pharm. Pharmacol. **36:** 70–72.
23. GRIFFIN, H. D. & J. N. HAWTHORNE. 1978. Biochem. J. **176:** 541–552.
24. BATTY, I., D. A. KENDALL & S. R. NAHORSKI. 1985. Br. J. Pharmacol. **84:** 108P.
25. JAFFE, J. H. 1984. *In* Pharmacological Treatments for Alcoholism. G. Edwards & J. Littleton, Eds. 463–490. Croom Helm. London.

DISCUSSION OF THE PAPER

A. SCARPA (*Case Western Reserve University, Cleveland, OH*): Suppose an animal is treated with ethanol but the diet is so that the total pool of specific phospholipid is altered. In fact, if the base line is higher or the isotopic equilibration of what is put inside is altered, then some of your data can be explained.

LITTLETON: Yes, I agree. We've looked to the concentrations of endogenous phosphotidylinositol bisphosphate, for example, and we don't find any great differences between control and ethanol-treated animals. I can't say whether that is the correct pool to look at, and of course the concentrations are very small anyway. So a difference between the two kinds of preparations would have to be rather great to reach significance. I agree entirely that I can't rule that out as a possible explanation for some of our results.

I. DIAMOND (*Ernest Gallo Clinic and Research Center, San Francisco, CA*): It's not often that people working in different systems have a chance to agree completely with each other, but we agree completely and can confirm what you've been finding in your studies. We have taken advantage of PC12 cells, which are chromaffin-like cells, and have found, at least in the cell culture system we use, that we had a very striking adaptive increase in the number of calcium channels and voltage-dependent calcium channels when measured either by calcium flux or by direct binding. The kinetics indicate that this 85% increase in calcium channels is not a change in affinity but actually in the number of calcium channels. This recovers completely during 16 hours of withdrawal to relatively normal levels. So at least in that cellular system it has been possible to confirm what you are showing, that is, that there's a dramatic change in the number of channels whether measured by the binding assay or calcium flux.

LITTLETON: The chromaffin cells are much easier to work with in many respects, because the majority of calcium channels on the surface seem to be sensitive to the dihydropyridines, whereas in neurones it's only a very small proportion of the calcium channels which are sensitive. I hope we can get the same results as you on our cell cultures.

D. MCCARTHY (*University of New Mexico School of Medicine, Albuquerque, NM*): Have you had any chance to look at the system with opiates and are they similarly effective on the inositol pathway? In particular, do you know if loperamide, which is also a calcium channel blocker, antagonizes the effects you've seen?

LITTLETON: I don't know anything about that drug. However, similar kinds of effects have been reported in relationship to opiate dependence in terms of changes in these dihydropyridine-sensitive calcium channels. Nevertheless, I don't think the changes that have been reported for opiates are as dramatic as those for ethanol dependence.

W. S. THAYER (*Hahnemann University School of Medicine, Philadelphia, PA*): Could you comment on what's known about the molecular nature of these calcium channels? In particular, do they involve proteins, and is there anything known about whether lipids modulate their activity, etc.?

LITTLETON: I can't answer that, I'm afraid. It's certainly thought that proteins are involved. I don't think anybody knows anything about the lipid modulation of this kind of calcium channel at the moment or indeed the lipid modulation of any kind of calcium channel. I know Dr. Harris has done experiments on calcium flux looking at different chain length alcohols on that. His results suggest that the channels can be modulated by the presence of alcohols in the lipid bilayer, but apart from that I don't know of any evidence of lipid modulation.

A. P. THOMAS (*Hahnemann University School of Medicine, Philadelphia, PA*): Could you tell us which inositol phosphates you are measuring?

LITTLETON: We're looking at the accumulation of inositol phosphates in the presence of lithium. We know that the inositol phosphate which accumulates most is inositol monophosphate, and we're at the moment trying to find out whether chronic ethanol treatment causes any change in the distribution of inositol phosphates between the trisphosphates, bisphosphate and the monophosphate. It looks as though it doesn't, so that the distribution is the same as it would be in preparations from control tissues.

THOMAS: That's especially important, though, in view of these two isomers of inositol trisphosphate, one of which in our hands is potentiated by lithium, the inactive isomer.

You suggested that acute ethanol treatment causes a minor inhibition of inositol lipid metabolism. Dr. Hoek in this volume discusses some measurements in the liver where there is clear evidence for a very rapid, acute change of inositol 1,4,5-trisphosphate in 10 to 20 seconds after acute ethanol addition. Have you looked at those sorts of time points?

LITTLETON: No, we haven't. One has to appreciate that the changes in brain are really rather small compared to a nice system like hepatocytes. So we haven't looked at that sort of time course. If we look at the effects of ethanol in brain we can sometimes see quite marked biphasic effects. At low concentrations of ethanol we get an enhanced inositol phosphate production, whereas at high concentrations we see an inhibition.

Membrane Structural Alterations Caused by Chronic Ethanol Consumption: The Molecular Basis of Membrane Tolerance[a]

THEODORE F. TARASCHI, JOHN S. ELLINGSON,
AND EMANUEL RUBIN

Department of Pathology
Thomas Jefferson University School of Medicine
Philadelphia, Pennsylvania 19107

INTRODUCTION

The effects of chronic ethanol consumption on the central nervous system, the liver, and a variety of other organs may be related in part to its interaction with biological membranes (reviewed in REF. 1). The acute presence of ethanol generally exerts a disordering (fluidizing) effect on the lipid bilayer of biological membranes. By contrast, chronic consumption of ethanol leads to altered membranes, which are resistant to this disordering (membrane tolerance).[2-4] In our animal model, rats ingest 14–16 g of ethanol/kg body weight for 35 days in a nutritionally adequate diet, while pair-fed controls consume the same diet, except that carbohydrate isocalorically replaces ethanol. Our criteria for a good animal model to investigate the membrane alterations caused by chronic ethanol consumption are: 1) the mode of ethanol administration is similar to that in human alcoholism, 2) the structural and chemical alterations are produced by physiologically relevant amounts of ethanol, and 3) the effects of acute or chronic ethanol treatment, measured by electron spin resonance (ESR) are large, easily discernible (*i.e.* "all or nothing"), and reproducible. The animal model we use meets all of these criteria. In this investigation we examined the structural properties of red blood cells and liver microsomes obtained from rats that had been chronically administered ethanol for 35 days and then withdrawn from ethanol for 1–10 days. In addition, using electron spin resonance, high pressure liquid chromatography (HPLC), and membrane reconstitution techniques, we identified the membrane component(s) responsible for membrane tolerance in the microsomal membrane for the first time.

RESULTS

Response to Withdrawal from Ethanol

Microsomes

The effect of ethanol *in vitro* on the membrane molecular order parameter was assessed by ESR for liver microsomes isolated from untreated rats, rats chronically fed

[a]Supported by United States Public Health Service Grants AA03442, AA05662, and AA00088. TFT is the recipient of a Research Scientist Development Award from the National Institute on Alcohol Abuse and Alcoholism.

ethanol, and alcoholic rats withdrawn from ethanol (FIG. 1). The molecular order parameter S provides a measure of membrane structural order, since the motionally averaged spectral hyperfine splittings are influenced by the angular amplitude and rate of motion experienced by the lipid hydrocarbon chains bearing the spin label group. The results were obtained from microsomes labeled with the phospholipid probe 1-palmitoyl-2-{12-[β-(4',4'-dimethyloxazolidinyl-N-oxyl)]stearoyl}phosphatidylcholine (PtdCho 12) (FIG. 1). Similar results were obtained with microsomes labeled with 12-[β-(4',4'-dimethyloxazolidinyl-N-oxyl)]stearic acid (Ste 12). In agreement with an earlier report from this laboratory,[4] the addition of various amounts of ethanol *in vitro*

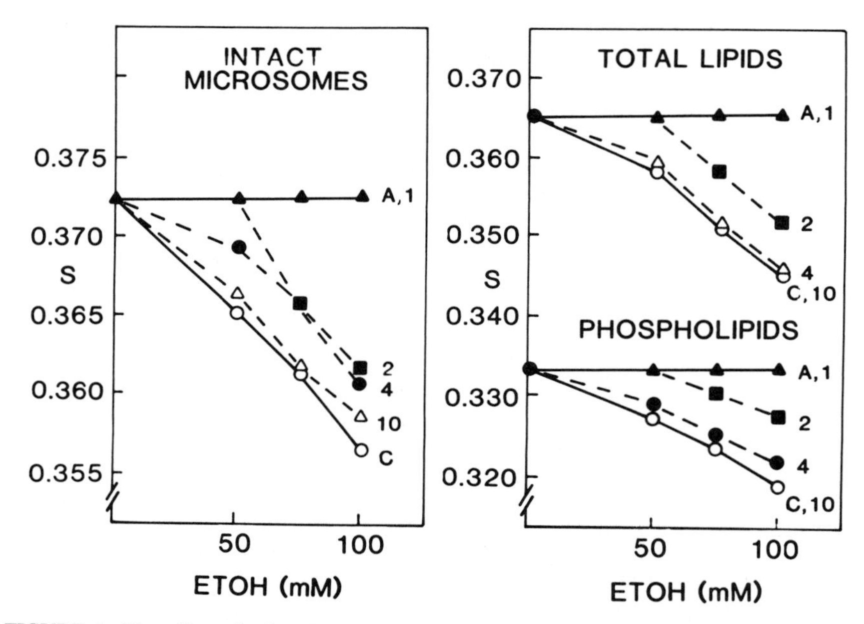

FIGURE 1. The effect of ethanol *in vitro* on the molecular order parameter S obtained for PtdCho 12-labeled intact rat liver microsomes and multilayer vesicles prepared from extracted microsomal total lipids and phospholipids. Microsomes were obtained from untreated rats (curve C, ○), alcoholic rats (curve A, ▲), and alcoholic rats withdrawn from ethanol feeding for 1 day (curve 1, ▲), 2 days (curve 2, ■), 4 days (curve 4, ●), and 10 days (curve 10, △). Each point represents the mean obtained from four pairs of animals. Errors in each point were never greater than 0.75%.

(50, 75, and 100 mM) caused a decrease in the order parameter from 0.372 to 0.356. Such behavior is indicative of significant membrane molecular disordering. Molecular disordering, characterized by an increase in the spectral splitting $2T_{\perp}$; reflects the greater angular displacement of the lipid fatty acid chains bearing the spin label group from the long molecular axis of the lipid in the membrane bilayer. Conversely, the order parameters determined for microsomes obtained from chronically alcoholic rats were unchanged over the same range of ethanol concentrations. This resistance to disordering by ethanol has been reported for alcoholic microsomes labeled with fatty acid or phospholipid spin probes.[4] No differences in baseline-order parameters (no ethanol added) were observed for microsomes obtained from control or ethanol-treated animals. Using the alcoholic liquid diet, we found that a period of 28–35 days was required for all of the animals treated chronically with ethanol to display membrane tolerance.

Following 1 day of ethanol withdrawal, the microsomes remained totally resistant to disordering by ethanol (up to 100 mM) *in vitro* (FIG. 1). After only 2 days of ethanol withdrawal, the microsomes were disordered by the addition of 75 mM ethanol (FIG. 1), as evidenced by a decrease in the order parameter. The extent of membrane disordering after 2 days of withdrawal was not as large (Δ S = 2.9%) as in the control membranes (Δ S = 4.6%) between 0 and 100 mM ethanol. Microsomes obtained from alcoholic animals that had been withdrawn from ethanol for 4 days had order-parameter profiles more similar to the controls. By 10 days of withdrawal, the order-parameter profiles were identical to those of the controls.

To locate the origin of the rapid loss of membrane tolerance, total lipid extracts were prepared from the microsomes used in the experiments described above. The results are summarized in FIGURE 1. Multilayer vesicles prepared from the control lipids were disordered by the addition of increasing amounts of ethanol (0 mM, S = 0.365; 75 mM, S = 0.358). Following 4 days of withdrawal, the order-parameter profile could be superimposed on that of the control lipids (FIG. 1). Between 0 and 100 mM ethanol, the bilayers from microsomal lipids obtained from rats withdrawn for 4 days experienced the same precent decrease in the order parameters as did bilayers from microsomes of control rats. This recovery was more rapid than in intact microsomes, which required a longer period of withdrawal (4–10 days) to display the full loss of membrane tolerance (FIG. 1).

To determine whether membrane resistance to disordering by ethanol is a characteristic of the phospholipids only, bilayers prepared from microsomal phospholipids were examined for membrane tolerance. Bilayers prepared from phospholipids of control microsomes were considerably disordered by ethanol *in vitro*. The baseline-order parameters obtained from the bilayers comprised solely of the phospholipids (S = 0.333) were lower than those from the intact microsomes (S = 0.372) and from the total lipid extracts (S = 0.365). Bilayers prepared from alcoholic phospholipids were resistant to disordering by ethanol. Furthermore, as was observed in the intact membranes and the total lipid extracts, resistance to disordering by ethanol was lost in the phospholipid bilayers from animals withdrawn for 2 days. In a fashion similar to that of the total lipid extracts, 4 days of ethanol withdrawal were required for the order-parameter profiles of those membranes to display disordering to the same extent as those from control phospholipids (FIG. 1).

There were no differences in the cholesterol-phospholipid ratios between control and ethanol-treated animals. The cholesterol-phospholipid ratio (0.10) obtained for control microsomes was in good agreement with reported values.[5,6]

Erythrocytes

The order-parameter profiles for Ste 12-labeled ghosts obtained from control, alcoholic, or ethanol-withdrawn animals are presented in FIGURE 2. Similar results were obtained with erythrocytes labeled with PtdCho 12. The erythrocytes were obtained from the same animals used in the studies of microsomes discussed above. Similar to microsomal membranes, control erythrocytes were considerably disordered by the addition of ethanol *in vitro* (0 mM, S = 0.565); 100 mM, S = 0.543), whereas erythrocytes from alcoholic rats were resistant to membrane disordering over the same range (0–100 mM) (FIG. 2). Interestingly, erythrocyte membranes from alcoholic rats displayed some loss of membrane tolerance after only 1 day of withdrawal. After 4 days the order-parameter profile of the ghosts could be superimposed on that of control membranes. This loss of tolerance was more rapid in lipid bilayers prepared from total erythrocyte lipids and from phospholipids only, beginning as early as day 1 (FIG. 2). It was observed that the baseline-order parameter in the erythrocyte phospholipid multilayers (S = 0.337) was considerably lower than that in bilayers composed of total lipids (S = 0.490) or in ghosts (S = 0.565).

The cholesterol-phospholipid ratio (0.68, mol/mol) was unchanged in the erythrocytes from alcoholic rats after withdrawal, a finding that indicates that alterations in the erythrocyte phospholipids, and not cholesterol content, are responsible for convey-

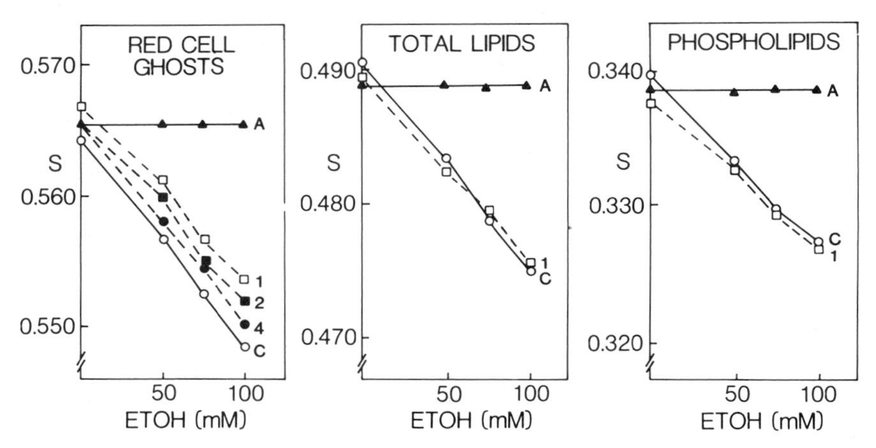

FIGURE 2. The effect of ethanol *in vitro* on the molecular order parameter S calculated for Ste 12-labeled ghosts and multilayer vesicles prepared from the erythrocyte total lipid and phospholipid extracts. The erythrocytes were obtained from the same animals used in the microsome studies presented in FIGURE 1. Resistance to disordering by ethanol was examined in ghosts isolated from untreated rats, (curve C, ○), alcoholic rats (curve A, ▲), and alcoholic rats withdrawn from ethanol feeding for 1 day (curve 1, □), 2 days (curve 2, ■), and 4 days (curve 4, ●). Points represent the mean from four pairs of animals. Errors in each point were never greater than 0.75%. Differences in baseline-order parameters (no ethanol added) are not statiscally significant.

ing membrane tolerance. The cholesterol-phospholipid ratio we obtained for the erythrocyte was in agreement with other reports.[7]

The Origin of Membrane Tolerance in Rat Liver Microsomes

Membrane tolerance in intact membranes is also observed in multilamellar vesicles composed of phospholipids extracted from membranes of ethanol-treated animals (synaptosomes,[8] hepatic mitochondria,[3] and microsomes and erythrocytes[9]). Thus, the property of membrane tolerance resides in the phospholipid portion of the membrane. Despite numerous studies of the effects of chronic ethanol consumption on the lipid composition of membranes (reviewed in REF. 1), no large changes have been reported, and no consistent pattern has emerged.

Since resistance to disordering by ethanol is a membrane property measured by physical techniques, *e.g.*, electron spin resonance (ESR) and fluorescence polarization, rather than chemical ones, it is unlikely that compositional studies alone will reveal the molecular origin of this response to chronic ethanol consumption. We, therefore, embarked upon a study to assess by ESR the ability of individual liver microsomal phospholipids from ethanol-fed rats to promote membrane tolerance in membrane vesicles composed of recombined individual phospholipids of the microsomal membrane. Microsomal membrane phospholipids were separated into classes by preparative HPLC. The membrane composition obtained with this procedure was 22 mole% phosphatidylethanolamine (PE), 8.5% phosphatidylinositol (PI), 4.0% phosphatidylserine (PS), and 65.5% of a phosphatidylcholine (PC)-sphingomyelin fraction. This composition is in agreement with other reported values for rat liver microsomes.[10] No significant differences in the molar proportions of individual phospholipid classes were observed as a result of chronic ethanol feeding.

The Effects of Ethanol on the Order Parameter of Recombined Microsomal Phospholipid Membranes

In vesicles comprised of varying mole fractions of microsomal phospholipids (before separation by HPLC) obtained from the livers of untreated and ethanol-fed rats, we determined that only 30–40 mole% of the phospholipids derived from the ethanol-fed animals was necessary to render the vesicles tolerant to ethanol-induced disordering. Thus, the phospholipids obtained from the ethanol-fed rats contained a component potent enough to confer membrane tolerance, even when diluted with a larger amount of phospholipids from untreated rats.

Vesicles were then prepared by recombining the HPLC-separated microsomal phospholipids in their naturally occurring molar proportions. Reconstituted membranes composed of phospholipids from untreated rats were disordered by the addition of ethanol, as evidenced by a significant decrease (4.6%) in the order parameter, from 0.314 to 0.299 (FIG. 3B). On the other hand, vesicles composed of microsomal phospholipids from ethanol-treated rats showed no change in the order parameter over the same range of ethanol concentrations (FIG. 3B). As in the intact membrane, chronic ethanol treatment did not alter the baseline-order parameters of the vesicles prepared from the phospholipids derived from the ethanol-fed rats. The response of the reconstituted vesicles to ethanol, *in vitro,* qualitatively paralleled that of the intact microsomal membranes, thereby providing the opportunity to identify the phospholipid(s) responsible for conferring membrane tolerance.

In an attempt to identify the component responsible for membrane tolerance, vesicles were made by recombining either all the individual liver microsomal phospholipids from the untreated animals or all the individual phospholipids from the ethanol-fed animals (in the same molar proportions as found in the microsomal membrane), except that in each preparation one different phospholipid class was omitted. The missing phospholipid in the untreated preparation was then replaced by the corresponding phospholipid from the ethanol-fed rats. Similarly, in the preparation of phospholipids from the ethanol-fed rats, a phospholipid was deleted and replaced by one from the control rats. Recombined membranes prepared from the phospholipids of ethanol-fed rats retained their resistance to fluidization by ethanol after substitution of PC (66.5%), PE (21%), PI (8.5%), or PS (4.0%) from untreated rats (FIG. 4). Recombined "untreated" membranes were still fluidized by ethanol after substitution of PC, PE, or PS from the ethanol-fed animals. By contrast, when PI (8.5%) from treated rats was substituted in the membranes made from phospholipids from untreated rats, the membranes were rendered resistant to disordering by ethanol (FIG. 4). To determine whether the effect of PI is peculiar to rat hepatic microsomes, we prepared vesicles that had the same proportions of phospholipid classes as the rat microsomal membrane, but were made from bovine liver PC, PE, and PI, and bovine brain PS and sphingomyelin. These reconstituted vesicles were disordered by ethanol to the same extent (S = 0.316, 0 mM; S = 0.301, 100 mM; % Δ S = 4.6) as vesicles made from the microsomal phospholipids of untreated animals. Vesicles in which bovine liver PI was replaced with microsomal PI from untreated rats were also disordered to the same extent. However, when the bovine liver PI was replaced by PI from the ethanol-fed rats, the bovine vesicles became resistant to disordering. Thus, 8.5% microsomal PI from the treated animals also conferred membrane tolerance to vesicles composed of 91.5% phospholipids from another animal species and another

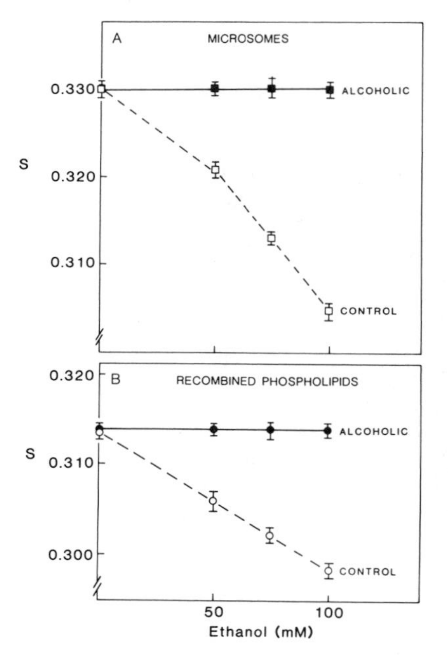

FIGURE 3. Typical order-parameter profiles showing the effect of ethanol *in vitro* on the molecular order parameters, S, obtained by ESR at 37° C. (**A**) Intact rat liver microsomes and (**B**) multilamellar vesicles prepared from recombined microsomal phospholipids that had been separated into individual classes by HPLC, were labeled with 12-doxylstearic acid. Each phospholipid class in the recombined vesicles was present in its naturally occurring mole fraction. Microsomes were obtained from untreated (control) rats (□----□) and rats chronically consuming ethanol (alcoholic) (■——■). The total control (untreated) (O----O) or alcoholic (ethanol-fed) (●——●) microsomal phospholipids used to prepare the reconstituted vesicles were isolated from the same microsomal preparations used to obtain the data in part (A). Points and bars represent the mean ± SD from four pairs of animals.

organ. The minimum amount of PI from ethanol-fed animals needed to confer tolerance to either microsomal phospholipids or bovine liver phospholipids was subsequently determined to be only 2.5%.

DISCUSSION

Withdrawal

The loss of membrane tolerance following ethanol withdrawal proceeds at different rates in the microsomal and erythrocyte membranes. Whereas microsomes exhibit a partial loss at 2 days and a total loss between 4 and 10 days, erythrocytes become partially sensitive to disordering by ethanol at 1 day and completely sensitive after 4

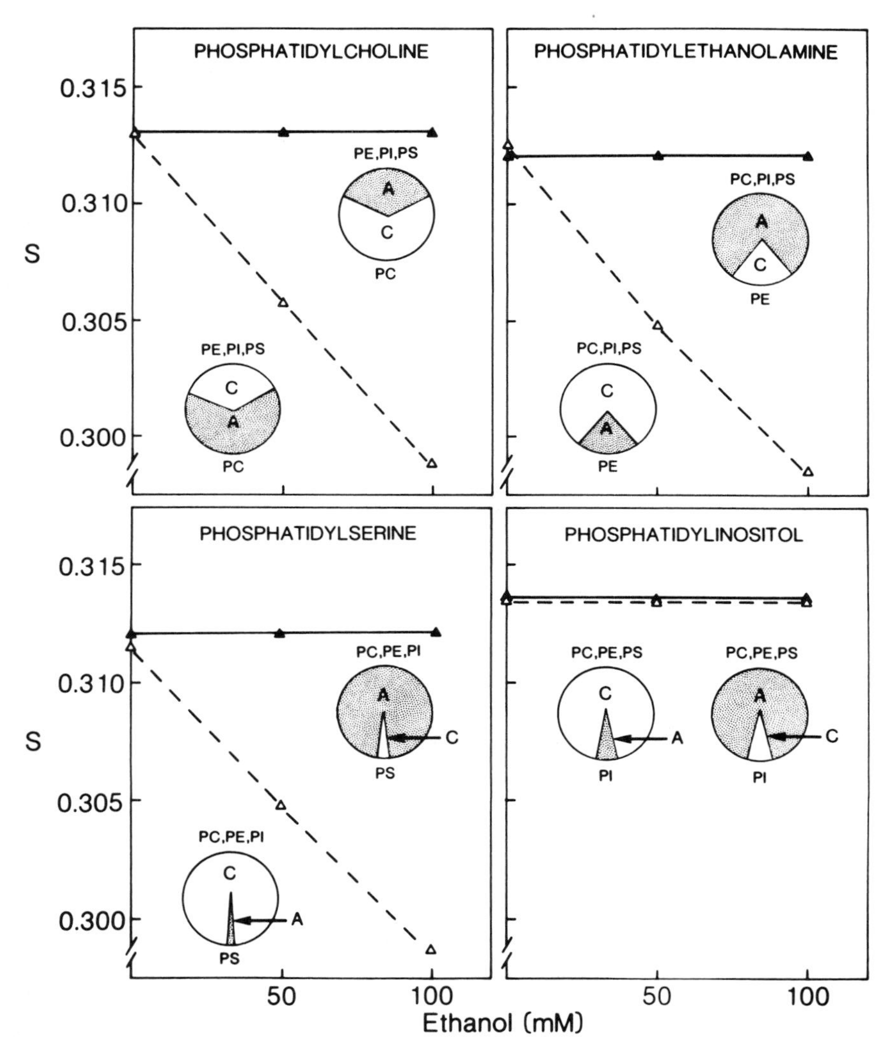

FIGURE 4. The capacity of individual microsomal phospholipid classes isolated from ethanol-fed rats to produce membrane tolerance in vesicles composed of microsomal phospholipids from untreated rats. The phospholipid compositions of each of the reconstituted membrane systems are indicated in the circle graphs. *Shaded* areas indicate the mole fraction of phospholipid(s) (A) from ethanol-fed rats and *unshaded* areas the mole fraction of phospholipid(s) (C) from untreated, control animals. ESR order parameters in the absence and presence of ethanol for recombined phospholipid vesicles comprising 3 phospholipid classes from ethanol-fed rats and 1 phospholipid class from untreated rats are shown as (▲—▲) and vesicles comprising 3 phospholipid classes from untreated rats and 1 phospholipid class from ethanol-fed rats as (△---△). The same results were obtained for 4 other sets of pair-fed animals. The error in each point was never greater than 0.75%.

days of ethanol withdrawal. Since these membranes differ in their phospholipid compositions, the source of phospholipids, and the rates of lipid turnover, these results are not unexpected. It seems that the factors responsible for the development of membrane tolerance are variable, depending on the membrane.

Tolerance does not solely result from ethanol-induced differences in the molecular order of the individual microsomal phospholipids. Using phospholipid spin probes, we found no differences in baseline molecular order (no ethanol present) among phosphatidylcholine, phosphatidylethanolamine, and phosphatidic acid in alcoholic or control microsomes.[4] Changes in cholesterol content are not responsible for the development of membrane tolerance in rat liver mitochondria, microsomes, or erythrocytes. This conclusion is supported by the fact that no changes in the cholesterol-phospholipid ratio are found in microsomes or erythrocytes before or after withdrawal from ethanol.

Since tolerance can be observed in phospholipids extracted from three different membranes, it appears that changes in protein-lipid interactions cannot be implicated as the cause of tolerance. However, such interactions may be involved to some degree, since loss of tolerance in reconstituted bilayers is demonstrable after shorter periods of time than in intact membranes. The considerably lower baseline-order parameters obtained for the phospholipid bilayers, compared to the intact membranes (see FIGS. 1 and 2), reflects greater "fluidity" in the absence of lipid-protein interactions and cholesterol, both of which tend to order the membrane. This difference is more pronounced in erythrocytes, because the cholesterol-phospholipid ratio is much higher (0.68) than in microsomes (0.10).

We have previously shown[9] that acquired membrane tolerance is associated with a decreased partitioning of ethanol and halothane into the alcoholic membranes. The decreased partition of ethanol into the alcoholic membrane may result from alterations in phospholipid molecular species, which might confer different molecular shapes to the phospholipids. Such alterations, which would go undetected using standard analytical techniques, could result in different packing properties of the membrane lipids, resulting in decreased partitioning of ethanol, and hence membrane tolerance. These alterations in lipid molecular species and packing properties would not necessarily cause any detectable change in membrane molecular order (fluidity) measured by ESR.

Membrane Tolerance

Although membrane tolerance has been observed by a number of investigators in a variety of membranes from different animal species, the molecular basis of this phenomenon has not been identified. Since the alterations in the physical properties of the vesicles prepared from microsomal phospholipids isolated from the ethanol-fed rats are the same as those observed in the intact membrane, the reconstituted systems described in this study provide a convenient model to identify the specific phospholipid that promotes membrane tolerance.

Interestingly, PI, which is a minor (8.5%) component of the microsomes, uniquely confers membrane tolerance, whereas PC or PE alone, which are present in much greater amounts (66% and 22%, respectively) are without effect. Even more surprising is the finding that as little as 2.5% content of PI from ethanol-fed animals renders the membrane resistant to fluidization. The ability of PI to promote membrane tolerance is not restricted to microsomal phospholipids; the inclusion of as little as 2.5% of PI from the ethanol-fed animals into vesicles of bovine phospholipids of brain and liver has the same effect. The observation that vesicles prepared from phospholipids of ethanol-fed animals, in which 8.5% PI from untreated animals replaced the PI from treated animals, remain resistant to disordering by ethanol (FIG. 4) suggests that minor ethanol-induced modifications in the other microsomal phospholipids make them weak promoters of tolerance, and that they can confer tolerance to reconstituted membranes when they comprise 90% of the vesicle phospholipids. Thus, while PI is a very strong promoter of membrane tolerance, combinations of other microsomal phospholipids (*e.g.* PC + PE) from ethanol-fed rats, when present in high concentrations, can also produce membrane tolerance.

In our studies, PI isolated from ethanol-fed rats exhibited a small decrease in the level of arachidonic acid and a slight increase in the quantity of oleic acid. These minor, ethanol-induced alterations in fatty acid composition may be related to the capacity of alcoholic PI to confer membrane tolerance. Chronic ethanol treatment may also cause changes in the phospholipid molecular species, although such alterations have not yet been sought. It may be that chemical alterations that occur in PI as a result of ethanol feeding change the packing properties of the microsomal phospholipids, such that the partition of ethanol into the membrane is reduced. The identification of the chemical and structural alterations in PI caused by chronic ethanol feeding is being actively pursued in our laboratory.

REFERENCES

1. TARASCHI, T. F. & E. RUBIN. 1985. Lab. Invest. **52:** 120–131.
2. CHIN, J. H. & D. B. GOLDSTEIN. 1977. Science **196:** 684–685.
3. WARING, A. J., H. ROTTENBERG, T. OHNISHI & E. RUBIN. 1981. Proc. Natl. Acad. Sci. USA **78:** 2582–2586.
4. TARASCHI, T. F., A. WU & E. RUBIN. 1985. Biochemistry **24:** 7096–7101.
5. RUGGIERO, F. M., C. LANDRISCINIA, G. V. GNOI & E. QUAGLIARIELLO. 1984. Lipids **19:** 171–178.
6. HOLLOWAY, C. T. & S. GARFIELD. 1981. Lipids **16:** 525–532.
7. NELSON, G. F. 1967. J. Lipid Res. **8:** 374–379.
8. JOHNSON, P. A., N. M. LEE, R. COOKE & H. H. LOH. 1979. Mol. Pharmacol. **15:** 739–746.
9. ROTTENBERG, H., A. WARING & E. RUBIN. 1981. Science **213:** 583–585.
10. WHITE, D. 1973. *In* Form and Function of Phospholipids. G. Ansell, J. Hawthorne & R. Dawson, Eds. 441–482. Elsevier Science Publishing Company. Amsterdam, The Netherlands.

DISCUSSION OF THE PAPER

C. C. CUNNINGHAM (*Wake Forest University, Winston-Salem, NC*): Your results for PI are very interesting to me. Of all the lipids that we analyzed in microsomes and mitochondria, PI was the one that increased in concentration as a result of ethanol feeding. Our numbers are not exactly the same as yours in terms of mole%, but the important thing is that we did see a significant increase in PI in microsomes, using the same feeding protocol as your group.

RUBIN: We have analyzed lipid composition by analytical HPLC and thin-layer chromatography (TLC) and have not found any major changes.

QUESTION: The microsomal membrane is very unusual in that it has very low cholesterol content as compared to other membranes and appears to be very sensitive to ethanol. Comparing your data to Dr. Goldstein's in her paper, you obtain a nice decrease in the order parameter at much lower ethanol concentrations than has been reported. I wonder if this is an unusual property of the microsomes, or is it a property of low-cholesterol-containing membranes?

RUBIN: The membrane fluidization we measure with 12-doxylstearic acid and the 12-doxyl phospholipid probes at room temperature in the microsomes is also obtained in red blood cells as well, which have a lot of cholesterol (0.68 mol phospholipid/mol cholesterol). I think the method is not restricted to cholesterol-poor membranes and the results that we have are really obtained from three types of membranes of different cholesterol content. The mitochondrial membrane, which has either no cholesterol or trivial amounts of cholesterol, the microsomal membrane, which has about 10%, and the red blood cell, which has much more. The question whether this can be modulated by cholesterol is one we have not addressed, but certainly the liposomes that we've used do not have any cholesterol. We doubt very much that cholesterol plays a role, because everything that we've done so far does not seem to involve cholesterol.

QUESTION: Can PI from other sources cause resistance?

RUBIN: In our hands, only the microsomal PI from the ethanol-fed rats that have been treated with alcohol for 35 days causes membrane resistance.

R. A. HARRIS (*University of Colorado School of Medicine, Denver, CO*): Does PI from other sources such as *E. coli* or plants that would have very different fatty acid chains have very different effects? In other words, if you think that the fatty acid composition and the lipid molecular species are very critical in the action of PI, then perhaps PI from a plant or a bacteria, which are very different from liver or brain PI might work.

RUBIN: We have not tested it; it's an interesting suggestion and I think we probably ought to pursue it.

The Effect of Chronic Ethanol Consumption on the Lipids in Liver Mitochondria[a]

CAROL C. CUNNINGHAM AND PRISCILLA I. SPACH

Department of Biochemistry
Bowman Gray School of Medicine of Wake Forest University
Winston-Salem, North Carolina 27103

INTRODUCTION

It has been established for many years that hepatic mitochondria are appreciably altered both in structure and function as a consequence of chronic ethanol consumption. In humans at the fatty-liver stage of ethanol-induced liver disease the mitochondria are large and misshapen, with disruption of the inner mitochondrial membrane, or cristae. These morphological changes are readily reproduced in experimental animals by maintaining them on an ethanol-containing diet for relatively short periods of time. In the rat, for example, both fatty-liver[1] and morphological changes in hepatic mitochondria[2] can be demonstrated after administration of an ethanol-containing diet for periods as short as three weeks. The morphological changes in the mitochondria, especially those occurring in the cristae, suggest that ethanol consumption affects the ability of the organelle to conserve energy, and it is now well established that hepatic mitochondria from ethanol-fed animals have diminished capacity for ATP synthesis via the oxidative phosphorylation process. These changes in energy conservation are accompanied by alterations in both the polypeptide and lipid composition of the mitochondrion. This paper focuses on those ethanol-elicited changes that occur in the lipids of the mitochondrion and on the alterations in the enzyme systems that catalyze their biosynthesis.

ETHANOL-ELICITED ALTERATIONS IN ENERGY-LINKED PROPERTIES OF HEPATIC MITOCHONDRIA; PREPARATION OF ORGANELLES FOR LIPID ANALYSES

Several laboratories have participated in characterizing the ethanol-elicited alterations in energy-linked parameters associated with hepatic mitochondria. While some of these investigations have been carried out with nonhuman primates,[3–5] most have been done with rats. In this presentation most of the studies to be discussed were implemented with the latter animal model. The ethanol-elicited alterations in the oxidative phosphorylation system in hepatic mitochondria are summarized in FIGURE 1. These measurements were carried out in our laboratory,[3,4,6–8] and are similar to the

[a]Supported by National Institute on Alcohol Abuse and Alcoholism (NIAAA) Grant 02887 and by the North Carolina Alcoholism Research Authority. CC was the recipient of NIAAA Research Scientist Development Award 00043 during the period in which these studies were carried out.

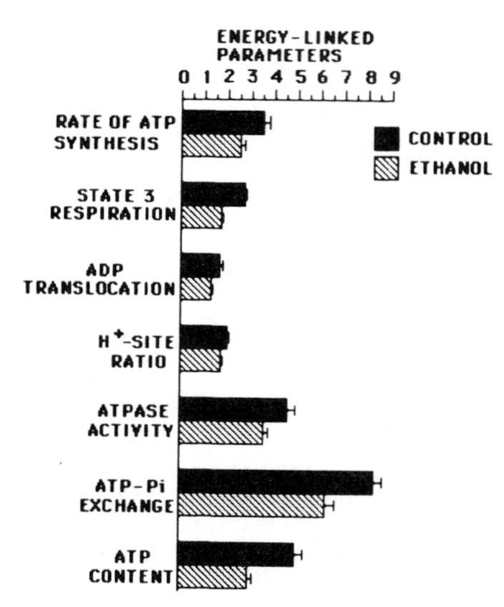

FIGURE 1. Ethanol-elicited alterations in parameters associated with mitochondrial energy metabolism. (The data in this figure were originally reported in REFERENCES 7 and 8.) The units for the following are: *rate of ATP synthesis,* μmol/min per mg of protein; *state 3 respiration,* μg-atoms of 0/min per mg of protein; *ADP translocation,* nmol/min per mg of protein; *H^+-site ratio,* protons translocated per coupling site; *ATPase activity,* μmol/min per mg of protein; *ATP-P_i exchange,* nmol ATP/min per mg protein × 10^{-1}; *ATP content,* nmol/mg of mitochondrial protein.

observations made by other investigators concerning changes occurring in energy-linked functions as a result of chronic ethanol consumption. An alteration that has the potential to affect the energy state of the liver is the lowered rate of ATP synthesis. Utilizing succinate as an energy source we observed a 30% decrease in the rate at which ADP was phosphorylated with radiolabeled inorganic phosphate. This was accompanied by significant decreases in 1) the respiratory rate in the presence of ADP, 2) the rate of ADP translocation, 3) proton translocation, and 4) activities associated with the oligomycin-sensitive ATP synthetase complex. We now have evidence that the ethanol-elicited decreases in the rate of ATP synthesis, state 3 respiration, ADP translocation, and the matrix concentrations of ATP are related to alterations in the functioning of the ATP synthetase, and we are investigating the possibility that ethanol consumption alters the concentration of one of the subunits of this enzyme complex. The importance of this decline in the rate of ATP synthesis cannot be minimized when one considers that chronic ethanol consumption results in lowered levels of hepatic ATP, as measured in both freeze-clamped liver[9] and isolated hepatocytes.[10] Possibly, the ethanol-elicited decreases in mitochondrial energy conservation have a strong negative impact on the energy state existing in fatty livers.

The above commentary reminds us of the physiological state of the organelle under investigation. In attempts to determine possible mechanisms by which ethanol elicits its effects on mitochondrial energy conservation, extensive studies have been carried out to characterize the changes occurring in both the polypeptide and lipid compositions of the mitochondrion. This discussion will be devoted to ethanol-related changes in membrane phospholipids.

A major concern in any investigation of the chemical composition of a cell organelle is its purity. How free is this structure from other cell components? The requirements for purity of the organelle are much more stringent when compositional studies are being carried out than when measurements of enzyme activities previously localized to the organelle are being implemented. When measuring mitochondrial phospholipids it is important to establish that they are not appreciably contaminated with those from lysosomes or endoplasmic reticulum, since these are the most likely contaminants. In our laboratory both the mitochondrial and microsomal preparations analyzed for phospholipid compositions have been purified by differential centrifugation techniques. We have included repeated washings to separate out contaminating membrane fractions that may have been trapped in the initial sediment. In addition, before microsomes are sedimented, lysosomes are removed from the post-mitochondrial supernatant by centrifuging at 17,300 × g for 10 minutes. This step improves microsomal purity appreciably.

The purity of mitochondria and microsomes has been assessed by measuring the distribution of marker enzymes in the isolated organelles. The results of such an analysis are shown in FIGURE 2. The cytochrome oxidase activity is dramatically reduced in mitochondria from ethanol-fed animals; this observation has been made in several laboratories. The low cytochrome oxidase activity in isolated microsomes demonstrates that they are minimally contaminated with mitochondria. Likewise, the low levels of NADPH-cytochrome c reductase activity in the preparations of mitochondria indicate that the mitochondria are relatively free of microsomes. This is emphasized by comparing the specific activities against that exhibited by purified microsomes. Lysosomal contamination of mitochondrial and microsomal preparations is low as is indicated by their levels of acid phosphatase specific activity. The microsomal preparations contain only 4% of the total homogenate acid phosphatase activity, whereas the mitochondrial preparations contain 15% of the total acid phosphatase activity (data not shown). Moreover, microsomes from ethanol-fed and control animals exhibit similar marker enzyme profiles.[4] This type of analysis provides a good measure of resolution of the organelles of interest from likely contaminants.

Another measure of organelle purity is the level of contamination by phospholipids previously established to be associated with certain cell compartments, but not others. For example, high amounts of sphingomyelin in liver mitochondrial preparations indicate significant contamination by other membrane fractions; likewise, measurable quantities of cardiolipin in microsomes establish the level of mitochondrial contamination. These considerations have been overlooked in some of the studies surveyed for this communication.

LIPID ANALYSES

Cholesterol

The cholesterol that would be expected to influence membrane-associated events occurring in the mitochondrion is the unesterified portion, since the degree to which cholesterol esters partition into phospholipid bilayers is minimal.[11] The concentration

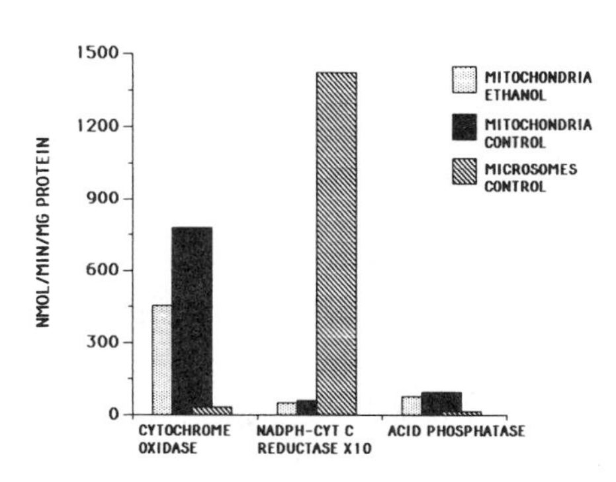

FIGURE 2. Distribution of marker enzymes in hepatic mitochondria and microsomes. (The distribution of marker enzymes in liver mitochondria was taken from REFERENCE 7.) The marker enzyme distributions in microsomes from *control* and *ethanol-fed* rats were not significantly different.

TABLE 1. Phospholipid and Unesterified Cholesterol Concentrations in Rat Liver Microsomes and Mitochondria[a]

Subcellular Fraction	Cholesterol (μg/mg Protein)	Phospholipid (μmol/mg Protein)	Cholesterol/Phospholipid (mol/mol)
Microsomes			
Control	23.4 ± 1.70	0.44 ± 0.02	0.137 ± 0.009
Ethanol-fed	22.9 ± 1.20	0.48 ± 0.03	0.123 ± 0.006
Mitochondria			
Control	2.67 ± 0.19	0.18 ± 0.01	0.038 ± 0.003
Ethanol-fed	3.08 ± 0.20	0.19 ± 0.01	0.042 ± 0.003

[a]Six pairs of animals; no statistically significant differences were observed. The unesterified cholesterol was determined as described in REFERENCE 4.

of unesterified cholesterol is quite low in mitochondria when compared with other subcellular organelles such as the endoplasmic reticulum (TABLE 1). Moreover, in the present study there were no significant ethanol-elicited differences in the unesterified cholesterol content of hepatic mitochondria when expressed on a per-mg mitochondrial protein or per-mol phospholipid basis. These results are consistent with those in other laboratories, and demonstrate that chronic ethanol consumption has no effect on the concentration of unesterified cholesterol, which is a minor constituent of hepatic mitochondrial membranes.

Mitochondrial Phospholipid-Protein Ratio

Many of the analyses presented in the remainder of this paper are summaries of several studies carried out in different laboratories. In evaluating the effect of chronic ethanol consumption on the amount of phospholipid in the mitochondrion 15 sets of data were utilized, including those from both rats and nonhuman primates.[3-6,12-17] These data were all analyzed together by statistical analysis after being converted to the unit, μmol phospholipid phosphorus/mg protein. Statistical analysis of the composite of these studies failed to verify any predictable effect of ethanol consumption on this chemical characteristic of mitochondrial membranes. Analysis of only those studies carried out with male rats fed a low-fat diet (< 12% of calories as fat) revealed, however, that chronic ethanol consumption elicited an average 18% decrease in the phospholipid-protein ratio (control—0.34 ± 0.05, ethanol-fed—0.28 ± 0.03 μmol phospholipid/mg protein). In contrast, in those studies carried out with animals fed a moderate-fat diet (> 30% of calories as fat) the mitochondrial phospholipid-protein ratio was not altered in any predictable manner by chronic ethanol consumption. Representative values for the phospholipid-protein ratios observed with animals on a moderate-fat diet are shown in TABLE 1. This comparison of the two dietary regimens emphasizes the interaction between the diet and ethanol in ethanol-elicited alterations in mitochondrial structure. This observation will be reinforced by several of the analyses to follow.

Mitochondrial Phospholipid Distribution

Approximately 95% of the total phospholipid in rat liver mitochondria from either control or ethanol-fed animals is comprised of phosphatidylcholine (PC) (~ 50%), phosphatidylethanolamine (PE) (~ 35%), and cardiolipin (CL) (~ 8%).[17] The remainder is made up of phosphatidylserine (PS), phosphatidylinositol (PI), and monoacyl species of PC and PE. FIGURE 3 illustrates that chronic ethanol consumption elicits no significant changes in the proportion of these phospholipids. These data, taken from our own rat studies, are in agreement with the composite analyses of the studies mentioned above[3–6,12–17] carried out with animals maintained on either low-fat or high-fat diets; the statistical analyses of those two groups of studies affirmed that ethanol feeding has no predictable effect on the distribution of the major phospholipids in hepatic mitochondria (FIGURE 4).

Acyl Chain Composition

While consumption of ethanol-containing diets has no effect on phospholipid distribution, it does influence significantly the acyl composition of both the total phospholipid complement and individual phospholipids. As shown in FIGURE 5, the fatty acid distribution of hepatic microsomes is also affected.[17] The most consistent alteration observed in our rat studies is a decrease in the palmitic acid (C16:0) level; this was demonstrated in the total phospholipid complement, PC, PE, and PI. Also noted were increases in eicosatrienoic (C20:3) and oleic (C18:1) acids. The same ethanol-related alterations are consistently noted in mitochondria but, in addition, significant increases in linoleic acid (C18:2) and decreases in arachidonic acid (C20:4) are noted in the two predominant phospholipids, PC and PE (FIGURE 6). The alterations in PC and PE are the most prominent quantitatively, because these two phospholipids comprise ~ 85% of the total phospholipid in the hepatic mitochondria from rats maintained on the moderate-fat diet.[17] Another significant ethanol-related alteration is a 17% decrease in C18:2 in cardiolipin.[17] Waring *et al.*[18] also demonstrated a 25% decrease in this predominant fatty acid of the cardiolipin molecule.

The dramatic change in C18:2 content of cardiolipin may reflect an alteration in phospholipid reacylation activity in the mitochondrion itself. Cardiolipin is synthesized in the mitochondrion on the inner membrane via the reaction: phosphatidylglycerol + CDP-diacylglycerol → cardiolipin + CMP. Interestingly, the precursor, rat liver mitochondria PG, contains low levels of C18:2;[19] moreover, there appears to be no acyl group specificity expressed by the cardiolipin synthetase for the CDP-diacylglycerol.[20] These observations indicate that either there is substantial reacylation of the cardiolipin once it is formed, or that reacylation of the precursors occurs as a preliminary step immediately preceding the condensation reaction designated above. In ethanol-fed rats the rate of this deacylation-reacylation process may be attenuated, resulting in a fatty acid composition with higher amounts of the saturated and monounsaturated fatty acids.[18] Both the deacylation[21] and reacylation[21–23] of endogenous phospholipids have

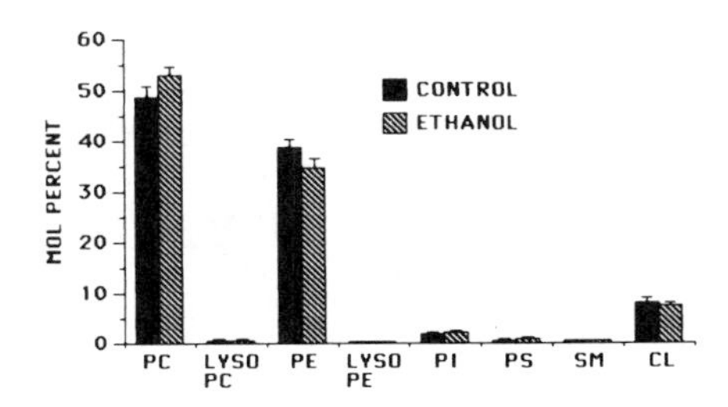

FIGURE 3. Phospholipid distribution in rat liver mitochondria from animals on a moderate-fat diet. (These data were taken from REFERENCE 17.)

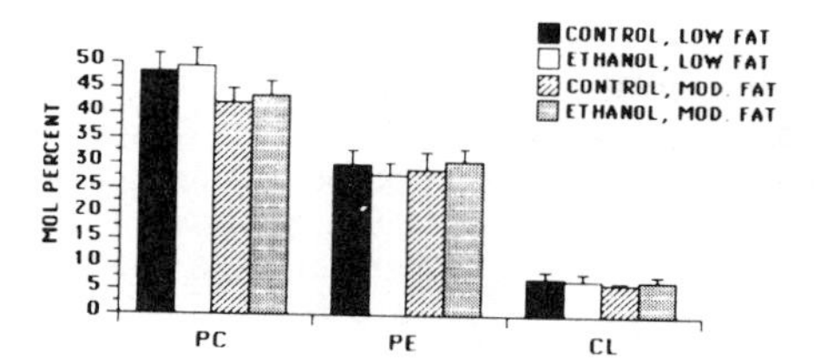

FIGURE 4. Composite analyses of the distribution of PC, PE, and CL in liver mitochondria from animals on either low-fat or moderate-fat diets. (The data upon which these analyses were performed were taken from REFERENCES 3–6 and 12–17.)

been demonstrated in hepatic mitochondria, and the degree to which reacylation takes place has been related to the energy state of the mitochondrion.[21] In mitochondria from ethanol-fed animals where the energy state is lowered (FIGURE 1), the reacylation of the glycerol moieties in cardiolipin may be depressed. This activity associated with mitochondrial lipid biosynthesis may therefore be affected by altered energy metabolism.

An alternative explanation which has not been rigorously ruled out is decreased cardiolipin synthetase activity in animals maintained on an ethanol-containing diet. This possibility gains credibility when it is recalled that other inner membrane activities are lowered as a result of ethanol consumption (FIGURES 1 and 2). If this were the case, then there might be higher PG concentrations in the mitochondria of ethanol-fed animals. Moreover, if the separation of PG from cardiolipin has been incomplete in the studies reported thus far, then the acyl composition of the cardiolipin might reflect the level of PG contamination, and indirectly the effect of ethanol on

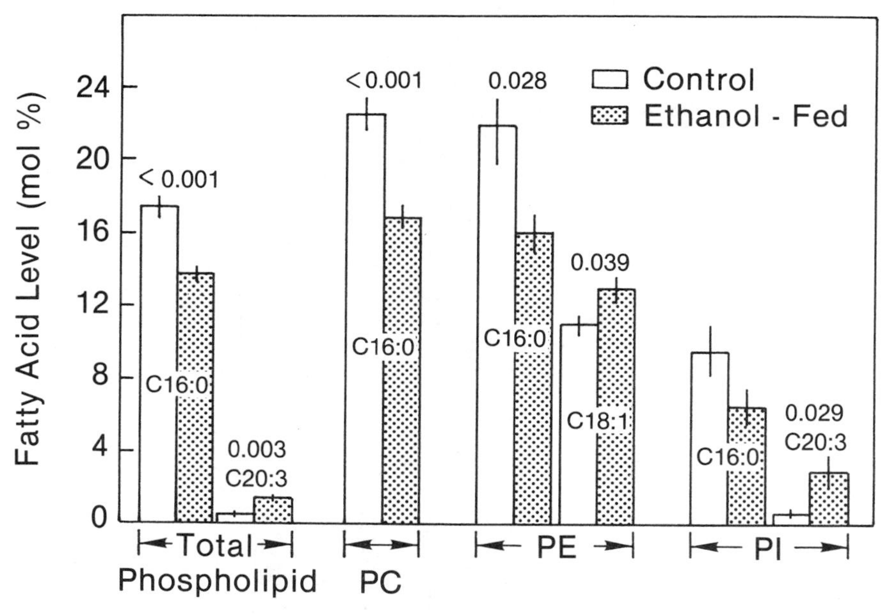

FIGURE 5. Effect of chronic ethanol consumption on acyl chain composition of rat liver microsomal phospholipids. (These data were taken from REFERENCE 17.)

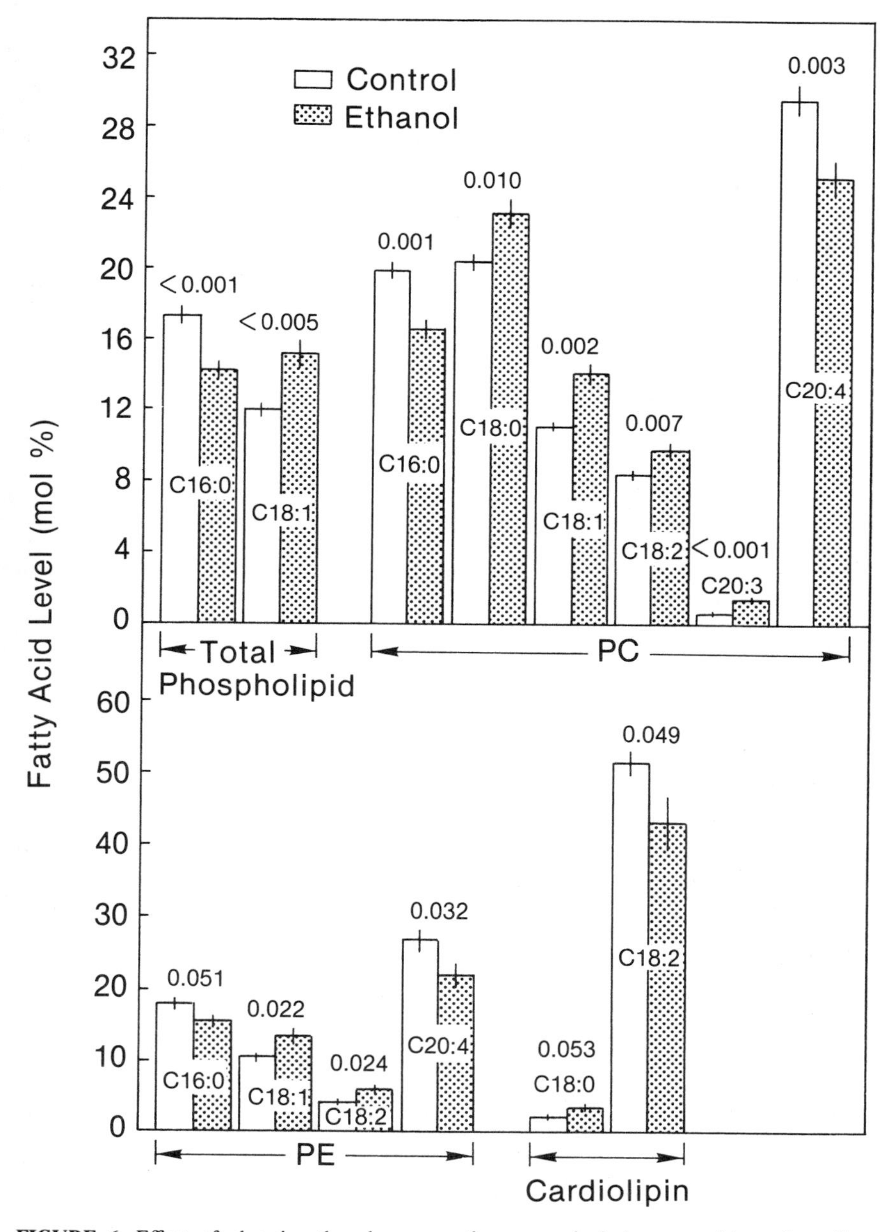

FIGURE 6. Effect of chronic ethanol consumption on acyl chain composition of rat liver mitochondrial phospholipids. (These data were taken from REFERENCES 6 and 17.)

cardiolipin synthetase. We are currently carrying out experiments to test this possibility.

Alterations in the mitochondrial phospholipid metabolism are also indicated when the ethanol-related changes in mitochondrial and microsomal acyl group composition are compared. As was illustrated in FIGURES 5 and 6, there were additional ethanol-related changes in the acyl composition of mitochondrial PC and PE that were not observed in the same microsomal phospholipids. For example, ethanol consumption resulted in a significant decrease in the C16:0 in microsomal PC; in the mitochondrial PC, however, there were significant alterations in C18:0, C18:1, C18:2, C20:3, and C20:4. Similar observations were made when comparing PE from microsomes and mitochondria. These differences are quantitatively important since PC and PE comprise ~ 85% of the total mitochondrial phospholipid. These data reaffirm earlier studies that suggest that after PC and PE are transferred from the endoplasmic reticulum to the mitochondrion,[24] their acyl group compositions are modified further by the mitochondrial deacylation-reacylation activities.[22,23] The greater number of ethanol-related alterations in the acyl composition of mitochondrial phospholipids suggests, therefore, that ethanol consumption depresses the normally occurring mitochondrial-associated reacylation activities. The functioning of phospholipid reacylation activities in liver is well established and provides a mechanism for modifying the acyl composition of phospholipids synthesized *de novo*.[25,26]

There appear to be patterns in the ethanol-related alterations of the acyl compositions of microsomal and mitochondrial phospholipids. These are observed most readily by comparing ratios of fatty acids that are in the same elongation-desaturation pathways. The two pathways affected by ethanol consumption are those that include the n-9 and n-6 fatty acids.[27] The n-9 fatty acids include C18:1 (oleic acid), which is derived from C16:0 that is elongated to C18:0 and then desaturated by a Δ^9 desaturase. The n-6 fatty acids begin with C18:2 which is desaturated by a Δ^6 desaturase, elongated, and desaturated again by a Δ^5 desaturase. We previously observed an increased C18:1-C16:0 ratio in both microsomal and mitochondrial phospholipids from livers of ethanol-fed rats, and also a decreased C20:4-C18:2 ratio in three mitochondrial phospholipids.[17] Our results encouraged us to look for these patterns among the investigations carried out on mitochondrial phospholipids in various laboratories including ours.[3,4,6,13–18,28–32] These studies were again subdivided into animal groups fed either low-fat or moderate-fat diets as designated above. The C18:1-C16:0 and C20:4-C18:2 ratios were calculated for the phospholipids analyzed in each study and these data assigned to either the low or the moderate dietary fat groups. These analyses, reported in TABLE 2, demonstrate that the fat content of the diet influenced the ethanol-related alterations in the C18:1-C16:0 and C20:4-C18:2 ratios. The C18:1-C16:0 ratio was not affected by ethanol in any predictable manner in mitochondrial phospholipids from animal groups maintained on low-fat diets. In contrast, there was an average 48% ethanol-related increase in this ratio in mitochondria from animal groups maintained on moderate-fat diets. While there may be a slight increase in the desaturation of C18:0 to C18:1 in the phospholipids from animals on the moderate-fat diets, it is apparent that the increase in the C18:1-C16:0 ratio observed is due to ethanol-related stimulation of the elongation step. There was an average 36% increase in the C18:0-C16:0 ratio in the mitochondrial phospholipids from ethanol-fed animals in the moderate-fat dietary groups (TABLE 2). This may be due to an increase in the concentration of NADPH, which results as a consequence of the shift in the cytoplasmic NADPH/NADP ratio that occurs during ethanol metabolism.[33] Moreover, the elongation may be stimulated only in the moderate-fat animal groups, since there would be lowered *de novo* fatty acid synthesis and thus less demand for the increased NADPH made available by the ethanol-induced shift in the hepatic oxidation-reduction state. In the low-fat animal groups increased concentrations of NADPH related to ethanol metabolism may be utilized primarily for *de novo* fatty acid synthesis. Confirmation of these suggestions will require more precise measurements of the NADPH concentrations in liver under the conditions utilized in the studies included in TABLE 2.

The effect of ethanol consumption on the C20:4-C18:2 ratio is also influenced by the level of fat in the diet. The data in TABLE 2 indicate that the conversion of C18:2 to C20:4 is depressed by including ethanol in a diet low in fat content. The lowered C20:4-C18:2 ratio observed in the mitochondrial phospholipids from animals on moderate-fat diets observed in some studies[17] was not confirmed in all investigations; thus the decrease reported in TABLE 2 for the studies utilizing a moderate-fat diet did not prove to be statistically significant. The lowered C20:4-C18:2 ratio suggests a decrease in either the elongation, Δ^6, or Δ^5 desaturation steps associated with the n-6 fatty acid family. The observation in some studies[4,15,17] that the C20:3 concentrations are elevated with ethanol feeding suggests that the Δ^5 desaturase activity becomes rate-limiting for this elongation-desaturation process. Normally this enzyme turns over

TABLE 2. Effect of Dietary Fat and Ethanol on the C18:1-C16:0 and C20:4-C18:2 Ratios in Mitochondrial Phospholipids[a]

	Ratios			
Diet Type	C18:1-C16:0	C18:1-C18:0	C18:0-C16:0	C20:4-C18:2
Low-fat				
Control	0.70 ± 0.09	0.69 ± 0.10	1.03 ± 0.07	2.61 ± 0.43
Ethanol	0.71 ± 0.08	0.68 ± 0.09	1.11 ± 0.13	1.99 ± 0.41
p	> 0.5	> 0.5	> 0.5	< 0.001
Moderate-fat				
Control	0.64 ± 0.05	0.50 ± 0.05	1.19 ± 0.12	2.51 ± 0.63
Ethanol	0.95 ± 0.07	0.57 ± 0.06	1.62 ± 0.21	2.20 ± 0.51
p	< 0.001	< 0.1	< 0.005	< 0.4

[a]Low-fat studies, n = 10 for C18:1-C16:0 ratios, and 13 for C20:4-C18:2 ratios; for the moderate-fat studies, n = 10 and 11 for the same two ratios. The p values were from paired t tests.

at a slower rate than do the Δ^9 and the Δ^6 desaturases,[27] and it can become rate-limiting for arachidonate synthesis if its activity is lowered by ethanol consumption. Administration of ethanol-containing diets to rats does appear to lower the activities of both the Δ^6 and Δ^5 desaturases,[34-36] which verifies that, indeed, the Δ^5 desaturase can become rate-limiting in the pathway for arachidonate biosynthesis. Since the Δ^6 and Δ^5 desaturases do not seem to be dramatically sensitive to the levels of carbohydrate in the diet, the decreases in this nutrient that result from an ethanol for carbohydrate substitution may not be a dominant factor in altering its activity. This suggests that the ethanol in the diet could be primarily responsible for the changes seen in Δ^6 and Δ^5 desaturase activity. This conclusion cannot be extended to the Δ^9 desaturase, however, which has been demonstrated to be more sensitive to both the fat and carbohydrate concentrations in the diet.[37]

Another consideration with respect to the deficit in C20:4 is that it occurs in the mitochondrial PC and PE, but not in their microsomal counterparts (FIGURES 5 and 6) according to our investigation.[17] In addition to reinforcing the conclusion that mitochondrial phospholipid reacylation may be depressed by ethanol consumption, this observation is also consistent with the possibility that the endoplasmic reticulum and mitochondria may compete for a limited amount of archidonate, but with the endoplasmic reticulum being more successful, since it is the site for synthesis of most of the C20:4.[27] In this regard it has been suggested that the synthesis of C20:4 and its subsequent esterification to phospholipids in the endoplasmic reticulum occurs in a concerted manner without accumulation of appreciable pools of free arachidonyl CoA.[27] Under this circumstance movement of arachidonate to the mitochondrion in livers of ethanol-fed animals may be limited somewhat due to reduced availability.

SUMMARY

The ethanol-related alterations in hepatic mitochondrial phospholipids are primarily changes in acyl chain composition. There are no alterations in the unesterified cholesterol content in the mitochondrion, as measured by the cholesterol-phospholipid ratio. Moreover, the distribution of mitochondrial phospholipids are not changed as a result of chronic ethanol consumption. There was a significant ethanol-related decrease (18%) in the phospholipid-protein ratio in mitochondria from rats maintained on a low-fat diet, which was not observed in studies where animals were fed diets containing a higher proportion of lipid. This effect of dietary composition on the phospholipid-protein ratio was also paralleled by the interaction between diet and ethanol in influencing the phospholipid acyl composition. The alterations in acyl chain distribution indicated that ethanol consumption stimulated elongation of palmitic acid, and depressed the Δ^5 desaturation step required for the formation of arachidonic acid. Elongation of palmitic acid was stimulated in studies where animals were fed diets with moderate amounts of fat, whereas depressed synthesis of arachidonate occurred more frequently, but not exclusively, in studies where low-fat diets were employed. These results indicate that there is a significant interaction between diet and ethanol in eliciting changes in hepatic mitochondrial phospholipids.

The significant decrease in the linoleic acid content of cardiolipin and the more prominent ethanol-associated alterations in mitochondrial phospholipids suggest that ethanol consumption depresses the phospholipid reacylation activities associated with the mitochondrion. The above observations indicate, therefore, that the alterations occurring in mitochondrial phospholipids are influenced by ethanol-related changes in mitochondrial enzymes involved in phospholipid metabolism. In addition, alterations in the availability of fatty acids due to ethanol-related changes in microsomal elongation and desaturation activities also appear to affect the fatty acid composition of phospholipids in mitochondria from ethanol-fed animals.

REFERENCES

1. LIEBER, C., D. JONES & L. DECARLI. 1965. J. Clin. Invest. **44:** 1009–1021.
2. PORTA, E., W. HARTROFT & F. DE LA IGLESIA. 1965. Lab. Invest. **14:** 1437–1455.
3. CUNNINGHAM, C., G. SINTHUSEK, P. SPACH & C. LEATHERS. 1981. Alcoholism: Clin. Exp. Res. **5:** 410–416.
4. CUNNINGHAM, C., R. BOTTENUS, P. SPACH & L. RUDEL. 1983. Alcoholism: Clin. Exp. Res. **7:** 424–430.
5. ARAI, M., M. LEO, M. NAKANO, E. GORDON & C. LIEBER. 1984. Hepatology **4:** 165–174.
6. SPACH, P., J. PARCE & C. CUNNINGHAM. 1979. Biochem. J. **178:** 23–33.
7. SPACH, P., R. BOTTENUS & C. CUNNINGHAM. 1982. Biochem. J. **202:** 445–452.
8. BOTTENUS, R., P. SPACH, S. FILUS & C. CUNNINGHAM. 1982. Biochem. Biophys. Res. Commun. **105:** 1368–1373.
9. BERNSTEIN, J., L. VIDELA & Y. ISRAEL. 1973. Biochem. J. **134:** 515–521.

10. MONTGOMERY, R., P. SPACH & C. CUNNINGHAM. 1984. Fed. Proc. **43:** 1878A.
11. JANIAK, M., C. LOOMIS, G. SHIPLEY & D. SMALL. 1974. J. Mol. Biol. **86:** 325–339.
12. LUNDQUIST, C., K. KIESSLING & L. PILSTROM. 1966. Acta Chem. Scand. **20:** 2751–2754.
13. FRENCH, S., B. IHRIG & R. MORIN. 1970. Quart. J. Stud. Alc. **31:** 801–809.
14. THOMPSON, J. & R. REITZ. 1978. Lipids **13:** 540–550.
15. SCHILLING, R. & R. REITZ. 1980. Biochim. Biophys. Acta **603:** 266–277.
16. GORDON, E., J. ROCHMAN, M. ARAI & C. LIEBER. 1982. Science **216:** 1319–1321.
17. CUNNINGHAM, C., S. FILUS, R. BOTTENUS & P. SPACH. 1982. Biochim. Biophys. Acta **712:** 225–233.
18. WARING, A., H. ROTTENBERG, T. OHNISHI & E. RUBIN. 1981. Proc. Natl. Acad. Sci. USA **78:** 2582–2586.
19. HOSTETLER, K. 1982. *In* Phospholipids. J. Hawthorne & G. Ansell, Eds. 215–261. Elsevier Biomedical Press. Amsterdam, The Netherlands.
20. HOSTETLER, K., J. GALESLOOT, P. BAER & H. VAN DEN BOSCH. 1975. Biochim. Biophys. Acta **380:** 382–389.
21. PARCE, J., C. CUNNINGHAM & M. WAITE. 1978. Biochemistry **17:** 1634–1639.
22. WAITE, M., P. SISSON & E. BLACKWELL. 1970. Biochemistry **9:** 746–753.
23. SARZALA, M., L. VAN GOLDE, B. DE KRUYFF & L. VAN DEENEN. 1970. Biochim. Biophys. Acta **202:** 106–119.
24. DAUM, G. 1985. Biochim. Biophys. Acta **822:** 1–42.
25. LANDS, W. & C. CRAWFORD. 1976. *In* The Enzymes of Biological Membranes. A. Martonosi, Ed. Vol. 2: 1–85. Plenum Press. New York, NY.
26. VAN GOLDE, L. & S. VAN DEN BERGH. 1977. *In* Lipid Metabolism in Mammals. F. Snyder, Ed. Vol. 1: 35–149. Plenum Press. New York, NY.
27. COOK, H. 1985. *In* Biochemistry of Lipids and Membranes. D. Vance & J. Vance, Eds. 181–212. Benjamin/Cummings Publishing. Menlo Park, CA.
28. FRENCH, S., A. SHEINBAUM & R. MORIN. 1969. Proc. Soc. Exp. Biol. Med. **130:** 781–783.
29. FRENCH, S. & R. MORIN. 1969. *In* Biochemical and Clinical Aspects of Alcohol Metabolism. V. Sardesai, Ed. 123–132. C. C. Thomas Publishing. Springfield, IL.
30. THOMPSON, J. & R. REITZ. 1976. Ann. N.Y. Acad. Sci. **273:** 194–204.
31. THOMPSON, J., J. BERGSTROM & R. REITZ. 1982. Biochim. Biophys. Acta **719:** 580–588.
32. ARAI, M., E. GORDON & C. LIEBER. 1984. Biochim. Biophys. Acta **797:** 320–327.
33. GUYNN, R. & J. PIEKLIK. 1975. J. Clin. Invest. **56:** 1411–1419.
34. NERVI, A., R. PELUFFO, R. BRENNER & A. LEIKEN. 1980. Lipids **15:** 263–268.
35. WANG, D. & R. REITZ. 1983. Alcoholism: Clin. Exp. Res. **7:** 220–226.
36. REITZ, R. 1984. Proc. West. Pharmacol. Soc. **27:** 247–249.
37. JEFFCOAT, R. 1979. *In* Essays in Biochemistry. P. Campbell & R. Marshall, Eds. Vol. 15: 1–36. Academic Press. New York, NY.

DISCUSSION OF THE PAPER

R. WRIGHT (*University of Nevada, Reno, NV*): A couple of years ago we reported that chronic alcohol inhibited the activity of all three of the acyl CoA desaturases—Δ^9, Δ^5, and Δ^6. We also find that if we bypass the Δ^6 desaturase by feeding the animals with primrose oil, which has a high content of the 18:2 or 18:3 fatty acids, we see a fairly marked increase in the 20:3 fatty acid in alcohol-treated animals, again indicating an effect on the Δ^5. The three desaturases, at least using a low-fat diet, are all affected fairly rapidly within one day and even out after 30 days of ethanol feeding.

E. RUBIN (*Thomas Jefferson University School of Medicine, Philadelphia, PA*): Just one comment about what constitutes a high-fat and low-fat diet for a rat. The chow diet is made low-fat because it's cheaper to make it that way and they have enough fat to thrive and do well. Presumably a rat that lives around a slaughter house is eating a high-fat diet, and if he's in the field he's eating a low-fat diet. It's hard to decide what's the "right" fat diet for the rat.

CUNNINGHAM: I agree. However, we have to recognize that there is an interplay between the diet and the ingested ethanol. When we see disagreement in data, one of the first things to look at is the diet composition. It may not only have effects on the lipids, it may also have an effect in other ethanol-related changes as well.

Y. ISRAEL (*University of Toronto, Toronto, Canada*): We have been able to confirm your experiments, when we give low-fat diets along with alcohol, that almost no mitochondrial damage occurs. When we give high-fat diets, we have massive mitochondrial damage. It seems that the fat composition of the diet is rather important in the final results you get.

Effects of Ethanol on Proteins of Mitochondrial Membranes[a]

WILLIAM S. THAYER

Department of Pathology and Laboratory Medicine
Hahnemann University School of Medicine
Philadelphia, Pennsylvania 19102

During the past few years, we have characterized the effects of chronic ethanol consumption on the structure and function of the hepatic mitochondrial membrane with respect to its protein composition using rats as an animal model. Studies at the cellular, organelle, and membrane level have been conducted using isolated hepatocytes, mitochondria, and submitochondrial particles. Studies at these three experimental levels of membrane organization have given consistent results and provide a comprehensive analysis of the effects of chronic ethanol consumption on most known components of the respiratory-phosphorylation system. Our studies have employed mainly biochemical and biophysical techniques, particularly optical and electron spin resonance (EPR) spectroscopy and enzyme activity measurements. Recent studies using a complementary immunochemical approach have provided a new insight into the actions of ethanol at the molecular level. In this communication, I will review some of the key findings from our earlier work and present new data having implications for the mechanisms underlying the effects of alcohol consumption on mitochondrial membrane proteins.

EFFECTS OF ETHANOL CONSUMPTION ON OXIDATIVE PHOSPHORYLATION

Alterations of mitochondrial morphology produced by alcohol consumption have been recognized for decades and described by many investigators. Studies of mitochondrial function by Cederbaum *et al.*[1] first showed that liver mitochondria isolated from rats fed ethanol chronically displayed decreased respiratory rates and decreased rates of ATP synthesis compared to pair-fed controls, particularly with NAD-linked substrates. To explore the structural basis for these effects, I studied the effects of ethanol feeding on oxidative phosphorylation using submitochondrial particles. These are vesicles of the inner mitochondrial membrane that have a sidedness opposite to that of intact mitochondria. This inverted orientation allows direct access of substrates and adenine nucleotides to the enzymes of the respiratory-phosphorylation apparatus independent of transport carriers. I devised a method for preparing submitochondrial particles from rat liver that retained high phosphorylation and energy coupling activities.[2] Marker enzyme and recovery studies demonstrated that submitochondrial particle preparations from either ethanol-fed or control rats were comparable with respect to both purity and yield.[2,3] Studies with these particles showed that respiratory rates, measured under both coupled and uncoupled conditions, were decreased after

[a]Supported by National Institute on Alcohol Abuse and Alcoholism Grants AA05662 and AA3442.

chronic ethanol treatment (FIG. 1). Decreases of 25–40% were observed with substrates donating reducing equivalents at each of the three classic sites of the respiratory chain. Under phosphorylating conditions, P/O ratios were decreased slightly (10–15%), but significantly, with all substrates. Steady state rates of ATP synthesis, the most critical parameter for energy metabolism, were lowered 35–40% with all substrates as a result of the combined effects of ethanol treatment on respiration and coupling efficiency.[2] With submitochondrial particles, the rate of ATP synthesis during oxidative phosphorylation appears to be limited mainly by the rate of electron transfer through the respiratory chain. Thus, results of these functional studies suggested that alterations in content or activities of respiratory chain components might account for the effect of ethanol consumption on oxidative phosphorylation.

EFFECTS OF ETHANOL CONSUMPTION ON RESPIRATORY CHAIN COMPONENTS

Cytochromes

The cytochrome composition of the membranes was quantified by optical spectroscopy measuring the characteristic heme absorbance differences between reduced and oxidized states.[3] Studies with both physiologic and chemical reductants revealed major decreases (about 50%) in amounts of cyts. aa_3 (oxidase) and b after ethanol treatment (TABLE 1). Similar changes were also found in both intact mitochondria and isolated hepatocytes. By contrast, the content of cyts. $c + c_1$ was essentially unaffected by chronic ethanol consumption. Although some decrease in cyts. $c + c_1$ was measured in submitochondrial particles, no significant change was found with intact mitochondria or isolated hepatocytes. The entire complement of cyt. c within the membrane of alcoholic animals was enzymatically functional and interacted with all of the cytochrome oxidase, even though the content of the latter was only half that of control membranes. As a consequence of the differential effect of ethanol consumption on these proteins, the cyts. $c + c_1$/cyt. aa_3 ratio was effectively doubled in membranes from ethanol-treated rats. In terms of redox interaction of these cytochromes, this change resulted in maintainance of cyts. $c + c_1$ at a higher level of steady state reduction during conditions of maximal electron flow.[3] The catalytic turnover number of cytochrome oxidase (electrons/s/nmole heme a) and apparent K_m for cyt. c were unchanged.[4] Thus, ethanol consumption decreases the membrane content of functional cytochrome oxidase without altering its kinetic properties.

The decrease in cyt. b produced by chronic ethanol consumption was determined from difference spectra recorded aerobically in the presence of the respiratory inhibitor antimycin. Such conditions promote reduction of the b cytochromes, while cytochromes $c_1 + c$, as well as aa_3, remain oxidized. Experimentally, this method allows more accurate measurement of cytochrome b content than spectra recorded under anaerobiosis or in the presence of dithionite, because spectral overlap due to absorbance of reduced cytochromes $c + c_1$ does not occur. The alteration of cytochrome b is complex, owing to the existence of at least two distinct species of cytochrome b, b-562 and b-566, in the electron transfer chain. More detailed spectral studies comparing the extent of cytochrome b reduction under substrate-reduced anaerobic conditions to that in the presence of antimycin suggested that little enzymatically-reducible cytochrome b-566 was present in membranes from ethanol-treated rats.[3] Thus, chronic ethanol consumption appears to decrease the content of cytochrome b-566 more extensively than that of cytochrome b-562.

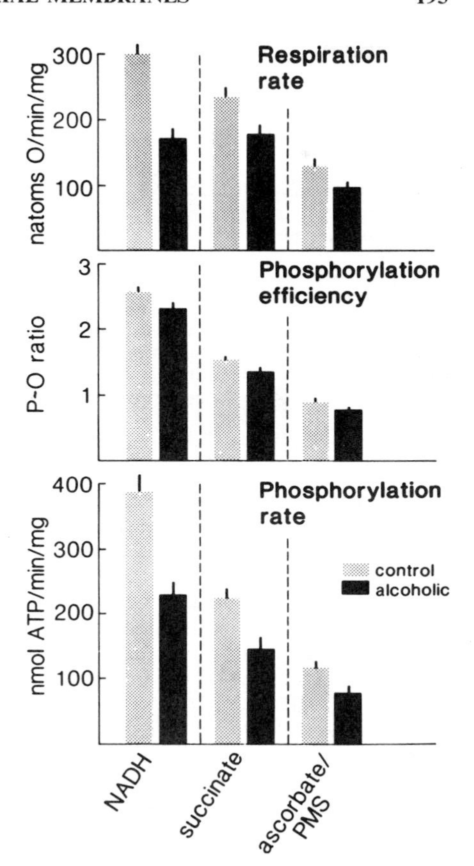

FIGURE 1. Effects of chronic ethanol consumption on bioenergetic activities of rat liver submitochondrial particles. Maximal respiratory rates were measured in the presence of an uncoupler. P/O ratios, nmoles ATP formed per natom O consumed, and steady state rates of ATP synthesis were determined in the absence of uncoupler. Data indicate mean and S.E. for 15 preparations. Control: *stipled bars;* alcoholic: *solid bars.* (Adapted from REF. 2.)

TABLE 1. Effect of Chronic Ethanol Treatment on the Cytochrome Content of Hepatic Mitochondrial Membranes[a]

Cytochromes	% Change		
	Hepatocytes	Mitochondria	Submitochondrial Particles
aa_3	−50	−47	−39
b	−42	−50	−45
$c + c_1$	0	−6	−22

[a]Cytochrome contents were determined by absorbance difference spectroscopy.[3] Data indicate the average % change of preparations from rats fed ethanol chronically (36% calories/40 days) compared to those from pair-fed control rats. (Adapted from REFS. 3 and 5.)

TABLE 2. Effect of Chronic Ethanol Treatment on the Antimycin Titer[a]

Experimental System	Antimycin Titer		
	Control	Ethanol-Treated	% Change
	(nmols/mg protein)		
Submitochondrial particles			
Respiratory inhibition	0.146 ± 0.010	0.098 ± 0.010	−33
Antimycin binding	0.15 ± 0.03	0.08 ± 0.02	−47
Hepatocytes			
Respiratory inhibition	0.020 ± 0.002	0.008 ± 0.02	−60

[a]The antimycin titer, defined as the minimum amount of inhibitor to produce maximum inhibition, was determined from antimycin titrations of respiration. With submitochondrial particles, the titer for antimycin binding to high-affinity sites was also determined from quenching of antimycin fluorescence. (Adapted from REF. 5.)

The decrease in cytochrome *b* content was confirmed by an alternative approach based on antimycin inhibition.[5] Antimycin binds stoichiometrically to a component of the cytochrome bc_1 complex, causing inhibition of electron flow in this region of the respiratory chain. Antimycin titrations of respiration with submitochondrial particles, mitochondria, and hepatocytes all indicated a decrease of approximately 50% in the antimycin titer (minimum amount of inhibitor producing maximal inhibition) in preparations derived from ethanol-treated rats compared to those from controls (TABLE 2). However, the maximum extent of inhibition in each of these systems was identical to that observed with preparations from control rats. Measurements of antimycin binding to submitochondrial particles by the quenching of antimycin fluorescence confirmed a similar decrease in the number of high-affinity antimycin binding sites.[5] These results demonstrate that the effect of chronic ethanol consumption on the cytochrome *b* content of the membrane is similar in magnitude to that on cytochrome aa_3.

Dehydrogenases

Chronic ethanol consumption affects the membrane content of proteins associated with NADH dehydrogenase. This enzyme contains FMN and several iron-sulfur clusters (designated by the prefix N-), which function in the transfer of electrons through this region of the respiratory chain. The iron-sulfur clusters of NADH dehydrogenase are paramagnetic in their reduced states. Thus, EPR spectrometry provided a method for estimating the contents of these clusters within membranes. In NADH-reduced submitochondrial particles, amplitudes of signals arising from clusters N-2, N-3, and N-4 were decreased 20% to 30% after ethanol treatment (TABLE 3).[6] Direct measurements of NADH dehydrogenase activity with exogenous electron acceptors indicated a 46% decrease in the Vmax,[6] confirming an effect of ethanol consumption on this enzyme. There was no change in acid-extractable flavin (non-covalently bound) after ethanol treatment.[3] However, since several flavoproteins are associated with the respiratory chain in liver submitochondrial particles, this determination is not specific for the flavin of NADH dehydrogenase. The decrease in enzyme activity and alteration in EPR spectra are more specific, and provide evidence for another alteration of mitochondrial membrane proteins induced by chronic ethanol consumption.

By contrast with the effects on NADH dehydrogenase, chronic ethanol treatment did not affect the membrane content of proteins associated with succinate dehydrogenase. This respiratory chain enzyme contains covalently bound FAD and specific iron-sulfur clusters (designated by the prefix S-). EPR spectra of submitochondrial particles showed no change in amplitude of signals attributable to clusters S-1 or S-3 after ethanol treatment (TABLE 3).[6] Measurement of covalently bound flavin, obtained by tryspin treatment of submitochondrial particles following acid extraction, showed no difference after ethanol treatment.[3] In this case, the flavin determination is specific for succinate dehydrogenase, since this enzyme is the only mitochondrial flavoprotein known to have a covalently attached flavin prosthetic group. These data indicate that chronic ethanol consumption does not affect the mitochondrial membrane proteins of succinate dehydrogenase.

Ubiquinone

The membrane content of ubiquinone, determined spectrophotometrically after extraction from submitochondrial particles, was unchanged by chronic ethanol consumption.[3] However, this finding cannot exclude a possible alteration in a minor fraction of the total ubiquinone that may be associated with a specific ubiquinone binding protein. The amount of ubiquinone associated with the electron transfer chain is much greater than that of the cytochromes and other electron carriers. The bulk of ubiquinone functions as a reducing equivalent pool linking the dehydrogenase enzymes with the cytochromes.

EFFECTS OF ETHANOL CONSUMPTION ON ATPase

The 10–15% decreases in phosphorylation efficiencies (P/O ratios) observed with submitochondrial particles, noted earlier, were suggestive of an effect of ethanol at the

TABLE 3. Effect of Chronic Ethanol Consumption on Iron-Sulfur Clusters in Rat Liver Submitochondrial Particles[a]

Cluster	Temperature of Observation (°K)	Characteristic *g* Values	% Change
NADH Dehydrogenase			
N-1a	30	2.03, 1.94	0
N-1b	30	2.03, 1.94, 1.91	0
N-2	12	2.05, 1.93	−30
N-3	8	2.04, 1.93, 1.86	−20
N-4	8	2.10, 1.94, 1.88	−20
Succinate Dehydrogenase			
S-1	30	2.08, 1.93, 1.91	0
S-3	8	2.01	0

[a]Contents of specific iron-sulfur clusters were determined by electron spin resonance spectroscopy. Data indicate the average % change in signal amplitudes at the characteristic g values for preparations from ethanol-fed rats compared to controls, as in TABLE 1. (Adapted from REF. 6.)

level of ATPase itself. Studies of ATP-dependent energy-linked functions, particularly $^{32}P_i$-ATP exchange and reverse electron transfer (reduction of NAD^+ by succinate), likewise indicated a comparable decrease in coupling efficiency.[2] Direct studies of ATPase activity showed that the maximal rate of ATP hydrolysis, measured under uncoupled conditions, was decreased 30% in preparations from ethanol-fed rats (FIG. 2).[2] This finding suggests that the membrane content of ATPase protein is also decreased substantially after chronic ethanol treatment.

TIME COURSE FOR REVERSAL OF CHARACTERISTIC EFFECTS OF ETHANOL ON MITOCHONDRIAL MEMBRANE PROTEINS

Data from our previous studies, outlined above, show that chronic ethanol treatment results in a selective, or characteristic, pattern of effects on proteins of the mitochondrial membrane. Four major membrane protein complexes are decreased substantially (30–50%) following alcohol consumption. These are cytochrome oxidase, cyts. *b*, NADH dehydrogenase, and ATPase.

To test the reversibility of these alterations, we have studied the time course for reversal of these characteristic changes following withdrawal of ethanol from the diet. For these studies, rats that had been fed ethanol chronically for the usual 40–45 days were placed on control liquid diet for 1–11 days. Liver submitochondrial particles were then prepared and assayed for cytochrome contents, NADH oxidase, and ATPase activities by our standard methods. As shown in FIGURES 3 and 4, the alcohol-induced protein changes gradually reverted to levels typical of control animals over the 11-day period. Interestingly, all four protein complexes exhibited similar time courses for the reversal. Approximately 11 days of non-alcohol feeding were necessary to achieve levels of these membrane proteins equivalent to those found in control animals. This length of time is compatible with the normal turnover time for mitochondrial proteins measured during studies of mitochondrial biogenesis. For example, radioisotope

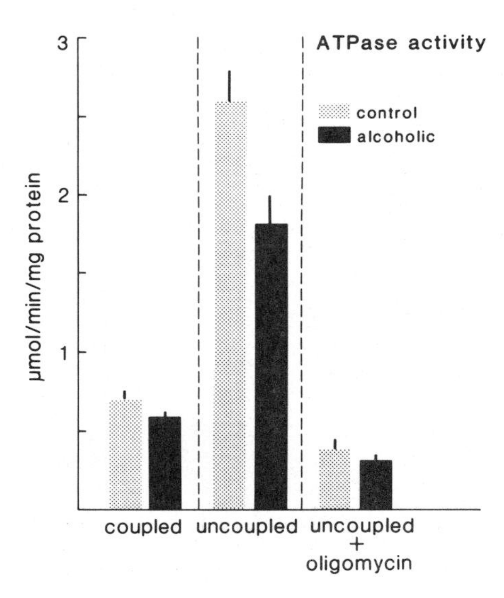

FIGURE 2. Effect of chronic ethanol consumption on ATPase activity of rat liver submitochondrial particles. The rate of ATP hydrolysis was measured in the absence (coupled) and presence (uncoupled) of carbonyl cyanide m-chlorophenyl hydrazone (1 μM). Rates observed under uncoupled conditions in the presence of the ATPase inhibitor oligomycin (2 μg/mg protein) are also shown. Data indicate mean and S.E. for 9 preparations. Control: *stipled bars;* alcoholic: *solid bars.* (Adapted from Ref. 2.)

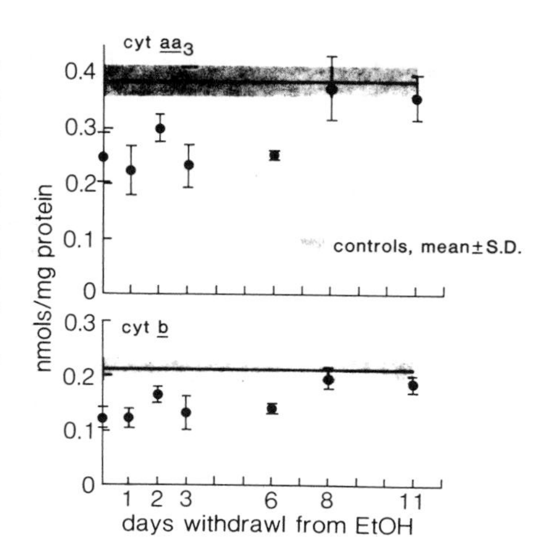

FIGURE 3. Time course for reversal of alcohol-induced cytochrome content alterations following ethanol withdrawal. Rats fed ethanol (36% calories) for 40–45 days (Day "0" of withdrawal) were subsequently fed control liquid diet without ethanol for periods as indicated. Liver submitochondrial particles were then prepared and cytochrome contents measured spectrophotometrically. Cyt. aa_3 content was determined from dithionite reduced minus oxidized difference spectra; cyt. *b* content was determined from succinate reduced plus antimycin (aerobic) minus oxidized spectra.[3] Each point indicates mean ± S.D. for 3 preparations from ethanol-withdrawn rats. Data for corresponding pair-fed control rats (never fed ethanol) are shown as the *shaded bar,* mean ± S.D. (n = 21).

labeling studies of protein degradation *in vivo* have recorded half-lives ranging from 3.8 to 10 days for mitochondrial inner membrane proteins.[7,8] Consequently, the time course data suggest that reversal of the alcohol-induced protein changes may involve a resynthesis and replacement of affected proteins.

IMMUNOCHEMICAL STUDIES OF THE EFFECT OF ETHANOL CONSUMPTION ON CYTOCHROME OXIDASE

In order to learn more about molecular mechanisms responsible for the effects of chronic ethanol consumption on mitochondrial proteins, we have focused our studies on cytochrome oxidase as a model. The biochemical data outlined above clearly indicate that the amount of enzymatically-active, or functional, cytochrome oxidase in membranes of alcoholic animals is only about half that of pair-fed controls. However, it is important to note that the physical content of the protein or apoprotein moiety, per se, was not determined in these previous studies. Both spectroscopy and activity measurements are dependent on the heme component of cytochrome oxidase. To address the fate of the protein portion of cytochrome oxidase, we have used an immunochemical approach.[9] We first purified cytochrome oxidase[10] from control rats and used it to raise a specific antibody in rabbits. The antiserum then served as a probe for detection of cytochrome oxidase protein in submitochondrial particles from alcoholic and control rats.

Several tests were used to characterize the antiserum obtained from rabbits that had been immunized with purified rat liver cytochrome oxidase. The antiserum inhibited cytochrome oxidase by precipitating the enzyme and thus removing it from solution. This occurred with both purified rat liver cytochrome oxidase preparations and triton-extracts of submitochondrial particles. Pre-immune serum did not cause precipitate formation and did not inhibit oxidase activity. Spectral analyses verified that heme aa_3 was removed from solution on formation of the immunoprecipitates. In double immunodiffusion studies, the antiserum formed a single precipitin line when tested against either purified cytochrome oxidase or triton-extracts of submitochondrial particles. Upon electrophoresis under dissociating conditions,[11] immunoprecipitates derived from purified cytochrome oxidase demonstrated the known subunit bands of the native enzyme, as well as IgG bands.

Our approach for measuring cytochrome oxidase was that of immunoinhibition titrations.[12] In this method triton-extracts of submitochondrial particles containing the antigen (Ag), cytochrome oxidase protein, were incubated overnight with the antiserum (Ab). Following centrifugation to remove the immunoprecipitate formed, the amount of antigen remaining in the supernatant was determined by measuring its cytochrome oxidase activity. Titration patterns predicted for extracts containing 1.0 and 0.5 units of Ag per unit of extract protein are shown in FIGURE 5. These would be analogous to those expected for control and alcoholic membranes, respectively, if enzymatically-active cytochrome oxidase was the only antigenic species present in the extracts. In the absence of antibody submitochondrial particles from alcoholic rats displayed only about one-half the apparent cytochrome oxidase activity per mg

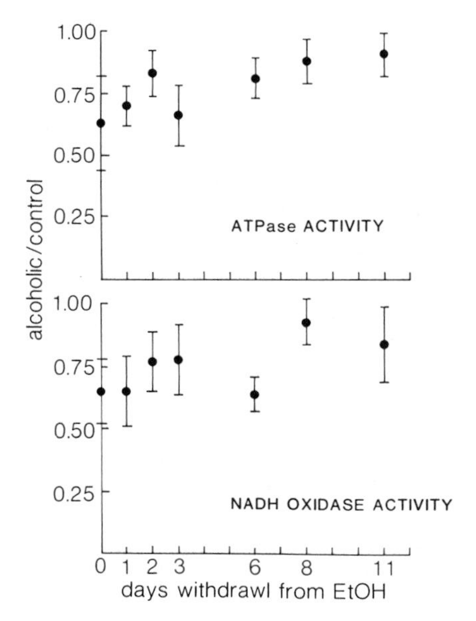

FIGURE 4. Time course for reversal of alcohol-induced enzyme activity alterations following ethanol withdrawal. Submitochondrial particles were prepared from alcoholic rats withdrawn from ethanol for various periods as described in FIGURE 3. Maximal rates of ATP hydrolysis and NADH respiration were measured in the presence of an uncoupler.[2] Data indicate the ratio of specific activities (nmol/min/mg protein) for the ethanol-withdrawn ("alcoholic") to the corresponding pair-fed control rats. Points indicate mean ± S.D. for 3 pairs of preparations at each time.

membrane protein as those from controls (FIG. 6), in agreement with previous findings. However, titrations of a fixed amount of anti-oxidase serum with increasing amounts of submitochondrial particle extract from alcoholic or control rats showed an upward deflection at equivalent amounts of membrane added to the incubation (FIG. 6). After incubation with antiserum, the plots of absolute activity versus amount of membrane protein were displaced toward higher protein concentration by equivalent amounts with both alcoholic and control rats. Titrations of a fixed amount of submitochondrial particles with increasing antiserum likewise showed that equivalent amounts of antiserum were necessary to inhibit the respective activity of each membrane preparation by 50% of its original value (FIG. 7). By contrast with predicted results, these titration patterns were indicative of a similar amount of immunologic reactivity per mg membrane protein in both preparations.

The absolute amounts of protein present in immunoprecipitates were comparable

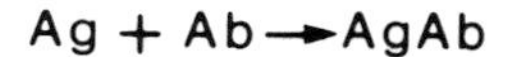

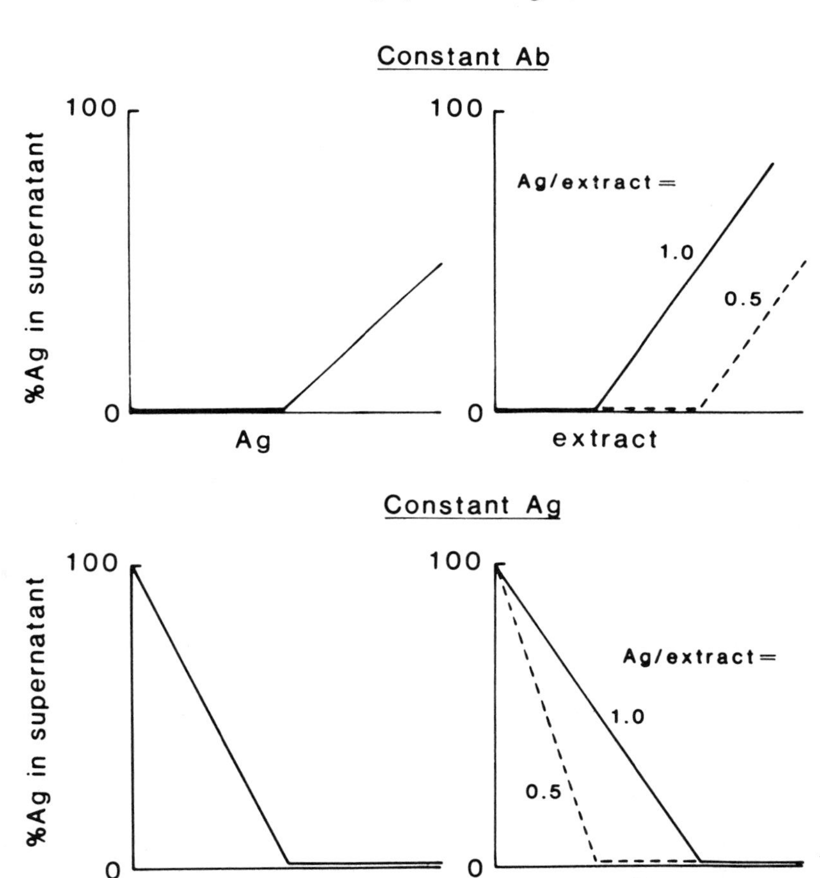

FIGURE 5. Theoretical immunotitration curves expected for reaction of a precipitating antibody (Ab) with an antigen (Ag) to form an insoluble immunoprecipitate. *Left:* General case; *Right:* Different amounts of Ag present in a constant amount of crude extract (*e.g.*, solubilized membrane preparation). When the Ag is an enzyme, it can be quantified by measuring its catalytic activity. The predicted results are based on the assumption that only active enzyme is present in the extract and recognized as antigen. Note that Ag/extract = 0.5 would be analogous to the approximately 50% decrease in heme aa_3 or cytochrome oxidase activity per mg submitochondrial particle protein found in alcoholic rats compared to controls having Ag/extract = 1.0.

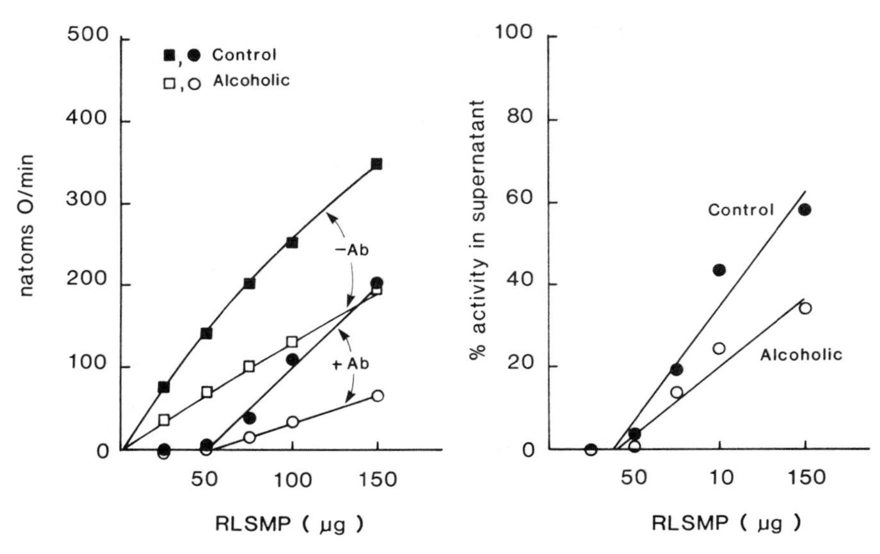

FIGURE 6. Membrane protein titration at fixed antibody concentration. *Left:* Rat liver submitochondrial particles (RLSMP) were solubilized in 0.5% triton X-100, 15 mM NaP_i, pH 7.4, and incubated overnight with 100 μl of antiserum (Ab) in a volume of 0.40 ml. Following centrifugation, aliquots of the supernatants were tested for cytochrome oxidase activity by a polarographic assay (*circles*). Activity in the absence of antiserum (*squares*) was measured similarly using an equivalent amount of protein for each incubation. *Right*: Results are expressed as the percentage of added activity recovered in the supernatants after immunoprecipitation (*i.e.*, natoms 0/min, $+Ab/-Ab \times 100\%$). Control: *solid symbols;* alcoholic: *open symbols.* (From Thayer and Rubin.[9] Reprinted by permission from *Biochimica et Biophysica Acta.*)

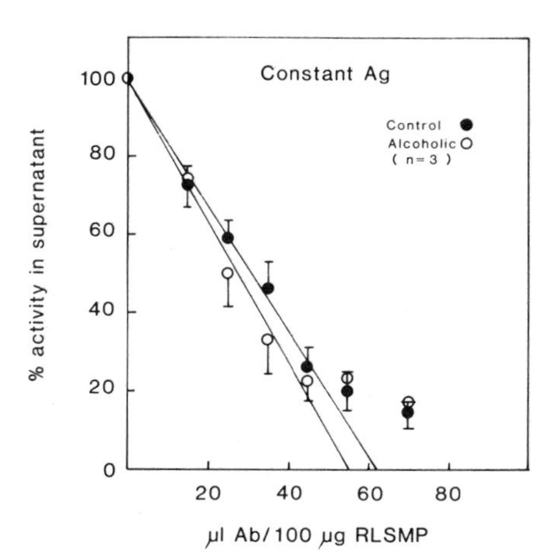

FIGURE 7. Antiserum titration at fixed membrane protein extract concentration. Rat liver submitochondrial particles (0.1 mg protein) were solubilized as described in FIGURE 6 and incubated overnight with varying amounts of antiserum (Ab) in a final volume of 0.10 ml. Following centrifugation, cytochrome oxidase activity remaining in the supernatants was measured by a spectrophotometric assay. Symbols indicate mean and S.D. for 3 pairs of submitochondrial particles from control (*solid circles*) and alcoholic (*open circles*) rats. (From Thayer and Rubin.[9] Reprinted by permission from *Biochimica et Biophysica Acta.*)

with both alcoholic and control preparations, in either fixed antiserum or fixed membrane protein titration experiments (FIG. 8). In addition, the electrophoretic profiles of immunoprecipitates from alcoholic and control rats were characteristic of cytochrome oxidase and similar with respect to both band locations and staining intensities. These findings corroborated the immunoinhibition titration data and provided additional evidence that both alcoholic and control membranes had quantitatively similar cytochrome oxidase immunologic reactivity.

These results indicate that the protein moiety of cytochrome oxidase recognizable by the antiserum must be present in similar amounts within submitochondrial particles of alcoholic and control rats. Together, the biochemical and immunochemical data

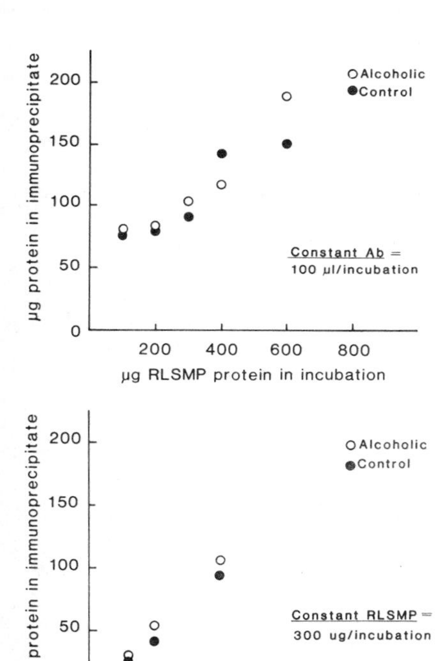

FIGURE 8. Comparison of amounts of protein in immunoprecipitates. Total protein of washed immunoprecipitates was measured by the Lowry method. *Upper panel:* Submitocondrial particle protein titration at fixed antiserum amount (100 μl) per 0.40 ml incubation. *Lower panel:* Antiserum titration at fixed submitochondrial particle protein concentration (300 μg) per 0.40 ml incubation. Control: *solid circles;* alcoholic: *open circles.*

suggest the presence of an inactive form of cytochrome oxidase in mitochondrial membranes of rats fed ethanol. That is, cytochrome oxidase biochemical reactivity is one-half but cytochrome oxidase immunologic reactivity is unchanged after alcohol treatment. The inactive form of oxidase is apparently devoid of heme, since the latter is undetectable spectrophotometrically.

IMPLICATIONS FOR MEMBRANE ASSEMBLY

The molecular mechanism underlying formation of the inactive form of cytochrome oxidase in ethanol-fed animals remains to be elucidated. One possibility is that ethanol consumption may interfere with the attachment of heme *a* to cytochrome oxidase. Alternatively, ethanol might cause an inactivation of pre-existing functional oxidase protein. Previous studies have demonstrated that chronic ethanol consumption decreases the ability of mitochondria to incorporate amino acids into membrane proteins, some of which have been suggested to be subunits of cytochrome oxidase.[13,14] The present findings do not rule out ethanol-related effects on mitochondrial protein synthesis as being involved in the alteration of cytochrome oxidase, but suggest that subunits of the oxidase are present in normal amounts in hepatic mitochondrial membranes of alcoholic rats. It is possible, however, that the oxidase subunits may be assembled incorrectly in alcoholic animals.

Studies of mitochondrial biogenesis have made considerable progress toward delineating the sequence of molecular events involved in the assembly of cytochrome oxidase and other mitochondrial proteins. For example, it is known that the cytoplasmically synthesized subunits of ATPase, and possibly of oxidase, are imported to the matrix as precursor proteins in an energy-dependent reaction prior to processing and assembly with the mitochondrially synthesized subunits.[15,16] In the case of cytochrome oxidase, subunits from both the mitochondrial and cytoplasmic translation systems appear to be assembled in a specific ordered sequence within the mitochondria.[17] In addition, recent evidence has shown that heme *a* promotes association of specific oxidase subunits during the assembly process.[18] However, many details in the assembly of cytochrome oxidase remain unknown. In particular, relationships between insertion of the protein into the lipid bilayer and heme attachment have not been clarified. Thus, alterations in membrane lipids associated with ethanol consumption may play a role in modulating the assembly of functional cytochrome oxidase. The existence of a heme-deficient form of oxidase in submitochondrial particles from alcoholic animals does suggest that during normal biogenesis oxidase subunits may be fully assembled and inserted into the membrane prior to heme attachment.

IMPLICATIONS FOR MECHANISM OF ACTION OF ETHANOL

The existence of an inactive form of cytochrome oxidase in ethanol-fed rats provides a fresh insight into the actions of alcohol. This finding may clarify at least one puzzling set of observations. Because by functional criteria several major mitochondrial membrane proteins appear decreased by 30–50% after chronic ethanol consumption, and other proteins do not appear compensatorily increased, one would anticipate a substantial decrease in total mitochondrial protein per gram liver. By contrast, the amount of total mitochondrial protein per gram liver is unchanged after chronic ethanol consumption.[19] However, if other membrane proteins are also present in inactive forms, analogous to the condition of cytochrome oxidase, then both sets of data would be readily understood.

These findings with cytochrome oxidase point toward a link between ethanol consumption and heme and/or iron metabolism as perhaps playing a key role in expression of the effects of ethanol on membrane proteins. It is noteworthy that another characteristic effect produced by chronic ethanol consumption is proliferation of the smooth endoplasmic reticulum and an increase in the cyt.-P450 content of microsomal membranes.[20,21] One study with isolated hepatocytes found that the total heme content of the rat liver was not altered by chronic ethanol consumption.[22] The decrease in mitochondrial hemoproteins was counterbalanced by the increase in microsomal hemoproteins.[22] Thus, chronic ethanol consumption appears to induce a redistribution of heme between mitochondrial and microsomal membranes. This could reflect some teleological priority of cyt.-P450, or simply its rapid rate of synthesis compared to mitochondrial cytochromes. However, other mechanisms are probably involved in the actions of ethanol, because the content of ATPase, which does not contain iron, is also affected by alcohol consumption. Cytochrome oxidase, cytochrome *b* (a component of ubiquinol-cytochrome c reductase), ATPase, and perhaps NADH dehydrogenase, the four protein complexes altered by alcohol consumption, are multisubunit enzymes composed of products derived from both mitochondrial and cytoplasmic translation.[15,16] This particular pattern of selective effects suggests a link between mitochondrial protein synthesis and the actions of ethanol. However, instead of direct inhibition of mitochondrial translation, this pattern might be explained by an effect of ethanol on a mitochondrial processing enzyme common to assembly of these protein complexes. The relatively long time required for reversal of the alcohol-induced protein changes following ethanol withdrawal also suggests the involvement of protein synthesis at some point in the mechanism of action of ethanol. Finally, the alternative possibility that mitochondrial membrane protein changes arise from inactivation of pre-existing functional proteins, perhaps linked in some way to ethanol metabolism, likewise deserves consideration.

In conclusion, the effects of ethanol consumption on mitochondrial membrane proteins appear to reflect subtle alterations in membrane assembly. We now know that alcoholic animals have markedly different mitochondrial membranes than controls with respect to both structure and function of their protein components. In future years, unraveling the details of the molecular mechanisms by which these differences come about should advance considerably our understanding of the pathophysiology of ethanol consumption.

ACKNOWLEDGMENTS

I thank Dr. Emanuel Rubin for giving me the opportunity to join the field of alcohol research and for his many stimulating discussions throughout the course of this work. I also thank the many research assistants and other colleagues who have contributed to the studies reported in this article.

REFERENCES

1. Cederbaum, A. I., C. S. Lieber & E. Rubin. 1974. Arch. Biochem. Biophys. **165:** 560–569.
2. Thayer, W. S. & E. Rubin. 1979. J. Biol. Chem. **254:** 7717–7723.
3. Thayer, W. S. & E. Rubin. 1981. J. Biol. Chem. **256:** 6090–6097.
4. Thayer, W. S. & E. Rubin. 1980. *In* Alcohol and Aldehyde Metabolizing Systems. R.G. Thurman, Ed. Vol. 4: 385–392. Plenum Publishing Corp. New York, NY.
5. Thayer, W. S. & E. Rubin. 1982. Biochim. Biophys. Acta **721:** 328–335.
6. Thayer, W. S., T. Ohnishi & E. Rubin. 1980. Biochim. Biophys. Acta **591:** 22–36.
7. Lipsky, N. G. & P. L. Pedersen. 1981. J. Biol. Chem. **256:** 8652–8657.
8. Beattie, D. S., R. E. Basford & S. B. Koritz. 1967. J. Biol. Chem. **242:** 4584–4586.
9. Thayer, W. S. & E. Rubin. 1986. Biochim. Biophys. Acta **849:** 366–373.
10. Kuboyama, M., F. C. Yong & T. E. King. 1972. J. Biol. Chem. **247:** 6375–6383.
11. Kadenbach, B., J. Jarausch, R. Hartmann & P. Merle. 1983. Anal. Biochem. **129:** 517–521.
12. Schimke, R. T. 1975. Methods Enzymol. **40:** 241–266.
13. Burke, J. P. & E. Rubin. 1979. Lab. Invest. **41:** 393–400.
14. Bernstein, J. D. & R. Penniall. 1978. Alcoholism: Clin. Exp. Res. **2:** 301–310.

15. ADES, I. Z. 1982. Mol. Cell. Biochem. **43:** 113–127.
16. HARMEY, M. A. & W. NEUPERT. 1985. *In* The Enzymes of Biological Membranes. 2nd edit. A. Martonosi, Ed. Vol. 4: 431–464. Plenum Press. New York, NY.
17. WIELBURSKI, A. & B. D. NELSON. 1983. Biochem. J. **212:** 829–834.
18. WIELBURSKI, A. & B. D. NELSON. 1984. FEBS Lett. **177:** 291–294.
19. CEDERBAUM, A. I., C. S. LIEBER, A. TOTH, D. S. BEATTIE & E. RUBIN. 1973. J. Biol. Chem. **248:** 4977–4986.
20. THURMAN, R. G. 1973. Mol. Pharmacol. **9:** 670–675.
21. RUBIN, E., P. BACCHIN, H. GANG & C. S. LIEBER. 1979. Lab. Invest. **22:** 569–580.
22. JONES, D. P., S. ORRENIUS & H. S. MASON. 1979. Biochim. Biophys. Acta **576:** 17–29.

DISCUSSION OF THE PAPER

QUESTION: Have you ever examined total mitochondrial protein content in response to chronic ethanol feeding? Does this change at all in your hands?

THAYER: Although I have not measured it, Drs. Cederbaum and Rubin have reported that the amount of mitochondrial protein per gram of liver is actually not changed in these alcoholic rats. This is consistent with my results. If all of the mitochondrial proteins are decreased so much, say 30 to 50%, and these are the major proteins of the inner mitochondrial membrane, you would think that there would probably be a decrease in total mitochondrial protein per gram liver. But it turns out that the data does not support this. So I think the idea that there could be inactive forms of cytochrome oxidase and possibly the other proteins might be an explanation of that if this holds up.

QUESTION: Cytochrome oxidase as I recall, has certain subunits synthesized in the cytoplasm and others in the mitochrondria itself. Have you examined the possibility that in your immunoprecipitates one or the other subunits might be decreased somewhat as a result of ethanol treatment?

THAYER: Essentially we see all the subunits. The alcoholic immunoprecipitates show all the subunits of cytochrome oxidase seen for the controls. There is one very intriguing difference that we are not absolutely certain about; in the alcoholic ones there's one particular subunit—subunit 4—that sometimes shows up as a double band, suggesting two forms. There may be some subtle alterations that we need to characterize in more detail.

Phospholipase Activity Is Enhanced in Liver Microsomes from Chronic Alcoholic Rats[a]

CELESTE L. PRYOR,[b] CHRISTOPHER D. STUBBS,[b]
AND EMANUEL RUBIN[b]

Hahnemann University
Philadelphia, Pennsylvania 19102

Chronic ethanol intake is known to give rise to a resistance in cell membranes to the disordering effect of the lipids, *in-vitro* addition of other hydrophobic agents as well as by ethanol itself (for reviews see REFS. 1, 2). Also alterations in the fatty acid profiles of the phospholipids have been demonstrated to occur. We have therefore initiated studies on the enzymes involved in the re-tailoring of the membrane phospholipid fatty acid profiles and here report on studies on phospholipase activities in rat liver microsomes.

TABLE 1. Phospholipase Activity in Liver Microsomes from Control and Ethanol-Fed Rats

Addition		Control		Alcoholic	
None		2.77 ± 0.08[a]	100%	1.38 ± 0.11	100%
Ethanol	200 nM	3.04 ± 0.05	110	1.48 ± 0.04	107
	500 nM	2.66 ± 0.01	96	1.38 ± 0.01	100
Butanol	50 nM	1.33 ± 0.05	48	0.71 ± 0.02	52
Propanol	50 nM	2.69 ± 0.20	97	1.38 ± 0.10	100

[a]nmol NBD-hexanoic acid released/min/mg protein at 37° C, +/− std. dev. (n = 3).

The activity of phospholipase in microsomes was measured using a new and highly sensitive fluorescence assay[3] that uses phosphatidylcholine (PC) with a fluorescent fatty acid at the sn-2 position of the glycerol backbone (1-palmitoyl,2-N-(4-nitrobenzo-2-oxa-1,3-diazole-aminohexanoyl)-PC; NBD-PC). On hydrolysis by phospholipase A_2, NBD-hexanoic acid is released. The reaction is stopped by the addition of chloroform/methanol and a lipid extraction performed according to the method of Bligh and Dyer.[4] The NBD-hexanoic acid is then recovered in the upper aqueous phase after the lipid extraction, whereas the unhydrolysed NBD-PC remains in the lower chloroform phase, from which it is therefore easily separated. In previous studies we compared the activity of purified pancreatic phospholipase A_2 against liposomes of 1-palmitoyl,2-oleoyl-PC (POPC) with varying mole % of NBD-PC.[3] It was found that

[a]Supported by National Institute on Alcohol Abuse and Alcoholism Grants AA3442, AA05662, and AA7457.
[b]Present address: Department of Pathology, Thomas Jefferson University, Philadelphia, PA 19107.

the activity was independent of the amount of NBD-PC present and was the same whether it was measured as the release of NBD-hexanoic acid or as the release of ^{3}[H] oleic acid from ^{3}[H]-oleoyl radiolabeled POPC. This shows that the phospholipase does not distinguish between the NBD-fatty acyl moiety and a natural fatty acid chain.

The NBD-PC was introduced into the microsomes as an aqueous dispersion, and it was found to incorporate within seconds at 37° C.[3] The NBD-hexanoic acid released was then quantified by measurement of the fluorescence emission at 530 nm on excitation at 472 nm. In FIGURE 1 the temperature dependence of the phospholipase activity of microsomes from control and alcoholic animals is shown. Chronic ethanol intake was found to result in a considerably elevated activity and the ratio of

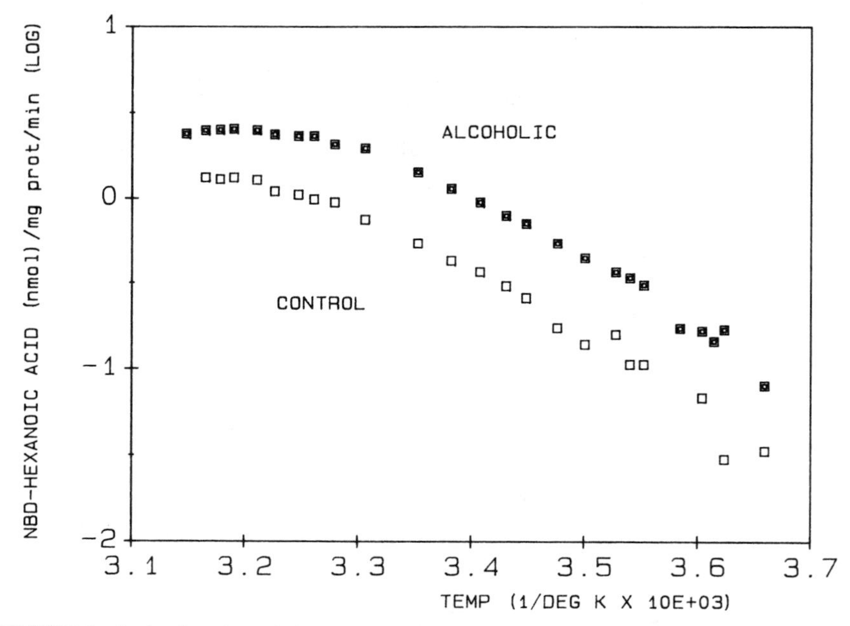

FIGURE 1. Arrhenius plots of the activity of microsomal phospholipases for liver microsomes from a control and ethanol-fed littermate pair of rats.

alcoholic/control was 2.15+/−0.04 (for four pairs of animals). The temperature dependences also show a difference, in terms of discontinuities in the Arrhenius plots, which may be attributable to an altered physical environment in the microsomes from the alcoholic animals. It is also possible that the levels of the different contributing phospholipases[5] present in microsomes (A_2, A_1, and lyso-) are affected by ethanol consumption in a different manner.

The *in-vitro* effect of several alcohols was also tested as shown in TABLE 1. At low concentrations, within the physiological range, ethanol was found to cause a 7–10% stimulation, which was lost at higher concentrations. In contrast, butanol was found to be an inhibitor, while propanol had little effect. In testing a range of membrane perturbants, apart from ethanol, so far no other stimulatory agents have been found.

REFERENCES

1. TARASCHI, T. F. & E. RUBIN. 1985. Lab. Invest. **52:** 120–131.
2. SUN, G. Y. & A. Y. SUN. 1985. Alcoholism **9:** 164–180.
3. STUBBS, C. D., C. L. PRYOR, J. B. HOEK, N. HARADA, M. TOMSHO & E. RUBIN. 1986. Submitted.
4. BLIGH, E. G. & W. J. DYER. 1959. Can. J. Biochem. Physiol. **37:** 911–917.
5. VAN DEN BOSCH, H. 1980. Biochim. Biophys. Acta **604:** 191–246.

Chronic Ethanol Ingestion Decreases the Partitioning of Hydrophobic Molecules into Cell Membranes Due to an Alteration in the Phospholipids[a]

YUSHENG NIE,[b] CHRISTOPHER D. STUBBS,[b]
AND EMANUEL RUBIN[b]

Hahnemann University
Philadelphia, Pennsylvania 19102

Chronic ethanol ingestion is known to induce adaptive changes in the structure, composition, and functioning of cell membranes. Previous studies (for recent reviews see REFS. 1, 2) have shown that the lipids of membranes from animals subjected to chronic ethanol intake are resistant to disordering or fluidizing by various hydrophobic agents, including ethanol itself. Other studies[3,4] have shown that a decreased partitioning into the membranes also results. The major components of cell membranes are the neutral lipids, phospholipids, and proteins. In the present study, we investigated which of these components might be altered as a result of chronic ethanol intake by measuring the partitioning of a model hydrophobic compound into intact membranes, liposomes of total lipids (including neutral lipids and phospholipids), and liposomes of phospholipids.

The effects of chronic ethanol intake on partitioning were determined using 5-doxyl-decane (5DD) as the model hydrophobic compound,[3,4] which distributes partially into the aqueous phase. First 1,6-diphenyl-1,3,5-hexatriene (DPH) was incorporated into the membranes or liposomes. The molar partition coefficient of 5DD was then calculated from the degree of quenching of the fluorescence of DPH that resides in the center of the lipid bilayer. The greater the degree of quenching the

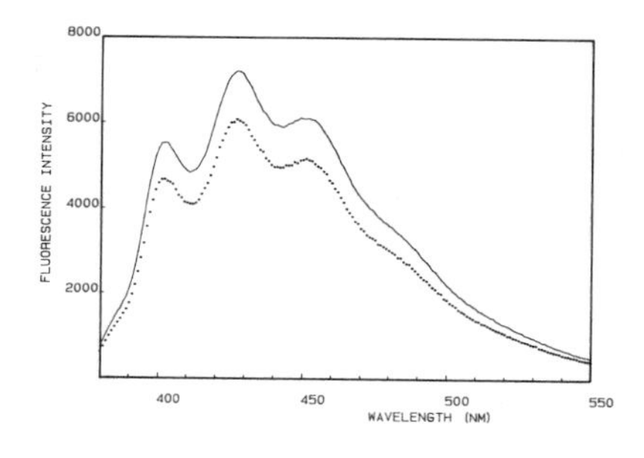

FIGURE 1. The emission spectra of DPH in liposomes of microsomal total lipids on excitation at 360 nm before (*solid line*) and after (*dashed line*) the addition of 5-doxyl decane (2.3 μM).

[a]Supported by National Institute on Alcohol Abuse and Alcoholism Grants AA3442, AA05662, and AA7457.

[b]Present address: Department of Pathology, Thomas Jefferson University, Philadelphia, PA 19107.

TABLE 1. Effect of Chronic Ethanol Intake on the Partition Coefficient of 5-Doxyl Decane into Rat Liver Microsomal Membranes

		Partition Coefficient	Temperature (°C)
Intact microsomes	Control	6210	25
	Alcoholic	5240 (84% of C)	
Liposomes of total lipids	Control	6235	25
	Alcoholic	5042 (83% of C)	
Liposomes of total lipids	Control	5560	37
	Alcoholic	4640 (83% of C)	
Liposomes of total phospholipids	Control	5853	37
	Alcoholic	5303 (91% of C)	

greater the partition coefficient. The quenching of DPH fluorescence is illustrated in FIGURE 1.

Microsomes from rats kept on liquid ethanol diets for 35 days were prepared as previously described[5] and total lipid extracts and total phospholipids obtained. The membranes or liposomes were labelled with DPH (at 0.25 mole % of the phospholipids) and the degree of fluorescence quenching determined for different quencher (5DD) concentrations determined, and the molar partition coefficient calculated,[6] according to the equations:

$$Fo/F = 1 + Kapp\,[Q] \qquad (1)$$

$$1/Kapp = Fm(1/K - 1/KP) + (1/KP) \qquad (2)$$

where Fo and F are the fluorescence intensities before and after quencher addition (concentration [Q]), Fm the volume fraction of the membrane, K the biomolecular quenching constant, P the partition coefficient, and $Kapp = K.P/(PFm + (1 - Fm))$. First Fo/F vs [Q] plots are constructed for different Fm (Equation 1) to obtain different Kapp. Then plots of 1/Kapp vs Fm are constructed from which P can be calculated (Equation 2).

The results (see TABLE 1) show that the partition coefficient of 5DD into intact membranes is markedly decreased after chronic ethanol intake compared to the control. The decrease is also seen in the reconstituted liposomes of microsomal total lipids and phospholipids showing that the primary course of the decrease in the partition coefficient is due to an alteration in the phospholipid composition. The difference between the partition coefficients of the phospholipids was less than for the total lipids showing that a neutral lipid component may also be involved in the decrease in partitioning caused by ethanol ingestion.

REFERENCES

1. TARASCHI, T. F. & E. RUBIN. 1985. Lab. Invest. **52:** 120–131.
2. SUN, G. Y. & A. Y. SUN. 1985. Alcoholism **9:** 164–180.
3. ROTTENBERG, H., A. WARING & E. RUBIN. 1981. Science **213:** 583–585.
4. KELLY-MURPHY, S., A. WARING, H. ROTTENBERG & E. RUBIN. 1984. Lab. Invest. **50:** 178.
5. PONNAPPA, B. C., A. WARING, J. B. HOEK, H. ROTTENBERG & E. RUBIN. 1982. J. Biol. Chem. **257:** 10141–10146.
6. LAKOWICZ, J. R., D. HOGEN & G. OMANN. 1977. Biochim. Biophys. Acta **471:** 401–411.

PART IV. EFFECTS OF ETHANOL ON TRANSPORT AND PROTEIN SYNTHESIS

Effects of Ethanol on Calcium Homeostasis in Rat Hepatocytes and Its Interaction with the Phosphoinositide-Dependent Pathway of Signal Transduction[a]

JAN B. HOEK

Department of Pathology
Thomas Jefferson University
Philadelphia, Pennsylvania 19107

CALCIUM HOMEOSTASIS IN HEPATOCYTES IN THE BASAL AND HORMONE-STIMULATED STATE

Concepts of the regulation of cellular calcium homeostasis have undergone marked revision in recent years. In large part, this development is due to the recognition of the widespread role of polyphosphoinositide turnover in signal transduction processes associated with calcium mobilization (see REFS. 1, 2 for reviews). In hepatocytes, as in many other cells, the activation of a polyphosphoinositide-specific phospholipase C in response to hormone-receptor interaction at the plasma membrane, results in the generation of at least two distinct second messenger molecules. One of these, inositol-1,4,5-trisphosphate (Ins-1,4,5-P_3) has been shown to open a Ca^{2+} channel in an intracellular storage site, presumably the endoplasmic reticulum.[2] This allows calcium to move rapidly down the concentration gradient maintained across the endoplasmic reticular membrane by the action of the ATP-dependent Ca^{2+} pump system. The resulting elevation of cytosolic calcium levels activates or potentiates a multitude of calcium-dependent protein kinases and other calcium-dependent enzyme activities, many of them mediated by the interaction of Ca^{2+} with calmodulin. One of these proteins is glycogen phosphorylase kinase, and the coversion of (inactive) phosphorylase *b* to the (active) phosphorylase *a* has been used extensively as a sensitive test for changes in cytosolic free Ca^{2+} in hepatocytes.

The elevation of cytosolic Ca^{2+} levels leads to increased Ca^{2+} transport processes in the mitochondrial and plasma membranes. An ATP-driven Ca^{2+} pump in the plasma membrane actively removes Ca^{2+} from the cytosol, and the cells are partly depleted of Ca^{2+} if this efflux is not balanced by an accompanying stimulation of calcium influx down the concentration gradient across the plasma membrane (see REFS. 2, 3 for review). There is evidence that hormone treatment indeed activates a plasma membrane Ca^{2+} channel.[2,3] Thus, the hormone-receptor interaction at the plasma membrane results not only in a redistribution of intracellular calcium stores,

[a]Supported by United States Public Health Service Grants AA05662, AA03422, and AM31086.

but also in an increased rate of cycling of calcium ions across the endoplasmic reticular membrane, and the plasma membrane as well as, presumably, across the mitochondrial membrane. On removal of the hormonal stimulus, the Ins-1,4,5-P_3 level falls, allowing the reaccumulation of calcium into the hormone-responsive endoplasmic reticular stores, the calcium influx across the plasma membrane is inactivated, and the cytosolic free Ca^{2+} levels return to basal.

EFFECT OF ETHANOL ON CYTOSOLIC FREE Ca^{2+} LEVELS IN INTACT HEPATOCYTES

Our interest in the potential interactions of ethanol with cellular calcium homeostasis was based on the hypothesis that membrane-bound ion transport activities may be sensitive to ethanol-associated changes in membrane ordering. Our earlier studies on isolated liver microsomes[4] supported this notion, and other authors have indicated

TABLE 1. Effect of Different Stimuli on Phosphorylase Activation and cAMP Levels in Intact Hepatocytes[a]

Addition	Phosphorylase *a* (nmol/min/mg prot)	cAMP (pmol/mg prot)
None	39 ± 5	0.37 ± 0.05
Ethanol (200 mM)	198 ± 35[b]	0.43 ± 0.07
Vasopressin (60 nM)	297 ± 23[b]	0.44 ± 0.15
Glucagon (10 nM)	210 ± 28[b]	1.79 ± 0.20

[a]Isolated hepatocytes (5 mg protein/ml) were preincubated for 30 min at 37° C in a Krebs-Ringer bicarbonate medium with lactate (10 mM) and pyruvate (1 mM) and 1% bovine serum albumin. Reactions were terminated 30 sec after addition of the stimulus, either with $HClO_4$ (for cAMP assay), or with an ice cold stopping medium containing 1 mM EDTA, 50 mM NaF, 0.2 mM digitonin, and 25 mM -glycerophosphate. Phosphorylase activity was measured by the method of Gilboe[22] and cAMP by the method of Gilman and Murad.[23] For further experimental details, see REF. 11.

[b]$p < 0.01$.

various degrees of modification of Ca^{2+} transport activities in other tissues as a consequence of the membrane-disordering effects of ethanol.[5] However, the revelance of these studies to the situation in intact hepatocytes remained to be elucidated.

In the work summarized in this report, we approached the study of calcium homeostasis at the level of the whole cell by examining the effect of ethanol treatment on calcium fluxes and intracellular calcium levels in intact, isolated hepatocytes. The analysis of calcium homeostasis in intact cells has been greatly facilitated since the introduction by Tsien and co-workers[7,8] of a series of fluorescent calcium indicators, of which quin 2 has so far been the most widely used. These indicators can be loaded into intact cells by preincubation with the tetraacetoxylmethylester, which readily penetrates across the cell membrane.[9] In the cytosol, the ester bonds are hydrolyzed by nonspecific esterases and the negatively charged indicator is trapped in the cell.[9] Much of the work presented here has relied on the use of quin 2-loaded hepatocytes.

The key observation that led us to this work is summarized in TABLE 1; ethanol, when added to isolated hepatocytes, caused a rapid activation of phosphorylase, without an accompanying rise in cAMP. This pattern is similar to that obtained with vasopressin or α_1-agonists, but different from glucagon. The activation of phosphorylase was comparable to that of calcium-mediated hormones in its rapid onset (maximal at < 30 sec) and in the lack of requirement for extracellular calcium (FIG. 1). However, the response to ethanol was more transient than that to the α_1-adrenergic agonist phenylephrine. The latter agent retained elevated phosphorylase levels for a prolonged period of time ($>$ 10 min) unless Ca^{2+} was removed from the incubation medium by the addition of EGTA (2.5 mM) immediately before the agonist. By contrast, the ethanol-induced phosphorylase activation decayed to baseline levels in approximately 5 min and was not markedly affected by the additon of an excess of EGTA (FIG. 1).

Further studies (see REF. 11) revealed that the activation of phosphorylase required a dose of ethanol in the range of 25–250 mM; this is the same range of

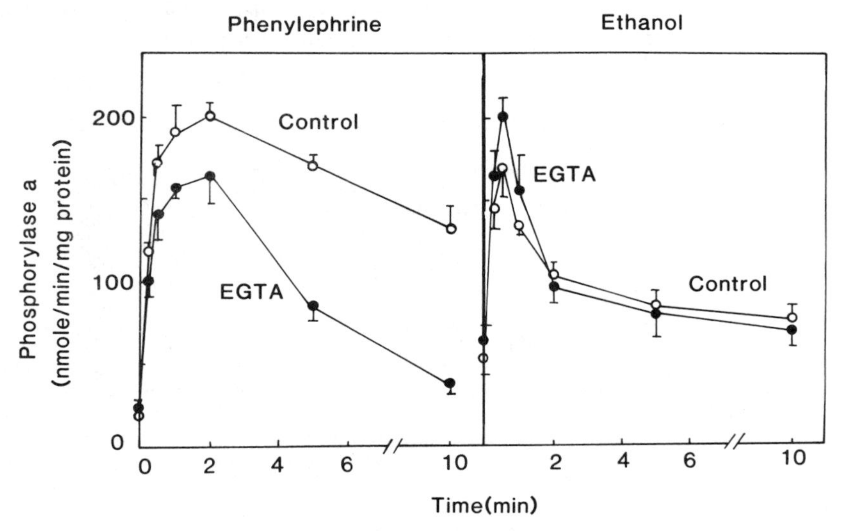

FIGURE 1. Time course of phosphorylase activation induced by ethanol (200 mM) and by phenylephrine (20 μM) in the absence and presence of EGTA. Hepatocytes isolated by a standard collagenase perfusion technique[10] were preincubated for 30 min at 37° C before addition of the agonist. For further experimental details, see REFERENCE 11. Where indicated, EGTA (2.5 mM) was added 5 min before the agonist.

concentrations where membrane disordering effects of ethanol become apparent. Other membrane disordering agents, including longer-chain alcohols and the general anesthetic halothane exerted similar effects at lower concentrations, whereas solvents such as dimethylsulfoxide and dimethylformamide, which do not markedly affect membrane ordering, had little effect on the phosphorylase activity. These characteristics are suggestive of a mechanism involving a disturbance of hydrophobic interactions (see below). A summary of these and other relevant characteristics of the ethanol-induced activation of phosphorylase is given in TABLE 2.

Direct evidence that ethanol interferes with cellular calcium homeostasis comes from studies with hepatocytes loaded with the fluorescent intracellular calcium indicator quin 2. A complication with this experimental system is the significant

TABLE 2. Characteristics of Ethanol-Induced Activation of Phosphorylase

1. Rapid and transient. Phosphorylase activation peaks after 20–30 sec and decays over a period of 5–10 min.
2. Requires ethanol levels in the "membrane-active" concentration range: half-maximal dose 75 mM ethanol, maximal (>90%) activation of phosphorylase at >250 mM ethanol.
3. Requires intact cells. Activation of phosphorylase kinase by ethanol in broken cells occurs only at concentrations of >0.3 m and requires >1 M ethanol for maximal activation.
4. Not sensitive to inhibition of alcohol dehydrogenase by 4-methylpyrazole.
5. Similar effects found with longer-chain alcohols, with other apolar solvents, and with the general anesthetic halothane.
6. Depends on intracellular calcium. Not affected by low extracellular calcium, but inhibited when cells are calcium-depleted.

overlap in the excitation and emission spectra of NADH and the quin 2-Ca complex. FIG. 2 illustrates two ways to distinguish between ethanol-induced NAD reduction and an ethanol-induced increase in cytosolic Ca^{2+} levels, detected by the fluorescence of the quin 2-Ca complex. In FIGURE 2a, a low concentration of ethanol (17 mM) was added to give maximal reduction of NAD. This concentration of ethanol is sufficient to saturate alcohol dehydrogenase, but has no detectable effect on phosphorylase activity. In FIGURE 2, the change in fluorescence induced by 17 mM ethanol was the same in quin 2-loaded cells and in unloaded control cells. A subsequent additon of 300 mM ethanol caused a reoxidation of the NADH in unloaded cells (FIG. 2b), due to substrate inhibition of alcohol dehydrogenase. In quin 2-loaded cells (FIG. 2a), transient increase of fluorescence upon addition of 300 mM ethanol is superimposed on the fluorescence change of NADH, indicating the increase in cytosolic free Ca^{2+} level. Digitonin is added subsequently to determine the fluorescence of Ca^{2+}-saturated quin 2. The fluorescence changes due to NADH can be effectively prevented by the addition of high levels (> 15 mM) of 4-methylpyrazole, an inhibitor of alcohol dehydrogenase. Under these conditions, only a minor increase in NADH is detectable upon addition of 300 mM ethanol. Thus, the ethanol-induced calcium mobilization is not dependent on the activity of alcohol dehydrogenase and in the quin 2-loaded cells a significant Ca^{2+}

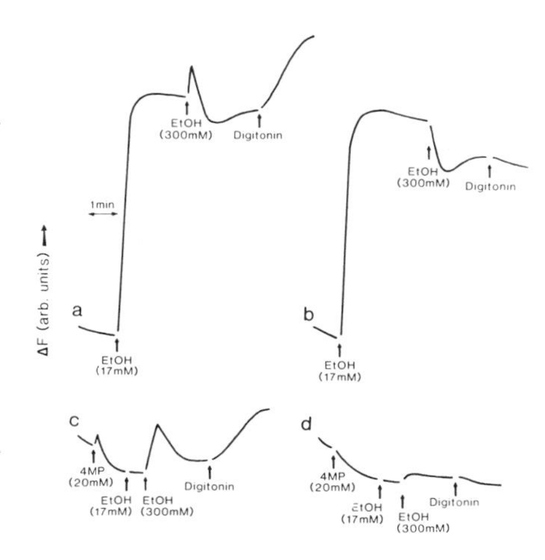

FIGURE 2. Effect of ethanol on fluorescence signal of quin 2-loaded hepatocytes (**a**) and (**c**) and unloaded cells (**b**) and (**d**). Conditions of quin 2 loading were described in detail elsewhere.[11] Batches of cells were loaded individually with 40 μM quin 2 acetoxylmethylester (from a stock solution in DMSO), washed twice, and preincubated for 5 min at 37° C before use. Quin 2-loaded cells were incubated in a bicarbonate-free salt medium, in the cuvet of a Perkin-Elmer MP-44B spectrofluorometer, using an excitation wavelength of 339 nm and an emission wavelength of 495 nm. Unloaded cells were treated similarly, but quin 2 AM was replaced by an equivalent volume of DMSO. 4 MP, 4-methylpyrazole.

increase can now be detected upon addition of 300 mM ethanol (FIG. 2c). At these concentrations, 4-methylpyrazole itself generates a small but transient increase in cytosolic Ca^{2+} levels, which, however, did not affect the subsequent response to ethanol. Correction for the contribution of NADH fluorescence is also possible by taking parallel measurements at excitation wavelengths of 339 nm and 357 nm (FIG. 3). At the latter wavelength, the quin 2-Ca^{2+} complex and free quin 2 have the same fluorescence. The fluorescence of NADH is essentially identical at these two wavelengths and the quin 2-Ca signal can be obtained by spectral subtraction. The difference spectra shown in FIGURE 3 with ethanol (300 mM) and phenylephrine (20 μM) as agonists, illustrate the kinetics of the changes in cytosolic free Ca^{2+}. The response to ethanol reached a maximum after 20–30 sec and decayed to baseline over a period of 2–4 min (FIG. 3a). The Ca^{2+} response to phenylephrine peaked after about 10 sec and decayed much slower (FIG. 3b). These kinetics are consistent with the changes in phosphorylase *a* activity obtained after ethanol addition (FIG. 1).

The ethanol-induced calcium mobilization in quin 2-loaded cells is dose-dependent. A detectable fluorescence response was obtained routinely with 50 mM ethanol; both the rate and the extent of the rise in cytosolic free Ca^{2+} increased with ethanol concentrations up to 500 mM. Higher levels of ethanol gave rise to a variable and inconsistent response, possibly reflecting interference with cellular integrity (data not shown). The cytosolic free Ca^{2+} concentration at the peak of an ethanol-induced Ca^{2+} response varies with the degree of quin 2 loading. In a series of experiments at different quin 2 levels, the peak free Ca^{2+} level attained after addition of 300 mM ethanol was 460 ± 47 nM at a quin 2 loading of 0.6 nmol/mg protein and 180 ± 18 nM at a loading of 3 nmol/mg protein (Mean ± S.E.M. of three experiments). In the same series a submaximal (2-μM) concentration of phenylephrine increased the cytosolic free Ca^{2+} level to 400 ± 21 nM and 207 ± 34 nM, respectively. Thus, the quin 2 loading may contribute a significant buffering capacity, which can suppress the changes in apparent cytosolic free Ca^{2+} with both ethanol and hormonal signals.

SOURCES OF THE CALCIUM MOBILIZED BY ETHANOL ADDITION

As with ethanol-induced phosphorylase activation, the rise and decay of cytosolic free Ca^{2+} levels on addition of ethanol detected with quin 2 is not significantly affected

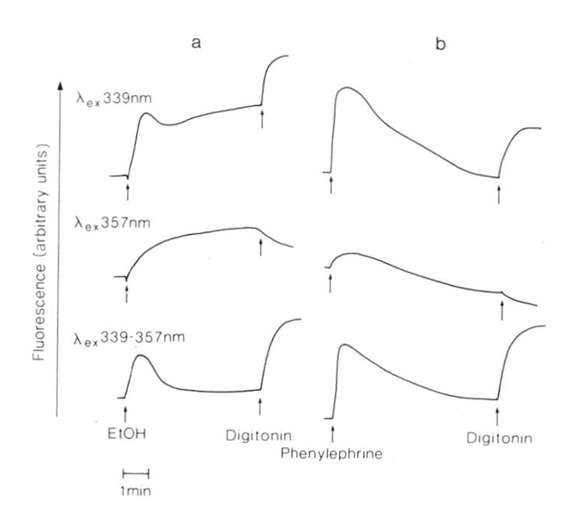

FIGURE 3. Fluoroescence changes induced by (**a**) ethanol (300 mM) and (**b**) phenylephrine (20 μM) in quin 2-loaded hepatocytes. The fluorescence excitation wavelength was manually alternated between 339 nm (peak of quin 2-Ca fluorescence) and 357 nm (isobestic point of quin 2-Ca complex and free quin 2).

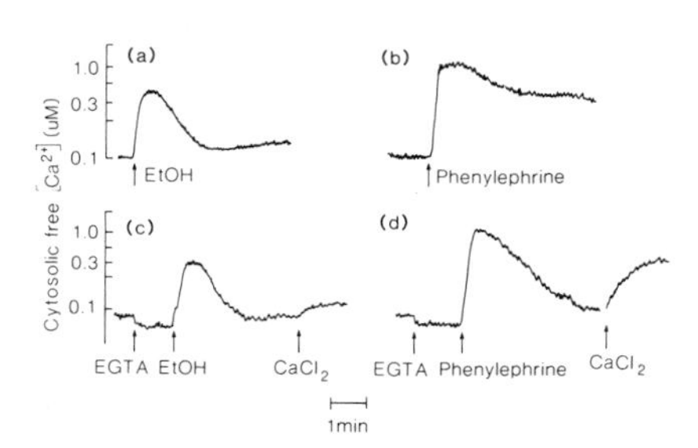

FIGURE 4. Effect of EGTA on calcium transients induced by ethanol (300 mM) or phenylephrine (20 μM). Cells were preincubated for 5–10 min in the standard incubation medium containing 1 mM $CaCl_2$ and 15 mM 4-methylprazole. Where indicated, 2 mM EGTA was added. Calibration of the cytosolic free Ca^{2+} scale was made, assuming a Kd for the quin 2-Ca^{2+} complex of 115 nM.[7] Estimates of the maximal and minimal quin 2-Ca^{2+} fluorescence were obtained after permeabilizing the cells by the addition of a minimal dose of digitonin (8 μg/ml), saturating intracellular quin 2 with Ca^{2+}, followed by an excess of Tris-EGTA, pH 8.5. Corrections were made for extracellular quin 2 on the basis of parallel incubations where an excess of EGTA was added before digitonin.

by decreasing the extracellular calcium level to the μM range (FIG. 4a and c). Readdition of $CaCl_2$ (1 mM excess over EGTA) after the signal had decayed to basal did not show evidence of a marked influx of Ca^{2+} into the cell from the medium (FIG. 4c). When cells were stimulated with phenylephrine (20 μM), the onset of the Ca^{2+} rise was also unaffected by EGTA, but the decay of the Ca^{2+} signal was much more pronounced in the absence of extracellular calcium (FIGS. 4b and d). The subsequent readdition of $CaCl_2$ now gave a substantial net increase in the cytosolic free Ca^{2+}. This observation is in agreement with the results of other workers (see REFS. 2, 3, 12) and represents the activation of a calcium influx pathway across the plasma membrane by hormone treatment. Thus, after hormonal stimulation the cytosolic calcium level remains elevated even after the intracellular Ins-1,4,5-P_3 sensitive calcium stores have been depleted. Taken together, the data of FIGURE 4 indicate that extracellular calcium does not contribute significantly to the rise in cytosolic Ca^{2+} level induced by ethanol.

The experiment of FIGURE 4 does not exclude the possibility that there is a transient opening of a plasma membrane Ca^{2+} channel after ethanol addition, which does not persist as it does after treatment with a hormone. FIGURE 5 shows that this is indeed possible. Extending the results of FIGURE 4, in this experiment we again used the method of Joseph *et al.*[12] to detect Ca^{2+} influx. Hepatocytes were loaded with quin 2, then washed in a calcium-free medium and stored on ice. Under these conditions the cells become largely depleted of calcium and the response to ethanol or low doses (1 nM) of vasopressin is minimal. Readdition of calcium 5 min after addition of either a saturating dose (40 nM) or a low dose (1 nM) of vasopressin causes an increase in quin 2 fluorescence (FIG. 5b and c), indicating an influx of Ca^{2+} into the cytosol. The initial rate of formation of the quin 2-Ca complex was taken as the net rate of Ca^{2+} transport across the plasma membrane. With ethanol as a stimulus, $CaCl_2$ addition after 5 min did not result in a calcium influx exceeding that observed in control incubations (FIG. 5a and d). However, when $CaCl_2$ was added within 1 min of the ethanol stimulus, a significantly higher rate of Ca^{2+} influx was observed. FIGURE 6 compares the time course of changes in the Ca^{2+} influx rate when $CaCl_2$ is added at different times after a stimulus of ethanol (300 mM) or vasopressin (1 nM). Both

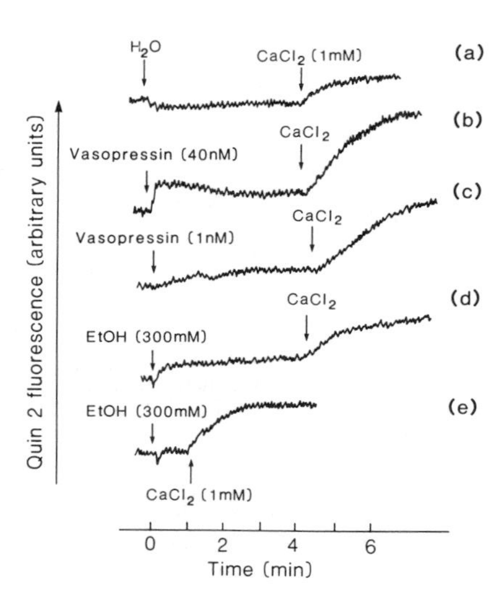

FIGURE 5. Activation of calcium influx by vasopressin and ethanol. Incubation conditions are described in the text and in the legend to FIGURE 4.

agents induce a rapid increase in calcium influx rate; but the response to ethanol decays to basal activity within 3 min, whereas after vasopressin an elevated Ca^{2+} influx rate persists. This finding emphasizes again the similarity between the responses to ethanol and hormones in the onset phase of the disturbance of cellular calcium homeostasis, and the dissimilarity in the rapid decay of the response of ethanol.

The experiment of FIGURE 7 provides evidence that calcium mobilized by hormones and by ethanol comes from the same intracellular source. Pretreatment of the hepatocytes with a saturating dose of vasopressin completely inhibited the capacity of ethanol to mobilize intracellular calcium (FIG. 7, top). The vasopressin treatment

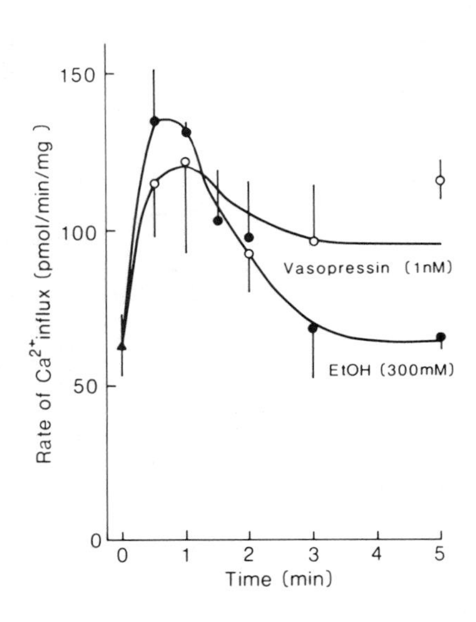

FIGURE 6. Time course of changes in calcium influx rate after stimulation with ethanol and vasopressin (1 nM). Experiments were carried out as shown in FIGURE 5, with additions of 1 mM $CaCl_2$ made at different times after the stimulus. The initial rate of formation of the quin 2-Ca complex was taken as the rate of calcium influx.

had completely depleted the Ins-1,4,5-P_3 sensitive stores of calcium, as indicated by the fact that phenylephrine was also incapable of mobilizing any further calcium. Similarly, depletion of the hormone-responsive calcium stores with phenylephrine prevented the ethanol-induced increase in cytosolic free Ca^{2+} (data not shown). These findings indicate that the Ins-1,4,5-P_3 sensitive calcium stores are the target of ethanol action in hepatocytes. By contrast, exposure of the cells to ethanol does not prevent, or diminish, the response to a subsequent dose of ethanol and, even after repeated doses of ethanol, phenylephrine addition still gives rise to a substantial calcium increase. Evidently, the transient calcium rise induced by ethanol treatment does not lead to a sustained depletion of the Ins-1,4,5-P_3 sensitive calcium stores as occurs with the hormones.

SUMMARY OF THE CHARACTERISTICS OF ETHANOL-INDUCED DISTURBANCE OF CELLULAR CALCIUM HOMEOSTASIS

The data presented in this report and elsewhere[11] demonstrate that ethanol addition causes a transient disturbance of cellular calcium homeostasis in intact

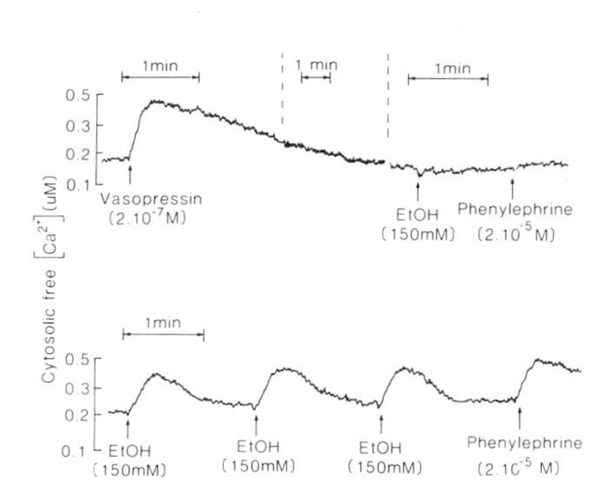

FIGURE 7. Inhibition of ethanol-induced calcium mobilization by depletion of the hormone-responsive calcium stores. Incubation conditions as described in the legend to FIGURE 4.

hepatocytes with characteristics that strongly resemble those induced by the glycogenolytic hormones acting through the phosphoinositide-linked signal transduction pathway. Calcium is mobilized from hormone-sensitive stores in the cell with a time course similar to that obtained at a physiological dose (1 nM) of vasopressin or at submaximal levels of the α_1-agonist phenylephrine. In additon to the mobilization of intracellular calcium, both ethanol and vasopressin also activate a Ca^{2+} influx pathway in the plasma membrane.[1] We have shown that ethanol also induces a small, but detectable efflux of calcium when cells are incubated in a calcium-free medium[11]; this presumably reflects the activation of the ATP-driven Ca^{2+} pump in the plasma membrane. Thus, ethanol may transiently activate the cycling of Ca^{2+} ions across the plasma membrane, but without inducing a net loss of Ca^{2+} from the cell in the presence of physiological extracellular Ca^{2+} levels. In contrast to the hormonally induced effects, the response to ethanol decays rapidly and calcium apparently re-enters the same intracellular stores from which it was mobilized. In this respect, the decay of the response to ethanol may resemble the pattern obtained after removal of a hormone or after addition of an antagonist.

SITE OF ACTION OF ETHANOL

The question remains where in this pathway ethanol exerts its effects. Studies on permeabilized hepatocytes carried out by Andrew P. Thomas in our department indicated that ethanol has no direct effect on the Ins-1,4,5-P_3 stimulated calcium efflux from the endoplasmic reticulum. Ethanol caused some inhibition of ATP-dependent calcium uptake, in agreement with our studies on isolated microsomes[4] and slightly elevated the calcium set-point maintained by the ATP-driven calcium uptake system (data not shown). However, these effects were too slow and too small to account for the calcium mobilization observed in intact cells. In other studies reported elsewhere in this volume[13–15] ethanol is shown to give a direct activation of the hormone-sensitive phospholipase C. Ethanol was found to give rise to a significant but transient elevation of both Ins-1,4,5-P_3 and Ins-1,3,4-P_3.[13] Quantitatively and kinetically the formation of Ins-1,4,5-P_3 is sufficient to account for the calcium mobilization observed in intact cells. Moreover, a study of the level of polyphosphoinositides and phosphatidic acid[14] demonstrated that ethanol addition causes a pattern of changes in these phospholipids compatible with an increased flux through the pathway consisting of phosphoinositide kinases and phospholipase C, similar to that induced by submaximal levels of vasopressin. These results provide strong evidence that the site of action of ethanol is at an early step in the hormone-responsive pathway of phosphoinositide turnover leading to the formation of the second messengers Ins-1,4,5-P_3 and diacylglycerol.

Further evidence emphasizing the similarity between the effects of ethanol and the calcium-mobilizing hormones is provided by the findings that the ethanol-induced calcium mobilization, like that caused by phenylephrine or low doses of vasopressin, is inhibited by pretreatment of the hepatocytes with the phorbol ester 12-0-tetradecanoyl-phorbol-13-acetate (TPA).[15] Other studies[16,17] have indicated that the phorbol esters may exert their action, in part, at the level of the receptor-phospholipase C interaction. Recent evidence[18] indicates that this interaction is regulated by a GTP-binding protein comparable, but not identical to the Gs and Gi (Ns and Ni) proteins that regulate the receptor-adenylate cyclase interaction. It is an attractive hypothesis, therefore, that ethanol acts at the level of receptor-phospholipase C coupling mediated by such regulatory G-proteins. This hypothesis is also appealing in view of the similarity with the adenylate cyclase system. Studies by Molinoff and co-workers[19,20] and others[21] have provided evidence that the G-proteins involved in receptor adenylate cyclase coupling are a potential site of action of ethanol in other membrane systems. Currently, we are working to obtain direct evidence for this hypothesis in our laboratory.

TRANSIENT NATURE OF THE RESPONSE TO ETHANOL

The experimental data discussed so far indicate that the exposure of hepatocytes to ethanol activates the complete pathway of signal transduction characteristic of calcium-mobilizing hormones. A major difference with the hormonally activated cells, however, is the transient nature of the response to ethanol. This differnce is apparent not only in the calcium response, shown, for instance in FIGURES 2 and 4, but also in the formation and degradation of $InsP_3$ isomers[13] and in the transient pattern activation of flux through the phosphoinositides.[14] It should be stressed that this transient response to ethanol is not a reflection of the disappearance of ethanol by oxidation or of the transient formation of a metabolic intermediate from ethanol. The calcium transients in quin 2-loaded cells were obtained in the presence of a high concentration of

4-methylpyrazole, a potent inhibitor of ethanol oxidation, and the kinetics of activation and inactivation of phosphorylase were identical in the absence or presence of the inhibitor. It is also unlikely that ethanol addition initiated a desensitization response to inhibit the initial activation; a further addition of either a hormone or a similar ethanol concentration generated an uninhibited response (FIG. 7). These features also make it unlikely that the membrane disordering effect of ethanol alone is sufficient to explain the activation of the hormone-sensitive phospholipase C. In agreement with that interpretation, we have observed that chronic ethanol exposure of the rat does not provide resistance to this specific effect of ethanol, even though membranes from these rats have acquired resistance to the disordering effect of ethanol (data not shown). However, the hormone-sensitive phospholipase C is a complex enzyme system that is probably regulated both by membrane-bound components responding to interactions in the hydrophobic milieu of the membrane core and by ionic interactions across the membrane surface. It is conceivable that the addition of ethanol interferes at the same time with the hydrophobic components through its membrane disordering effects and with the ionic interactions through its effect on the solvent properties of the aqueous phase. The system may respond to these disturbances by an adjustment of the interactions between individual components that would allow the system as a whole to return to its basal activity. Further study of the factors involved in the regulation of the receptor-coupled phospholipase C will be needed to assess the validity of such a model.

The physiological implications of such transient effects of ethanol on the phosphoinositide-linked signal transduction system also remain to be elucidated. It should be emphasized that a transient effect on second messenger signals may well result in a metabolic response that is sustained for a much longer period of time. From the data in this and accompanying reports,[13] it is evident, for instance, that the Ins-1,4,5-P_3 signal decays to baseline levels within 2 min, well before the Ca^{2+} signal, whilst the phosphorylase activity may remain significantly elevated for up to 10 min. It is not unusual, in biological systems, that relatively short-lived signals have metabolic consequences that persist long after the initial message has disappeared. The concentrations of ethanol (25–500 mM) used in the present study are within the range that may be of physiological relevance. The liver is routinely exposed to concentrations of ethanol in the portal blood that are much higher than are what would ever be reached in the circulation. Repeated intake of ethanol is likely to result in a recurrent activation of the phosphoinositide-specific second messenger system, with a consequent repeated elevation of cytosolic calcium levels. The metabolic consequences of such transient events are likely to have a significant longer-term effect on the liver and possibly on other tissues.

ACKNOWLEDGMENTS

Expert technical assistance in these studies was provided by William A. Habib, Janet B. Smith, and Michelle Tomsho. I gratefully acknowledge support from and helpful discussions with Drs. Emanuel Rubin, Andrew P. Thomas, and Raphael Rubin.

REFERENCES

1. BERRIDGE, M. 1984. Biochem. J. **220:** 345–360.
2. WILLIAMSON, J. R., R. H. COOPER, S. K. JOSEPH & A. P. THOMAS. 1985. Am. J. Physiol. **248:** C203–C216.

3. REINHART, P. H., W. M. TAYLOR & F. L. BYGRAVE. 1984. Biochem. J. **223:** 1–3.
4. PONNAPPA, B. C., A. J. WARING, J. B. HOEK, H. ROTTENBERG & E. RUBIN. 1982. J. Biol. Chem. **257:** 10141–10146.
5. YAMAMOTO, H. A. & R. A. HARRIS. 1983. Biochem. Pharmacol. **32:** 2787–2791.
6. Reference deleted.
7. TSIEN, R. Y., T. POZZAN & T. J. RINK. 1982. J. Cell Biol. **94:** 325–339.
8. GRYNKIEWICZ, G., M. POENIE & R. Y. TSIEN. 1985. J. Biol. Chem. **260:** 3440–3450.
9. TSIEN, R. Y. 1981. Nature (Lond.) **290:** 527–528.
10. MEYER, A. J., J. A. GIMPEL, G. DELEEUW, J. M. TAGER & J. R. WILLIAMSON. 1975. J. Biol. Chem. **250:** 7728–7738.
11. HOEK, J. B., A. P. THOMAS & R. RUBIN. 1987. J. Biol. Chem. **262:** 682–691.
12. JOSEPH, S. K., K. E. COLL, A. P. THOMAS, R. RUBIN & J. R. WILLIAMSON. 1985. J. Biol. Chem. **260:** 12508–12515.
13. THOMAS, A. P., J. B. HOEK & E. RUBIN. 1987. Ann. N.Y. Acad. Sci. This volume.
14. RUBIN, R. & J. B. HOEK. 1987. Ann. N.Y. Acad. Sci. This volume.
15. HOEK, J. B., R. RUBIN & A. P. THOMAS. 1987. Ann. N.Y. Acad. Sci. This volume.
16. COOPER, R. H., K. E. COLL & J. R. WILLIAMSON. 1985. J. Biol. Chem. **260:** 3281–3288.
17. LYNCH, C. J., R. CHAREST, S. B. BOCCHINO, J. H. EXTON & P. F. BLACKMORE. 1985. J. Biol. Chem. **260:** 2844–2851.
18. UHING, R. J., V. PRPIC, H. JIANG & J. H. EXTON. 1986. J. Biol. Chem. **261:** 2140–2146.
19. RABIN, R. A. & P. B. MOLINOFF. 1983. J. Pharmacol. Exp. Ther. **227:** 551–556.
20. SAITO, T., J. M. LEE & B. TABAKOFF. 1985. J. Neurochem. **44:** 1037–1044.
21. KISS, Z. & V. A. THACHUK. 1984. Eur. J. Biochem. **142:** 323–329.
22. GILBOE, D. P., K. L. LARSON & F. Q. NUTTAL. 1972. Anal. Biochem. **47:** 20–27.
23. GILMAN, A. G. & F. MURAD. 1974. Meth. Enzymol. **38:** 49–61.

DISCUSSION OF THE PAPER

E. NOBLE (*University of California, Los Angeles, CA*): When you added alcohol to your system there was a very rapid rise of calcium as measured by the quin 2 method, followed by a very rapid fall to the basal level. What is happening to that calcium? Do you think it's being sequestered?

HOEK: Our experiments suggest that most of the calcium is being sequestered again into the same stores. When we give repeated additions of ethanol we can remobilize calcium again. There is, however, some depletion of the calcium stores; after three sequential ethanol additions the amount of calcium mobilized by phenylephrine is lower than without the ethanol pretreatment. The calcium that is resequestered after ethanol addition may have gone into calcium stores that are not sensitive to inositol-trisphosphate. There may also have been some extrusion of calcium. When there is no calcium outside and we have EGTA in the system, repeated additions of ethanol also give repeated bursts of calcium mobilization. Under these conditions, however, there is a decay of this response, indicating that the calcium stores are now being depleted due to the loss of some calcium from the cells.

NOBLE: Compared to stimulation with an agonist like vasopressin, is the response to ethanol generally slower?

HOEK: With a maximally saturating concentration of vasopressin, the increase in cytosolic calcium is much more rapid than after addition of ethanol. At the same time, the inositol-trisphosphate level goes up twentyfold or so, an order of magnitude more than we see with ethanol. We have selected a 1-nM vasopressin concentration specifically to give the same rate of increase of cytosolic calcium as we see with ethanol. When the calcium is mobilized at the same rate, we observed that inositol-triphosphate is formed at the same rate, suggesting that it can, in fact, account for all the calcium mobilization with either vasopressin or ethanol.

A. SCARPA (*Case Western Reserve University, Cleveland, OH*): What I find intriguing is that ethanol is present at 150 mM and you can challenge again with the same concentration. Is there an adaptation to the presence of ethanol in one minute?

HOEK: I agree, there is a mechanistic problem for which I have no adequate explanation. All I can say is, our observations are that calcium is repeatedly mobilized with sequential additions of ethanol, even though the ethanol is still present. In other words, it appears to be the addition of ethanol that triggers the system, rather than the concentration present. Even if a desensitization system is activated, I find it hard to visualize that the second addition of ethanol doesn't induce a slower rate of calcium mobilization. We may have to understand in much more detail what ethanol is doing at the level of the membrane before we can expect an answer to this kind of question.

SCARPA: Can you induce phospholipid degradation by ethanol in a simpler system?

HOEK: Dr. Raphael Rubin has been working on that problem during the past few months. We have a permeabilized liver cell system that exhibits quite a good activation of the phospoinositide-specific phopholipase C and responds to GTP-α-S with stimulation of the phospholipase C. In that system, we have so far not been able to demonstrate a stimulatory effect of ethanol. We are still studying that system further.

R. A. HARRIS (*University of Colorado School of Medicine, Denver, CO*): Whenever you add ethanol in a concentration of 150 mM or more you increase the osmotic strength quite a bit. Of course, this osmotic gradient will slowly dissipate over a few seconds as the ethanol diffuses from outside to inside. A trivial and hopefully incorrect explanation for these data would be that it is simply due to a change in osmotic strength which dissipates over a few seconds, and the effect, the change in calcium level, disappears. Have you tried adding a 150 mM sucrose or some substance other than ethanol?

HOEK: We have not tested the effect of 150 mM sucrose, but we have compared the response to ethanol with that of an equivalent amount of water. Ethanol has quite a rapid rate of penetration and it is unlikely that any osmotic effect would last for 20 seconds. The effects of DMSO or methanol are much less than those of ethanol, but I have no information on the relative permeabilities. With the longer chain alcohols n = propanol or n = butanol we find the same effects on calcium mobilization at much lower concentrations, which is not compatible with an explanation based on osmotic effects.

Effects of Ethanol on Rat Placental and Fetal Hepatocyte Transport of Amino Acids[a]

GEORGE I. HENDERSON, DAVID W. HEITMAN, AND STEVEN SCHENKER

Departments of Medicine and Pharmacology
The University of Texas Health Science Center at San Antonio
San Antonio, Texas 78284
and
Research Service, Audie L. Murphy Memorial Veterans' Hospital
San Antonio, Texas 78284

Numerous studies, including our own, have provided evidence that, in animal model systems, maternal ethanol consumption impairs fetal and subsequently neonatal growth.[1-3] Evidence to date suggests that the origin of this impaired development is likely multifactorial[4] with one potential mechanism being impaired transfer of nutrients to the fetus. Previous studies in this laboratory and others[5-8] have indicated that ethanol may alter placental transport of various amino acids following acute and, especially chronic, exposure. More recent work has investigated the possibility that ethanol may alter the flux of nutrients across the fetal hepatic plasma membrane.

METHODS

Placental Transport—Animal Models

Acute Ethanol Exposure In Vivo. Rats (200–225 g, Sprague Dawley, Harlan Sprague Dawley, Inc., Indianapolis, IN) were administered ethanol by gavage (4 g/kg, 25% v/v) 2 hr prior to sacrifice on the 20th day of gestation. Day 1 was taken as the day of appearance of sperm in vaginal smears. Maternal body temperatures averaged 36.1° C ± 0.09 S.E.M. and maternal blood ethanol levels averaged 2.58 mg/ml ± 0.7 S.E.M. at sacrifice. Controls received isocaloric dextrose by gavage and body temperatures averaged 38.2° C ± 0.08 S.E.M. Both ethanol-fed and control rats were fasted overnight prior to the morning study.

Chronic Ethanol Exposure In Vivo. The chronic ethanol regimen was that previously outlined.[2] Basically, Sprague-Dawley rats (200-225 g) were maintained on the Lieber-DeCarli liquid diet (Bio Serv, Frenchtown, NJ) containing either 6% (v/v) ethanol or isocaloric maltose dextrins for at least 30 days prior to and throughout gestation. Mean daily ethanol intake was 3.7 ml of absolute ethanol/rat. Maternal blood ethanol levels averaged 1.43 mg/ml ± 0.01 S.E.M. and maternal body

[a]Supported by Veterans Administration (Research Service, Audie Murphy Memorial Veterans' Hospital) and National Institute on Alcohol Abuse and Alcoholism Grant 7R01AA05814-01.

temperatures averaged 37.7° C ± 0.06 S.E.M. at sacrifice on the 20th day of gestation.

In-Vitro *Placental transport.* Basically, placental uptake of amino acids was determined in villous fragments isolated on the twentieth day of gestation.[5,6] Rate of net uptake was studied following a 30-minute preincubation and accumulation of labeled amino acid in the intracellular water (ICW) was noted at time intervals following its addition to the incubation media. The ratio of ICW to extracellular water (ECW) was not altered by *in-vivo* or *in-vitro* ethanol exposure.

In-Vivo *Placental Transport.* Maternal/fetal exchange of amino acids, *in vivo,* were determined following injection of labeled amino acid (2 μ Ci, 0.04 μmole/animal) into the femoral vein of the dams.[9] Subsequently, at appropriate time intervals, accumulation of the identified label in the placentas and fetuses was established.

Fetal Hepatocyte Primary Cultures. Fetal hepatocytes were obtained from Sprague-Dawley rats on the 20th day of gestation. Techniques used for isolation were similar to those used by Leffert and co-workers.[10,11] William's E medium (without arginine) was utilized for isolation and plating, the latter supplemented with 10% dialyzed fetal bovine serum and insulin (8 μg/ml). These two substances were subsequently omitted after a 24-hour "plating" period. The growth media was supplemented with varying concentrations of EGF and/or ethanol. Cells were thereafter maintained in culture for 48 to 72 hours and a Coulter Counter used to determine cell numbers. Net uptake of amino acids utilized labeled compounds with concentrations adjusted to 10^{-4} M. Initially, time courses for net uptake determined periods of linearity.

RESULTS

Amino Acid Uptake by Villous Fragments

Acute In-Vivo *Ethanol Exposure.* FIGURE 1 illustrates the effects of a single *in-vivo* dose of ethanol (4 g/Kg, PO) administered 2 hr prior to sacrifice (day 20 of gestation) on subsequent amino acid net uptake by rat placental villous fragments. Lysine net uptake (panel E) was entirely unaffected while that for α-aminoisobutyric (A) and L-leucine (D) were reduced by 20% and 18% respectively, although statistical significance was not quite attained ($p > 0.05$). Net uptake of cycloleucine (B) and L-alanine (C) were reduced by 29% ($p < 0.05$) by prior acute ethanol exposure. FIGURES 2A and 2B indicate a similar inhibitory effect (40 and 31%) on val uptake by placentas exposed to a single dose of ethanol on days 20 and 16 of gestation, respectively. Additionally, maintenance of maternal body temperature at 38° C did not reverse this inhibitory response, indicating that ethanol-induced hypothermia is not a significant causal factor. The effects of 3-day *in-vivo* ethanol exposure regimens on *in-vitro* val uptake are shown in FIGURE 3. If placentas were exposed to ethanol early in gestation (days 8–10, FIG. 3A) no subsequent effect on val uptake (on day 20) was apparent. However, exposure on days 11–13 and 14–16 caused significant ($p < 0.05$) reductions in net uptake of 24% and 28%, respectively, determined *in vitro* on day 20 of gestation.

Chronic Ethanol Exposure. If rats were maintained on the ethanol diet throughout gestation and placental amino acid uptake determined *in vitro* on day 20, transport of

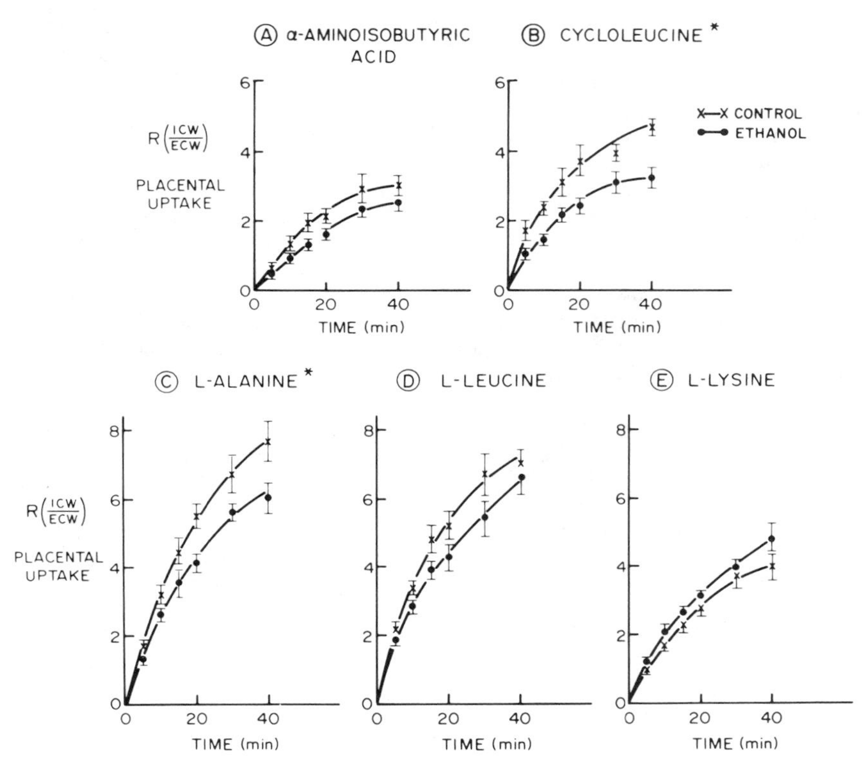

FIGURE 1. Effect of a single *in-vivo* dose of ethanol on placental amino acid uptake measured *in vitro*. The vertical axis represents the ratio of accumulated amino acid [(**A**) AIB, (**B**) cycloleucine, (**C**) L-Ala, (**D**) L-Leu, (**E**) L-Lys] in the ICW to that in the ECW. The horizontal axis indicates the time of incubation of villous fragments after addition of labeled amino acid (10^{-6} M). Ethanol (4 g/kg, 25% solution) or isocaloric dextrose was administered by intubation 2 hr before sacrifice on day 20 of gestation. Placentas from each pregnancy were pooled and the resulting fragments divided into four incubation tubes, one for ICW determinations and three for amino acid uptake studies. n = 10 dams for each group, and 91 to 115 placentas each for ethanol (●) and control (x) experiments. Vertical bars represent S.E. and an asterisk (*) next to the name of the amino acid indicates that the ethanol values differ significantly ($p < 0.05$) from the corresponding control. (From Henderson *et al.*[5] Reprinted by permission from *Alcoholism: Clinical Experimental Research.*)

all amino acids tested was impaired ($p < 0.05$). FIGURE 4 indicates the following reductions in amino acid net uptake: AIB—38%, Ala—35%, cycloleucine—45%, Leu—25%, and Lys—34%.

Amino Acid Uptake In Vivo. FIGURES 5 and 6 illustrate the effect of a single dose of ethanol on maternal-fetal exchange of valine *in vivo*. A 2-g/Kg dose of ethanol reduced placental accumulation of val by 50%, while the 4-g/Kg dose caused a 63% impairment of uptake. Fetal accumulation of val (FIG. 6) was also reduced in a dose-dependent fashion. Chronic exposure (FIG. 7) impaired both placental and fetal retention of val by 47% and 46%, respectively.

Effect of Ethanol on Amino Acid Net Uptake by Cultured Fetal Hepatocytes. Recent studies in our laboratory have utilized monolayer cultures of rat fetal hepatocytes. These cells were found to replicate only in the presence of EGF at concentrations greater than 1 ng/ml. This replication was completely and irreversibly blocked by exposure to ethanol (1.7 or 3.9 mg/ml) for periods in excess of 9 hr (unpublished data). In addition to impaired replication, a 48-hr exposure to ethanol caused a 3- to 4-times increase in cellular water and total protein ($p < 0.05$). Amino acid net uptake per cell was stimulated by 54% to 55% for AIB and methyl AIB and by 128% for cycloleucine (TABLE 1). Further studies utilizing AIB as the marker amino acid indicated that uptake, not efflux, is affected by ethanol and that the component(s) of transport that is increased is not dependent on the presence of sodium in the uptake media.

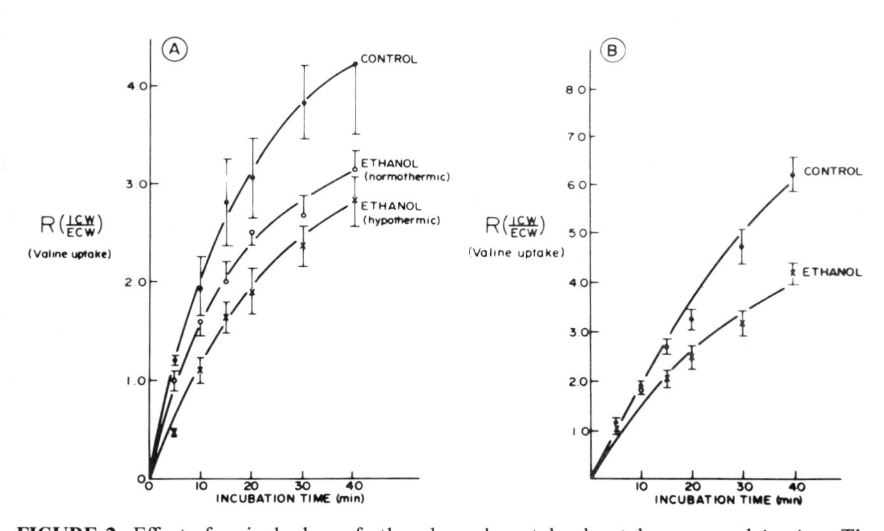

FIGURE 2. Effect of a single dose of ethanol on placental val uptake measured *in vitro*. The vertical axis represents the ratio of accumulated [^{14}C] val in the ICW to that in the ECW. The horizontal axis indicates the time of incubation of villous fragments after addition of [^{14}C] val (10^{-6} M). Ethanol (4 g/kg, 25% solution) or isocaloric dextrose was administered by intubation 2 hr before sacrifice on either day 20 (**A**) or day 16 (**B**) of gestation. Ethanol induced a 2° C drop in body temperature (x, ethanol hypothermic) as compared to normothermic controls (●, control). Twenty-day pregnant dams labeled ethanol (O, normothermic) were treated in a manner identical to ethanol (hypothermic) rats except that their body temperature was maintained at 38° C throughout the 2-hr ethanol exposure period. Placentas from each pregnancy were pooled and the resulting fragments divided into four incubation tubes, one for ICW determinations and three for val uptake studies; n = 12 dams and 126 placentas for 20-day control, n = 11 dams and 113 placentas for 20-day ethanol (hypothermic), n = 11 dams and 116 placentas for ethanol (normothermic), n = 21 dams and 232 placentas for 16-day control, and n = 26 dams and 282 placentas for 16-day ethanol experiments. Vertical bars represent S.E. The ethanol and control values differed statistically ($p < 0.05$). (From Henderson *et al.*[6] Reprinted by permission from *Journal of Pharmacology and Experimental Therapeutics*.)

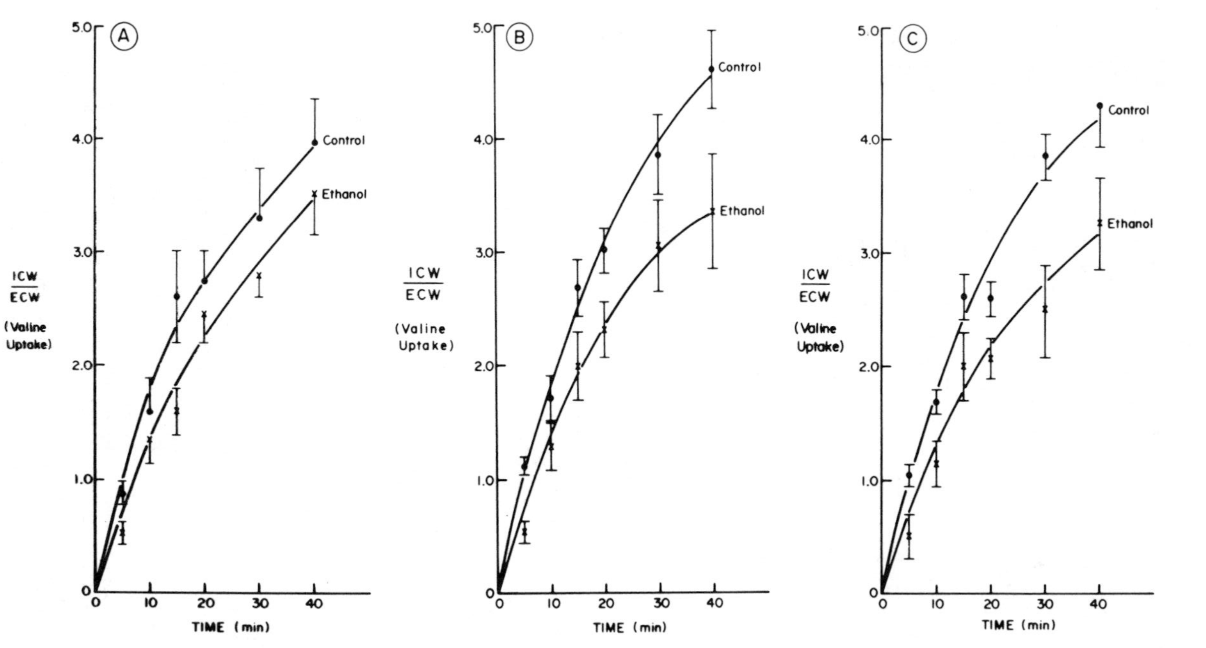

FIGURE 3. Effect of 3-day maternal ethanol intake regimens on placental val uptake measure *in vitro*. The vertical axis represents the ratio of accumulated [^{14}C] val in the ICW to that in the ECW. The horizontal axis indicates the time (min) after addition of [^{14}C] val (10^{-6} M) to the incubation media. Ethanol was administered by intubation (4 g/kg, 25% solution) twice a day for 3 consecutive days [(**A**) days 8–10, (**B**) days 11–13, and (**C**) days 14–16] during gestation and the rats were sacrificed on day 20 of gestation (x). Control (●) were pair-fed throughout pregnancy and were isocalorically intubated with dextrose. Placentas from each pregnancy were pooled and the resulting fragments divided into four incubation tubes; n = 11 control dams (107 placentas) for days 8–10 and n = 11 ethanol dams (114 placentas), n = 8 control dams (71 placentas) for days 11–13 and n = 8 ethanol dams (84 placentas), n = 10 control dams (109 placentas) for days 14–16 and n = 10 ethanol dams (108 placentas). Vertical bars represent S.E.s. Ethanol and control values differed significantly ($p < 0.05$) for placentas exposed to ethanol on days 11–13 and days 14–16 of gestation. (From Henderson *et al.*[6] Reprinted by permission from *Journal of Pharmacology and Experimental Therapeutics.*)

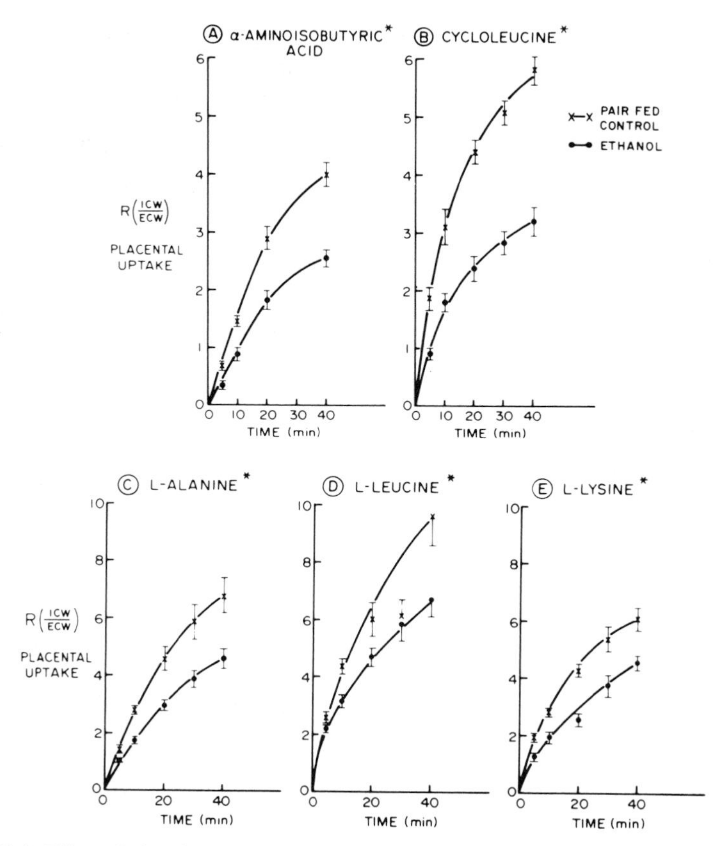

FIGURE 4. Effect of chronic ethanol consumption on placental amino acid uptake measured *in vitro*. The vertical axis represents the ratio of labeled amino acid [(**A**) AIB, (**B**) cycloleucine, (**C**) L-Ala, (**D**) L-Leu, (**E**) L-Lys] in the ICW to that in the ECW. The horizontal axis indicates the time (min) after addition of amino acid (10^{-6} M) to the incubation media. Rats were maintained on the Lieber-DeCarli liquid diet containing either 6 percent ethanol (●) or isocaloric maltose dextrins (x) for at least 30 days prior to and throughout gestation. At sacrifice on day 20 of gestation, maternal blood ethanol levels averaged 1.45 mg/ml. Placentas (an average of nine) from each pregnancy were pooled and the resulting fragments divided into four incubation tubes. The number of paired pregnancies for each amino acid study was: (AIB) = 11, cycloleucine = 10, L-Ala = 8, L-Leu = 11, and L-Lys = 13. Vertical bars represent S.E. All ethanol values differed significantly ($p < 0.05$) from the corresponding pair-fed controls (*). (From Henderson *et al.*[5] Reprinted by permission from *Alcoholism: Clinical Experimental Research.*)

DISCUSSION

The preceding studies indicate that acute and especially chronic maternal ethanol consumption can impair placental transport of a variety of amino acids. Other investigators have provided similar evidence.[7,8] While the precise mechanisms responsible for this effect remain to be elucidated, *in-vitro* studies not included in this brief synopsis have provided some insight into the observed phenomena. *First,* acute ethanol

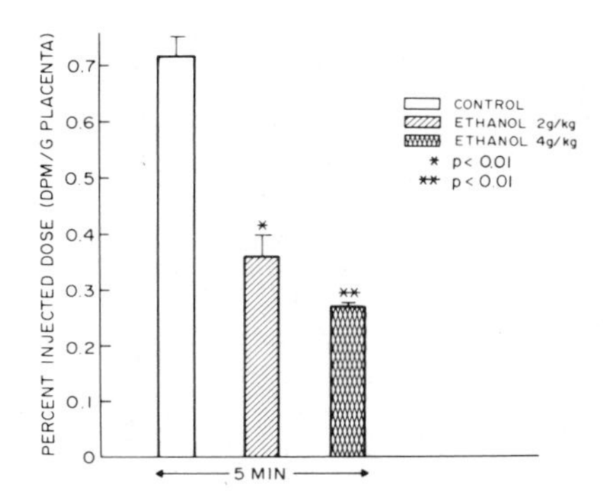

FIGURE 5. Effect of short-term ethanol ingestion on the placental uptake of [^{14}C] valine (mean ± S.E.; n = 40–50 placentas in 5–6 animals) ($p < 0.05$). *Compared to control; **compared to both control and 2 gm/kg ethanol groups. (From Patwardhan *et al.*[9] Reprinted by permission from the *Journal of Laboratory and Clinical Medicine.*)

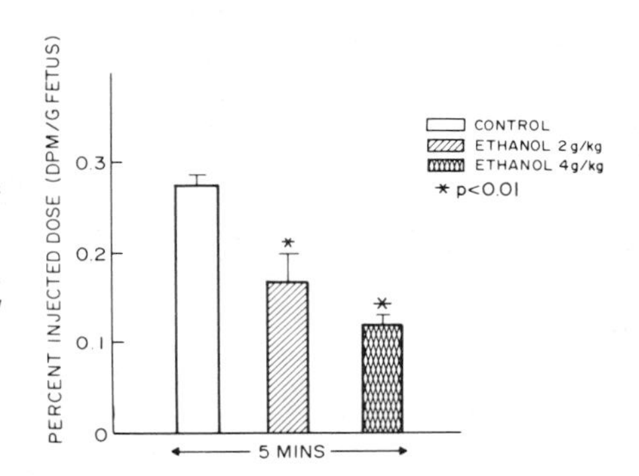

FIGURE 6. Effect of short-term ethanol ingestion on the fetal uptake of [^{14}C] valine (mean ± S.E.; n = 40–50 fetuses in 5–6 animals) ($p < 0.05$). *Compared to control. (From Patwardhan *et al.*[9] Reprinted by permission from the *Journal of Laboratory and Clinical Medicine.*)

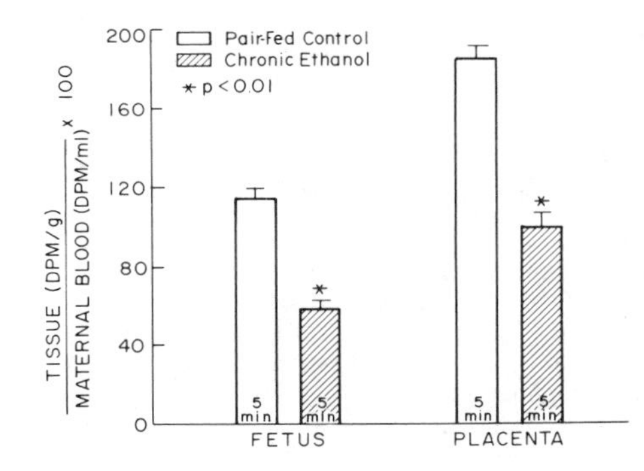

FIGURE 7. Effect of long-term ethanol administration on fetal and placental uptake of [^{14}C] valine (mean ± S.E.; n = 35–50 fetuses and placentas in 6 animals) ($p < 0.05$). *Compared to control. (From Palwardhan *et al.*[9] Reprinted by permission from the *Journal of Laboratory and Clinical Medicine.*)

TABLE 1. Effect of Ethanol on AIB, Methyl AIB, and Cycloleucine Uptake by Rat Fetal Hepatocytes[a]

		Net Uptake (nM/Cell/ Min × 10^{-6})	
	AIB	Methyl AIB	Cycloleucine
Control	0.270 ± 0.011[b]	0.158 ± 0.004	0.267 ± 0.008
Ethanol[c]	0.419 ± 0.020	0.243 ± 0.010	0.609 ± 0.016
Increase from Control	55%[d]	54%[d]	128%[d]

[a]Cells cultured in the presence of 2 ng/ml EGF.
[b]Mean ± S.E.M.
[c]Mean ethanol concentration ± S.E.M. – 3.9 ± 0.2 mg/ml.
[d]$p < 0.05$.

exposure appears to affect influx of amino acids into placental cells and has no effect on efflux. *Second,* it appears that the observed changes in influx are caused by ethanol rather than acetaldehyde. This is based on *in-vitro* studies illustrating that, acutely, the minimum concentration of acetaldehyde needed to elicit an effect on amino acid uptake is in excess of 300 μM, while ethanol at 2–3 mg/ml impairs transport. However, an effect from long-term (chronic) exposure to lower concentrations of acetaldehyde cannot be ruled out. *Third,* ethanol inhibition of placental amino acid uptake appears to primarily involve sodium-dependent transport processes.

Functional impacts of the observed effects of ethanol on fetal hepatocyte amino acid transport are more difficult to assess at this stage. It is possible that the increased cellular uptake of amino acids following ethanol exposure is an expression of increased metabolic need caused by lesions in normal metabolic function. Thus, this response could have a positive functional significance with respect to cell survival. On the other hand, this increased transport could be a reflection of ethanol-induced cellular pathology, the latter resulting in hepatocyte swelling, among other parameters. This could, in turn, cause more sodium-independent amino acid transport sites to become available on the cell surface. There are likely to be other explanations, but further elucidation of this phenomenon will have to follow future mechanistic probes.

The role of placental transport dysfunction in offspring growth and developmental delays associated with maternal ethanol consumption remains to be confirmed. Additionally, substantial species differences in placental structure, and possibly transport processes, complicate extrapolation to the human. Given the multitude of cellular responses to ethanol, including direct effects on cell protein accrual machinery, it seems likely that the detrimental fetal responses to ethanol reflect a multifactorial event, one component of which could be malfunctioning placental transport processes for amino acids.

REFERENCES

1. ABEL, E. L. & B. A. DINTCHEFF. 1978. Effects of prenatal alcohol exposure on growth and development in rats. J. Pharmacol. Exp. Ther. **207:** 916–921.
2. HENDERSON, G. I., A. M. HOYUMPA, C. MCCLAIN & S. SCHENKER. 1979. The effects of chronic and acute alcohol administration on fetal development in the rat. Alcoholism: Clin. Exp. Res. **3:** 99–106.
3. ABEL, E. L. & H. B. GREIZERSTEIN. 1979. Ethanol-induced prenatal growth deficiency: changes in fetal body composition. J. Pharmacol. Exp. Ther. **211:** 668.
4. HENDERSON, G. I., R. V. PATWARDHAN, A. M. HOYUMPA, JR. & S. SCHENKER. 1981. Fetal alcohol syndrome: overview of pathogenesis. Neurobehav. Toxicol. Teratol. **3:** 73–80.
5. HENDERSON, G. I., R. V. PATWARDHAN, S. MCLEROY & S. SCHENKER. 1982. Inhibition of placental amino acid uptake in rats following acute and chronic ethanol exposure. Alcoholism: Clin. Exp. Res. **6:** 495–505.
6. HENDERSON, G. I., D. TURNER, R. V. PATWARDHAN, L. LUMENG, A. M. HOYUMPA & S. SCHENKER. 1981. Inhibition of placental valine uptake after acute and chronic maternal ethanol consumption. J. Pharmacol. Exp. Ther. **216:** 465–472.
7. LIN, G. W.-J. 1981. Effect of ethanol feeding during pregnancy on placental transfer of alpha-aminoisobutyric acid in the rat. Life Sci. **28:** 595–601.
8. FISHER, S. E., M. ATKINSON, D. VANTHIEL, E. ROSENBLUM, R. DAVID & I. HOLZMAN. 1981. Selective fetal malnutrition: the effect of ethanol and acetaldehyde upon *in vitro* uptake of alpha amino isobutyric acid by human placenta. Life Sci. **29:** 1283–1288.
9. PATWARDHAN, R. V., S. SCHENKER, G. I. HENDERSON, N. N. ABOU-MOURAD & A. M. HOYUMPA, JR. 1981. Short-term and long-term ethanol administration inhibits the placental uptake and transport of valine in rats. J. Lab. Clin. Med. **98:** 251–262.
10. LEFFERT, H. L. & D. PAUL. 1972. Studies in primary cultures of differentiated fetal liver cells. J. Cell Biol. **52:** 559–568.

11. LEFFERT, H. L. & D. PAUL. 1973. Serum dependent growth of fetal rat hepatocytes in arginine-deficient medium. J. Cell Physiol. **81:** 113–124.

DISCUSSION OF THE PAPER

E. NOBLE (*University of California, Los Angeles, CA*): Since you can dissociate the effect of ethanol on amino acid uptake from the effect on growth, have you looked at the next obvious question, the effect of ethanol on protein synthesis?

SCHENKER: We haven't yet done that; it's number two on the agenda, as soon as we find out what the sodium influx is into those cells. We want to use a smaller molecule and first prove that this is not some artifact within this culture system. Because we are concerned about this possibility, these data have not yet been published.

D. MCCARTHY (*University of New Mexico School of Medicine, Albuquerque, NM*): Usually we like to keep fibroblasts out of the cultures, but obviously an interesting question is, what is the effect of alcohol on the uptake of amino acids into hepatic fibroblasts?

SCHENKER: It's enhanced. Interesting that you should mention that. There are two systems in which it is enhanced. In an old paper a marked enhancement was reported of the uptake of amino acids by fibroblasts that are exposed to alcohol. The only other system that I know of that may have an enhancement are hepatoma cells that Dr. Simon in Denver has in culture, which also seem to show an enhancement of AIB uptake with alcohol. The obvious question is, do rapidly growing cells behave differently than other cells?

D. B. GOLDSTEIN (*Stanford University School of Medicine, Stanford, CA*): I was interested in the effect of ethanol on the EGF—simulated growth of the hepatocytes. Did you study whether the alcohol changed the affinity of the EGF for its receptor?

SCHENKER: We are just about to do this. It is a very exciting question, because it may also have implications for hepatic regeneration. Could it be, for example, that cells that are damaged by continuing alcoholism are not regenerated as well, because there's some problem in the affinity for various growth-promoting factors or with the internalization of those factors within the hepatocyte? The answer is, we don't know, but it's a very interesting possibility.

Effects of Ethanol on Protein Synthesis[a]

MARCUS A. ROTHSCHILD,[b] MURRAY ORATZ,[c]
AND SIDNEY S. SCHREIBER[b]

*[b]Nuclear Medicine Service,
Veterans Administration Medical Center
Departments of Medicine and Radiology
New York University Medical Center
New York, New York 10010*
and
*[c]Department of Biochemistry
New York University College of Dentistry
New York, New York 10010*

Alcohol is known to affect protein synthesis and/or protein secretion.[1-18] In the case of hepatic-made proteins destined for secretion as extracellular proteins there is conflicting evidence as to which aspect is affected, synthesis or secretion, since these extracellular proteins are the net result of both processes.[16-23] In the case of intracellular proteins secretion is not a factor. However, both intracellular proteins and those destined for export share common pathways of synthesis and intracellular transport.

Protein synthesis begins in the nucleus with: 1) gene transcription into mRNA; 2) post-transcriptional processing of the mRNA including the excission of nucleotides, the addition of a polyadenylate sequence, and the formation of 5′ methylated termini ("cap"); and, finally, 3) secretion from the nucleus into the cytoplasm.

It is in the cytoplasm that de facto protein synthesis occurs. Here, the small ribosomal subunit, which contains some initiation factors, forms a complex with the initiator tRNA and several other factors to form a pre-initiation complex. This is followed sequentially by the binding of mRNA and the large ribosomal subunit to form the active 80S initiation complex. Chain elongation occurs as the ribosome moves along the mRNA one codon at a time. At each codon specific tRNAs with the appropriate anticodons bind to the mRNA, and the amino acid they carry is covalently linked to the growing polypeptide chain. Chain elongation continues until a termination codon is reached. Each step requires several protein factors as well as energy.[24]

In most types of eukaryotic cells there are two major classes of ribosomes: bound ribosomes and free ribosomes. The bound ribosomes are bound to an internal network of lipoprotein membranes—the endoplasmic reticulum. There is no structural difference between a free and a bound ribosome. Binding to the membrane occurs after protein synthesis has started, probably by means of a leader polypeptide chain ("signal hypothesis").[25] This hypothesis can be satisfied by a number of possible mechanisms: binding to a receptor, or insertion across the lipid bilayer without another membrane-protein participation. Whether a ribosome binds or does not bind to the membrane depends upon the protein synthesized.

Most proteins destined to be secreted or to be stored in intracellular vesicles are synthesized by ribosomes attached to the endoplasmic reticulum. These proteins are

[a]Supported in part by the Louise and Bernard Palitz Research Fund.

primarily found in the cisternae of the endoplasmic reticulum. In contrast, those proteins destined to be free-floating in the cytoplasm are made on free ribosomes.

The signal for the binding of the ribosome to the endoplasmic reticulum is an amino terminal sequence about 15–30 amino acids long containing at least 11 very hydrophobic residues. It is this portion of the growing polypeptide chain as it emerges from the channel in the large ribosomal subunit that is responsible for the binding of the ribosome-mRNA complex to the membrane and for translocating the nascent protein across the membrane into the cisternal space of the endoplasmic reticulum. There it is removed from the mature protein by proteolysis. If the protein synthesized is destined for membrane incorporation this signal peptide is responsible for this protein's translocation across the membrane except that polypeptide transfer is arrested leaving the polypeptide anchored to the membrane by means of the hydrophobic amino acid sequence. The binding of the signal peptide portion of the growing protein chain to the membrane is a two-step process consisting of binding of the signal peptide to a ribonucleoprotein complex [signal recognition particle (SRP)] arresting further protein synthesis until this complex binds to a receptor on the endoplasmic reticulum [SRP receptor, or docking protein (DP)]. Binding to this receptor displaces the SRP and translocation proceeds coincidentally with renewed chain elongation. The SRP and DP would be recycled as the peptide is inserted into the membrane.[26–28] Other concepts have the important signal sequence buried in the elongating peptide with the peptide folded on itself, and thus portions may be outside the membrane. This concept suggests that some proteins destined for export may have unexcised residues of leader segment (FIG. 1).[29]

It is now recognized that the Golgi apparatus plays a central role in directing the various protein molecules to their proper final destination. It is imperative that secretory proteins and lysosomal enzymes appear to function as recognition markers that direct these enzymes to lysosomes, whereas sialic acid residues appear to be the markers for secretory proteins. As yet unexplained, albumin a secretory protein does not have any glycosyl markers. However, albumin is initially synthesized with a 24 amino acid residue extension at its amino terminus, preproalbumin. The first octadecapeptide functions as the signal for endoplasmic membrane insertion where it is proteolytically removed in the lumen. The subsequent 6 amino acid residues are cleaved off in the Golgi. Within the Golgi, there is intercompartmental transfer, not necessarily moving from one stack to another but random encapsulation in vesicles. Removal of the hexapeptide is necessary for the secretion of albumin and there is some evidence that this hexapeptide sequence may function as a negative feedback inhibitor of albumin synthesis.[30–34]

In addition to those points, the liver cell has been classified as a constitutive cell, one which transports all secretory proteins by bulk flow; intracellular transport and secretion are independent of recognition signals. Associated with this concept is one for regulated cells—that is those cells which have a storage capacity for proteins whose ultimate destination is outside of the cell. Secretion for these proteins requires an extracellular signal, and larger amounts of protein will be secreted than can be accounted for by synthesis. This storage-secretion concept would require a means whereby the storage vesicle would respond to an appropriate extracellular stimulus. Perhaps different from albumin, these regulated secretion processes are not inhibited if proteolysis of the precursor peptide is inhibited.[32]

Another point should be mentioned concerning the hepatocyte's protein secretion. Albumin has been claimed to be secreted only at the sinusoidal surface. This implies another type of cellular secretory control—movement of this bulk flow only along certain intracellular pathways or tubules or at least only in one direction. Microtubules may only be involved in transport in polarized cells.[32]

We initially present this data on protein cellular secretion and synthesis for two reasons. First, this is a highly complex and clearly integrated and organized system, and, second, to our knowledge there is not a single piece of datum to pinpoint any ethanol effect on any of these steps; in fact we are uncertain whether any such studies are even being conducted at present.

Let us turn now from the single hepatocyte to the whole human organism. The normal human liver produces about 15 grams of albumin a day and it is now believed that each cell is capable of producing albumin. If each cell were operating at its normal rate all day long much more albumin would be synthesized than is secreted and intracellular degradation would be excessive—a wasteful use of the cell's energy.[35]

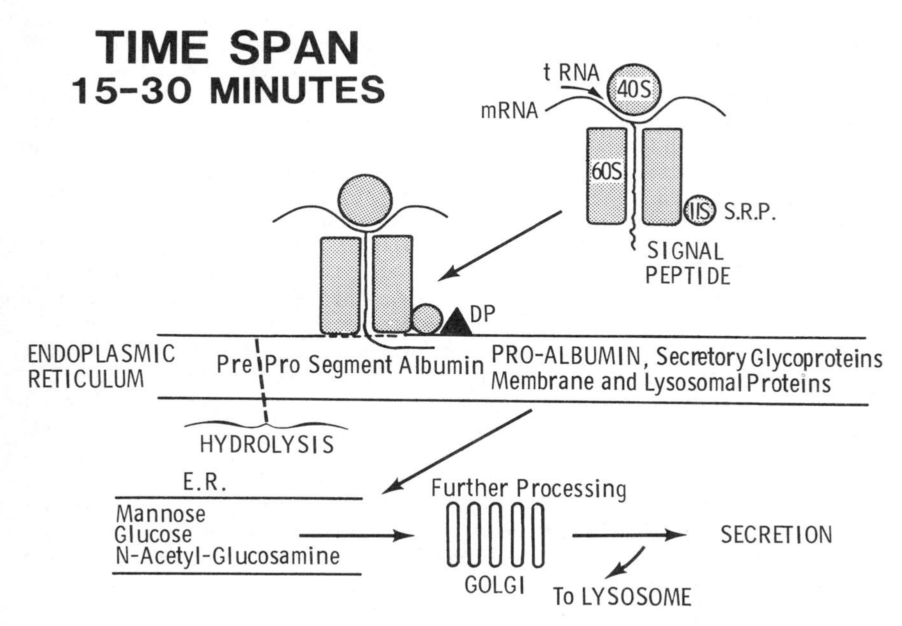

FIGURE 1. A schematic drawing depicting the time sequence of some of the major steps in cytosolic eukaryotic protein synthesis. Each step has many as yet unsolved components and depends on other factors such as energy sources, cellular integrity, and microtubular availability. Numerous signal sequences must be synthesized and recognized by other synthesized proteins or receptors and internal regulation and degradation must be taken into account.

When the human liver is exposed to the chronic intake of ethanol, cirrhosis of the liver may develop with its consequent perturbations on blood supply, cellular organization, and metabolism. The serum albumin level is most frequently depressed, and the serum albumin concentration has been used as a prognostic sign at least in nonascitic patients. The total exchangeable albumin pool has been measured in these patients and in contrast to the serum albumin level, is not depressed in many or most, but in most of the cirrhotic-ascitic subjects the exchangeable albumin pool was elevated.[36–37] Of course, this includes the amount of albumin in the ascitic volume, and this albumin achieves equilibrium slowly with the remaining exchangeable albumin.[38] Thus, in the face of severe liver disease, persistent ethanol intake, and altered total nutrition, the liver is capable of synthesizing and secreting large amounts of albumin. Micronutrition must also be considered, for with the development of portal hypertension, the hepatic acinus, distorted and hemmed in by fibrous bands, will receive even less in the way of nutrients and oxygen. There is considerable interest in how disease and blood flow alters the enzyme activity of the periportal and perivenous areas with fasting, diabetes, and the direction of blood flow into the liver, *i.e.*, portal vs. retrograde, all playing separate roles.[39–42] Alcohol will acutely increase portal flow, but the long-term effects on cellular enzyme activity are still in dispute.

The next step was to try to determine if altered nutrition or if ethanol per se was affecting albumin synthesis in any fashion. A short-term fast results in a loss of hepatic RNA, a disaggregation of the polysome bound to the endoplasmic reticulum and a decrease in albumin synthesis *in vivo* and in many different *in-vitro* systems.[43] The mRNA for albumin under these circumstances seems to be sequestered in a ribonucleoprotein particle in the cytosol.[44] Upon refeeding, albumin synthesis returns to normal or even to excessive levels. When ethanol is added to an *in-vitro* system capable of metabolizing ethanol and not previously exposed to ethanol *in vivo* then most data indicate that albumin synthesis is lowered and the bound polysome is disaggregated. That is, alcohol added *in vitro* appears to act as a pharmacologic fast. RNA synthesis and RNA polymerases have been shown to be inhibited in regenerating livers exposed to ethanol for only 24 hours. There is a decrease in the incorporation of nucleotides into RNA *in vitro.* When acute ethanol is studied *in vivo,* changes in protein synthesis are not necessarily seen. While the acute administration of ethanol has been shown to decrease tracer incorporation into albumin in some studies, others using similar but not identical techniques have shown no difference. When livers from ethanol-treated donors are studied in the presence of chronic ethanol treatment, again no difference in the incorporation of tracer activity into protein was noted. One of the important factors that could contribute directly to protein synthesis is the availability of amino acids. The evidence that alcohol per se or alcohol metabolism alters the available supply of amino acids separate from that which will accompany hepatic disease is not consistent. Aspartate and glutamate have been observed to increase and ethanol was observed to have no major effect on gluconeogenesis. A decrease in the net formation of alanine has been reported with an increase in free proline. On the other hand changes in aspartate and glutamate levels have been reported to be ethanol dose dependent, and no consistent changes in amino acid levels have also been reported. A decrease in the hepatic uptake of alpha-amino isobutyric acid was noted, which was not altered when ethanol metabolism was blocked with pyrazole. Certainly the microenvironmental concentration of amino acids may change around the liver cell during ethanol ingestion. However, a consistent trend has not been observed, and the effects of specific changes are not known.[43–54]

Associated with those acute *in-vitro* effects are changes in the lactate-pyruvate ratio indicating a change in the redox state of the liver.[55] When livers from chronic donors are studied their redox changes are moderated.[16] Therefore, the metabolism of ethanol may be the effective mechanism depressing albumin synthesis.[56] This aspect was investigated employing the isolated perfused rabbit liver from naive donors. When the donor was fed ad lib, the administration of 4-methylpyrazole, which prevented 95% of hepatic ethanol metabolism, did not improve albumin synthesis, which was still depressed to about 50% of the control-fed rate. However, the polysomes bound to the endoplasmic reticulum were not disaggregated. Thus, it seems that ethanol per se is toxic, but in a fashion differing from that occurring during active ethanol metabolism.[57]

The end product of ethanol metabolism is acetaldehyde, which has been considered to be the toxic agent of ethanol metabolism. However, while acetaldehyde reduces albumin synthesis in the perfused liver obtained from a fed donor, again the polysomes bound to the endoplasmic reticulum are not altered. Similar observations have been made for heart muscle protein synthesis.[58] Perfusing the liver with acetaldehyde in the presence of 4-methylpyrazole and disulfiram or of disulfiram alone, albumin synthesis was not affected but remained at a control-fed rate.[59]

When we add another parameter to this picture, namely, fasting, the effects of ethanol exposure are changed again (FIG. 2). Here, too, the data are not in agreement. Hepatocytes, for example, demonstrated no effect of ethanol on protein incorporation of tracer activity when the donor was fed but a significant effect when fasted cells were studied.[22] Other data show that fasting plus ethanol is not as readily reversible as when one or the other stress is studied separately using protein synthesis as an end point.[5] Acetaldehyde on the other hand has been shown not to depress albumin synthesis in livers from fasted donors.[57] Excess amino acids can reverse the toxic effects of ethanol when the liver is derived from a fed donor but not when the donor is fasted.[45] Recently, clinical data have been reported that suggest that excess amino acid supplementation improves the rate of recovery of acute alcoholic hepatitis, a finding that is in agreement with the experimental data cited above.[60] Again, to point up how much we do not understand, *in-vivo* administration of ethanol results in an acute increase in the portal blood flow as well as a decrease in the hepatic extracellular volume.[61,62] However, while

	STARVATION	ETHANOL ACUTE
Hepatic RNA	↓	unchanged
Albumin Synthesis	↓	↓
Urea Synthesis	↑	↓
Bound Polysome	disaggregated	disaggregated
Reversed by specific Amino Acids	yes	yes

FIGURE 2. The contrasting effects of fasting and perfusate ethanol 200 mgs% on albumin synthesis employing the perfused rabbit liver. Alcohol, in this regard, may be considered to act as an acute pharmacologic fast.

these vascular and volume changes are occurring for the liver as a whole we are unaware of the specific changes in the microenvironment. It would seem possible that an acute increase in blood flow might result in the delivery of a "protective" quantity of amino acids as well as an increase in ethanol.

Not all amino acids have been shown to be protective or restorative. Also, of interest, the branched chain amino acids, valine and leucine, were not effective in preventing polysome disaggregation or stimulating albumin synthesis, whereas isoleucine was effective.[48] Tryptophan has been shown to be essential in this regard, for its absence in an additive protein solution fails to reverse stress-induced polysome disaggregation.[63] In other studies in which tryptophan was added, urea synthesis was observed to increase; this association of improved albumin synthesis plus augmented urea synthesis was observed for each stimulatory amino acid.[48] This observation suggested a relationship between the urea cycle and the protein synthesis (TABLE 1). Further evidence for this possible association was that one of the stimulating amino acids was ornithine, which is not incorporated into protein.[48] Ornithine, however, is the precursor of the polyamines putrescine, spermidine, and spermine. These amines have been shown to play pivotal roles in many stages of cell division and protein synthesis. The addition of spermine at a physiological concentration resulted in a marked and rapid reaggregation of the endoplasmic-reticulum bound polysomes. An increase in

TABLE 1. Albumin Synthesis and Amino Acids[a]

Perfusate from Fasted Donors	Urea Synthesis	Albumin Synthesis
ARG. ORN. TRP. LYS. PHE. ALA. THR. PRO. GLN.	Increased	Increased
LEU. VAL. MET. HIS.	Unaffected	Unaffected

[a]Effects of specific amino acids in 10 mM concentrations on the rate of albumin and urea synthesis employing perfused livers derived from fasted rabbit donors. Ornithine, an amino acid not incorporated into albumin, is one of the stimulating amino acids suggesting a possible role of the polyamines, synthesized from ornithine, in the increase in albumin production.

albumin synthesis in fasted studies, however, required the addition of excess amino acids in the perfused liver. The addition of spermine also prevented and actually reversed the diminished protein synthetic capacity of livers exposed to ethanol *in vitro* and to CCl_4 *in vivo,* and electron microscopy demonstrated polysome reaggregation.[5,64] Further studies employing an inhibitor of the enzyme ornithine decarboxylase demonstrated that the protective actions of these amino acids could be blocked. Thus, the urea cycle in its role of providing a source of ornithine (FIG. 3) may play a more important role in influencing protein metabolism than simply that of providing the route for nitrogen excretion.[65,66] In this regard ethanol depresses the urea cycle, though by what means is not yet clarified.

So far we have been concerned with the effects of ethanol on protein synthesis; however, there is a large body of evidence to suggest that ethanol may inhibit the secretion of proteins from the cell.[16–19,67] Again it is necessary to consider both acute and chronic effects of ethanol. A basic observation is that the hepatocyte contains an increased amount of cellular nitrogen. At this point it is important to recall that the hepatocyte is not supposed to have the capacity to store protein and that secretion of protein from the hepatocyte may well be governed by the amount synthesized. There is no doubt that ethanol exposure disrupts the microtubules, and acetaldehyde has been shown to covalently bond to tubulin preventing polymerization.[68] A putative role for the intracellular microtubular structure is its participation in secretory vesicle transport to the plasma membrane for subsequent exocytosis. This may explain the data that have been developed from liver slice studies. In these alcohol experiments whether an amino acid label was used (leucine) or a carbohydrate label (fucose), a decreased rate of secretion of albumin and glycoproteins was noted. Also an increased content of carbohydrate label remained within the Golgi apparatus in slices of liver exposed to ethanol.[67] There are other conditions that could modify the covalent binding of

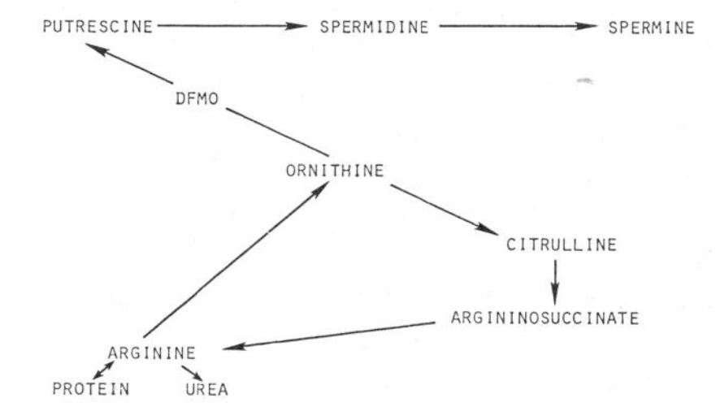

FIGURE 3. A schematic representation of the Krebs urea cycle. The ornithine generated is the precursor of the polyamines putrescine, spermidine, and spermine. The rate-limiting enzyme ornithine decarboxylase can be irreversibly inhibited by DFMO (difluoromethyl ornithine, generously supplied by Merrell Dow, Inc.).

acetaldehyde to liver proteins and to tubulin in particular. These include the level of acetaldehyde and its time of contact, other drugs, altered redox state, and the presence of lysine residues, which appear to make proteins more susceptible to adduct formation with subsequent loss of activity. These points are discussed in more detail in the paper in this volume by Dr. Sorrell and Dr. Tuma. A similar inhibition was not noted with ethanol-exposed hepatocytes obtained from naive rats. In the liver, albumin is released normally at the sinusoidal surface; perhaps where there is no cellular contact (hepatocyte preparations), polarization may not be apparent, and agents that cause transient microtubular interruption might be less effective (*e.g.* ethanol). Against this hypothesis is the fact that a known tubule-disrupting agent, colchicine, does impede albumin and total protein secretion in this type of liver cell preparation.[69] The hepatocyte, however, is a very special preparation. It does not metabolize ethanol as rapidly as do other preparations and may be more sensitive to redox changes. Thus, strict comparison between these models may not be possible.

Before proceeding further we must introduce another word of caution. In future discussions many different models will provide the data base. Obviously an *in-vivo,* adult male or female, nourished or starved, and genetically predisposed would not necessarily be expected to show a casual response reaction—an identifying reaction—as would an *in-vitro* model. Any stress (and alcohol is the most common self-induced environmental stress) will evoke *in vivo* a massive integrated response. There are many factors such as hormone levels, temperature,[70] amino acid delivery, cell permeability, and alterations in carbohydrate and fat metabolism that can influence protein synthesis indirectly. Therefore, that chronic ethanol consumption does not confirm all of the effects of ethanol or acetaldehyde that are observed *in vitro* is not surprising. Why should the polysome removed from an animal who rarely chooses alcohol mimic the *in-vivo* liver—or, in fact, the intact cell? Does a liver perfused with oxygenated blood through the portal low-oxygen venous system mimic the liver slice or the collagenase-treated liver cell for that matter? What does alcohol do to the specific activity of the precursor amino acid at the site of protein synthesis? What does the previously synthesized hepatic interstitial protein do to protein synthesis? The feeding of alcohol results in a marked enlargement of the liver, and an increase in hepatic nitrogen has been observed. This observation coupled with the measurement of retained proteins destined for export suggested that the retention was the cause of the increased nitrogen. However, the retention of albumin and transferrin accounts for only about 1% of the increment in cellular nitrogen, and so this degree of retention does not explain the increase in cellular nitrogen.[17] The specific proteins causing this nitrogen increment have not been identified. It has been postulated that the larger cells may compress the hepatic extracellular space resulting in an increase in intrahepatic pressure. Associated with this change (and it must be recalled that both acute and chronic ethanol exposure result in a decrease in hepatic extracellular volume)[71–72] is an increase in collagen synthesis, which must further compromise the available interstitial volume. Such changes could result in increases in colloid oncotic pressure within this volume. Osmotically reciprocal depression in serum albumin levels have been noted with rises in globulin levels in chronic liver disease.[73] A colloid osmotic regulator has been proposed and supported by numerous *in-vitro* studies.[74] One of the mechanisms whereby chronic ethanol exposure could alter hepatic albumin synthesis may, in fact, involve this mechanism. The acute administration of ethanol impedes albumin synthesis within minutes as does cycloheximide, whereas altered hepatic interstitial colloid requires hours. Therefore, the acute ethanol effect is probably not mediated by this osmotic mechanism.[75]

Another factor involved in this protein retention has been acetaldehyde. It has been pointed out that the electrophilic carbonyl carbon of acetaldehyde binds to the nucleophilic groups of peptides in a covalent bond. These bonds may be stable and do occur during ethanol oxygenation. Furthermore, areas of decreased oxygen availability (zone 3 of the hepatic acinus) would favor the development of a stable adduct. Decreased enzyme activities have been demonstrated and long-term effects on tubulin polymerization have been reported.

Finally it is clear that ethanol exposure has an effect on hepatic protein synthesis. It appears that this effect may well be moderated during chronic exposure, can be overcome, and affects different proteins in different ways. A stimulus to collagen synthesis may occur simultaneously with a depression of albumin and transferrin synthesis and release. Cellular transport and secretion may also be altered, but whether these changes are preceded or accompanied by alterations in synthesis remains to be clarified. Protein synthesis can be influenced by so many *in-vivo* variables that the consequences of long-term ethanol exposure *in vivo* on protein production cannot be understood unless other models are employed. The applications of state-of-the-art microbiological techniques to hepatic protein synthesis and secretion will be necessary to clarify the effects of ethanol and to determine if the effects of ethanol on protein metabolism have any relationship to cellular damage.

REFERENCES

1. Rothschild, M. A., M. Oratz, J. Mongelli, *et al.* 1971. Alcohol induced depression of albumin synthesis: reversal by tryptophan. Clin. Invest. **50:** 1812–1818.
2. Kirsch, R. E., L. O. Frith, R. H. Stead, *et al.* 1973. Effect of alcohol on albumin synthesis by the isolated perfused rat liver. Amer. J. Clin. Nutr. **26:** 1191–1194.
3. Chamber, J. W. & V. J. Piccirillo. 1973. Effects of ethanol on amino acid uptake and utilization by the liver and other organs of rats. Quart. J. Stud. Alcohol **34:** 707–717.
4. Morland, J. 1975. Incorporation of labelled amino acids into liver protein after acute ethanol administration. Biochem. Pharmacol. **24:** 439–442.
5. Oratz, M., M. A. Rothschild & S. S. Schreiber. 1976. Alcohol, amino acids, and albumin synthesis. II. Alcohol inhibition of albumin synthesis reversed by arginine and spermine. Gastroenterology **71:** 123–127.
6. Perin, A., G. Scalabrino, A. Sassa, *et al.* 1974. In vitro inhibition of protein synthesis in rat liver as a consequence of ethanol metabolism. Biochem. Biophys. Acta **366:** 101–108.
7. Sorrell M. F., D. J. Tuma, E. C. Schafter, *et al.* 1977. Role of acetaldehyde in the ethanol-induced impairment of glycoprotein metabolism in rat liver slices. Gastroenterology **73:** 137–144.
8. Jeejeebhoy, K. N., A. Bruce-Robertson, J. Ho, *et al.* 1975. The effect of ethanol on albumin and fibrinogen synthesis in vitro and in hepatocyte suspensions. *In* Alcohol and Abnormal Protein Synthesis. M. A. Rothschild, M. Oratz & S. S. Schreiber, Eds. Pergamon Press. New York, NY.
9. Morland, J. & A. Bessesen. 1977. Inhibition of protein synthesis by ethanol in isolated rat liver parenchymal cells. Biochem. Biophys. Acta **474:** 312–320.
10. Renis, M., A. Giovine & A. Bertolino. 1975. Protein synthesis in mitochondrial fractions from rat brain and liver after acute and chronic ethanol administration. Life Sci. **16:** 1447–1458.
11. Rubin, E., D. S. Beattie & C. S. Lieber. 1970. Effects of ethanol on the biogenesis of mitochondrial functions. Lab. Invest. **23:** 620–627.
12. Burke, J. P., M. E. Tumbleson, K. W. Hicklin, *et al.* 1975. Effects of chronic ethanol ingestion on mitochondrial protein synthesis in Sinclair (S-a) miniature Swine. Proc. Soc. Exp. Biol. Med. **149:** 1051–1056.
13. Kuriyam, K., P. Y. Sze & G. E. Rauscher. 1971. Effects of acute and chronic ethanol administration on ribosomal protein synthesis in mouse brain and liver. Life Sci. **10:** 181–189.
14. Khawaja, J. A. & D. B. Lindholm. 1978. Differential effect of ethanol ingestion on the

protein synthetic activities of free and membrane-bound ribosomes from liver of the weanling rat. Res. Commun. Chem. Pathol. Pharmacol. **19:** 129–139.
15. MORLAND, J., A. BESSESEN, A. SMITH-KIELLAND, *et al.* 1983. Ethanol and protein metabolism in the liver. Pharmacol. Biochem. Behav. **18:** 251–256.
16. BARAONA, E., P. PIKKARAINEN, M. SALASPURO, *et al.* 1980. Acute effects of ethanol on hepatic protein synthesis and secretion in the rat. Gastroenterology **79:** 104–111.
17. BARAONA, E., A. LEO, S. A. BOROWSKY, *et al.* 1977. Pathogenesis of alcohol-induced accumulation of protein in the liver. J. Clin. Invest. **60:** 546–554.
18. TUMA, D. J., R. B. JENNETT & M. F. SORRELL. 1981. Effect of ethanol on the synthesis and secretion of hepatic secretory glycoproteins and albumin. Hepatology **6:** 590.
19. VOLENTIN, G. D., D. J. TUMA & M. F. SORRELL. 1984. Acute effects of ethanol on hepatic glycoprotein secretion in the rat in vivo. Gastroenterology **86:** 225–229.
20. WALLIN, B., A. BESSESEN, A. M. FIKKE, *et al.* 1984. No effect of acute ethanol administration on hepatic protein synthesis and export in the rat in vivo. Alcoholism: Clin. Exp. Res. **8:** 191–195.
21. PRINCEN, J. M. G., G. P. B. M. MOL-BACKX & S. H. YAP. 1981. Acute effects of ethanol intake on albumin and total protein synthesis in free and membrane-bound polyribosomes of rat liver. Biochim. Biophys. Acta **655:** 119–127.
22. MORLAND, J., M. A. ROTHSCHILD, M. ORATZ, *et al.* 1981. Protein secretion in suspensions of hepatocytes: no influence of acute ethanol administration. Gastroenterology **80:** 159–165.
23. ZERN, M. A., P. R. CHAKRABORTY, N. RUIZ-OPAZO, *et al.* 1983. Development and use of a rat albumin cDNA clone to evaluate the effect of chronic ethanol administration on hepatic protein synthesis. Hepatology **3:** 317–322.
24. MOLDAVE, K. 1985. Eukaryotic protein synthesis. Annu. Rev. Biochem. **54:** 1109–1149.
25. BLOBEL, G. & B. DOBBERSTEIN. 1975. Transfer of proteins across membranes. I. Presence of proteolytically processed and unprocessed nascent immunoglobulin light chains on membrane-bound ribosomes of murine myeloma. J. Cell Biol. **67:** 835–851.
26. WALTER, P., R. GILMORE & G. BLOBEL. 1984. Protein translocation across the endoplasmic reticulum. Cell **38:** 5–8.
27. HORTSCH, M., D. AVOSSA & D. I. MEYER. 1985. Factors mediating protein translocation in the endoplasmic reticulum: the docking protein and beyond. *In* Protein Transport and Secretion. Cold Spring Harbor Conference on Protein Transport and Secretion. M. J. Gething, Ed. 24–32. Cold Spring Harbor Laboratory. Cold Spring Harbor, NY.
28. WALTER, P. & G. BLOBEL. 1981. Translocation of proteins across the endoplasmic reticulum. III. Signal recognition protein (SRP) causes signal sequence-dependent and site-specific arrest of chain elongation that is released by microsomal membranes. J. Cell Biol. **91:** 557–561.
29. WICKNER, W. T. & H. F. LODISH. 1985. Multiple mechanisms of protein insertion into and across membranes. Science **230:** 400–407.
30. GLAUMANN, H. & J. L. E. ERICSSON. 1970. Evidence for the participation of the Golgi apparatus in the intracellular transport of nascent albumin in the liver cell. J. Cell Biol. **47:** 555–567.
31. FARQUHAR, M. G. & G. E. PALADE. 1981. The Golgi apparatus complex—(1954–1981)—from artifact to center stage. J. Cell Biol. **91:** 77s–103s.
32. KELLY, R. B. 1985. Pathways of protein secretion in eukaryotes. Science **230:** 25–32.
33. ROTHMAN, J. E., R. L. MILLER & L. J. URBANI. 1984. Intercompartmental transport in the Golgi complex is a dissociative process: facile transfer of membrane protein between two Golgi populations. J. Cell Biol. **99:** 260–271.
34. WEIGAND, K., M. SCHMID, A. VILLRINGER, *et al.* 1982. Hexa- and pentapeptide extension of proalbumin: feedback inhibition of albumin synthesis by its propeptide in isolated hepatocytes and in the cell-free system. Biochemistry **21:** 6053–6059.
35. ROTHSCHILD, M. A. & M. ORATZ. 1976. Albumin synthesis and degradation. *In* Structure and Function of Plasma Proteins. A. C. Allison, Ed. Vol. 2: 79–105. Plenum Press. London.
36. ROTHSCHILD, M. A., M. ORATZ, D. ZIMMON, *et al.* 1969. Albumin synthesis in cirrhotic subjects with ascites studied with carbonate-^{14}C. J. Clin. Invest. **48:** 344–350.

37. ROTHSCHILD, M. A., M. ORATZ & S. S. SCHREIBER. 1970. Changing concepts of albumin metabolism and distribution in cirrhosis of the liver. Scand. J. Gastro. Suppl. **7:** 17–23.
38. BERSON, S. A. & R. S. YALOW. 1954. The distribution of ^{131}I-labeled human serum albumin introduced into ascitic fluid. Analysis of the kinetics of a three compartment catenary transfer system in man and speculations on possible sites of degradation. J. Clin. Invest. **33:** 377–387.
39. ORREGO, H., L. M. BLENDIS, I. R. CROSSLEY, *et al.* 1981. Correlation of intrahepatic pressure with collagen in the Disse space and hepatomegaly in humans and in the rat. Gastroenterology **80:** 546–556.
40. MIETHKE, H., B. WITTIG, A. NATH, *et al.* 1985. Metabolic zonation in liver of diabetic rats. Zonal distribution of phosphoenolpyruvate carboxykinase, pyruvate kinase, glucose-6-phosphatase, and succinate dehydrogenase. Biol. Chem. Hoppe Seyler **366:** 493–501.
41. THURMAN, R. G. & F. C. KAUFFMAN. 1985. Sublobular compartmentation of pharmacologic events (SCOPE): metabolic fluxes in periportal and pericentral regions of the liver lobule. Hepatology **5:** 144–151.
42. JUNGERMANN, K. & N. KATZ. 1982. Functional hepatocellular heterogeneity. Hepatology **2:** 385–395.
43. ORATZ, M. & M. A. ROTHSCHILD. 1975. The influence of alcohol and altered nutrition on albumin synthesis. *In* Alcohol and Abnormal Protein Biosynthesis. M. A. Rothschild, M. Oratz & S. S. Schreiber, Eds. 343–372. Pergamon Press. New York, NY.
44. YAP, S. H., R. K. STRAIR & D. A. SHAFRITZ. 1978. Identification of albumin mRNPs in the cytosol of fasting rat liver and influence of tryptophan or a mixture of amino acids. Biochem. Biophys. Res. Comm. **83:** 427–433.
45. ROTHSCHILD, M. A., M. ORATZ & S. S. SCHREIBER. 1974. Alcohol, amino acids, and albumin synthesis. Gastroenterology **67:** 1200–1213.
46. JEFFERSON, L. S. & A. KORNER. 1969. Influence of amino acid supply on ribosomes and protein synthesis of perfused rat liver. Biochem. J. **111:** 703–712.
47. WANNEMACHER, R. W., JR., C. F. WANNAMACHER & M. B. YATVIN. 1971. Amino acid regulation of synthesis of ribonucleic acid and protein in the liver of rats. Biochem. J. **124:** 385–392.
48. ROTHSCHILD, M. A., M. ORATZ, J. MONGELLI, *et al.* 1969. Amino acid regulation of albumin synthesis. J. Nutr. **98:** 395–403.
49. ROSA, J. & E. RUBIN. 1980. Effects of ethanol on amino acid uptake by rat liver cells. Lab. Invest. **43:** 366–372.
50. KREBS, H. A., R. HEMS & P. LUND. 1973. Accumulation of amino acids by the perfused rat liver in the presence of ethanol. Biochem. J. **134:** 697–705.
51. PETIT, M. A. & I. BARRAL-AIX. 1979. Effect of ethanol dose on amino acid and urea concentrations in the fed rat liver in vivo. Biochem. Pharmacol. **28:** 2591–2596.
52. HAKKINEN, H. M. & E. KULONEN. 1974. Effect of ethanol on the metabolism of alanine, glutamic acid, and proline in rat liver. Biochem. Pharmacol. **24:** 199–204.
53. STUBBS, M. & H. A. KREBS. 1975. The accumulation of aspartate in the presence of ethanol in rat liver. Biochem. J. **150:** 41–45.
54. PICCIRILLO, V. J. & J. W. CHAMBER. Inhibition of hepatic uptake of alpha amino isobutyric acid by ethanol: effects of pyrazole and metabolites of ethanol. Res. Comm. Chem. Path. Pharm. **13:** 297–308.
55. LIEBER, C. S. 1984. Alcohol and the liver: 1984 update. Hepatology **4:** 1243–1260.
56. PERIN, A., G. SCALABRINO, A. SESSA, *et al.* 1974. In vitro inhibition of protein synthesis in rat liver as a consequence of ethanol metabolism. Biochim. Biophys. Acta **366:** 101–108.
57. ORATZ, M., M. A. ROTHSCHILD & S. S. SCHREIBER. 1978. Alcohol, amino acids, and albumin synthesis. III. Effects of ethanol, acetaldehyde, and 4-methylpyrazole. Gastroenterology **74:** 672–676.
58. SCHREIBER, S. S., M. ORATZ & M. A. ROTHSCHILD. 1976. Alcoholic cardiomyopathy: the effect of ethanol and acetaldehyde on cardiac protein synthesis. *In* Recent Advances in Studies on Cardiac Structure and Metabolism. P. Harris, R. J. Bing & A. Fleckenstein, Eds. Vol. **7:** 431–442. University Park Press. Baltimore, MD.
59. ROTHSCHILD, M. A., M. ORATZ, S. S. SCHREIBER, *et al.* 1980. The effects of acetaldehyde and disulfiram on albumin synthesis in the isolated perfused rabbit liver. Alcoholism: Clin. Exp. Res. **4:** 30–33.

60. DIEHL, A. M., J. K. BOITNOTT, H. F. HERLONG, *et al.* 1985. Effect of parenteral amino acid supplementation in alcoholic hepatitis. Hepatology **5:** 57–63.
61. CARMICHAEL, F. J., J. P. MCKAIGNEY, V. SALDIVIA, *et al.* 1986. Ethanol induced increase in splanchnic blood flow: role of ethanol metabolism. Am. J. Physiol. In press.
62. VILLENEUVE, J. P., J. POMINER & P. M. HUET. 1981. Effect of ethanol on hepatic blood flow in unanesthetized dogs with portal and hepatic vein catherization. Can. J. Physiol. Pharmacol. **59:** 598–603.
63. MCGOWN, E., A. G. RICHARDSON, L. M. HENDERSON, *et al.* 1973. Effect of amino acids on ribosome aggregation and protein synthesis in perfused rat liver. J. Nutr. **103:** 109–116.
64. ORATZ, M., M. A. ROTHSCHILD, S. S. SCHREIBER, *et al.* 1980. Spermine stimulation of CCl_4 depressed protein synthesis in rabbits. Gastroenterology **79:** 1165–1173.
65. ROTHSCHILD, M. A., M. ORATZ, S. S. SCHREIBER, *et al.* 1983. Putrescine synthesis: essential for increased albumin synthesis. Hepatology **3:** 821.
66. ORATZ, M., M. A. ROTHSCHILD & S. S. SCHREIBER. 1983. The role of the urea cycle and polyamines in albumin synthesis. Hepatology **3:** 567–571.
67. VOLENTINE, G. D., D. J. TUMA & M. F. SORRELL. 1986. Subcellular location of secretory proteins retained in the liver during the ethanol-induced inhibition of hepatic protein secretion in the rat. Gastroenterology **90:** 158–165.
68. JENNETT, R. B., E. L. JOHNSON, M. F. SORRELL, *et al.* 1985. Preferential covalent binding of acetaldehyde to the alpha-chain of tubulin. Hepatology **5:** 1056.
69. REDMAN, C. M., D. BANERJEE, K. HOWELL, *et al.* 1975. Colchicine inhibition of plasma protein release from rat hepatocytes. J. Cell Biol. **66:** 42–59.
70. HENDERSON, G. I., A. M. HOYUMPA, JR., M. A. ROTHSCHILD, *et al.* 1980. Effect of ethanol and ethanol-induced hypothermia on protein synthesis in pregnant and fetal rats. Alcoholism: Clin. Exp. Res. **4:** 165–177.
71. MONGELLI, J., M. ORATZ, M. A. ROTHSCHILD, *et al.* 1984. Ethanol induced decrease in hepatic interstitial volume—an osmotic regulator of albumin synthesis. Hepatology **4:** 1027.
72. VIDINS, E. I., R. S. BRITTON, A. MEDLINE, *et al.* 1985. Sinusoidal caliber in alcoholic and nonalcoholic liver disease: diagnostic and pathogenic implications. Hepatology **4:** 408–414.
73. BJORNEBOE, M. 1946. Studies on the serum proteins in hepatitis. I. The relation between serum albumin and serum globulin. Acta Med. Scand. **123:** 393–401.
74. ORATZ, M., M. A. ROTHSCHILD & S. S. SCHREIBER. 1977. Albumin-osmotic function. *In* Albumin Structure, Function and Uses. V. M. Rosenoer, M. Oratz & M. A. Rothschild, Eds. 275–282. Pergamon Press. New York, NY.
75. ROTHSCHILD, M. A., M. ORATZ & S. S. SCHREIBER. 1985. The effects of hyperosmotic perfusates on hepatic interstitial volume and albumin release and synthesis. Hepatology **5:** 968.

DISCUSSION OF THE PAPER

E. RUBIN (*Thomas Jefferson University School of Medicine, Philadelphia, PA*): You referred to the interstitial space as a concept. What about the real interstitial space? For instance, the portal tracts contain interstitial space.

ROTHSCHILD: I assume that sucrose and DTPA, which were equilibrated with the liver for 30 minutes to an hour, would get into these spaces. These are physiologic spaces, not anatomic ones. We simply took the liver from the perfused system, gently patted both surfaces, cut it up into half a dozen sections, dropped them into counting tubes, and counted them. Thus, we counted everything.

RUBIN: What about the chronic effects of ethanol in the patients? The patients who show deficits in protein synthesis are those who have ascites and, presumably, are no different from patients who have cirrhosis from any other cause. Do you relate decreased protein synthesis to alcohol intake or to chronic liver disease?

ROTHSCHILD: That has not been conclusively determined.

Y. ISRAEL: (*University of Toronto, Toronto, Canada*): Alcohol may swell the hepatocytes, and this may lead to a reduction in extracellular space.

ROTHSCHILD: I agree with you.

Phorbol Esters Inhibit Ethanol-Induced Calcium Mobilization and Polyphosphoinositide Turnover in Isolated Hepatocytes[a]

JAN B. HOEK, RAPHAEL RUBIN, AND ANDREW P. THOMAS

Department of Pathology
Thomas Jefferson University
Philadelphia, Pennsylvania 19107

Ethanol (50–500 mM) causes a transient activation of the hormone-sensitive phosphoinositide-specific phospholipase C in intact hepatocytes, generating two second messenger products, inositol-1,4,5-trisphosphate (Ins-1,4,5-P_3), which mobilizes Ca^{2+} from endoplasmic reticular storage sites, and diacylglycerol (detected as phosphatidic acid), which can activate protein kinase C.[1–4] The phorbol ester 12-0-tetradecanoate

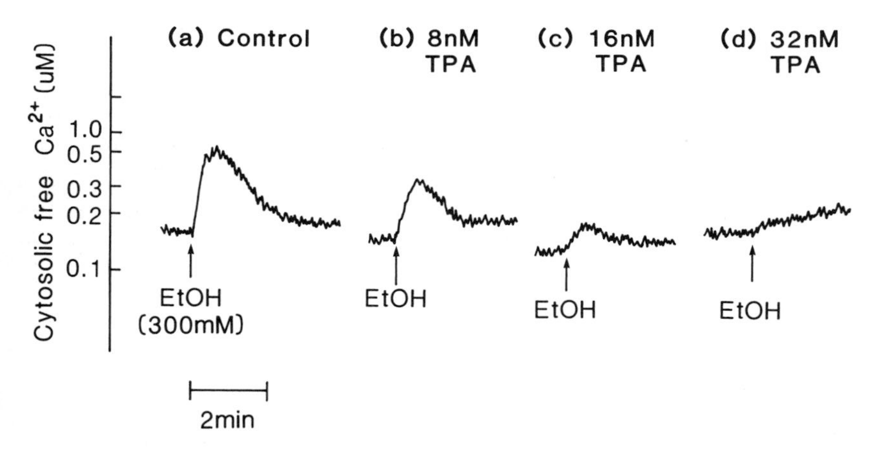

FIGURE 1. Inhibition by phorbol ester of ethanol-induced calcium mobilization in quin 2-loaded hepatocytes. Isolated hepatocytes were loaded with quin 2 and incubated at 37° C in the cuvet of a Perkin-Elmer MPF44B spectrofluorometer, as described previously.[1] TPA was added from a stock solution in DMSO. Wavelength pair: λ ex 339 nm, λ em 500 nm.

phorbol-13-acetate (TPA), a membrane permeant protein kinase C activator, has been used extensively as a tool to study the interrelationships between the calcium-mediated and the protein kinase C-mediated branches of the phosphoinositide-linked signal transduction pathway. In liver cells, TPA pretreatment prevents the activation of

[a]Supported by United States Public Health Service Grants AA05662, AA03422, AM31086, and AA06502.

phospholipase C by α_1-adrenergic agonists and certain other stimuli.[5,6] In this study, we have investigated the effect of TPA pretreatment on the ethanol-induced activation of phospholipase C and the consequent increase in cytosolic free Ca^{2+} levels.

FIGURE 1 illustrates the effect of different doses of TPA on the ethanol-induced calcium mobilization in hepatocytes loaded with the fluorescent calcium indicator quin 2. The phorbol ester inhibited the transient Ca^{2+} mobilization induced by the addition of 300 mM ethanol. Half-maximal inhibition was obtained at doses of approximately 10 nM, and complete inhibition occurred at TPA concentrations of 30 nM and above. Similar levels of TPA inhibited calcium mobilization induced by phenylephrine (20 μM) and glucagon (100 nM) but not by a saturating dose of vasopressin (100 nM) (data not shown), in agreement with earlier reports.[5,6] TPA was maximally effective after 2 min preincubation, but had no effect when added together with, or after the ethanol. Oleyl-acetylglycerol, a membrane-permeant diglyceride, partially inhibited ethanol-induced calcium mobilization. Phorbol esters that do not activate protein kinase C did not block this effect of ethanol. These results indicate that the inhibition of ethanol-induced calcium mobilization by TPA is mediated by protein kinase C.

TABLE 1. Effect of Phorbol Ester Pretreatment on the Ethanol-Induced Changes in the Level of Polyphosphoinositides, Inositol Phosphates, and Phosphatidate[a]

Metabolite	Ethanol	Ethanol + TPA	p
	(% of Control)		
a. Ptd Ins-4-P	110 ± 1.2	99 ± 0.9	< 0.05
b. Ptd Ins-4,5-P_2	103 ± 1.7	102 ± 0.8	n.s.
c. Phosphatidic Acid	111 ± 1.0	105 ± 2.1	< 0.05
d. Ins-1,4,5-P_3	157 ± 17.2	91 ± 7.0	< 0.01
e. Ins-1,3,4-P_3	197 ± 19.4	101 ± 7.3	< 0.01
f. Ins-P_2	139 ± 11.0	100 ± 22.0	< 0.05

[a]Isolated hepatocytes were preincubated for 90 min to label inositol phospholipids, either with ^{32}P (a–c) or with [^3H] inositol (d–f) as described elsewhere.[2,3] Samples were taken at 20 sec (d–f) or 30 sec (a–c) after the addition of ethanol (300 mM). Where present, TPA (100 nM) was added 5 min before ethanol. For analysis of inositol phosphates and phospholipids, see REFERENCES 2,3. Results are mean ± S.E.M. for 5 (a–c) or 3 (d–f) experiments. Statistical significance was estimated using Student's t test.

The data in TABLE 1 demonstrate that TPA pretreatment also inhibits the ethanol-induced changes in the levels of polyphosphoinositides and inositol phosphates. The ethanol-induced increase in phosphatidylinositol-4-phosphate and phosphatidic acid, which reflect the increased flux through the system of phosphoinositide kinases and phospholipase C,[3] was blocked by TPA, as was the formation of Ins-1,4,5-P_3, the intermediate directly responsible for Ca^{2+} mobilization.[4]

These results indicate that the effect of phorbol esters is exerted at an early step in the signal transduction pathway. Other studies[5,6] suggested the α_1-adrenergic receptor as a potential target. However, ethanol-induced calcium mobilization does not require receptor occupancy and is not sensitive to the α-adrenergic antagonist phentolamine. An alternative potential target site of TPA could be one of the putative GTP-binding regulatory proteins that mediate receptor-phospholipase C interaction. This would imply that the site of action of ethanol on this signal transduction pathway may be localized within the intramembrane coupling system that links hormone receptors to phospholipase C.

REFERENCES

1. HOEK, J. B., A. P. THOMAS, R. RUBIN & E. RUBIN. 1987. J. Biol. Chem. **262:** 682–691.
2. HOEK, J. B. 1987. Ann. N.Y. Acad. Sci. This volume.
3. RUBIN, R. & J. B. HOEK. 1987. Ann. N.Y. Acad. Sci. This volume.
4. THOMAS, A. P., J. B. HOEK & E. RUBIN. 1987. Ann. N.Y. Acad. Sci. This volume.
5. COOPER, R. H., K. E. COLL & J. R. WILLIAMSON. 1985. J. Biol. Chem. **260:** 3281–3288.
6. LYNCH, C. J., R. CHAREST, S. B. BOCCHINO, J. H. EXTON & P. F. BLACKMORE. 1985. J. Biol. Chem. **260:** 2844–2851.

Ethanol-Induced Activation of Polyphosphoinositide Turnover in Rat Hepatocytes[a]

RAPHAEL RUBIN AND JAN B. HOEK

Department of Pathology
Jefferson Medical College
Philadelphia, Pennsylvania 19107

Ethanol activates glycogen phosphorylase in rat hepatocytes by releasing calcium from hormone-sensitive intracellular stores.[1] Pretreatment of hepatocytes with vasopressin or the adrenergic agonist phenylepherine can eliminate further calcium-mobilizing effects of ethanol.[1] In liver, calcium-mobilizing hormones such as vasopressin and phenylepherine activate a phosphoinositide-specific phospholipase C, leading to the rapid breakdown of phosphatidylinositol 4,5-bisphosphate and the production of inositol trisphosphate and diacylglycerol.[2,3] Inositol trisphosphate has been demonstrated to directly mobilize calcium from a nonmitochondrial store, presumably endoplasmic reticulum.[4] Diacylglycerol is known to activate protein kinase C.[2] In this study, we have examined the effect of ethanol on hormone-sensitive phospholipase C.

Rat hepatocytes were isolated by collagenase perfusion[5] and incubated in Krebs-Ringer solution containing 10 uCi/ml ^{32}P for 90 min. In cells treated this way, changes in polyphosphoinositide-associated ^{32}P-radioactivity over short periods reflect changes in lipid mass.[3] Addition of 200 mM ethanol resulted in a small (5–10%) decline in phosphatidylinositol 4,5-bisphosphate and a 10–20% rise in phosphatidylinositol 4-phosphate and phosphatidic acid. All changes were maximal by 30 sec (FIG. 1). Phosphatidylinositol and phosphatidylcholine levels were unchanged. The effects of ethanol on phosphatidylinositol 4-phosphate and phosphatidic acid were concentration-dependent up to 500 mM ethanol (data not shown). Comparable changes in polyphosphoinositides and phosphatidic acid were obtained with a low concentration (1 nM) of vasopressin, which resembled ethanol in the rate and extent of calcium mobilization.

The time course and concentration dependency of the ethanol-induced changes in polyphosphoinositides and phosphatidic acid closely parallel the elevation of cytosolic calcium, as measured either by quin-2 fluorescence or by activation of phosphorylase.[1] The decrease in phosphatidylinositol 4,5-bisphosphate is compatible with the results of Thomas *et al.*,[6] who demonstrated an increased production of inositol trisphosphates by ethanol under similar conditions. The production of phosphatidic acid is presumed to reflect formation of diacylglycerol, which, in liver cells, is rapidly phosphorylated by diacylglycerol kinase.

We conclude that the changes in polyphosphoinositides and phosphatidic acid in response to ethanol addition are compatible with an activation of the hormone-sensitive, phosphoinositide-specific phospholipase C and a concomitant increase in the flux through the phosphatidylinositol kinases, similar to the effects observed with low doses of calcium-mobilizing hormones. The generation of the second messengers

[a]Supported by United States Public Health Service Grants AA05662, AA03422, AM31086, and AA06502.

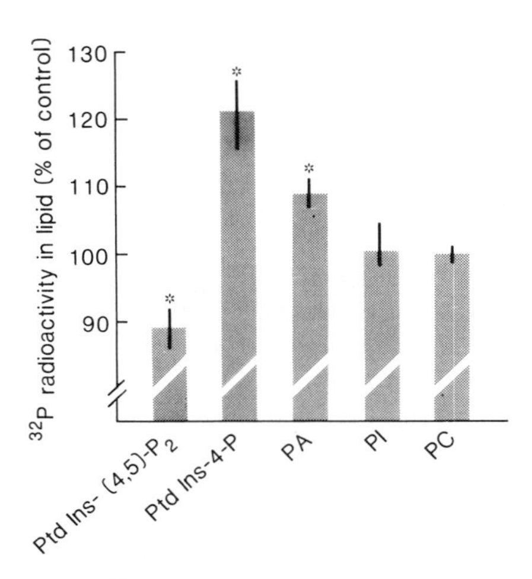

FIGURE 1. Effect of ethanol on ^{32}P-radioactivity in lipids of isolated rat hepatocytes. Isolated hepatocytes were prepared from male Sprague-Dawley rats by collagenase perfusion.[5] Cells were incubated for 90 min in a shaking water bath at 37° C in stoppered 25-ml flasks in Krebs-Ringer bicarbonate buffer, pH 7.4, containing 2% dialyzed bovine serum albumin and 10 uCi/ml ^{32}P. The gas phase contained 95% 0_2/5% CO_2. 30 seconds after the addition of 200 mM ethanol, 0.15-ml samples were quenched in 0.56 ml chloroform/methanol/HCL (200:100:5). The ^{32}P-labeled lipids were extracted into chloroform after the addition of 0.188 ml of 2M KCL/10 mM EDTA and 0.188 ml chloroform, and then the chloroform was evaporated in a Rotovac Evaporator. The lipids were separated by thin-layer chromatography as described in REFERENCE 3. Data are mean ± S.E.M. of triplicate determinations from 1 of 5 similar experiments and are expressed as a percentage of the appropriate controls. Significance of differences from the control calculated by paired T test were:*, significant at $P < 0.05$.

inositol 1,4,5-trisphosphate and diacylglycerol can lead to the mobilization of intracellular calcium and the activation of protein kinase C, respectively, which could have multiple effects on the regulation of cellular metabolism.

The mechanism(s) of ethanol-induced phospholipase C activation has not yet been identified. Ethanol may act on the membrane to affect the interaction of phospholipase C with its substrate phosphatidylinositol 4,5-bisphosphate, or with a regulatory component, such as the putative G-protein mediating receptor-phospholipase C interaction.

REFERENCES

1. HOEK, J. B., A. P. THOMAS, R. RUBIN & E. RUBIN. 1987. J. Biol. Chem. **262:** 682–691.
2. WILLIAMSON, J. R., R. H. COOPER, S. K. JOSEPH & A. P. THOMAS. 1985. Am. J. Physiol. **248** (Cell Physiol. **17**):C203–C216.
3. THOMAS, A. P., J. S. MARKS, K. E. COLL & J. R. WILLIAMSON. 1983. J. Biol. Chem. **258:**5716–5725.
4. JOSEPH, S. K., A. P. THOMAS, R. J. WILLIAMS, R. F. IRVINE & J. R. WILLIAMSON. 1984. J. Biol. Chem. **259:**3077–3081.
5. BERRY, M. N. & D. S. FRIEND. (1969). J. Cell Biol. **43:**506–520.
6. THOMAS, A. P., J. B. HOEK & E. RUBIN. 1987. Ann. N.Y. Acad. Sci. This volume.

Elevation of Inositol 1,4,5-Trisphosphate Levels after Acute Ethanol Treatment of Rat Hepatocytes[a]

ANDREW P. THOMAS, JAN B. HOEK,
AND EMANUEL RUBIN

Department of Pathology
Thomas Jefferson University
Philadelphia, Pennsylvania 19107

Acute treatment of isolated rat liver cells with ethanol (25–300 mM) results in a rapid transient elevation of cytosolic free Ca^{2+} ($[Ca^{2+}]_i$) and a consequent activation of phosphorylase.[1] Similar changes are brought about by glycogenolytic hormones such as vasopressin and α-adrenergic agonists in liver. These hormones increase $[Ca^{2+}]_i$ by activating a phospholipase C that hydrolyzes phosphatidylinositol 4,5-bisphosphate to yield diacylglycerol and inositol 1,4,5-trisphosphate ($Ins(1,4,5)P_3$), the latter product being a second messenger that causes Ca^{2+} mobilization from intracellular Ca^{2+} storage sites.[2,3] In this report we show that ethanol also elevates $Ins(1,4,5)P_3$ levels, and we therefore propose that acute ethanol treatment may directly activate the signal transduction mechanism utilized by Ca^{2+} mobilizing hormones.

$Ins(1,4,5)P_3$ was measured in hepatocytes whose inositol lipids were prelabelled by incubation with myo-[2-^{3}H]inositol for 90 min prior to treatment with hormones or ethanol. FIGURE 1A shows an HPLC separation where a ^{3}H-labelled standard of $Ins(1,4,5)P_3$ was added to an extract of unlabelled cells. The major absorbance peak is ATP, and $Ins(1,4,5)P_3$ elutes shortly after this. Panels B–G of FIGURE 1 show the inositol trisphosphate region of HPLC ^{3}H elution profiles from samples of ^{3}H-inositol-labelled cells incubated under different conditions. In addition to $Ins(1,4,5)P_3$ eluting at 35 min, there is a second isomer eluting at 32–33 min, which has been identified as $Ins(1,3,4)P_3$.[4] This second isomer, the function of which is unknown, appears to change in concert with $Ins(1,4,5)P_3$ but with a somewhat slower time

FIGURE 1. Separation of inositol phosphates by HPLC. Isolated rat hepatocytes (10 mg protein/ml) were incubated at 37° C for 90 min in Krebs-Ringer bicarbonate buffer supplemented with 15 mM glucose, 2% dialysed bovine serum albumin, and 50 uCi/ml of myo-[2-^{3}H]inositol (American Radiolabelled Chemicals). At the end of this time the excess ^{3}H-inositol was removed by washing and the cells incubated for a further 10 min in fresh buffer before the addition of agonists as required. Incubations were terminated by quenching 0.6 ml of cell suspension into 0.2 ml of 14% $HClO_4$. The precipitated protein and lipid was removed by centrifugation, and the supernatant was then neutralized using KOH. After removal of the $KClO_4$ precipitate the extracts were stored frozen for HPLC analysis.[4] The HPLC column used was a Partisil SAX-10 anion exchange column. The flow rate was 1.6 ml/min throughout the run. After injecting the sample, the column was washed through with H_2O for 7 min followed by a linear gradient to 0.8 M NH_4COOH (pH adjusted to 3.7 with H_3PO_4). This was followed by a reduced gradient over the range where the inositol trisphosphate isomers eluted. The elution

[a]Supported by National Institute on Alcohol Abuse and Alcoholism Grants AA05622 and AA03422.

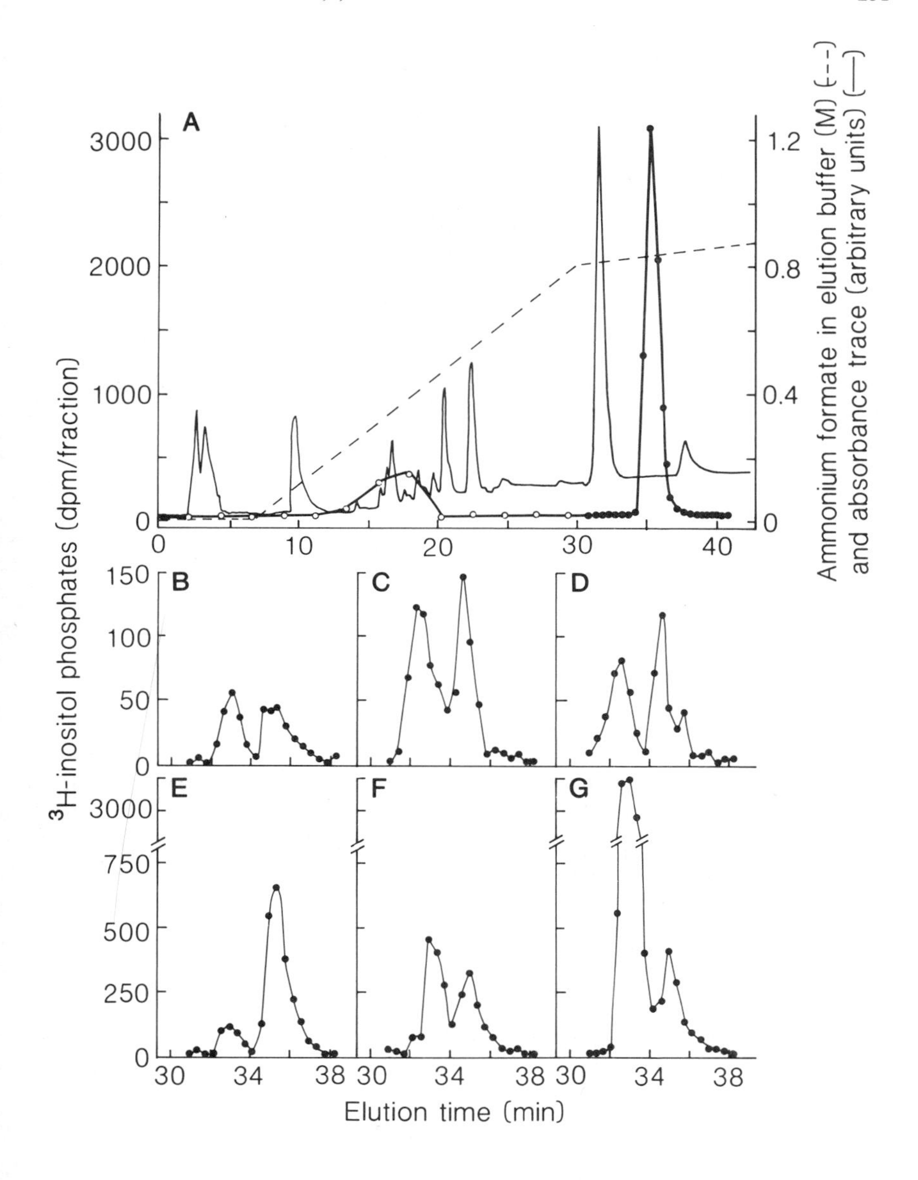

gradient is shown by the *broken line* in panel (**A**). Panel (**A**) also shows the absorbance trace at 254 nm for liver cell extract (*continuous line*) and the elution profile of an added ^{3}H-Ins(1,4,5)P_3 standard with fractions collected for liquid scintillation counting at intervals of 2 min (○) and 0.4 min (●). The remaining panels show the ^{3}H-inositol trisphosphates in the 30–40-min elution period of HPLC runs using extracts of cells prelabelled with ^{3}H-inositol and then treated prior to extraction as follows: (**B**), no further additions; (**C**), 300 mM ethanol for 15 sec; (**D**), 1 nM vasopressin for 15 sec; (**E**), 100 nM vasopressin for 15 sec; (**F**), 100 nM vasopressin for 12 min; (**G**), 10 mM LiCl + 100 nM vasopressin for 12 min. All of these traces are representative of results from at least 3 separate experiments.

course. FIGURE 1B shows that the basal levels of Ins(1,3,4)P_3 and Ins(1,4,5)P_3 are rather similar. After 15 sec incubation with 300 mM ethanol (a near maximal dose) both isomers increased by about twofold (FIG. 1C). With 1 nM vasopressin (FIG. 1D), which gives a $[Ca^{2+}]_i$ increase equivalent to that with 300 mM ethanol,[1] the increases of Ins(1,4,5)P_3 and to a lesser extent Ins(1,3,4)P_3 were of a similar magnitude to those observed with ethanol. However, at supraphysiological concentrations (FIG. 1E) vasopressin was about 5 times more effective than ethanol in elevating Ins(1,4,5)P_3 levels. FIGURE 1F shows that after prolonged incubation of hepatocytes with vasopressin, the predominant isomer was Ins(1,3,4)P_3, and, interestingly, the accumulation of this isomer can be specifically potentiated by the addition of LiCl (FIG. 1G).

Time courses for changes of Ins(1,4,5)P_3 and $[Ca^{2+}]_i$ following ethanol treatment are shown in FIGURE 2. It is apparent that Ins(1,4,5)P_3 increases sufficiently rapidly to account for the rise in $[Ca^{2+}]_i$. Furthermore, the decay of Ins(1,4,5)P_3 clearly precedes the decline in $[Ca^{2+}]_i$. The transient nature of the ethanol response with respect to both inositol phosphates and $[Ca^{2+}]_i$ is in marked contrast to the changes observed with

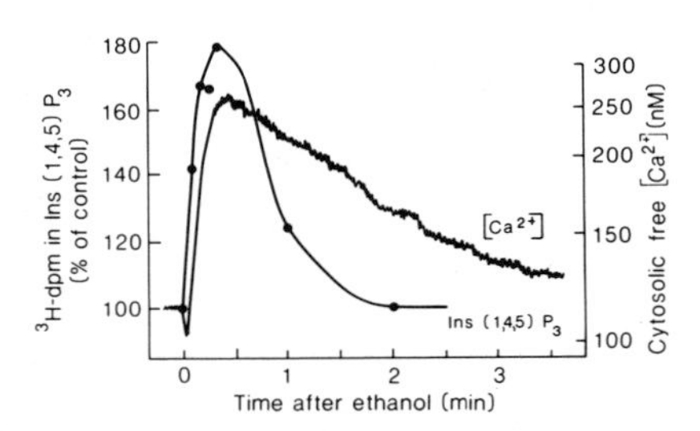

FIGURE 2. Time courses for changes of Ins(1,4,5)P_3 and $[Ca^{2+}]_i$ after treatment of hepatocytes with ethanol. Experimental procedures for the separation and measurement of ^{3}H-Ins(1,4,5)P_3 were as described in the legend to FIGURE 1. Samples were taken at the times indicated after addition of 300 mM ethanol (●). Each point represents the mean of values from 5 separate experiments. Cytosolic free Ca^{2+} was measured by using the intracellular fluorescent Ca^{2+} indicator, quin 2, as described elsewhere.[1]

vasopressin,[2] a difference that may have some bearing on the mechanism by which ethanol activates phospholipase C to break down phosphatidylinositol 4,5-bisphosphate. In conclusion, these data indicate that ethanol can interact acutely with the intracellular signalling mechanism that mediates the metabolic effects of several important hormones in liver.

ACKNOWLEDGMENT

We thank Richard Hager for excellent technical assistance.

REFERENCES

1. HOEK, J. B. 1987. Ann. N.Y. Acad. Sci. This volume.
2. THOMAS, A. P., J. ALEXANDER & J. R. WILLIAMSON. 1984. J. Biol. Chem. **259:** 5574–5584.
3. WILLIAMSON, J. R., R. H. COOPER, S. K. JOSEPH & A. P. THOMAS. 1985. Am. J. Physiol. **248:** C203–C216.
4. IRVINE, R. F., E. E. ANGGARD, A. J. LETCHER & C. P. DOWNES. 1985. Biochem. J. **229:** 505–511.

Hepatic Protein Catabolism During Chronic Ethanol Administration

TERRENCE M. DONOHUE, JR., JOHN M. PARR, MICHAEL F. SORRELL, DEAN J. TUMA, AND ROWEN K. ZETTERMAN

Liver Study Unit
Veterans Administration Medical Center
and
Department of Internal Medicine and Biochemistry
University of Nebraska College of Medicine
Omaha, Nebraska 68105

An early consequence of chronic alcohol consumption is enlargement of the liver; this enlargement precedes the more serious disturbances of alcoholic liver disease.[1] Alcoholic hepatomegaly results partially from protein accumulation in the liver, and an understanding of the pathogenesis of this net protein gain remains unclear. Our aim in these investigations was to examine the effects of chronic ethanol (EtOH) administration on the catabolism of both short-lived and long-lived hepatic proteins to

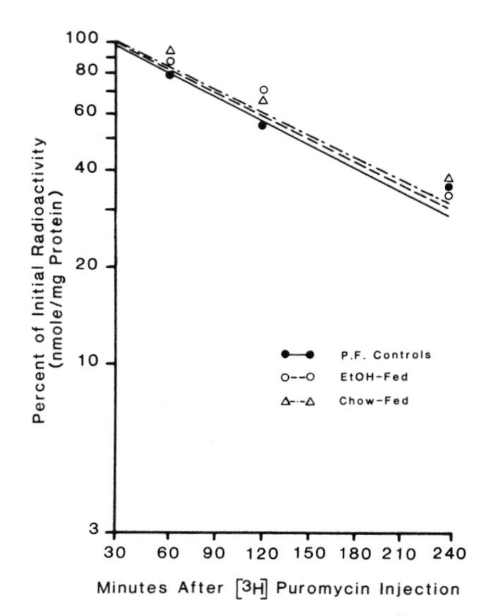

FIGURE 1. Determination of degradation rates of short-lived [^{3}H]puromycinyl peptides in livers of EtOH-fed, pair-fed control, and chow-fed rats. Rats were injected with [^{3}H]puromycin (4 μmoles/100 g body wt) 6 days after initiation of pair-feeding and were killed at one of the indicated time points. Each data point is the average value from 4 to 7 rats. Protein half-lives are given in the text and are calculated from regression analyses of the depicted curves.

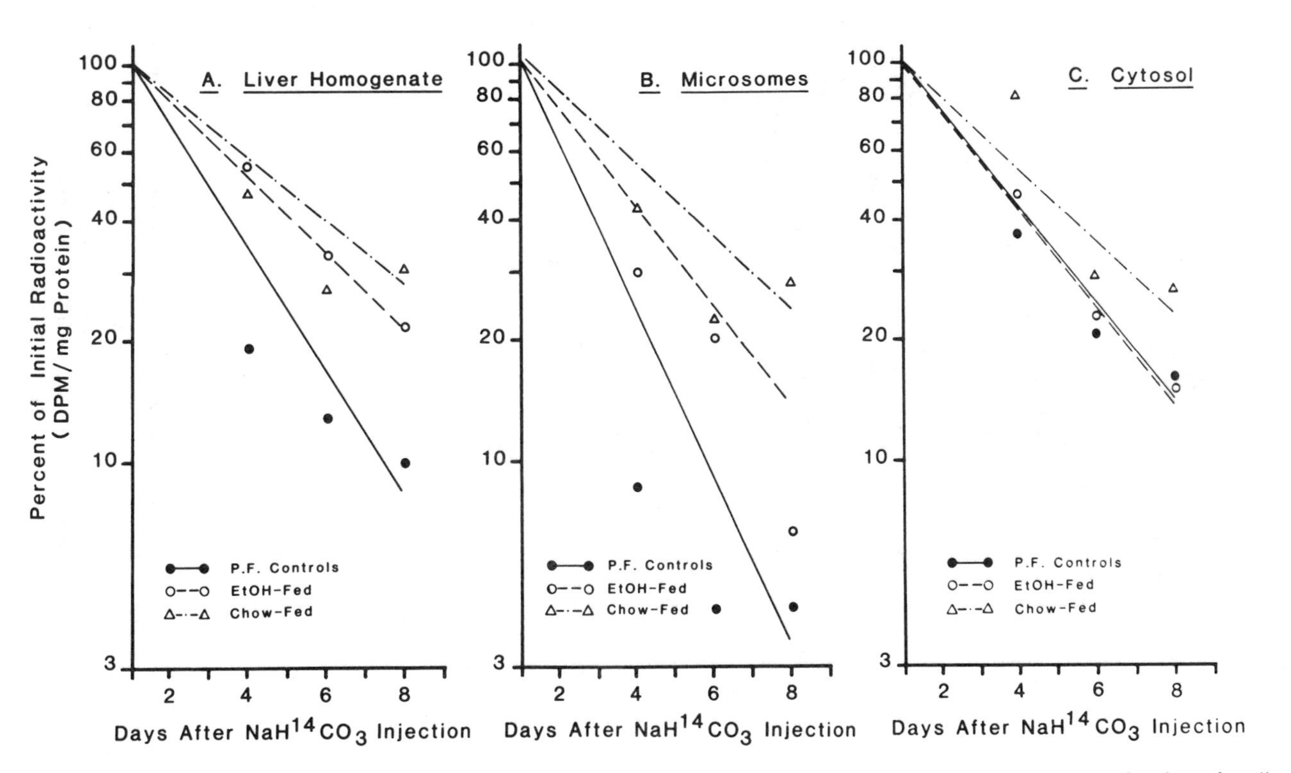

FIGURE 2. Determination of degradation rates of long-lived proteins in (**A**) liver homogenates, (**B**) microsomes, and (**C**) cytosolic fractions of rat liver. Rats were injected with NaH $^{14}CO_3$ (250 uCi/100 g body wt) one day after initiation of pair-feeding. Each data point is the mean value from 4 to 7 rats, which were killed at the indicated time points. Protein half-lives are given in the text and are calculated from regression analyses of the depicted curves.

determine whether an EtOH-induced alteration in protein degradation may be a factor that contributes to the increase in hepatic proteins.

Male Sprague-Dawley rats (170–180 g) were fed the Lieber-DeCarli liquid diet[2] containing EtOH, and were pair-fed with rats given an isocaloric control diet. A third group of chow-fed rats served as an additional control group. We previously observed (unpublished) that within 2–4 days after commencement of EtOH feeding, hepatic protein levels begin to rise appreciably. Therefore, degradation rates of both short-lived and long-lived hepatic proteins were measured within the first nine days after initiation of pair-feeding. Catabolism of short-lived proteins was measured by following the rate of decay (over 4 hr) of proteins prelabeled with [^{3}H]puromycin.[3] Degradation rates of long-lived proteins were measured by following the loss of radioactivity (over 8 days) in proteins prelabeled with [^{14}C]NaH CO_3.

Decay rates of short-lived ([^{3}H]puromycin-labeled) proteins in liver homogenates of EtOH-fed, pair-fed control, and chow-fed rats are depicted in FIGURE 1. Calculation of degradation rates by regression analysis revealed half lives of 135, 140, and 140 min for EtOH-fed, pair-fed control, and chow-fed rats, respectively, with no apparent difference in the degradation rates of short-lived proteins among the three groups.

Degradation rates of long-lived proteins ($NaH^{14}CO_3$-labeled) in the three groups of animals are shown in FIGURE 2. From the figure, half-lives of proteins in the liver homogenate and the isolated microsomal and cytosolic fractions of EtOH-fed rats were calculated to be 3.2, 2.7, and 2.5 days, respectively. Half-lives of proteins in corresponding fractions of pair-fed controls were 2.1, 1.8, and 2.5 days, and in chow-fed rats 3.9, 3.5, and 3.2 days. The results indicate that, compared with their pair-fed controls, EtOH-fed rats showed overall a decreased protein catabolic rate. This relative decrease is apparently confined mainly to the microsomal fraction. Comparison of protein degradation rates among all three groups shows that the slowest rates of hepatic protein degradation were found in the chow-fed rats.

In summary, the results indicate the following: 1) Degradation rates of short -lived hepatic proteins were the same in all three groups of rats, suggesting that degradation of short-lived proteins is unaffected by EtOH and proceeds by a pathway different from that of long-lived proteins.[4] 2) Long-lived proteins in EtOH-fed animals were degraded more slowly than those in their pair-fed controls, suggesting that EtOH consumption may impair protein degradation and thereby contribute to the net protein gain in the liver. 3) Slower protein catabolism observed in livers of chow-fed rats suggests that factors in addition to EtOH (*e.g.* dietary composition and/or feeding regimen) also play an important role in the regulation of protein catabolism in the liver.

REFERENCES

1. ISRAEL, V., R. BRITTON & H. ORREGO. 1982. Clin. Biochem. **15:** 189–192.
2. LIEBER, C. S. & L. M. DECARLI. 1970. Am. J. Clin. Nutr. **23:** 474–478.
3. LAVIE, L., A. Z. RESNICK & D. GERSHON. 1982. Biochem. J. **202:** 47–51.
4. AMENTA, J. S. & S. C. BROCHER. 1981. Life Sci. **28:** 1195–1208.

The Effect of Ethanol on Hepatic mRNA Regulation in Baboon and Man

MARK J. CZAJA,[a] FRANCIS R. WEINER,[a] MARIE-ADELE GIAMBRONE,[a] MARIE A. LEO,[b,c] CHARLES S. LIEBER,[b,c] AND MARK A. ZERN[a,b,c]

[a]Marion Bessin Liver Research Center
and
Department of Medicine
Albert Einstein College of Medicine
1300 Morris Park Avenue
Bronx, New York 10461

[b]Bronx Veterans Administration Medical Center
Bronx, New York 10468

[c]Mount Sinai School of Medicine
City University of New York
New York, New York 10029

Serum albumin levels are frequently depressed in patients with Laennec's cirrhosis, and it is thought that alcohol has a toxic effect on hepatic protein synthesis. However, the experimental findings are conflicting (see our review).[1] With the exception of our prior work on ethanol-induced fatty liver in rats,[2] no studies have addressed the effects of chronic ethanol administration on the molecular mechanisms of hepatic protein synthesis. Therefore, the primary objective of our study was to employ molecular hybridization techniques to evaluate protein synthesis in chronic alcoholic liver disease in a baboon model and then to extend our findings to man. Our second objective was to evaluate the molecular mechanisms responsible for increased collagen content in the fibrotic liver. It has been shown that in the immunological hepaptic injury of murine schistosomiasis there is a striking increase in type I collagen,[3] accompanied by an increase in type I procollagen mRNA content.[4] In the present study we extended our previous work on the hepatotoxic effects of ethanol exposure by investigating the changes in type I procollagen mRNA in the livers of baboons exposed to chronic alcohol administration and in man.

Using minor modifications of the procedure of Chirgwin *et al.*[5] we extracted total RNA from percutaneous liver biopsies of five baboons who were chronically fed an ethanol-rich liquid diet, and their pair-fed controls. The RNA was then used in *in-vitro* protein synthesis assays[4] or in RNA-DNA hybridization studies.[6] Chronic alcohol administration in baboons with liver fibrosis and a normal serum albumin increased *in-vitro* protein synthesis as measured by [^{35}S]-methionine incorporation ($15.3 \pm 1.8 \times 10^4$ cpm/μg RNA, in ethanol-fed vs 8.43 ± 1.4 in controls; $P < 0.01$). Alcohol-fed baboons also had increased albumin mRNA content ($180\% \pm 21$ of control; $P < 0.05$) and type I procollagen mRNA content ($183\% \pm 28$ of control; $P < 0.02$). There was no difference in the β-actin mRNA content (a constitutive protein).

We were then able to apply to man the techniques of RNA extraction developed in the baboon model. RNA was extracted from unused portions of percutaneous liver biopsies from humans having this procedure done for clinically indicated reasons. The

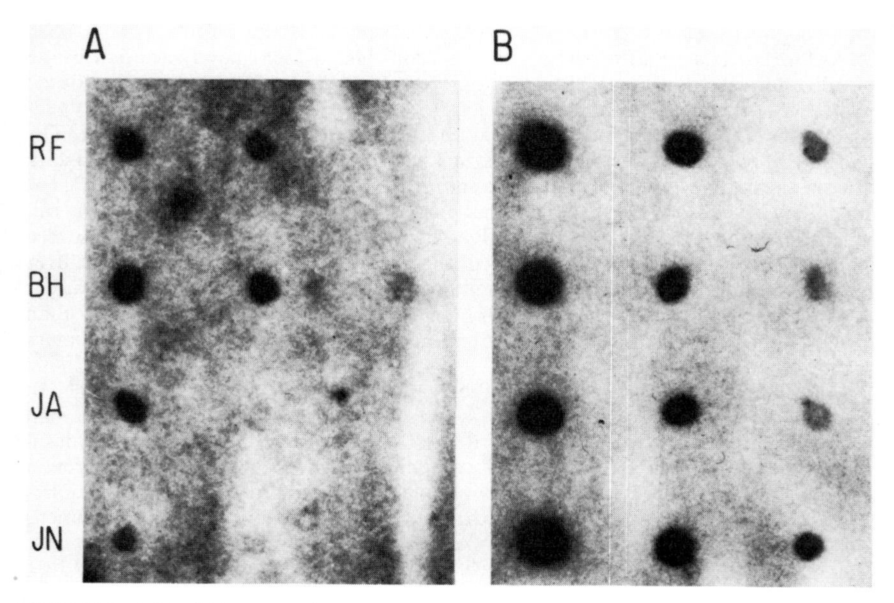

FIGURE 1. A "dot" blot assay of type I procollagen and albumin mRNA in total RNA extracted from human liver biopsies. Total RNA was extracted, spotted in decreasing concentration on Gene Screen filters, and hybridized with (**A**) a human type I pro α 1 collagen cDNA probe or (**B**) an albumin cDNA probe. *Each horizontal row* represents RNA from a different patient.

TABLE 1. Clinical Information Correlated with Densitometry Tracings of Total RNA Hybridized with Albumin and Collagen cDNA Probes

Patient	Clinical Data	Liver Histology	Serum Albumin mg/dl	Albumin mRNA (Densitometry Tracing % of Control)[a]	Collagen mRNA (Densitometry Tracing % of Control)[a]
JN	Psoriasis, methotrexate ETOH−	Mild fatty changes	4.4	100	100
JA	HBcAb+ ETOH+	Fatty changes, inactive cirrhosis	3.3	57	149
RF	Sarcoidosis ETOH−	Granulomatous hepatitis, slight fibrosis	4.3	80	195
BH	HBsAg− ETOH− chronic active hepatitis	Chronic active hepatitis, fibrosis	3.7	34	267
RJ	HBsAg+ ETOH−	Chronic active hepatitis, fibrosis	4.2	81	698

[a]Control considered to be JN with essentially normal biopsy.

RNA was then used for hybridization studies as previously described for the baboon RNA. FIGURE 1 is an autoradiograph of a "dot" blot, which shows variation in the hybridization signals when RNA from different human biopsies was serially diluted (with decreasing concentrations from left to right) and hybridized with a type I procollagen cDNA probe on the left and an albumin cDNA probe on the right. The quantitative representation of this data is in TABLE 1, which compares the clinical data in five of the patients with the molecular findings for albumin and procollagen mRNA. The patient JN with an essentially normal biopsy is considered the control. In general, the hybridization signals for albumin mRNA tend to reflect the serum albumin levels. The procollagen mRNA levels tend to correlate with the activity of the chronic liver disease and the active formation or accumulation of fibrosis. For example, the patient JA with "inactive" cirrhosis had levels of procollagen mRNA only slightly higher than normal, whereas the patients with chronic active hepatitis had substantially more procollagen mRNA.

Many studies of hepatic protein synthesis and fibrogenesis have been performed in cell culture or in isolated perfused organ systems. The advantage of evaluating human liver tissue directly is clear. The use of RNA-DNA hybridization in man provides a useful tool for evaluating the effects of pathophysiological states on hepatic protein synthesis and the future likelihood for developing fibrosis. In addition, the techniques we describe have great potential application in evaluating small human specimens from other organ systems for specific RNA species.

REFERENCES

1. ZERN, M. A., D. A. SHAFRITZ & D. SHIELDS. 1982. *In* The Liver: Biology and Pathobiology. I. M. Arias, H. Popper, D. Schacter & D. A. Shafritz, Eds. 103–122. Raven Press. New York, N.Y.
2. ZERN, M. A., P. R. CHAKRABORTY, N. RUIZ-OPAZO, S. H. YAP & D. A. SHAFRITZ. 1983. Hepatology **3:** 317–322.
3. TAKAHASHI, S., M. DUNN & S. SEIFTER. 1980. Gastroenterology **78:** 1425–1431.
4. ZERN, M. A., M. A. SABER & D. A. SHAFRITZ. 1983. Biochemistry **22:** 6072–6076.
5. CHIRGWIN, J. M., A. E. PRZBYLA, R. J. MACDONALD & W. J. RUTTER. 1979. Biochemistry **18:** 5294–5299.
6. ZERN, M. A., E. SCHWARTZ, M.-A. GIAMBRONE & O. BLUMENFELD. 1985. Exp. Cell Res. **160:** 307–318.

The Acute *In-Vitro* Effect of Ethanol and Acetaldehyde on Ion Flux in Isolated Human Leukocytes and Erythrocytes

RACHEL J. GREEN AND D. N. BARON

Department of Chemical Pathology and Human Metabolism
Royal Free Hospital and School of Medicine
Pond Street
London NW3 2QG, England

Acute exposure to ethanol and its primary metabolite acetaldehyde *in vitro* has been demonstrated to inhibit sodium and potassium transporting adenosine triphosphatase (Na^+,K^+-ATPase E C 3.6.1.37).[1,2] We assessed the inhibitory effect of ethanol and acetaldehyde on this enzyme, using isolated human leukocytes and erythrocytes as

TABLE 1. ^{86}Rb Influx with Ethanol in Leukocytes and Erythrocytes[a]

	Rubidium Influx ($mmol \cdot kg^{-1}$ Pr $\cdot h^{-1}$)		
	Total	Ouabain-Sensitive	Ouabain-Insensitive
Leukocytes			
Control	389	240	143
	(355–469)	(200–293)	(105–179)
8 $mmol \cdot l^{-1}$	405	248	144
	(349–461)	(222–273)	(115–173)
20 $mmol \cdot l^{-1}$	400	245	141
	(310–445)	(219–300)	(98–174)
40 $mmol \cdot l^{-1}$	388	235	148
	(339–433)	(204–266)	(116–165)
80 $mmol \cdot l^{-1}$	360*	208*	145
	(319–413)	(180–255)	(103–178)
160 $mmol \cdot l^{-1}$	303***	185***	123*
	(283–341)	(152–207)	(97–152)
Erythrocytes			
Control	7.55	5.30	2.25
	(6.60–8.10)	(4.00–6.05)	(1.25–2.85)
8 $mmol \cdot l^{-1}$	7.40	5.25	2.15
	(6.65–8.00)	(4.30–5.70)	(1.65–2.75)
20 $mmol \cdot l^{-1}$	6.95*	4.70*	2.10
	(5.85–7.80)	(3.40–5.30)	(1.50–2.70)
40 $mmol \cdot l^{-1}$	6.30**	4.45**	1.75*
	(5.50–7.95)	(3.30–5.60)	(1.40–3.00)
80 $mmol \cdot l^{-1}$	6.70***	4.50***	1.95**
	(2.95–7.60)	(3.85–5.30)	(1.30–2.70)
160 $mmol \cdot l^{-1}$	5.80***	4.30***	1.35***
	(4.75–6.80)	(3.45–5.05)	(1.05–2.10)

[a]Wilcoxon matched pairs signed ranks test. Results are medians and ranges n = 8–10 per group.
*$p < 0.05$.
**$p < 0.02$.
***$p < 0.01$.

model cells, by measuring the active fluxes of rubidium (Rb: equivalent to K) and sodium (Na).

Mixed leukocytes were isolated from heparinized venous blood from young healthy volunteers, using the dextran sedimentation method.[3] Flux rates in leukocytes were measured according to the method of Hilton and Patrick[4] and modified for erythrocytes. Concentrations of up to 40 mmol · l^{-1} ethanol produced no significant change in ^{86}Rb influx in leukocytes (TABLE 1). 80 mmol · l^{-1} inhibited total and ouabain-sensitive ^{86}Rb influx; at 160 mmol · l^{-1} ouabain-insensitive ^{86}Rb influx was also inhibited, suggesting that mechanisms other than active Rb(K) entry into leukocytes may also be inhibited by ethanol. In erythrocytes both total and ouabain-sensitive ^{86}Rb influx were inhibited at 20 mmol · l^{-1} ethanol (TABLE 1). A dose-related effect was found at higher concentrations, as for leukocytes. 80 and 120 mmol · l^{-1} ethanol significantly inhibited total and ouabain-sensitive sodium efflux rate and rate constant in leukocytes ($p < 0.02$ and < 0.01, respectively); and to a greater extent in erythrocytes ($p < 0.01$ in both cases). There was no decrease in the ouabain-insensitive efflux. The concentrations of acetaldehyde used in these experiments were approximately 100 times higher than those found in the peripheral circulation after drinking ethanol. Despite this neither 0.1 nor 0.2 mmol · l^{-1} acetaldehyde inhibited ^{86}Rb influx in leukocytes. In erythrocytes all fractions of ^{86}Rb influx were significantly inhibited (TABLE 2).

TABLE 2. ^{86}Rb Influx with Acetaldehyde[a]

	^{86}Rubidium Influx (mmol · kg^{-1} Pr · h^{-1})		
	Total	Ouabain-Sensitive	Ouabain-Insensitive
Leukocytes			
Control	433 (330–463)	273 (233–307)	155 (100–185)
0.1 mmol · l^{-1}	430 (356–481)	275 (243–315)	160 (105–190)
0.2 mmol · l^{-1}	432 (312–483)	280 (232–295)	158 (100–198)
Erythrocytes			
Control	6.65 (6.17–7.28)	4.64 (4.35–5.09)	1.84 (1.50–2.88)
0.1 mmol · l^{-1}	6.16** (5.92–6.62)	4.38** (3.79–4.84)	1.84* (1.42–2.13)
0.2 mmol · l^{-1}	5.68*** (5.02–7.16)	4.04*** (3.68–4.26)	1.76** (1.25–2.17)

[a]Results are medians and ranges n = 8 per group.
*$p < 0.05$.
**$p < 0.02$.
***$p < 0.01$.

In conclusion, inhibition of active cation fluxes occurs in erythrocytes at much lower concentrations of ethanol and acetaldehyde than in leukocytes, where the concentrations would be potentially fatal in man. In uncomplicated alcoholism there are no gross changes in plasma electrolytes, though this would occur if general body cells behaved as erythrocytes. Leukocytes, therefore, appear a better model by which to assess the response to ethanol. We are currently undertaking experiments to corroborate our *in-vitro* findings with *in-vivo* studies, by raising blood ethanol concentrations in volunteers above 20 mmol · l^{-1}.

REFERENCES

1. GONZALEZ-CALVIN, J. L., J. B. SAUNDERS & R. WILLIAMS. 1983. Biochem. Pharmacol. **32:** 1723–1728.
2. ERWIN, V. G., J. KIM & A. D. ANDERSON. 1975. Biochem. Pharmacol. **24:** 2089–2095.
3. BARON, D. N. & S. A. AHMED. 1969. Clin. Sci. **37:** 205–219.
4. HILTON, P. J. & J. PATRICK. 1973. Clin. Sci. **44:** 439–445.

Cytoskeletal Pathology Induced by Ethanol[a]

S.W. FRENCH,[b] Y. KATSUMA,[b] M.B. RAY,[c] AND S.H.H. SWIERENGA[d]

[b]*Department of Pathology*
University of Ottawa
Ottawa, Ontario, Canada K1H 8M5

[c]*Department of Pathology*
University of Cincinnati
Cincinnati, Ohio 45267

[d]*Toxicology Division*
Department of Health and Welfare
Ottawa, Ontario, Canada K1A 0L2

INTRODUCTION

Chronic ethanol ingestion alters cells in many different organ systems including the nervous system, muscle, pancreas, liver, hematopoietic, and digestive tract.[1] One of the cellular components that is altered by ethanol is the cytoskeleton including the microfilaments, microtubules, and intermediate filaments.

Most of the studies on the effect of ethanol on the cytoskeletal elements have been done on the liver of experimental ethanol-fed animals or chronic alcoholic patients. These effects may be acute while ethanol is in the blood stream or chronic after ethanol has been consumed for long periods. Much of the evidence that the cytoskeleton has been effected by ethanol is indirect in studies of changes in a cytoskeleton-dependent function, but direct morphologic measurements of the cytoskeletal elements have also been made. The evidence for microtubule involvement in chronic ethanol ingestion is both morphologic and functional. The evidence that the intermediate filaments are altered is morphologic only. Whether microfilaments are affected by ethanol or not depends on how ethanol affects microfilaments in the microvesicular transhepatic transport and the bile canalicular contraction systems.

MICROTUBULES

The first indication that hepatocytic microtubules might be affected by ethanol was that the yield of the Golgi fraction isolated from rat hepatocytes was increased if the rat was first given ethanol orally (6 g/kg, 90 mins before sacrifice).[2] Ethanol treatment causes a dramatic increase in the number of VLDL particles found in the Golgi apparatus involving several cisternae and along the sinusoidal cell front. The

[a]Supported by a grant from the Canadian Medical Research Council.

effect on the Golgi was similar to that of colchicine,[3] an antimicrotubule agent. Acute ethanol treatment, like colchicine[4] also inhibited release of lipoprotein from the rat liver *in vivo* presumably by disrupting the microtubular dependent exocytosis of VLDL-laden secretory vesicles.[5] The volume of hepatic Golgi increases after both acute and chronic ethanol treatment, and this is inversely proportional to the amount of polymerized tubulin.[6] Secretion of hepatic export proteins is inhibited by ethanol[7–10] similar to colchicine,[11,12] and the effect of ethanol and colchicine are additive in this regard.[12] The defect in secretion, presumably, is due to a decrease in polymerized tubulin[6,8] and a decrease in microtubules measured morphometrically in both experimental and human alcoholic liver disease,[13–15] although the valildity of the method used to measure polymerized tubulin has been disputed.[16–19] The decrease in polymerized microtubules is due to an alteration in tubulin and is reversed to normal by a 24-hr

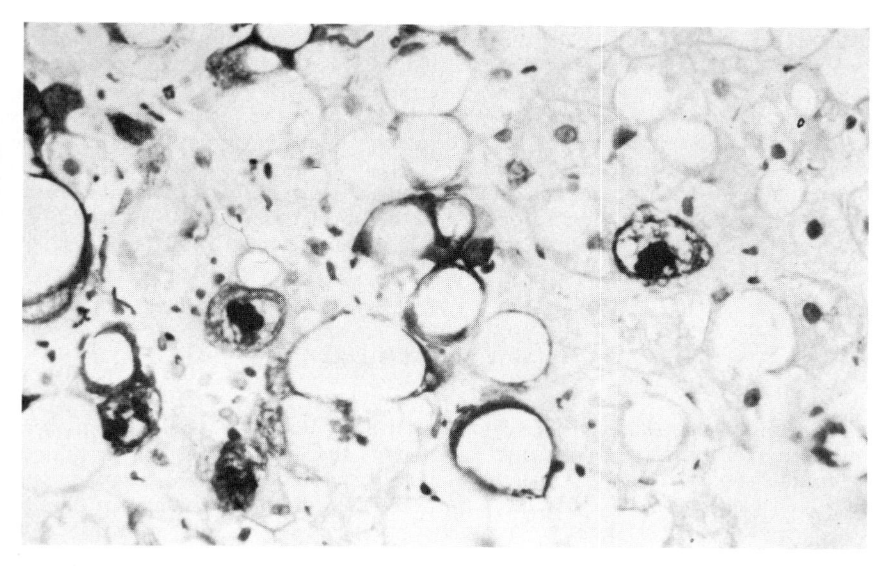

FIGURE 1. Human liver biopsy from a patient with alcoholic hepatitis stained with antihuman keratin monoclonal antibodies (AE1 and 3). Note that the centrolobular hepatocytes stain with horseradish peroxidase product as do Mallory bodies. Avidin-biotin peroxidase with hematoxylin counterstain. (Photograph provided by Dr. Ray.)

fast.[17] The net effect of ethanol ingestion is a retention of export proteins by hepatocytes,[6,8–10] which contributes to the enlargement of the liver and the ballooning of hepatocytes in rats and man.[20,21] The increase in liver cell protein content is associated with an increased polyploidy caused by chronic ethanol ingestion.[22]

MICROFILAMENTS

The microfilament system is primarily composed of F actin and actin-associated proteins that are contractile in nature. In the case of the liver, microfilaments provide the motive force for bile flow and transhepatic transport.[23,24] Acute ethanol ingestion inhibits bile flow, possibly by inhibiting microfilament contraction similar to that seen with chlorpromazine treatment.[25] Alcohol reduced and caused a delay in the peak of transhepatic transport of horseradish peroxidase (HRP) from blood to bile *in vivo*.[25] This diminution in the microvesicular transport and secretion of HRP into bile caused by ethanol may be due to its antimicrotubule effect, since cytochalasin D, a microfilament inhibitor, does not affect transhepatic transport of HRP.[26] However, the function of both microtubules and microfilaments may be involved in transhepatic transport of vesicles from the sinusoid to the bile canaliculus.[27]

Transport of sinusoidal IgA by hepatocytes into bile is a useful measure of transhepatic transport. IgA transport from blood to bile through a microvesicular nonlysosomal pathway is inhibited both by antimicrotubule[28] and antimicrofilament agents.[27] Surface binding of IgA is not affected by these agents. Cytochalasin B impaired transcellular transport of IgA apparently by paralyzing uptake and translocation, even though IgA binds to secretory IgA on the cell surface normally.[27] These observations are of interest, since IgA accumulates on the surface of hepatocytes in the livers of patients with alcoholic liver disease (ALD). IgA deposits in a continuous or pericellular pattern along the hepatic sinusoidal surface in ALD.[29–30] The presence of the continuous pattern correlates with progression of ALD.[31] It appears that IgA of intestinal origin, which is normally cleared by the liver and secreted in the bile, is sequestered on the surface of the sinusoids and liver cells but is not taken up by the hepatocyte and transported into the bile in ALD. This failure to transport IgA to the bile could result from either the antimicrotubular or the antimicrofilament effect of ethanol. It could also be the result of some other mechanism such as increased IgA synthesis.[32] These IgA deposits seen in ALD do not correlate with hepatocellular necrosis or cholestasis.[32]

INTERMEDIATE FILAMENTS

The intermediate filament (IF) cytoskeleton of hepatocytes has only recently been fully characterized morphologically,[33–37] although the composition and the antigen determinants of the cytokeratin polypeptide has been established in liver and in tissue culture of hepatocytes.[38–41] This data forms the basis for understanding how chronic ethanol ingestion affects the IFs of hepatocytes. Rats fed ethanol for 3 months show no morphological alterations in IFs.[16] However, livers from chronic alcoholic patients with liver disease show profound alterations in IFs in centrolobular hepatocytes using immunoperoxidase localization of cytokeratin polypeptide antigen determinants.[42] Using a mixture of two monoclonal antibodies (Hybritech, San Diego, CA) to human epidermal cytokeratins (acidic, 50 K and 56.5 K and basic, 58 K, 65 K, and 67 K), Dr. Ray demonstrated that centrolobular hepatocytes in livers from alcoholic liver disease (ALD) bound antibody (FIG. 1) whereas only bile duct epithelium bound antibody in normal livers (FIG. 2). Antigen was localized by the avidin-biotin antibody peroxidase complex method on paraffin sections digested with trypsin. The two monoclonal antibodies were characterized by Eichner *et al.*[43] who found that they recognized almost all keratins expressed by human epidermal cells.

The finding that centrolobular hepatocytes express IF polypeptides not normally expressed by hepatocytes gains significance from the fact that Mallory bodies (MBs), also found in ALD, likewise bind these monoclonal-antikeratin antibodies (FIG. 1). The implication of this finding is that the MB phenomenon results from a change in the cytokeratin gene family expression in hepatocytes when ethanol is ingested in large amounts for many years. The AE1 and AE3 monoclonal antibodies recognize human epidermal mutually exclusive subfamilies of cytokeratins[43] not normally expressed by hepatocytes.[44] The normal hepatocyte expresses the 45- and 52-kd proteins (45–52-kd pair).[38,45] In addition, experimentally induced MBs include a high molecular weight cytokeratin (65-kd band).[46] By two-dimensional gel electrophoresis separation, the 65-kd band is an acidic polypeptide with variant isoelectric points at pH 5.4, 5.38, and 5.2.[46] The isoelectric variants of the normal mouse lower molecular weight cytokeratins differed slightly from those found in MBs. The 65-kd cytokeratin in MBs differs from the large prekeratin polypeptides of mouse epidermis and other stratified squamous epithelia, which have all been shown to have isoelectric values above pH 7.2.[47] Antibodies raised against human epidermis prekeratin stain MBs in mice livers using immunofluorescent technique. These antibodies stain positively the 55- and 65-kd bands derived from isolated MB polypeptides on nitrocellulose immunoblots.[46] These

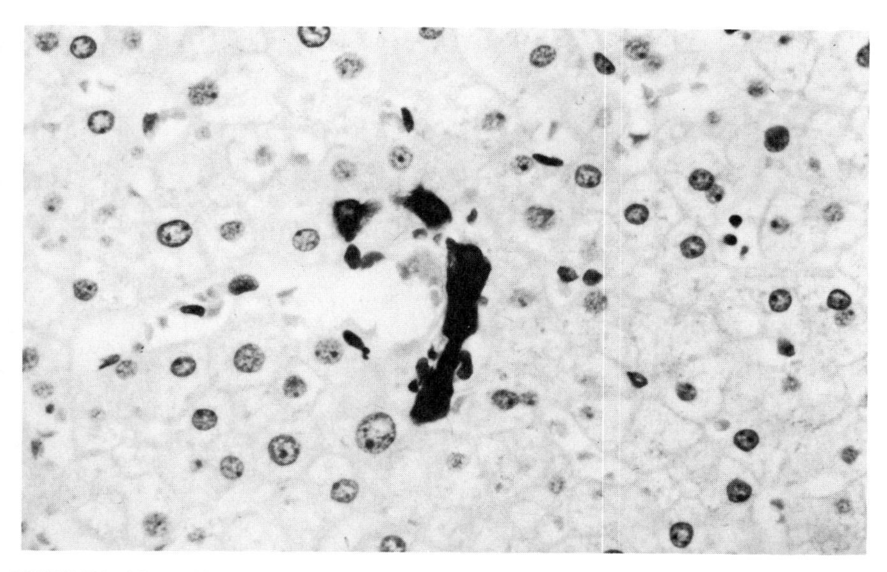

FIGURE 2. Normal human liver stained with monoclonal antibodies to human epidermal keratin as in FIGURE 1. Note that the bile duct epithelium stains but the hepatocytes do not. (Photograph provided by Dr. Ray.)

observations lend credence to the notion that MBs are composed of cytokeratin polypeptides. Some polypeptides are of lower molecular weight and are found in normal hepatocytes, and one is a large molecular weight polypeptide that has antigen determinants similar to epidermal cytokeratins. What remains to be established is whether the 65-kd cytokeratin in the MB is the same polypeptide expressed in the cytoplasm of the centrolobular hepatocytes in ALD. Cross reactivity of antibodies to the various cytokeratins is commonly observed.[46]

What are the implications of the observation that alcohol ingestion induces the expression of epidermal-like cytokeratin polypeptides in hepatocytes that culminate in the formation of abnormal MB filament aggregates? This phenomenon may help explain why an alcoholic must ingest large amounts of alcohol for many years before alcoholic hepatitis and cirrhosis develops. Regulation of cytokeratin gene expression by environmental changes can have a profound influence on the polypeptide composition and IF rearrangement based on observations of cellular differentiation of a variety of cell types.[41,48–58] In the case of MB formation, a type of pathological rearrangement of IFs, there are a large number of inducers including chronic drug treatment, preneoplastic and neoplastic transformation, and cirrhosis formation.[34,59,60] From these observations we have postulated that alcohol-induced MB formation represents a preneoplastic transformation of centrolobular hepatocytes in response to ethanol-induced increase in microsomal P-450 conversion of xenobiotics to carcinogenic intermediates.[61] This hypothesis finds some support in the fact that preneoplastic foci and hepatomas induced by griseofulvin in mice and hepatoma cells in man express both gamma glutamyl transferase and MB formation.[62,63] Similar observations have been made in rat hepatoma cells in tissue culture.[64]

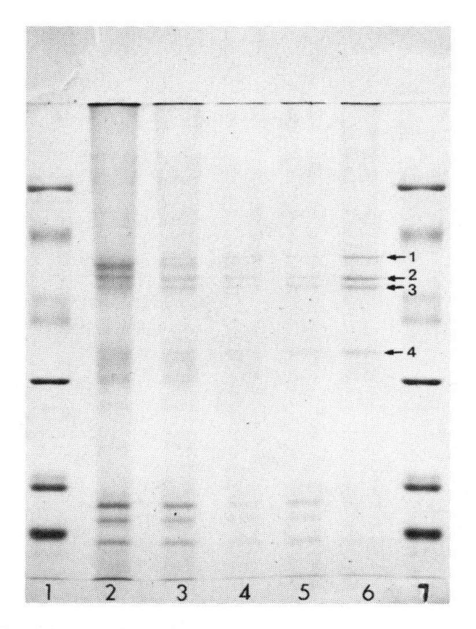

FIGURE 3. Track **1.** BioRad low MW stds. **2.** Unextracted slices. **3.** CSK extracted. **4.** CSK-AS extracted. **5.** CDM extracted. **6.** CDM-AS extracted. **7.** BioRad low MW stds. Method: Normal rat liver, frozen in isopentane at $-155°$ C was sectioned in a cryostat; slices were cut (14 μm) and adhered to coverslips previously coated with poly-L-lysine. The coverslips were washed twice with Tris buffered saline (TBS). All solutions contained 0.1 mM iodoacetamide and 1.2 mM phenylmethyl sulfonyl fluoride (PMSF). The following serial procedure was used for the extractions. Tissue slices were exposed first to cytoskeleton buffer (CSK) containing 300 mM sucrose, 0.5% Triton X-100, 100 mM NaCl, 3 mM $MgCl_2$ in 10 mM PIPES pH 6.8 at 0° C for 10 min. This extraction was followed by a similar one in which the $MgCl_2$ was replaced with 0.25 M $(NH_4)_2SO_4$ (CSK-AS). The slices were then exposed to chromatin digestion medium (CDM), which is similar to CSK, but contains only 50 mM NaCl and includes 100 μg/ml each of ribonuclease and deoxyribonuclease at 22° C. Finally, the tissues were exposed for 5 min to CDM-AS buffer, which is the same as CDM buffer except it contains an additional 0.25 M $(NH_4)_2SO_4$. After extraction, the pellets were homogenized in a glass-teflon homogenizer in 0.2 ml SDS sample buffer—0.5% SDS, 5 mM 2-mercaptoethanol, 62.5 mM Tris, pH 6.8—heated to 100° C for 5 min before electrophoresis. SDS-PAGE was performed on 10% gels according to the method of Laemmli (Nature 227: 680–685, 1970), then stained with Coomassie blue. Residual cytoskeletal protein bands after extraction (Track 6, *arrows*): Band 1 is 60kd, 2 is 55kd, 3 is 49 kd, 4 is 36 kd.

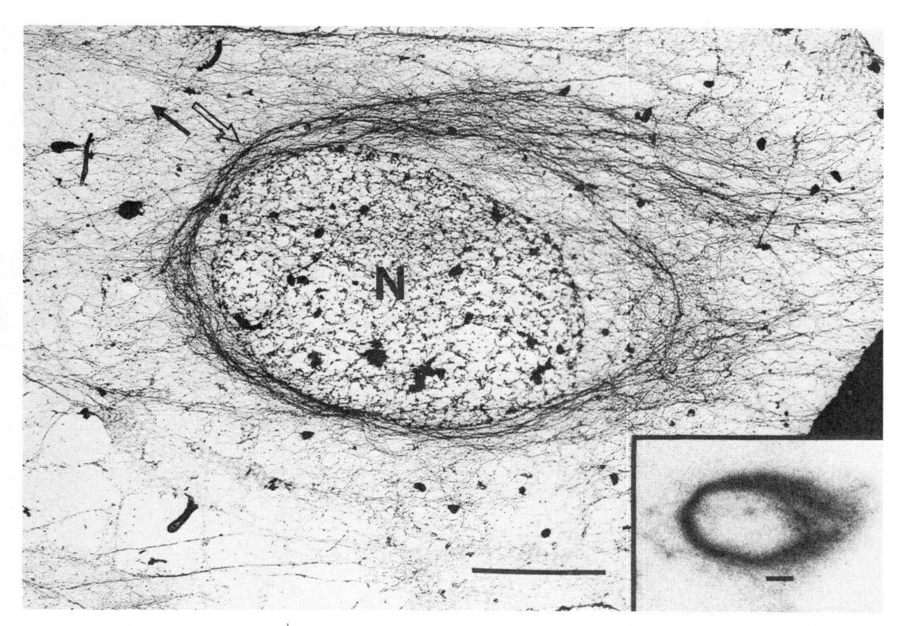

FIGURE 4. Direct comparison of ultrastructure and immunofluorescent staining (*inset*) of the same T51B cell using a monoclonal antibody against rat liver 55-kd cytokeratin. Cultured cells were extracted as in FIGURE 3, fixed with 2% paraformaldehyde, rinsed by phosphate-buffered saline (PBS), and processed for indirect immunofluorescent staining. After immunofluorescent photographs were taken, cells on the grid were rinsed by PBS, dehydrated by ethanol, dried through the CO_2 critical point, and coated with carbon for unembedded whole mount electron microscopic observation. Immunofluorescence shows characteristic ring-shaped perinuclear zone and filaments extending to the cell periphery (*inset*). At the ultrastructural level, cytokeratin IFs, which mainly surround the nucleus, are heavily decorated by immunoglobulin of the 1st and 2nd antibody, so that the diameter of cytokeratin (*open arrows*) increased up to twofold to threefold compared with the original filaments. Vimentin IFs, which were unstained (*closed arrow*) were identified by the same method using monoclonal antibody against vimentin instead of cytokeratin (not shown). Because of the morphological similarity, these two types of IFs cannot be identified without immunocytochemical procedure. **N,** nucleus. *Bar,* 5 μm.

The MB rearrangement of IFs seen in ALD may be due to or dependent on vitamin A deficiency. Decreased vitamin A serum levels are commonly observed in alcoholics and levels of vitamin A in the liver of ALD and in baboons and rats fed ethanol are decreased.[65–66] Mice fed griseofulvin to induce MB formation have depleted levels of vitamin A in their livers.[67] Based on these observations, we fed mice a vitamin-A deficient diet for 18–24 months and observed variable numbers of MBs in a high percentage.[68] To our surprise, one control mouse fed adequate vitamin A levels in the diet also had a few MBs in the liver. It was concluded that a vitamin-A deficient environment may enhance MB formation but is not essential. This observation may be important because vitamin A regulates differentiation of epithelia and the expression of groups or families of cytokeratin polypeptides.[43,56,69,70] Removal of vitamin A from the culture medium leads to synthesis of a 67-kd keratin and reduced synthesis of 52-kd and 40-kd keratins by epidermal or conjunctival cells in tissue culture. This process is reversed by adding retinyl acetate to the culture.[56] Thus, it is possible that hepatocytes in an ethanol-induced environment low in vitamin A may express high molecular weight polypeptide, *i.e.,* 65-kd, which causes derangement of the IF cytoskeleton with MBs resulting.

To gain further insight into the formation of MBs, we have been studying MB-related IF morphology in primary cultures of griseofulvin-induced neoplasms[35] and also hepatocytes extracted *in situ* by triton X-100 perfusion. Extracted hepatocytes were viewed by transmission[33] and scanning EM.[34,71] The *in-vivo* extracts showed a rich network of IFs in normal as well as MB-containing hepatocytes. The MBs were intimately connected and suspended by this filament network, which radiates from the nuclear surface and plasma membranes. The IF system appeared as an elaborate continuous scaffolding within the cell interior. MBs appeared to be composed of tightly compressed or collapsed IFs matted into an irregular-shaped mass.[71] The IFs associated with MBs showed focal condensation or areas where IFs were less dense, which indicated that there was a maldistribution of the filaments within the cell. This contrasted with the uniform spacing of filaments within adjacent cells that did not contain MBs.

In morphometric studies of *in-situ* detergent-extracted mouse livers, we failed to

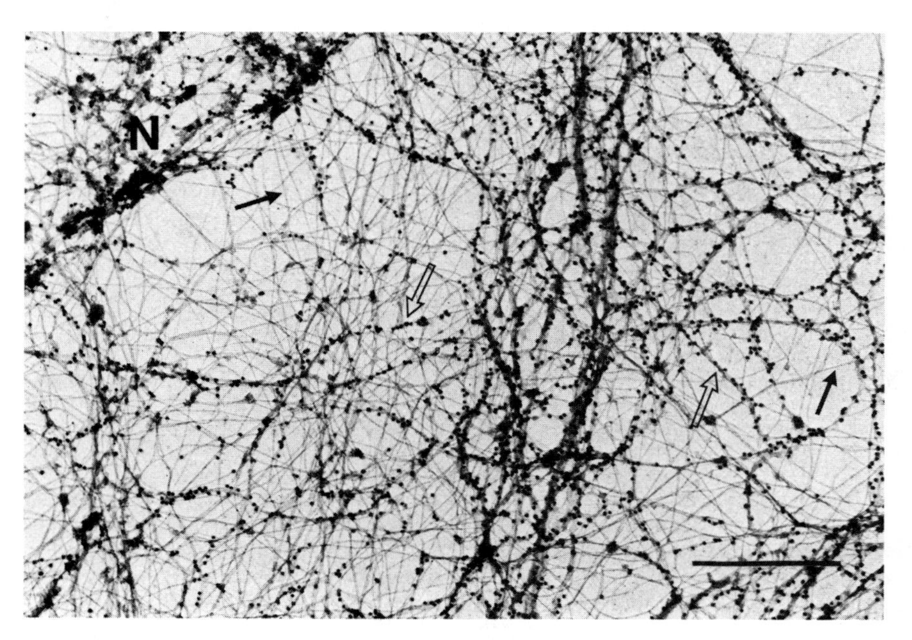

FIGURE 5. Immunogold labelling for cytokeratin in T51B cell. Cultured cells on the gold grid were extracted as in FIGURE 3, fixed with 0.1% glutaraldehyde for 1 min, rinsed with PBS, and incubated with monoclonal antibody against rat liver cytokeratin for 40 min at room temperature. After rinsing with PBS, cells were incubated with 10 nm colloidal gold-conjugated goat anti-mouse IgG for 40 min at room temperature. Cells were rinsed with PBS, fixed with 2% glutaraldehyde and 1% osmium tetroxide, dehydrated by ethanol, dried through the CO_2 critical point, and coated with carbon. 10-nm gold particles were detected mainly on the IFs encircling the nucleus with some stained IFs extending into cytoplasm (*open arrows*). Vimentin IFs remained unstained (*closed arrows*). **N**, nucleus. *Bar,* 0.5 μm.

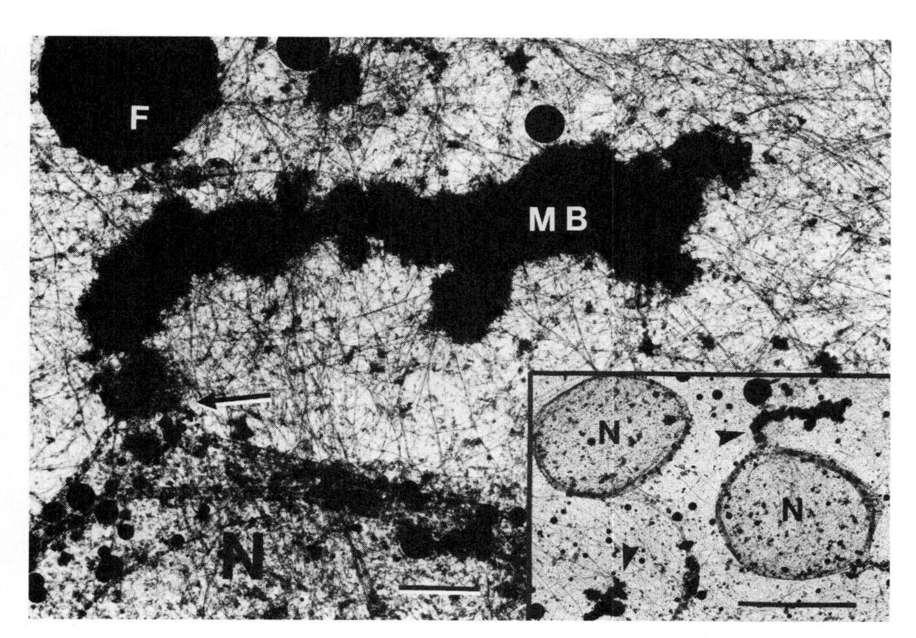

FIGURE 6. Electron micrograph of whole mount extracted hepatocyte with MB in primary culture using griseofulvin-fed C3H mouse liver. The MB, which appears as electron-dense material with fuzzy or filamentous edge, is anchored by numerous IFs. This MB is directly connected with the nuclear envelope cytoskeletal filaments; the former appears to pull the latter to form a tent, indicating that the two structures are attached (*arrow*). *Inset:* The micrograph at a lower magnification shows that the binuclear hepatocyte contains two MBs, which have an elongate contour (*arrowheads*). **MB,** Mallory body. **N,** nucleus. **F,** fat droplet. *Bar,* 1 μm. *Bar in inset,* 10 μm.

find a significant change in the volume of either microtubules or IFs in hepatocytes containing MBs compared to neighboring or control liver hepatocytes.[72] This was at variance with the results of anti-MB antibody and anti-prekeratin polyclonal antibody staining of hepatocytes containing MBs in alcoholic hepatitis and mice fed griseofulvin where MB-containing cells showed no staining of cytokeratin in the ectoplasm and protoplasm.[73,74] These findings were interpreted to mean that the loss of IFs from the cytoplasm was due to incorporation of IF polypeptides into the MB.[73] The results of Ray's studies[42] (FIG. 1) explain this apparent contradiction in observations. Denk *et al.*[73] and Kimoff and Huang[74] used antibodies that stained cytokeratins normally found in hepatocytes, whereas Ray used a mixture of two monoclonal antibodies, which do not stain normal hepatocytes but rather stain large molecular weight cytokeratins present in MBs and epidermal cells. This means that there are IFs present in MB-containing cells, but these IFs are composed of polypeptides not normally expressed by hepatocytes. What must happen to explain these observations is that the normal cytokeratin proteins are incorporated into the MB and are absent elsewhere in the MB, whereas newly expressed cytokeratin polypeptides occupy both the MB and the rest of the cytoplasm. The normal IF proteins persist within the MB, because the newly synthesized MB-forming cytokeratins make the peptides within the MB resistant to degradation possibly by calcium-activated neutral protease.[75] This hypothesis needs to be tested further.

To further determine the spacial relationship of MBs induced by griseofulvin feeding, we studied whole mount nonembedded detergent and nuclease-extracted cytoskeletal preparations[76,77] of primary cultures and frozen sections of hepatoma cells containing MBs. The removal of soluble proteins and the persistence of insoluble intermediate filaments can be seen both morphologically and by gel electrophoresis of the cytoskeletal proteins at each stage of extraction (FIG. 3). The cells were observed by immunofluorescence using a monoclonal antibody to rat liver 55-kd cytokeratin[35,78] (a gift from Dr. N. Marceau, Laval University, Quebec City, Canada). This antibody stains the IF cytoskeleton of the extracted T51B cell line of nonneoplastic rat liver epithelial cells (FIG. 4). Two types of IFs can be distinguished by immunofluorescence and immunogold techniques,[79] *i.e.,* 55-kd cytokeratin staining filaments ((FIG. 5) and vimentin (not shown). MBs in primary culture of mouse hepatomas were located by immunofluorescence[35] and by toluidine blue staining metachromasia after alcohol fixation. A MB-containing, extracted, nonembedded hepatocyte in primary culture is shown in FIGURE 6. When viewed as a stereo pair, the three-dimensional framework of the MB suspended by a rich network of IFs can be seen (FIG. 7). Frozen sections of the gris-fed mouse liver extracted by the same method as used with primary culture revealed identical findings, showing that the MB-containing cells had an intact IF cytoskeleton (FIG. 8). This finding confirms our previous observations by transmission and scanning EM that there is little diminution in the IFs seen in MB-containing cells.

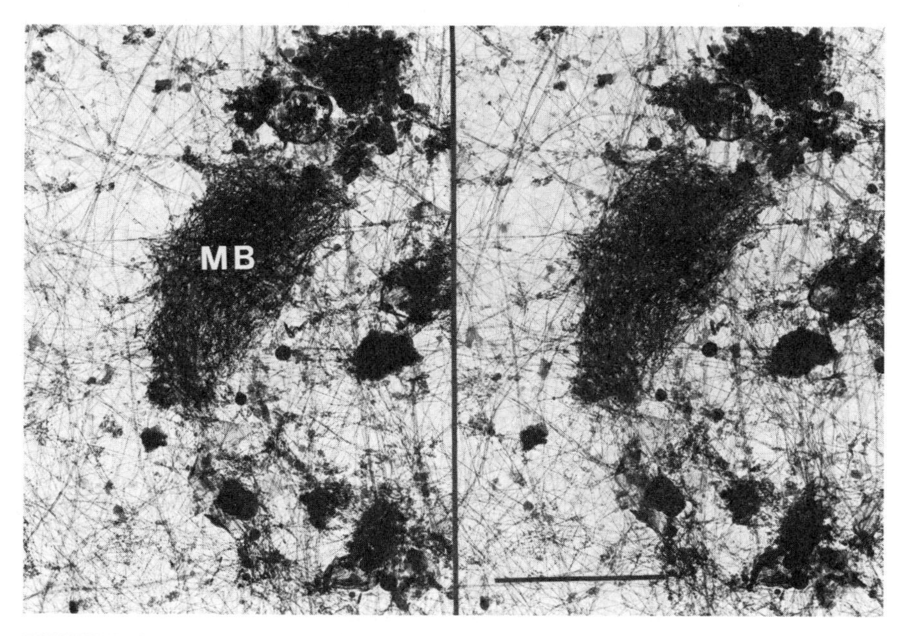

FIGURE 7. Stereoscopic finding of a small MB in a hepatocyte from a primary culture that was prepared from griseofulvin-fed C3H mouse liver. Cells were treated for whole mount extraction as in FIGURES 5 and 6. In the small MB shown, a filamentous structure is seen. The aggregated thick and short filaments are consistent with a Yokoo's type II MB. *Bar,* 1 μm.

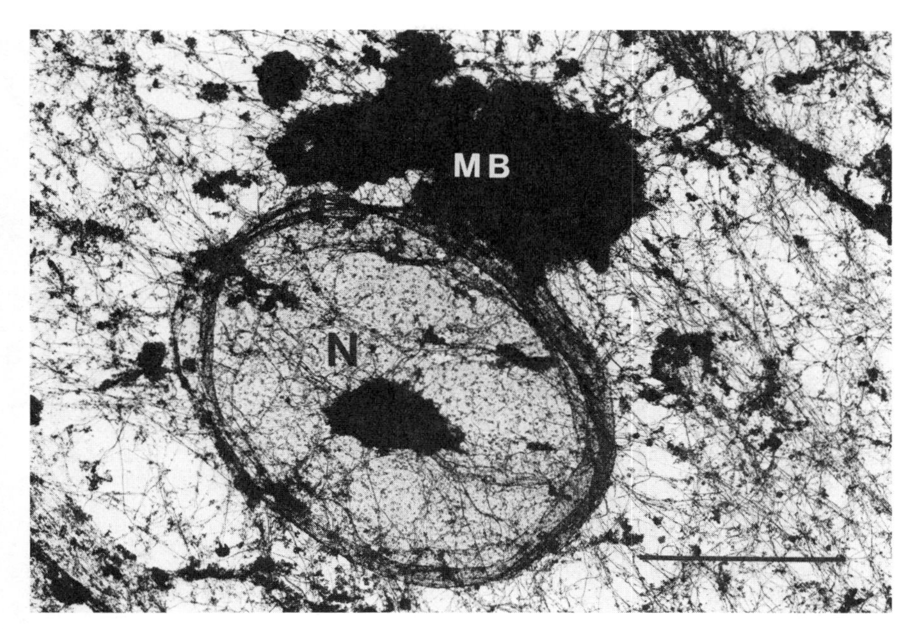

FIGURE 8. Frozen, unembedded, extracted section of a griseofulvin-fed C3H mouse liver preparation. A small piece of liver tissue was frozen in isopentane at −150° C precooled by liquid nitrogen. Frozen sections 8 μm thick were cut in a cryostat and mounted on formvar-coated copper grids; then an unembedded, extracted preparation was made as in FIGURE 7. A MB is seen adjacent to the nucleus. A portion of the MB connects directly to the nuclear envelope cytoskeletal filament. Note that numerous IFs coexist with and anchor the MB. **MB,** Mallory body. **N,** nucleus. *Bar,* 5 μm.

ACKNOWLEDGMENTS

The authors wish to acknowledge the valuable assistance of Kim Wong in EM studies and Douglas Mauldin in preparation of serial-extracted frozen sections of liver polyacrylamide electrophoresis.

REFERENCES

1. LIEBER, C. S. 1982. Medical disorders of alcoholism. Pathogenesis and treatment. *In* Major Problems in Internal Medicine. L. H. Smith, Jr., Ed. Vol. 22: 1–589. W. B. Saunders. Philadelphia, PA.
2. FARQUHAR, M. G., J. J. M. BERGERON & G. E. PALADE. 1974. Cytochemistry of Golgi fractions prepared from rat liver. J. Cell Biol. **60:** 8–25.
3. STEIN, O., L. SANGER & Y. STEIN. 1974. Colchicine-induced inhibition of lipoprotein and protein secretion into the serum and lack of interference with secretion of biliary phospholipids and cholesterol by rat liver in vivo. J. Cell Biol. **62:** 90–103.
4. STEIN, O. & Y. STEIN. 1973. Colchicine-induced inhibition of very low density lipoprotein release by rat liver in vivo. Biochim. Biophys. Acta **306:** 142–147.
5. MADSEN, N. P. 1969. Reduced serum very low-density lipoprotein levels after acute ethanol administration. Biochem. Pharmacol. **18:** 261–262.

6. BARAONA, E., Y. MATSUDA, P. PIKKARAINEN, F. FINKELMAN & C. S. LIEBER. 1981. Effects of ethanol on hepatic protein secretion and microtubules. Possible mediation by acetaldehyde. *In* Currents in Alcoholism. M. Galanter, Ed. Vol. 3: 421–434. Grune & Stratton, Inc. New York, N.Y.
7. BARAONA, E., M. A. LEO, S. A. BOROWSKY & C. S. LIEBER. 1975. Alcoholic hepatomegaly: accumulation of protein in the liver. Science **190:** 794–795.
8. BARAONA, E., M. A. LEO, S. A. BOROWSKY & C. S. LIEBER. 1977. Pathogenesis of alcohol-induced accumulation of protein in the liver. J. Clin. Invest. **60:** 546–554.
9. TUMA, D. J. & M. F. SORRELL. 1982. Effects of ethanol on glycoprotein synthesis and secretion during inflammation-induced stimulation of hepatic glycoprotein secretion. Toxicol. Appl. Pharmacol. **63:** 303–311.
10. TUMA, D. J., R. B. JENNETT & M. F. SORRELL. 1981. Effect of ethanol on the synthesis and secretion of hepatic secretory glycoproteins and albumin. Hepatology **1:**590–598.
11. REDMAN, C. M., D. BANERJEE, K. HOWEL & G. E. PALADE. 1975. Colchicine inhibition of plasma protein release from rat hepatocytes. J. Cell Biol. **66:** 49–59.
12. REDMAN, C. M., D. BANERJEE, K. HOWELL & G. E. PALADE. 1975. The step at which colchicine blocks the secretion of plasma proteins by rat liver. Ann. N.Y. Acad. Sci. **253:** 780–788.
13. MATSUDA, Y., E. BARAONA, M. SALASPURO & C. S. LIEBER. 1979. Effects of ethanol on liver microtubules and Golgi apparatus. Possible role in altered hepatic secretion of plasma protein. Lab. Invest. **41:** 455–463.
14. MATSUDA, Y., A. TAKADA, R. KANAYAMA & S. TAKASE. 1983. Changes of hepatic microtubules and secretory proteins in human alcoholic liver disease. Pharmacol. Biochem. Behav. **18** (Suppl. 1): 479–482.
15. OKANOUE, T., O. ONGYOKU, M. OHTA, J. YOSHIDA, M. MORISHI, T. YUKI, T. OKUNO & T. TAKINO. 1984. Effect of chronic ethanol administration on the cytoskeleton of rat hepatocytes-inducing morphometric analysis. Acta Hepatal. Jap. **25:** 210–213.
16. BERMAN, W. J., J. GIL, R. B. JENNETT, D. TUMA, M. F. SORRELL & E. RUBIN. 1983. Ethanol, hepatocellular organelles, and microtubules. A morphometric study in vivo and in vitro. Lab. Invest. **48:** 760–767.
17. BARAONA, E., F. FINKELMAN & C. S. LIEBER. 1984. Reevaluation of the effects of alcohol consumption on rat liver microtubules. Effects of feeding status. Res. Commun. Chem. Pathol. Pharmacol. **44:** 265–278.
18. JENNETT, R. B., D. A. BURNETT, M. F. SORRELL & D. J. TUMA. 1985. Colchicine-binding properties of hepatic tubulin. The role of "time-decay." Lab. Invest. **53:** 111–115.
19. TANIKAWA, K., H. MOTORI, S. SAKISAKA, K. NOGUDRI, K. YAMAUCHI & H. ABE. 1984. Distribution of microtubules in cultured hepatocytes and effects of colchicine, cold and ethanol. Hepatology **4:** 773.
20. MATSUDA, Y., S. TAKASE, A. TAKADA, H. SATO & M. YASUHARA. 1985. Comparison of ballooned hepatocytes in alcoholic and non-alcoholic liver injury in rats. Alcohol **2:** 303–308.
21. MATSUDA, Y., A. TAKADA, H. SATO, M. YASUHARA & S. TAKASE. 1985. Comparison between ballooned hepatocytes occurring in human alcoholic and non-alcoholic liver diseases. Alcoholism: Clin. Exp. Res. **9:** 366–370.
22. GAUB, J., L. FAUERHOLDT, S. KEIDING, J. KONDRUP, P. PETERSEN & G. LANGEWANTZIN. 1981. Cytophotometry of liver cells from ethanol-fed rats: ethanol causes increased polyploidization and protein accumulation. Europ. J. Clin. Invest. **11:** 235–237.
23. PHILLIPS, M. J., C. OSHIO, M. MIYAIRI, S. WATANABE & C. R. SMITH. 1983. What is actin doing in the liver cell? Hepatology **3:** 433–436.
24. FRENCH, S. W. 1985. Role of canalicular contraction in bile flow. Lab. Invest. **53:** 245–249.
25. OKANOUE, T., I. KONDO, T. J. IHRIG & S. W. FRENCH. 1984. Effect of ethanol and chlorpromazine on transhepatic transport and biliary secretion of horseradish peroxidase. Hepatology **4:** 253–260.
26. KACICH, R. L., R. H. RENSTON & A. L. JONES. 1983. Effects of cytochalasin D and colchicine on the uptake, translocation and biliary secretion of horseradish peroxidase and (^{14}C)sodium taurocholate in the rat. Gastroenterology **85:** 385–394.
27. GEBHARDT, R. 1984. Participation of microtubules and microfilaments in the transcellular

biliary secretion of immunoglobulin A in primary cultures of rat hepatocytes. Experientia **40:** 269–271.

28. GOLDMAN, I. S., A. L. JONES, G. T. HRADEK & S. HULING. 1983. Hepatocyte handling of immunoglobulin A in the rat: the role of microtubules. Gastroenterology **85:** 130–140.
29. KATER, L., A. C. JOBSIS, E. H. BAART DE LA FAILLE-KUYPER, A. J. M. VOGTEN & R. GRIJM. 1979. Alcoholic hepatic disease. Specificity of IgA deposits in liver. Am. J. Clin. Pathol. **71:** 51–57.
30. SWERDLOW, M. A., L. N. CHOWDHURY & T. HORN. 1982. Patterns of IgA deposition in liver tissues in alcoholic liver disease. Am. J. Clin. Pathol. **77:** 259–266.
31. SWERDLOW, M. A. & L. N. CHOWDHURY. 1984. IgA deposition in liver in alcoholic liver disease. Arch. Pathol. Lab. Med. **108:** 416–419.
32. COPPO, R., S. ARICO, G. PICCOLI, B. BASOLO, D. ROCCATELLO, A. AMORE, M. TABONE, M. DE LA PIERRE, A. SESSA, D. L. DELACROIX & J. P. VERMAN. 1985. Presence and origin of IgA_1- and IgA_2-containing circulating immune complexes in chronic alcoholic liver diseases with and without glomerulonephritis. Clin. Immunol. Immunopathol. **35:** 1–8.
33. FRENCH, S. W., I. KONDO, T. IRIE, T. J. IHRIG, N. BENSON & R. MUNN. 1982. Morphologic study of intermediate filaments in rat hepatocytes. Hepatology **2:** 29–38.
34. OKANOUE, T., M. OHTA, S. FUSHIKI, O. OU, K. KACHI, T. OKUNO, T. TAKINO & S. W. FRENCH. 1985. Scanning electron microscopy of the liver cell cytoskeleton. Hepatology **5:** 1–6.
35. FRENCH, S. W., T. OKANOUE, S. H. H. SWIERENGA & N. MARCEAU. 1987. The cytoskeleton of hepatocytes in health and disease. *In* The Pathogenesis of Liver Disease. E. Farber & M. S. Phillips, Eds. 95–112. Williams and Wilkins Publ. New York, NY.
36. YOKOTA, S. & H. D. FAHIMI. 1979. Filament bundles of prekeratin type in hepatocytes: revealed by detergent extraction after glutaraldehyde fixation. Biol. Cellulaire **34:** 116–119.
37. JAHN, W. 1982. Electron microscopic characterization of intermediate filaments in rat liver parenchymal cells. Eur. J. Cell Biol. **26:** 259–264.
38. DENK, H., R. KREPLER, E. LACKINGER, V. ARTLIEB & W. W. FRANKE. 1982. Biochemical and immunocytochemical analysis of the intermediate filament cytoskeleton in human hepatocellular carcinomas and in hepatic neoplastic nodules in mice. Lab. Invest. **46:** 584–595.
39. FRANKE, W. W., E. SCHMID, J. KARTENBECK, D. MAYER, H.-J. HACKER, P. BANNISH, M. OSBORN, K. WEBER, H. DENK, J.-C. WANSON & P. DROCHMANS. 1979. Characterization of the intermediate-sized filaments in liver cells by immunofluorescence and electron microscopy. Biol. Cellulaire **34:** 99–110.
40. FRANKE, W. W., H. DENK, R. KALT & E. SCHMID. 1981. Biochemical and immunological identification of cytokeratin proteins present in hepatocytes of mammalian liver tissue. Exp. Cell Res. **131:** 299–318.
41. MARCEAU, N., R. GOYETTE, G. PELLETIER & T. ANTAKLY. 1983. Hormonally-induced changes in the cytoskeleton organization of adult and newborn rat hepatocytes cultured on fibronectin precoated substratum: effect of dexamethasone and insulin. Cell. Mol. Biol. **29:** 421–435.
42. RAY, M. B. 1986. Distinctive distribution patterns of cytokeratin filaments in alcoholic and non-alcoholic liver disease. Lab. Invest. **54:** 52A.
43. EICHNER, R., P. BONITZ & T. T. SUN. 1984. Classification of epidermal keratins according to their immunoreactivity, isoelectric point, and mode of expression. J. Cell Biol. **98:** 1388–1396.
44. TSENG, S. C. G., M. J. JARVINEN, W. G. NELSON, J.-W. HUANG, J. WOOKCOCK-MITCHELL & T.-T. SUN. 1982. Correlation of specific keratins with different types of epithelial differentiation: monoclonal antibody studies. Cell **30:** 361–372.
45. FRANKE, W. W., E. SCHMID, C. GRUND & B. GEIGER. 1982. Intermediate filament proteins in non-filamentous structures: transient disintegration and inclusions of subunit protein in granular aggregates. Cell **30:** 103–113.
46. DENK, H., R. KREPLER, E. LACKINGER, U. ARTLIEB & W. W. FRANKE. 1982. Immunological and biochemical characterization of the keratin-related component of Mallory bodies: a pathological pattern of hepatocytic cytokeratins. Liver **2:** 165–175.
47. FRANKE, W. W., D. SCHILLER, R. MOLL, S. WINTER, E. SCHMID, H. DENK, R. KREPLER &

B. PLATZER. 1981. Diversity of cytokeratins: differentiation specific expression of different cytokeratins in epithelial cells and tissues. J. Mol. Biol. **153:** 933–959.

48. WEISS, R. A., G. Y. A. GUILLET, I. W. FREEDBERG, E. R. FARMER, E. A. SMALL, M. M. WEISS & T.-T. SUN. 1983. The use of monoclonal antibody to keratin in human epidermal disease: alterations in immunohistochemical staining pattern. J. Invest. Dermatol. **81:** 224–230.
49. SUN, T.-T., R. EICHNER, A. SCHERMER, D. COOPER, W. G. NELSON & R. A. WEISS. 1984. Classification, expression, and possible mechanisms of evolution of mammalian epithelial keratins: a unifying model. *In* Cancer Cells. The Transformed Phenotype. Vol. 1: 169–176. A. Levine, W. Topp, G. Vande Woude & J. D. Watson, Eds. Cold Spring Harbor Laboratory. Cold Spring Harbor, NY.
50. GREEN, H., E. FUCHS & F. WATT. 1982. Differentiation structural components of the keratinocytes. Cold Spring Harbor Symp. Quart. Biol. **46:** 293.
51. MOLL, R., W. W. FRANKE, D. L. SCHILLER, B. GEIGER & R. KREPLER. 1982. The catalog of human cytokeratins: patterns of expression in normal epithelia, tumors and cultured cells. Cell **31:** 11–24.
52. WOODCOCK-MITCHELL, J., R. EICHNER, W. G. NELSON & T.-T. SUN. 1982. Immunolocalization of keratin polypeptides in human epidermis using monoclonal antibodies. J. Cell Biol. **95:** 580–588.
53. WEISS, R. A., R. EICHNER & T.-T. SUN. 1984. Monoclonal antibody analysis of keratin expression in epidermal diseases: a 48- and 56-k dalton keratin as molecular markers for hyperproliferative keratinocytes. J. Cell Biol. **98:** 1397–1406.
54. NELSON, W. G. & T.-T. SUN. 1983. The 50 kd and 58 kd keratin classes as molecular markers for stratified squamous epithelia: cell culture studies. J. Cell Biol. **97:** 244–251.
55. BREITKREUTZ, D., A. BOHNERT, E. HERZMANN, P. E. BOWDEN, P. BOUKAMP & N. E. FUSENIG. 1984. Differentiation specific functions in cultured and transplanted mouse keratinocytes: environmental influences on ultrastructure and keratin expression. Differentiation **26:** 154–169.
56. FUCHS, E. & H. GREEN. 1981. Regulation of terminal differentiation of cultured human keratinocytes by vitamin A. Cell **25:** 617–625.
57. GEIGER, B., T. E. KREIS, O. GIGI, E. SCHMID, S. MITTNACH, J. L. JORCANO, D. B. VON BASSEWITZ & W. W. FRANKE. 1984. Dynamic rearrangements of cytokeratin in living cells. *In* The Transformed Genotype. 201–215. A. J. Levine, G. F. Vande Woude, W. C. Topp & J. D. Watson, Eds. Cold Spring Harbor Laboratory. Cold Spring Harbor, NY.
58. SWIERENGA, S. H. H., R. GOYETTE & N. MARCEAU. 1984. Differential effects of calcium deprivation on the cytoskeleton of non-tumorigenic and tumorigenic rat liver cell culture. Exp. Cell Res. **153:** 39–49.
59. FRENCH, S. W. 1983. Present understanding of the development of Mallory's body. Arch. Pathol. Lab. Med. **107:** 445–450.
60. NAKANUMA, Y. & G. OHTA. 1985. Is Mallory body formation a preneoplastic change? A study of 181 cases of liver bearing hepatocellular carcinoma and 82 cases of cirrhosis. Cancer **55:** 2400–2404.
61. FRENCH, S. W. 1981. Nature, pathogenesis and significance of the Mallory body. Seminars Liver Dis. **1:** 217–231.
62. TAZAWA, J., T. IRIE & S. W. FRENCH. 1983. Mallory body formation runs parallel to -glutamyl transferase induction in hepatocytes of griseofulvin-fed mice. Hepatology **3:** 989–1001.
63. NAKANUMA, Y. & G. OHTA. 1986. Expression of Mallory bodies in hepatocellular carcinoma in man and its significance. Cancer **57:** 81–86.
64. BORENFREUND, E., P. J. HIGGINS & A. BENDICH. 1980. In vivo-in vitro rat liver carcinogenesis: modifications in protein synthesis and ultrastructure. Ann. N.Y. Acad. Sci. **349:** 357–372.
65. SATO, M. & C. S. LIEBER. 1981. Hepatic vitamin A depletion after chronic ethanol consumption in baboons and rats. J. Nutr. **111:** 2015–2023.
66. LEO, M. A. & C. S. LIEBER. 1982. Hepatic vitamin A depletion in alcoholic liver injury in man. N. Engl. J. Med. **37:** 597–601.

67. DENK, H., W. W. FRANKE, D. KERJASCHI & R. ECKERSTORFER. 1979. Mallory bodies in experimental animals and man. Rev. Exp. Pathol. **20:** 77–121.
68. AKEDA, S., K. FUJITA, Y. KOSAKA & S. W. FRENCH. 1986. Mallory body formation and amyloid deposition in the liver of aged mice fed a vitamin A deficient diet for a prolonged period. Lab. Invest. **54:** 228–233.
69. GREEN, H. & F. M. WATT. 1982. Regulation by vitamin A of envelope cross-linking in cultured keratinocytes derived from different human epithelia. Mol. Cell. Biol. **2:** 1115–1117.
70. GILFIX, B. M. & R. L. ECKERT. 1985. Coordinate control by vitamin A of keratin gene expression in human keratinocytes. J. Biol. Chem. **260:** 14026–14029.
71. OKANOUE, T., O. MASAHARU, O. OU, K. KACHI, K. KAGAWA, T. YUKI, T. OKUNO, T. TAKINO & S. W. FRENCH. 1985. Relationship of Mallory bodies to intermediate filaments in hepatocytes. A scanning electron microscopy study. Lab. Invest. **53:** 534–540.
72. IRIE, T., N. C. BENSON & S. W. FRENCH. 1982. Relationship of Mallory bodies to the cytoskeleton of hepatocytes in griseofulvin treated mice. Lab. Invest. **47:** 336–345.
73. DENK, H., W. W. FRANKE, B. DRAGOSICS & I. ZEILER. 1981. Pathology of cytoskeleton of liver cells: demonstration of Mallory bodies (alcoholic hyalin) in murine and human hepatocytes by immunofluorescence microscopy using antibodies to cytokeratin polypeptides from hepatocytes. Hepatology **1:** 9–20.
74. KIMOFF, R. J. & S.-N. HUANG. 1981. Immunocytochemical and immunoelectron microscopic studies on Mallory bodies. Lab Invest. **45:** 491–503.
75. IRIE, T., N. C. BENSON & S. W. FRENCH. 1984. Electron microscopic study of the in vitro calcium-dependent degradation of Mallory bodies and intermediate filaments in hepatocytes. Lab. Invest. **50:** 303–312.
76. FEY, E. G., D. G. CAPCO, G. KROCHMALNIC & S. PENMAN. 1984. Epithelial structure revealed by chemical dissection and unembedded electron microscopy. J. Cell Biol. **99:** 203s–208s.
77. FEY, E. G., K. M. WAN & S. PENMAN. 1984. Epithelial cytoskeletal framework and nuclear matrix-intermediate filament scaffold: Three dimensional organization and protein composition. J. Cell Biol. **98:** 1973–1984.
78. SWIERENGA, S. H. H. 1985. Use of low calcium medium in carcinogenesity testing: studies with rat liver cells. *In* In Vitro Models for Cancer Research. M. M. Weber, Ed. Vol. 2: 61–89. CRC Press. Boca Raton, FL.
79. BENDAYAN, M. 1984. Protein A-gold electron microscopic immunocytochemistry: methods, applications and limitations. J. Electron Microsc. Tech. **1:** 243–270.

DISCUSSION OF THE PAPER

QUESTION: Could increased cytosolic calcium play a role in stimulating the degradation of intermediate filaments?

FRENCH: Calcium does activate the degradation of the intermediate filaments in the experimental model *in vitro*. What happens in the intact cell is a more difficult question. We speculate that there is no change in turnover of the normal filaments, but that the hyaline bodies are resistant to the protease.

E. RUBIN: (*Thomas Jefferson University School of Medicine, Philadelphia, PA*): An alcoholic can drink for many years without the appearance of Mallory bodies. Then suddenly he develops alcoholic hepatitis, with many Mallory bodies there. This is not consistent with a slow accumulation of Mallory bodies or with slow phenotypic change. Do you have any ideas on that subject?

FRENCH: What turns on the cell to produce Mallory bodies is uncertain. In mice fed griseofulvin it takes more than a month to clear the liver of the Mallory bodies. They reaccumulate in a matter of two to three days if you put griseofulvin back into their diet. It seems that hyaline is quickly inducible once the cell is preconditioned to express that phenomenon. It's very slow to be reversed, at least *in vivo*.

E. MEZEY: (*Johns Hopkins School of Medicine, Baltimore, MD*): Could you speculate on the mechanism of the accumulation of Mallory bodies in vitamin A deficiency? Does vitamin A or its deficiency increase microtubule formation and bile flow?

FRENCH: We found some Mallory bodies in the control mice after two years, so that it's not strictly a vitamin A phenomenon. Yet there is a quantitative switch in the various types of cytokeratin messenger RNA's in the presence or absence of vitamin A.

The Interaction of Acetaldehyde With Tubulin[a]

DEAN J. TUMA, RICHARD B. JENNETT, AND MICHAEL F. SORRELL

Liver Study Unit
Veterans Administration Medical Center
and
Departments of Internal Medicine and Biochemistry
University of Nebraska Medical Center
Omaha, Nebraska 68105

INTRODUCTION

Alcoholic liver disease is a major health problem in the United States. Despite considerable research efforts in this area, the mechanism(s) of ethanol-induced liver injury have remained elusive. In order to formulate an adequate pathogenic model for alcoholic liver damage, we must take into account several known factors: (i) The liver is both the principal site of ethanol metabolism as well as the principal site of ethanol-induced toxicity in the body. (ii) Ethanol oxidation generates acetaldehyde, which is a very reactive compound capable of undergoing chemical reactions with a variety of substances found in biological material. (iii) Genetic, environmental, and nutritional factors are capable of influencing the hepatotoxicity of alcohol.[1-8] In this regard, we have recently proposed that ethanol may exert some of its toxic effects by virtue of its metabolic conversion to acetaldehyde, which subsequently covalently binds to tissue proteins.[9] Although the binding of acetaldehyde to proteins and the formation of acetaldehyde adducts with hepatic proteins during ethanol oxidation have been established,[10-13] the functional consequences of such binding have not been thoroughly studied. Therefore, in this report we shall describe the interaction of acetaldehyde with the model protein, tubulin, and shall relate this interaction or binding to changes in the structure and function of this protein.

CHEMISTRY OF ACETALDEHYDE-PROTEIN INTERACTIONS

Studies in our laboratory[10,13] and others[5,8] have demonstrated that acetaldehyde can bind to a variety of proteins *in vitro* to form both stable and unstable adducts. This reactivity of acetaldehyde with proteins appears to be mainly due to the reaction of the carbonyl carbon of the aldehyde with nucleophilic groups on the polypeptide chain especially the ϵ-amino group of lysine.[14,15] The majority of adducts formed when acetaldehyde reacts with proteins for short time periods (1–3 hr) are unstable and have been identified to be Schiff bases, which are unsaturated addition products that readily dissociate.[10] Treatment with reducing agents, such as sodium borohydride, stablizes Schiff bases by reduction to secondary amines and, thereby, allows the detection of these adducts. Stable adducts, on the other hand, are relatively irreversible and survive

[a]Supported by National Institute on Alcohol Abuse and Alcoholism Grant AA04961 and Postdoctoral Fellowship IF32AA05220-01 and by the Veterans Administration.

acid or base treatment, gel filtration, and exhaustive dialysis.[10,13] Both types of adducts mainly involve participation of lysine residues of the protein.[9,10,15] Our studies[13] have also demonstrated that physiological reducing agents, such as ascorbate, can increase the formation of stable adducts, presumably by reducing Schiff bases, although other mechanisms and reaction products have not been ruled out.

Recent studies in our laboratory[11,12] have demonstrated the formation of both unstable and stable acetaldehyde-protein adducts in the liver during ethanol oxidation. These observations as well as the chemistry of acetaldehyde-protein interactions have led us to propose a reaction scheme that describes acetaldehyde adduct formation in the liver during ethanol oxidation.[9] As indicated in FIGURE 1, oxidation of ethanol, via alcohol dehydrogenase in the liver, yields acetaldehyde, which initially reacts with free ε-amino groups of lysyl residues on the polypeptide chain to form unstable Schiff base adducts. In addition, excess formation of reducing equivalents (increased NADH) accompanies the oxidative reaction, and could potentially be available to promote the reduction of Schiff bases to stable protein adducts.

Although acetaldehyde adduct formation in the liver during ethanol oxidation seems very likely, the functional consequences of adduct formation have not been systematically explored. This latter aspect of adduct formation appears to be essential

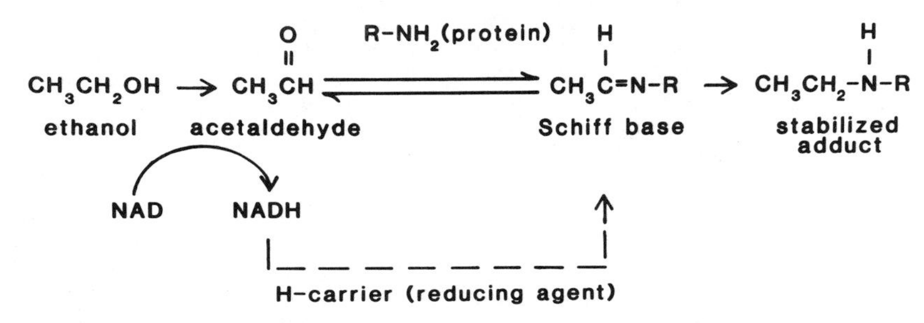

FIGURE 1. Formation of acetaldehyde-protein adducts during ethanol oxidation in the liver.

in clarifying the role of acetaldehyde adducts in ethanol-induced liver injury. However, in order to incriminate acetaldehyde-protein adducts as being involved in liver injury, it is necessary to first show and characterize the structural or functional alterations of a specific protein by adduct formation. Such an approach should yield valuable initial information that could be applied to studies that examine the role of adducts in the more complicated processes of cell injury. Therefore, we have characterized the *in-vitro* interaction of acetaldehyde with the specific protein, tubulin.

TUBULIN AS A MODEL PROTEIN TO STUDY THE FUNCTIONAL CONSEQUENCES OF ADDUCT FORMATION

Tubulin was chosen as a model protein for investigating the functional consequences of acetaldehyde adduct formation for several reasons. (i) Tubulin is a biologically important protein, which is found in one form or another in virtually all eukaryotic cells. Tubulin generally exists in cells both as the free dimer or assembled into filamentous structures known as microtubules. Microtubules are thought to be involved in mitosis, cell motility phenomena, and secretory events, and, in addition, may serve a general cytoskeletal role.[16,17] (ii) Tubulin is located in the cytoplasmic compartment of the liver cell where most of the acetaldehyde generation takes place, thus facilitating local interaction. (iii) The tubulin=microtubule system has been characterized in detail and lends itself especially well to *in-vitro* study because of several unique properties of tubulin. Tubulin has the ability to polymerize (*in vitro*) to form microtubules under proper conditions. In addition, the specific binding of colchicine and GTP can be used as sensitive probes of the conformation of tubulin.[18–20] (iv) Tubulin is known to be a globular protein, which is composed of two polypeptides each of about 55,000 Daltons, known as α and β.[21] Each polypeptide chain contains 27 ± 1 lysines and therefore has multiple ϵ-amino groups available to react with acetaldehyde.[22] Recent elegant work by Sternlicht and co-workers[20,23] has shown that there are 1 or 2 lysine residues on the α-chain, which are both uniquely important to the polymerization process and exhibit an enhanced reactivity toward formaldehyde. This interesting observation led us to wonder whether tubulin could be a specific target of acetaldehyde adduct formation in the cell by virture of certain key lysine residues on the α-chain with enhanced reactivity toward aldehydes in general. (v) Finally, tubulin can be easily isolated in good yield and purity from bovine brain. Tubulin so purified is in its native state as evidenced by its ability to bind colchicine and other ligands as well as by its ability to form microtubules of normal morphology (*in vitro*).[24] Brain tubulin was used for these experiments rather than liver tubulin because of the relative abundance of tubulin in the brain and its ease of isolation compared to the liver. The limited amount of information available in the literature suggests that brain and liver tubulin are structurally very similar.[18]

Further rationale for the use of tubulin as a model target protein to study the effects of acetaldehyde adduct formation is based upon previous data that suggests that acetaldehyde blocks the secretion of protein from the liver.[25–27] Microtubules are believed to be prominently involved in hepatic protein secretion, and impaired microtubule formation (tubulin polymerization) has been advanced as an explanation of this effect.[28] It should be emphasized, however, that protein secretion is a complex and not fully understood event, with multiple possible sites of impairment other than microtubules. In addition, there is some uncertainty in the literature concerning the extent to which microtubules are actually affected by ethanol (acetaldehyde),[29] probably reflecting methodologic difficulties involved in the measurement of liver microtubules.[30,31] Nonetheless, tubulin seems to be an excellent candidate protein for assessing acetaldehyde adduct-induced functional and structural changes.

COVALENT BINDING OF ACETALDEHYDE TO TUBULIN

Like most other proteins, tubulin binds acetaldehyde to form both stable and unstable adducts. In short-term incubations (up to 2 hr), the majority of adducts (80–90%) formed are unstable, but the percentage of total adducts that are stable gradually increases with time. When initial experiments were conducted to assess the binding of exogenously added [^{14}C]acetaldehyde to cycle-purified tubulin, the detectable protein-bound radioactivity increased in a time- and concentration-dependent manner as shown in FIGURE 2. The inclusion of sodium cyanoborohydride, a specific reducing agent of Schiff bases, in the incubation mixture resulted in a dramatic increase in detectable binding.[32,33] Cyanoborohydride treatment provides a means to greatly increase the amount of stable adduct formation under conditions which are nondestructive to the protein and, in addition, it strongly implicates the involvement of

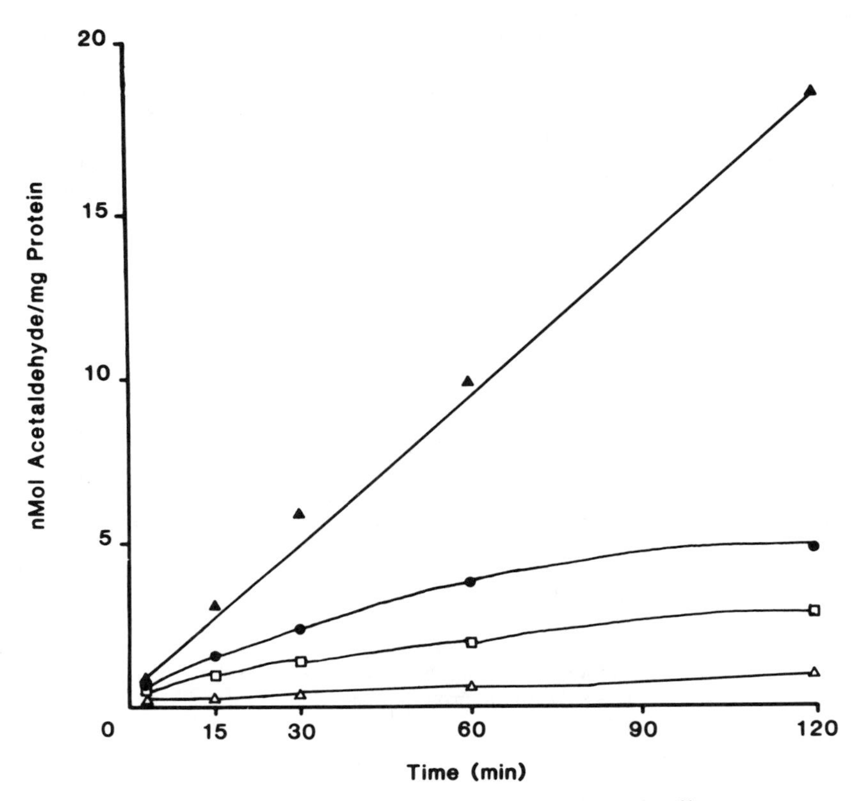

FIGURE 2. Time and concentration dependence of covalent binding of [^{14}C]acetaldehyde to tubulin. Bovine neurotubulin was cycle-purified as previously described.[36] For these experiments tubulin 2.75 mg/ml was incubated at 37° (pH 6.7) with [^{14}C]acetaldehyde at the following concentrations: 1 mM (△), 2.5 mM (□), 5.0 mM (●); additional samples contained 1 mM [^{14}C]acetaldehyde plus 6 mM sodium cyanoborohydride (▲). At designated times samples were processed to determine protein bound radioactivity after an extensive washing-dialysis procedure.

Schiff base intermediates in stable adduct formation. Jentoft and Dearborn,[34] who carried out extensive studies in an analagous system (the methylation of proteins by formaldehyde in the presence of cyanoborohydride), found that the only groups modified by this method were free amino groups of the protein (ϵ-amino of lysine and α-terminus). Additional experiments were conducted in our laboratory with tubulin that was more extensively purified using phosphocellulose chromatography and comparable formation of stable and unstable acetaldehyde adducts was observed.[32,35] In general agreement with Szasz *et al.*,[20] we observed that MAPs (microtubule-associated proteins) could also bind acetaldehyde (unpublished observation). Therefore, at first consideration, it appeared that the binding of acetaldehyde to tubulin resembled, in most respects, the binding reported for a variety of other proteins.

Since detailed studies by Sternlicht and co-workers[23] demonstrated that the α-chain of tubulin has 1 or 2 lysine residues that possess enhanced reactivity toward formaldehyde (reductive methylation), we undertook studies to further characterize the binding of acetaldehyde to tubulin. We were especially interested in determining whether tubulin exhibited similar α-chain reactivity toward acetaldehyde. In these experiments,[35] cycle-purified beef brain tubulin was further purified using phosphocellulose chromatography to eliminate the microtubule-associated proteins that copurify with tubulin. Tubulin was then incubated with [^{14}C]acetaldehyde and subjected to

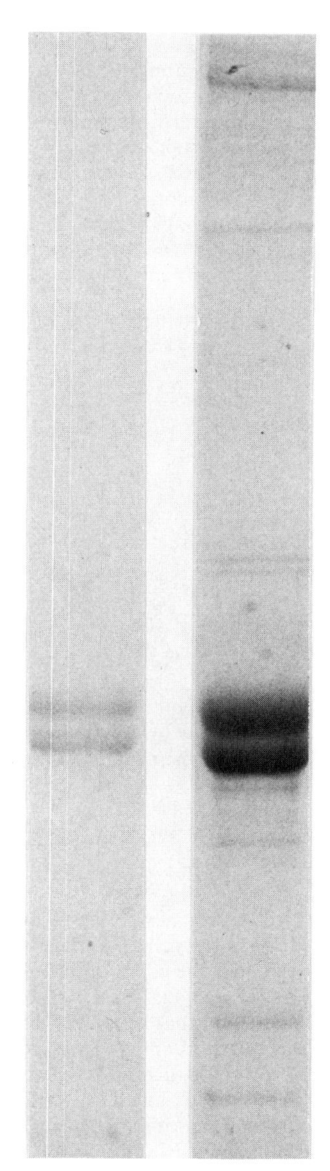

FIGURE 3. Electrophoretic resolution of tubulin into its α- and β-chains. SDS-PAGE (7.5% gels) was conducted by a modification of the procedure described by Szasz *et al.*[20] The *right lane* is cycle-purified tubulin that contains microtubule-associated proteins (MAPs) and other impurities, whereas the *left lane* represents more highly purified tubulin prepared by phosphocellulose chromatography.[20]

SDS-PAGE to resolve the α- and β-chains of tubulin. Resolution of tubulin into its α- and β-chains is shown in FIGURE 3. As indicated in FIGURE 4 the α-chain of tubulin bound considerably more [^{14}C]acetaldehyde than the β-chain despite an approximately equal lysine content. As expected, this selectivity was more pronounced at low acetaldehyde-to-protein ratios (FIG. 4A) where the more reactive α-chain effectively competed for a limited amount of acetaldehyde. At higher acetaldehyde-to-protein ratios (FIG. 4B) bulk ethylation of lysine became more prominent. The tubulin (1 mg/ml) and acetaldehyde (50 μM) concentrations are comparable to those that are believed to exist in the liver during ethanol oxidation. The enhanced reactivity of the α-chain of tubulin was also apparent when competitive binding experiments were conducted in which the α-chain of tubulin was found to preferentially compete with serum albumin for binding to acetaldehyde (unpublished data). Taken as a whole, these results show that acetaldehyde preferentially binds to certain key lysine amino groups on the α-chain of tubulin and further indicate that the magnitude of this specificity increases at "physiologic" concentrations of acetaldehyde and tubulin.

The findings of these binding studies suggest that tubulin could be a selective target for acetaldehyde binding in the liver. Since the specific lysine residues on the α-chain, which exhibit enhanced reactivity toward aldehydes, have been shown to be intimately

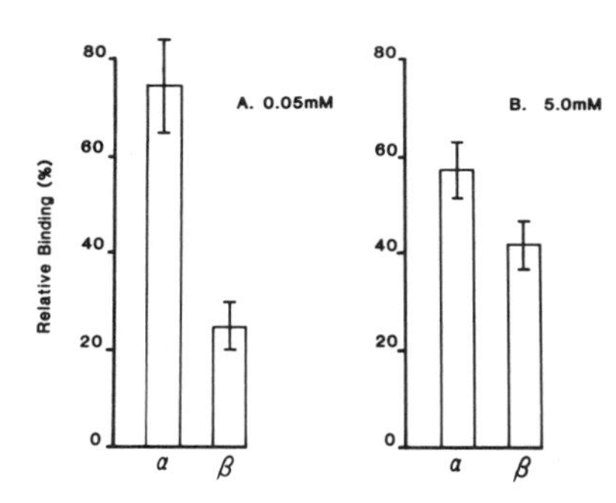

FIGURE 4. Effect of acetaldehyde concentration on subunit binding specificity. For these experiments, phosphocellulose-purified brain tubulin (1 mg/ml) was incubated at 37° with 50 μM (*panel A*), and 5.0 mM (*panel B*) [^{14}C]acetaldehyde for 2 hr. Dialyzed reaction mixtures were then subjected SDS-PAGE as described in FIGURE 3 to resolve the α- and β-chains. Chain specific radioactivity was then determined after digestion of appropriate gel slices. As shown, the specificity of α-chain labeling was predominant at the lower acetaldehyde concentration.

involved in polymerization (tubulin function),[23] these results raise the possibility that relatively low levels of specific acetaldehyde binding to tubulin may interfere with the function of the tubulin-microtubule system in the liver.

FUNCTIONAL CONSEQUENCES OF ACETALDEHYDE BINDING TO TUBULIN

The polymerization of tubulin to form microtubules is both an important cellular function as well as an easily measured property of this protein. For this reason, we measured the effect of acetaldehyde adduct formation on *in-vitro* tubulin polymerization, employing an assay that takes advantage of the increase in optical density (turbidity) that accompanies microtubule formation. In order to verify the sensitivity and specificity of our assay system, we confirmed that colchicine, a known inhibitor of microtubule formation, caused a concentration-dependent inhibition of polymerization with levels as low as 1 μM showing significant effects (FIGURE 5). These data along with electron microscopic evidence showing the formation of microtubules of normal morphology established the validity of our assay system.[36]

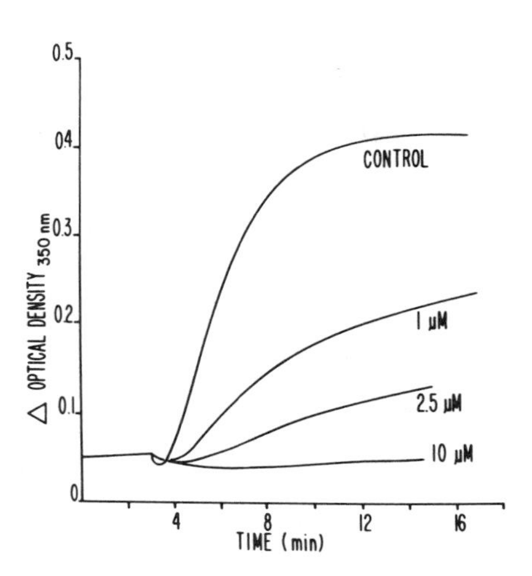

FIGURE 5. Inhibition of *in-vitro* tubulin assembly by colchicine. Tubulin polymerization was measured by a well established spectrophotometric technique.[36] To verify the sensitivity of our assay system, colchicine was added to selected samples. As shown, colchicine exhibited a potent antimicrotubule effect at low concentrations in our model.

When tubulin was preincubated with acetaldehyde (5 mM), a time-dependent increase in polymerization inhibition was observed. After 5 min of preincubation a 20% inhibition of polymerization was noted, whereas after 120 min complete inhibition occurred.[32] Addition of sodium cyanoborohydride resulted in increased stable adduct formation and dramatically decreased both the time required and the concentration of acetaldehyde necessary to inhibit microtubule assembly.[32] As shown in FIGURE 6, 1-hr pretreatment of tubulin with acetaldehyde (1 mM) in the presence of cyanoborohydride resulted in a 50% inhibition of tubulin assembly. Complete inhibition of microtubule formation was observed after 90–120 min of incubation of tubulin with

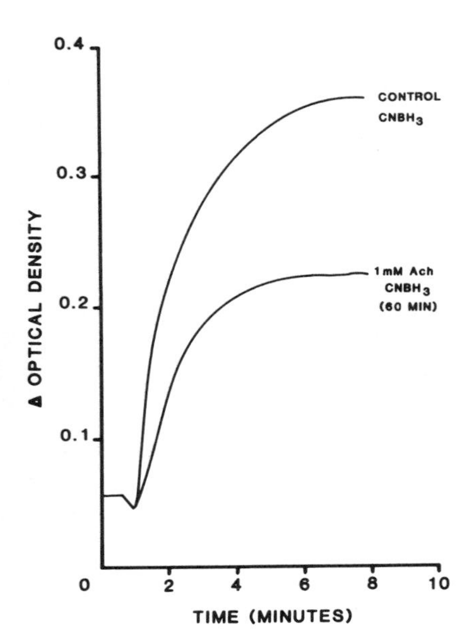

FIGURE 6. Inhibition of tubulin assembly by pretreatment with acetaldehyde. Depolymerized tubulin was pre-incubated for 1 hr at 37° with 1 mM acetaldehyde in the presence of 6 mM sodium cyanoborohydride. Samples were dialyzed versus the assembly assay buffer to remove unreacted reagents. Tubulin polymerization was then assayed spectrophotometrically after addition of GTP (guanosine-5′-triphosphate). As shown, under these conditions, assembly of acetaldehyde-treated tubulin was inhibited compared to control tubulin, which was treated with sodium cyanoborohydride alone.

1 mM acetaldehyde in the presence of cyanoborohydride.[32] Under these same conditions, lower concentrations of acetaldehyde partially inhibited tubulin polymerization. In more prolonged incubations carried out at 0° (conditions necessary to maintain tubulin stability) concentrations of acetaldehyde as low as 200 μM completely inhibited tubulin assembly after 4 days of pretreatment (unpublished observation). Inhibition of tubulin assembly by low concentrations of acetaldehyde was also previously reported by others.[37] In addition, aldehydes have been shown to be antimitotic agents, a characteristic property of antimicrotubular agents.[38] On the other hand, Zeeberg *et al.*[39] studied the effects of reductive ethylation on tubulin assembly and came to the different conclusion that tubulin so modified could co-polymerize with normal tubulin. It should be noted, however, that they used a lightly modified protein under different experimental conditions. However, most of the data do support the conclusion that the covalent binding of acetaldehyde to tubulin is associated with impaired microtubule formation.

Since the presence of cyanoborohydride increased stable adduct formation as well as the acetaldehyde-induced inhibition of polymerization, it would appear that stable rather than unstable adducts were responsible for inhibition of tubulin assembly. Furthermore, reaction mixtures that were dialyzed prior to assembly assay to dissociate unstable adducts did not influence the effects of acetaldehyde pretreatment on polymerization[32] (unpublished observations). To further investigate this point, we compared the time course and α-subunit binding specificity of stable and unstable adduct formation with the time course of assembly impairment.[40] For these experiments, tubulin was incubated with [^{14}C]acetaldehyde in the absence of cyanoborohydride for various times followed by determination of tubulin polymerization and measurement of stable and unstable adducts. The ratios of $\alpha : \beta$ subunit [^{14}C]radioactivity of both types of adducts were also determined. The gradual increase of stable adduct formation with time paralleled the steady decrease in the ability of tubulin to form microtubules much more closely than did the formation of unstable adducts, which formed rapidly and then remained constant throughout the preincubation period. In addition, the $\alpha : \beta$ binding ratios were >1.5 for stable adducts, whereas the ratios were near 1.0 for unstable adducts.[40] These results suggest that stable adducts rather than unstable adducts are responsible for the acetaldehyde-induced decrease in tubulin polymerization. In addition, the increased $\alpha : \beta$ ratio observed for stable adducts suggests α-chain specificity of this effect. These results, however, do not completely exclude the possibility that unstable adducts or even noncovalent interactions of acetaldehyde with tubulin could also contribute to altered tubulin function.

SUMMARY AND CONCLUSIONS

Acetaldehyde covalently binds to purified tubulin *in vitro* to form both stable and unstable adducts. The formation of stable adducts can be greatly facilitated by the inclusion of the relatively gentle and Schiff base specific reducing agent, sodium cyanoborohydride. Although the tubulin molecule has multiple lysine resides available to react with acetaldehyde, certain key lysine residues on the α-chain appear to be selective targets for adduct formation. The formation of α-chain specific stable acetaldehyde-tubulin adducts results in functional impairment of the ability of tubulin to polymerize. Under relatively physiologic conditions where acetaldehyde-to-protein ratios are low, α-chain specific binding is prominent. These results, coupled with the studies presented in another report in this volume,[41] raise the possibility that low levels of adduct formation may be detrimental to the structure or function of certain proteins (*e.g.* tubulin) in the liver. The alteration of this or other biologically important proteins by sustained low levels of adduct formation may contribute to the pathogenesis of alcoholic liver injury.

REFERENCES

1. LIEBER, C. S. 1984. Alcohol and the liver: 1984 update. Hepatology **4:** 1243–1260.
2. ORREGO, H., Y. ISRAEL & L. M. BLENDIS. 1981. Alcoholic liver disease: information in search of knowledge. Hepatology **1:** 267–283.
3. MEZEY, E. 1985. Metabolic effects of alcohol. Fed. Proc. **44:** 134–138.
4. SORRELL, M. F. & D. J. TUMA. 1979. Effects of alcohol on hepatic metabolism: selected aspects. Clin. Sci. **57:** 481–489.
5. STEVENS, V. J., W. J. FANTL, C. B. NEWMAN, R. V. SIMS, A. CERAMI & C. M. PETERSON. 1981. Acetaldehyde adducts with hemoglobin. J. Clin. Invest. **67:** 361–369.
6. KENNY, W. C. 1982. Acetaldehyde adducts of phospholipids. Alcoholism: Clin. Exp. Res. **6:** 412–416.
7. GAINES, K. C., J. M. SALHANY, D. J. TUMA & M. F. SORRELL. 1977. Reactions of acetaldehyde with human erythrocytes membrane proteins. FEBS Lett. **75:** 115–119.
8. NOMURA, F. & C. S. LIEBER. 1981. Binding of acetaldehyde to rat liver microsomes: enhancement after chronic alcohol consumption. Biochem. Biophys. Res. Commun. **100:** 131–137.
9. SORRELL, M. F. & D. J. TUMA. 1985. Hypothesis: alcoholic liver injury and the covalent binding of acetaldehyde. Alcoholism: Clin. Exp. Res. **9:** 306–309.
10. DONOHUE, T. M., D. J. TUMA & M. F. SORRELL. 1983. Acetaldehyde adducts with proteins: binding of [^{14}C]acetaldehyde to serum albumin. Arch. Biochem. Biophys. **220:** 239–246.
11. DONOHUE, T. M., D. J. TUMA & M. F. SORRELL. 1983. Binding of metabolically derived acetaldehyde to hepatic proteins in vitro. Lab. Invest. **49:** 226–229.
12. MEDINA, V. A., T. M. DONOHUE, M. F. SORRELL & D. J. TUMA. 1985. Covalent binding of acetaldehyde to hepatic proteins during ethanol oxidation. J. Lab. Clin. Med. **105:** 5–10.
13. TUMA, D. J., T. M. DONOHUE, V. A. MEDINA & M. F. SORRELL. 1984. Enhancement of acetaldehyde-protein adduct formation by L-ascorbate. Arch. Biochem. Biophys. **234:** 377–381.
14. O'DONNELL, J. P. 1982. The reaction of amines with carbonyls: its significance in the nonenzymatic metabolism of xenobiotics. Drug Metab. Rev. **13:** 123–159.
15. TUMA, D. J. & M. F. SORRELL. 1985. Covalent binding of acetaldehyde to hepatic proteins: role in alcoholic liver injury. *In* Aldehyde Adducts in Alcoholism. M. A. Collins, Ed. Prog. Clin. Biol. Res. Vol. **183:** 3–17. Alan R. Liss, Inc. New York, NY.
16. OLMSTED, J. B. & G. G. BORISY. 1983. Microtubules. Annu. Rev. Biochem. **42:** 507–540.
17. TIMASHEFF, S. N. & L. M. GRISHAM. 1980. In vitro assembly of cytoplasmic microtubules. Annu. Rev. Biochem. **49:** 565–591.
18. PATZELT, C., A. SINGH, Y. LEMARCHAND, L. ORCI & B. JEANRENAUD. 1975. Colchicine-binding protein of liver. J. Cell Biol. **66:** 609–620.
19. JENNETT, R. B., D. J. TUMA & M. F. SORRELL. 1980. Colchicine-binding properties of the hepatic tubulin/microtubule system. Arch. Biochem. Biophys. **204:** 181–190.
20. SZASZ, J., R. BURNS & H. STERNLICHT. 1982. Effects of reductive methylation on microtubule assembly. J. Biol. Chem. **257:** 3697–3704.
21. LUDUEÑA, R. F., E. M. SHOOTER & L. WILSON. 1977. Structure of the tubulin dimer. J. Biol. Chem. **252:** 7006–7014.
22. LU, R. C. & M. ELZINGA. 1978. The primary structure of tubulin. Sequences of the carboxyl terminus and seven other cyanogen bromide peptides from the α-chain. Biochem. Biophys. Acta **537:** 320–328.
23. SHERMAN, G., T. L. ROSENBERRY & H. STERNLICHT. 1983. Indentification of lysine residues essential for microtubule assembly. J. Biol. Chem. **258:** 2148–2156.
24. LEE, Y. C., F. E. SAMSON, L. L. HOUSTON & R. H. HIMES. 1974. The in vitro polymerization of tubulin from beef brain. J. Neurobiol. **5:** 317–330.

25. SORRELL, M. F., D. J. TUMA, E. C. SCHAFER & A. J. BARAK. 1977. Role of acetaldehyde in the ethanol-induced impairment of glycoprotein metabolism in rat liver slices. Gastroenterology **73:** 137–144.
26. VOLENTINE, G. D., D. K. KORTJE, D. J. TUMA & M. F. SORRELL. 1985. Role of acetaldehyde in the ethanol-induced inhibition of hepatic glycoprotein secretion in vivo. Hepatology **5:** 1056.
27. SORRELL, M. F., D. J. TUMA & A. J. BARAK. 1977. Evidence that acetaldehyde irreversibly impairs glycoprotein metabolism in liver slices. Gastroenterology **73:** 1138–1141.
28. MATSUDA, Y., E. BARAONA, M. SALASPURO & C. S. LIEBER. 1979. Effects of ethanol on liver microtubules and Golgi apparatus. Lab. Invest. **41:** 455–463.
29. BERMAN, W. J., J. GIL, R. B. JENNETT, D. J. TUMA, M. F. SORRELL & E. RUBIN. 1983. Ethanol, hepatocellular organelles and microtubules. Lab. Invest. **48:** 760–767.
30. BARAONA, E., F. FINKELMAN & C. S. LIEBER. 1984. Reevaluation of the effects of alcohol consumption on rat liver microtubules: effects of feeding status. Res. Commun. Chem. Pathol. Pharmacol. **44:** 265–278.
31. JENNETT, R. B., D. A. BURNETT, M. F. SORRELL & D. J. TUMA. 1985. Colchicine-binding properties of hepatic tubulin, the role of "time-decay." Lab. Invest. **53:** 111–115.
32. JENNETT, R. B., E. L. JOHNSON, M. F. SORRELL & D. J. TUMA. 1985. Covalent binding of acetaldehyde to tubulin is associated with impaired polymerization. Hepatology **5**(5)A: 436.
33. JENNETT, R. B., M. F. SORRELL & D. J. TUMA. Covalent binding of acetaldehyde to tubulin. Manuscript in preparation.
34. JENTOFT, N. & D. G. DEARBORN. 1979. Labeling of proteins by reductive methylation using sodium cyanoborohydride. J. Biol. Chem. **254:** 4359–4365.
35. JENNETT, R. B., E. L. JOHNSON, M. F. SORRELL & D. J. TUMA. 1985. Preferential covalent binding of acetaldehyde to the α-chain of tubulin. Hepatology **5**(5)A: 437.
36. JENNETT, R. B., D. J. TUMA & M. F. SORRELL. 1980. Effect of ethanol and its metabolites on microtubule formation. Pharmacology **21:** 363–368.
37. BARAONA, E., Y. MATSUDA, P. PIKKARAINEN, F. FINKELMAN & C. S. LIEBER. 1980. Acetaldehyde-mediated disruption of liver microtubules after alcohol consumption. Alcoholism: Clin. Exp. Res. **4:** 209.
38. SENTEIN, P. 1975. Action of glutaraldehyde and formaldehyde in segmentation mitoses. Exp. Cell Res. **95:** 233–246.
39. ZEEBERG, B., J. CHEEK & M. CAPLOW. 1980. Preparation and characterization of [^{3}H]ethyltubulin. Anal. Biochem. **104:** 321–327.
40. JENNETT, R. B., S. L. SMITH, M. F. SORRELL & D. J. TUMA. 1986. Inhibition of tubulin assembly by stable acetaldehyde adducts. Alcoholism: Clin. Exp. Res. **10**(1): 58.
41. SORRELL, M. F. & D. J. TUMA. 1987. The functional implications of acetaldehyde binding to cell constituents. Ann. N.Y. Acad. Sci. This volume.

DISCUSSION OF THE PAPER

Y. ISRAEL (*University of Toronto, Toronto, Canada*): Do you see lysine binding in these stable adducts?

TUMA: In the case of albumin we observed that 80–90% of the adducts involved lysine residues and the remaining 10% some other amino acid.

K. ISSELBACHER (*Harvard Medical School, Boston, MA*): What do you mean by the term "key lysine residues"?

TUMA: They're key from two points of view. First they show enhanced reactivity. Secondly, they're uniquely important to the assembly of tubulin to form microtubules.

Effects of Ethanol on Endocrine Cells: Testicular Effects

DAVID H. VAN THIEL, JUDITH S. GAVALER, ELAINE ROSENBLUM, AND PATRICIA K. EAGON

University of Pittsburgh School of Medicine
Pittsburgh, Pennsylvania 15261

Ethanol has been shown to produce a wide variety of endocrine effects involving such organs as the hypothalamus and the pituitary, gonadal, adrenal, and thyroid glands. Of these the best studied are those that involve the hypothalamic-pituitary-gonadal axis and particularly the testes. Thus ethanol ingestion by man and experimental animals has been shown to reduce plasma testosterone levels and if continued in a chronic fashion to be associated with the development of overt testicular atrophy. The following is a presentation of the morphologic and biochemical consequences of alcohol ingestion by rats initiated in an effort to understand the mechanisms responsible for alcohol-associated testicular dysfunction observed clinically in chronically alcoholic men.

MATERIALS AND METHODS

Animals

Weanling male rats aged 20 days were obtained from Charles River Breeding Laboratories, Wilmington, MA. They were individually caged after pairing for body size and pair-fed either an alcohol-containing liquid diet, in which ethanol accounted for 36% of the total calories, or an identical diet in which the ethanol was isocalorically replaced with dextrimaltose (Bioserve, Diet 711-A and 711-C, Frenchtown, NJ) for 6 weeks. The animals were sacrificed by exsanguination and autopsied. The testes were removed, decapsulated, weighed, and sectioned into 1-mm cubes and were fixed in 3% glutaraldehyde (Ladd Research Industries, Inc., Burlington, VT) in 0.1 M cacodylate buffer (pH 7.4) for 1 hr at 4°C. After primary aldehyde fixation, the tissue samples were washed for 1 hr in 0.1 M cacodylate buffer and post-fixed in 1% osmium tetroxide in 0.1 M cadodylate buffer for 2 hr at room temperature. After fixation, the tissue was rinsed, dehydrated in graded ethanol series, and embedded in Epon-Araldite. Thin sections with a gold interference were cut with glass or diamond knives on a Sorvall MT 1 microtome (Ivan Sorvall, Inc., Norwalk, CT), mounted on scored 200 to 300 mesh grids, and stained with 3% uranyl acetate and lead citrate. Grids were examined and photographed at 60 kv with an electron microscope (Phillips Electronic Instruments, Mount Vernon, NY), using a double condenser system.

[a]Supported in part by National Institute on Alcohol Abuse and Alcoholism Grants RO–AA04425–05 and RO1–AA06772–01.

Morphometric Analysis

To guarantee that sampling was random, fields were examined in a predetermined sequence in the four corners and center of the grid squares. From each grid, five photomicrographs of Leydig cells at an original magnification of X2,970, X5,300 and X10,800 were taken. Prints were made at an enlargement of 2.65. The high-power photomicrographs were used for determination of volume densities of smooth endoplasmic reticulum and mitochondria. All measurements were made using a Graphic Tablet System attached to an Apple II computer (Apple Computer, Inc., Cupertino, CA 95014). A sheet of transparent acetate was placed upon each photomicrograph to prevent electrostatic interference and the detecting pen was carefully passed upon each of the Leydig cells and their organelles. The data thus obtained was interpreted by the computer and translated into square and cubic microns. Prior to use, the graphics tablet was standardized using appropriate test points (standards).

To demonstrate that the number of measurements available for any given parameter satisfied requirements of sampling adequacy for sterological analysis, the method of Bolender was used.[1] The adequacy of the number of measurements was determined by grouping data from increasing numbers of samples until the standard error of the mean achieved a plateau at a level less than 10% of the mean. For each set of measurements used in the analyses presented here, the number available exceeded the number required to satisfy this criterion.

Hormone Determinations

Plasma testosterone[2] was determined in duplicate. All samples were measured in a single assay. The intra-assay variation for known standards was less than 8%. The detection limit was 0.1 ng. Plasma gonadotropins were measured utilizing specific radioimmunoassays and reagents supplied by the NIAMDD, Rat Pituitary Hormone Distribution Program. NIAMDD Rat FSH-RP-1 and NIAMDD Rat LH-RP-1 were used as reference preparations. The hormones were iodinated with 125I using the chloramine-T method.[3] Iodinated hormones were purified by polyacrylamide gel electrophoresis using 9% acrylamide with 2% crosslinking.[4] Other details of the assays have been described previously.[5] The detection limit for both assays was 4.0 ng with an intra-assay variation of less than 8%. Potency estimates were calculated using the computer program of Rodbard.[6]

Chemicals

NAD+, NADH, 4-methylpyrazole, retinol, dithiothreitol Trisma base, and a variety of antipeptidases including phenylmethylsulfonyl fluoride (PMSF), *L*-tosylamide-2-phenylethylchloromethyl ketone (TPCK), *N*--*p*-tosyl-lysine chloromethyl ketone HCI (TLCK), *N*--benzoyl-L-arginine methyl ester (BAME), leupeptin, pepstatin A, aprotinin, egg white trypsin inhibitor and benzamidine-HCl were purchased from Sigma Chemical Co., St. Louis, MO. Eagle's minimum essential medium with Earle's salt was obtained from Gibco Laboratories, Grand Island, NY.

Purification of Rat Testicular Alcohol Dehydrogenase

The testes were removed within 60 sec of sacrificing the animals and were decapsulated and homogenized in 3 volumes of ice-cold 0.25 mM sucrose containing 50 mM Tris-HCl, pH 7.4, dithiothreitol (1 mM), benzamidine-HCl (25 mM), leupeptin (4 μg/ml), pepstatin A (50-μg/ml), aprotinin (5 μg/ml), TPCK (80 μg/ml), TLCK (20 μg/ml), BAME (80 μg/ml), egg white trypsin inhibitor (100 μg/ml), EDTA (5 mM), EGTA (1 mM), and PMSF (5 mM) with the aid of a polytron homogenizer (Brinkman Instruments, Inc., Westbury, NY. The homogenate was centrifuged at 12,000 × g for 60 min in a Beckman L5–50B ultracentrifuge. This supernatant was designated as cytosol. Unless otherwise stated, all procedures were performed at 4°C.

The cytosol was subfractionated with solid ammonium sulfate and the enzyme was precipitated between 35% and 70% fractional saturation. The precipitate was collected by centrifugation at 38,000 × g for 20 min and was resuspended in the minimum volume of 50 mM Tris-HCl buffer, pH 7.4, containing dithiothreitol, benzamidine-HCl, leupeptin, pepstatin A, aprotinin, TPCK, TLCK, BAME, egg white trypsin inhibitor, EDTA, EGTA, and PMSF, each at the concentration used for the initial homogenate preparation. The resultant solution was dialyzed overnight with two buffer changes and the contents of the dialysis tubing were cleared of insoluble debris by centrifugation at 38,000 × g for 10 min in a Sorvall RC-2B centrifuge. The specific activity of the various preparations obtained ranged from 0.4–1.0 units/mg of protein and compares favorably with those reported for the rat liver enzyme.

Enzyme Assays

Alcohol dehydrogenase activity was determined by the method of Bonnichsen[7] using a Cary spectrophotometer (model 118) at 340 nm. The assay mixture contained the following components (in μmol/1.0 ml): Tris-HCl buffer, pH 8.5, (200); NAD+, (20); ethanol, (10). Enzyme activity was expressed as μmol of NAD+ reduced per min at 22°C. When retinol was used as substrate, the retinol dehydrogenase activity was assayed according to the method of Mezey and Holt[8] in which the retinal generated is determined spectrophotometrically by using its absorption maximum at 510 nm.

Isolation of Testicular Leydig Cells

Leydig cells were separated from seminiferous tubules from rats' testes using a modification of the method of Oeltman *et al.*[9] The decapsulated testes were homogenized in a loose-fitting hand-held homogenizer containing 4 ml of Eagle's minimum essential medium with Earle's salt. Homogenates were filtered through cheese cloth and the Leydig-cell rich filtrate was recovered. The Leydig-cell filtrate was centrifuged at 150 × g for 15 min to separate intact cells from cellular debris. The resultant cell pellet was resuspended in Eagle's medium and was "washed" twice more. This procedure resulted in a final Leydig-cell enriched preparation that contained 10–15% non-Leydig cells, the majority of which were spermatozoa.

Tissue Preparation for Lipid Analysis

The decapsulated testes were homogenized in 6 volumes of 0.25-M sucrose containing 1 mM magnesium chloride and 10 mM Tris-HCl buffer, pH 7.4. Subcellular fractions were separated by differential centrifugation. The 600 × g supernatant was centrifuged at 9,700 × g for 30 min. The resultant mitochondrial pellet was washed in 0.15-M potassium chloride–0.15-M potassium phosphate buffer, pH 7.4, and centrifuged for an additional 30 min. The 9,700 × g supernatent was centrifuged at 95,000 × g for 60 min. The resultant microsomal pellet was washed in a 0.15-M

KCl–0.15-M potassium phosphate buffer for 30 min. The final mitochondrial or microsomal pellet was resuspended in 0.15-M KCl–0.15-M potassium phosphate buffer, pH 7.4. Fifty percent of this preparation was frozen for assay of protein content utilizing the method of Lowry *et al.*[10] and for assay of malonaldehyde according to the method of Bieri and Anderson;[11] the remainder was extracted for subsequent lipid analysis.

Lipid Extraction and Diene Conjugate Determination

Total lipids were extracted with chloroform/methanol (2/1) (v/v) according to the method described by Folch *et al.*[12] with the modification of Bligh and Dyer.[13] Prior to extraction, an antioxidant, 2,6, di-tert-butyl-p-cresol, and an internal standard, heptadecanoic acid (17:0) were added to the mitochondrial pellet. The resultant chloroform layer was rotoevaporated and was treated briefly with ethereal diazomethane to form the methyl derivatives of the contained free fatty acids.

Diene conjugates were measured by ultraviolet spectroscopy using a Cary 118 spectrophotometer according to the method of Hashimoto and Recknagel.[14] The results of diene conjugation assays were expressed as optical density readings at 233 nm adjusted per milligram of extracted lipid.

Fatty Acid Analysis

Separation and quantification of the fatty acid methyl esters were accomplished using a Packard 430 gas chromatograph. The chromatographic module was equipped with a hydrogen flame detector and a glass column (6 ft by 4 mm) packed with 10% Silar 10C on 100/120 Gas-Chrom Q (Applied Science). The assay was carried out with a programmed oven temperature rise of 10°/min between 150°C and 230°C, with an initial hold of 2 min, and a final hold of 30 min. The flash heater and detector temperatures were 270°C and the carrier gas flow of nitrogen was 30 ml/min. The areas under the individual peaks were quantitated by an electronic integrator. The structures of all of the isolated methyl esters were confirmed using gas chromatography/mass spectrometry (GC/MS) on a Hewlett-Packard 5992 mass spectrometer fitted with a 6-ft by 2-mm glass column, and operated under conditions optimized by the Hewlett-Packard program AUTOTUNE.

Silica high-pressure liquid chromatography (HPLC) was used to separate the triglycerides from the other neutral lipids and various polar lipids. Chromatograms were obtained with a Perkin Elmer liquid chromatograph fitted with a 30-cm by 3.9-mm Porasil column. Triglycerides were eluted with hexane/tetrahydrofuran at a flow rate of 1.0 ml/min. After acid hydrolysis and methanolysis of the triglycerides, their fatty acid methyl esters were quantified by gas-liquid chromatography as described above. Trimyristin was used as the internal standard.

Malonaldehyde Assay

One-tenth of the testicular mitochondrial and microsomal extract were peroxidized *in vitro* by incubation at 37°C for 60 min. The incubation system contained 2.0 ml of 0.15-M potassium phosphate–0.15-M potassium chloride buffer, pH 7.4, 1.0 ml of the mitochondrial extract prepared above, 0.5 ml of 0.25-nM ferrous sulfate, and 0.5 ml of 1.25-nM sodium ascorbate. Protein concentrations ranged from 0.2 to 0.6 mg/ml. The reaction was stopped by the addition of 1.0 ml of cold 25% trichloroacetic acid and centrifuged at 1,000 × g to remove the precipitated protein. A 2.0-ml aliquot of the resultant deproteinized extract was then reacted with 2.0 ml of 0.67%-thiobarbituric acid and 0.5 ml of 1.0-N HCl. The color was developed by heating at 100°C for 10 min. The resultant absorbance of the assay mixture was read at 535 nm after cooling. Absorbance was directly proportional to the malonaldehyde standard that was produced by the acid hydrolysis of 1,1,3,3 tetraethoxypropane. Absorbance was linear over the concentration range of 0.8–4.0-μg tetraethoxypropane.

Reduced Glutathione (GSH) Assay

Determination of rat testicular GSH content was accomplished by homogenizing 300–500 mg of testicular tissue in 3.5 ml of 0.1-M sodium phosphate–0.005-M EDTA buffer, pH 8.0. Subcellular fractions were removed by differential centrifugation at 17,000 × g for 20 min and at 100,000 × g for 60 min. The resultant supernatant was deproteinized by the addition of 0.5 ml of 25%-m phosphoric acid (HPO) (w/v) followed by centrifugation at 3,000 × g for 10 min. A tenfold dilution of the resultant protein-free supernatant was made to yield a final solution containing 0.1 to 10 μg GSH/ml. The GSH content of the solution was determined fluorometrically according to a modification of the method described by Hissin and Hilf.[15] The assay mixture contained 1.8 ml phosphate-EDTA buffer, 100 μl of opthalaldehyde (1 μg/μl) and 100 μl of the deproteinized supernatant described above. The assay mixture was incubated at room temperature for 20 min prior to analysis. Fluorescence at 420 nm was determined following activation at 350 nm. Fluorescence was directly proportional to the GSH concentration and was linear over the concentration range of 0.1 to 1.0 μg. The protein content of each preparation was assayed by the method of Bradford.[16]

Statistical Analysis

Data were analyzed using the paired t test or linear regression. All results are expressed as mean values ±S.E.M. A p value of <0.05 was considered to be significant.

RESULTS

Gross Anatomy

The testes of the alcohol-fed animals weighed 1.47 ± 0.32 g (mean ± S.E.M.), which was 50% less than that of the isocaloric controls (2.98 ± 0.17 g) ($p < 0.01$).

Hormone Levels

Plasma testosterone concentrations were reduced in the alcohol-fed animals (2.4 ± 0.2 ng/ml) compared to those in the isocaloric control animals (4.7 ± 0.1) ($p < 0.01$). Plasma FSH levels in the alcohol-fed animals (761 ± 133 ng/ml) were twice those of the isocaloric controls (378 ± 57 ng/ml) ($p < 0.05$). In contrast, and despite the marked reduction in testosterone levels present in the alcohol-fed animals,

LH levels did not differ between the two groups (alcohol-fed 50.8 ± ng/ml vs. 59.2 ± 4.1 ng/ml for the isocaloric controls).

Light Microscopy

As previously reported, the testes of the alcohol-fed animals contained smaller seminiferous tubules with a reduced number of germ cells, compared to those of the isocaloric controls.

Electron Microscopy

Ultrastructural examination of rat gonadal interstitial tissue disclosed the presence of loose connective tissue with a population of polygonal, fusiform, and clear cells, and a rich network of blood vessels. The large, polygonal cells (Leydig cells) were found to be irregularly grouped around capillaries. They had a single, large, eccentric nucleus with a dense rim of chromatin attached to the inner membrane of the nuclear envelope. They had a prominent nucleolus which was composed of a distinct pars amopha and pars reticularis. As noted above, the Leydig cells were intimately associated with each other with contiguous cells being separated by a regular space. The Leydig cells present in the testes of the alcohol-fed rats exhibited numerous microvilli varying in length and shape and protruding into this intercellular space or interdigitating with microvilli of neighboring interstitial cells. This was not so in testes obtained from the isocalorically fed control rats. Mitochondria varying in size and shape were present in moderate numbers within the Leydig cells. Their inner structure consisted of numerous tubular cristae of uniform size. The mitochondrial matrix was moderately dense. Occasionally small, dense granules were encountered within the matrix. Cup-shaped and elongated mitochondria were most conspicuous in the Leydig cells of alcohol-fed rats. The membranes of the Golgi complex appeared as flattened cisternae in a juxtanulcear location. A granular endoplasmic reticulum was present in all of the interstitial cells examined.

Morphometric Analysis

The Leydig cell volume of the alcohol-fed animals was less (2071.0 ± 117.0 μ^3) than that of the isocaloric controls (2382.5 ± 148.5) ($p < 0.05$). In contrast, no difference in Leydig cell nuclear volume was seen between the two groups (alcohol-fed 786.5 ± 57.5 μ^3 vs. 846.5 ± 65.0 for the isocaloric controls). As a result, the cytoplasmic volume of Leydig cells of alcohol-fed animals was less (1280.5 ± 71.5 μ^3) than that of the isocaloric controls (1535.5 ± 99.5; $p < 0.05$). The volume of the smooth endoplasmic reticulum (SER) was reduced in the Leydig cells of the alcohol-fed animals (422.0 ± 27.5 μ^3) as compared to that of the isocaloric controls (497.5 ± 37.5; $p < 0.05$); but, when the volume of SER was corrected for the reduced volume of the cytoplasm, no difference existed between the two groups. In contrast, the mitochondria of the Leydig cells of alcohol-fed animals were larger (225.0 ± 15.0 μ^3) than that of the isocaloric controls (161.0 ± 13.5; $p < 0.005$). Moreover, when corrected for the reduced cytoplasmic volume, this increase in mitochondrial volume was even more apparent (alcohol-fed 17.83 ± 0.77% cytoplasmic volume vs. isocaloric controls 10.28 ± 0.55) ($p < 0.005$).

TABLE 1. Purification of Rat Testicular Alcohol Dehydrogenase

Fraction	Volume (ml)	Protein (mg)	Total Activity (units)	Specific Activity (units/mg)	Purification (-fold)
Cytosol	240	3,980	6.0	0.0015	1
Ammomium Sulfate Fractionation	25	910	4.6	0.005	3

Isolation of Alcohol Dehydrogenase Activity from the Testes

A. Yield and Stability. By using ammonium sulfate precipitation, the testicular enzyme was prepared over a period of 48 hr with an activity yield of 76% (TABLE 1). Following its relative concentration, the enzyme was found to be relatively unstable, losing 50% of its activity over a 2-week period, despite its being maintained at 4°C. In contrast, by using the same procedure, hepatic alcohol dehydrogenase can be purified with an activity yield of 97%. The better yield obtained with the hepatic enzyme probably reflects the fact that the content of enzyme present in testes is only about 1% of that present in the liver.

B. pH Optimum. The effect of varying the pH and the assay buffer used on the activity of rat testicular ADH was determined using ethanol as the substrate (FIG. 1). As can be seen, testicular ADH was found to have two pH optima: one at pH 7–7.5 and a second at pH 10–10.5. In contrast, hepatic ADH demonstrated a maximum activity at pH 11.0 in glycine-NaOH buffer and a broad pH range of enzymatic activity with a maximum at 8.5 when using the Tris-HCl buffer.

C. Substrate Specificity. Rat testicular ADH has a broad substrate specificity including various primary and secondary, but not tertiary, saturated aliphatic and aromatic alcohols (TABLE 2). 2,3-Unsaturated alcohols and polyenic alcohols, such as farnesol or retinol, are also good substrates for the enzyme. Moreover, rat testicular ADH was also found to catalyze the dehydrogenation of some diols.

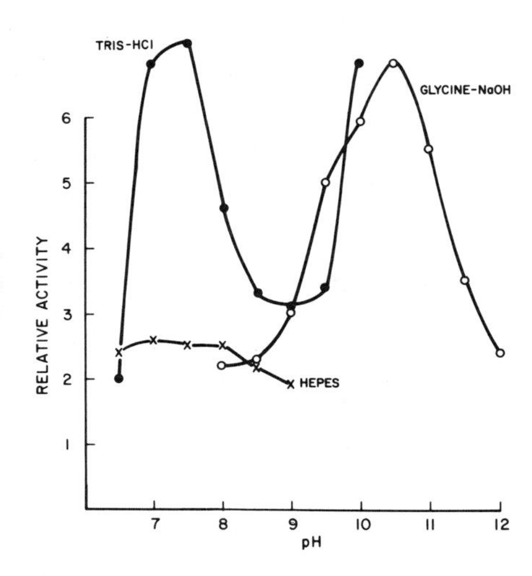

FIGURE 1. Activity curves of rat testicular ADH at different pH values and using three different buffers—C Tris-HCl, glycine-NaOH, and HEPES.

TABLE 2. Substrate Specificity of Rat Testicular Alcohol Dehydrogenase[a]

Substrate	Concentration mM	Relative Activity (%)	Substrate	Concentration mM	Relative Activity (%)
Methanol	1240	0	1.2-Ethanediol	895	12
Ethanol	860	100	1.2-Propanediol	680	55
n-Propanol	670	85	1.3-Butanediol	560	90
n-Butanol	550	112	1.4-Butanediol	560	155
n-Pentanol	230	177	1.5-Pentanediol	475	137
n-Hexanol	60	70	1.6-Hexanediol	1	90
n-Heptanol	7	81	2-Butene-1.4-diol	600	65
n-Octanol	6	109	2-Butene-1-01	585	75
2-Propanol	660	80	2-Hexene-1-01	425	149
2-Butanol	550	110	Benzyl alcohol	485	25
tert-Butanol	530	0	Cinnamyl alcohol	1.6	203
2-Pentanol	460	133	Farnesol	200	102
			Retinol	4	83

[a]The rates of oxidation of alcohols were compared with that of ethanol. Saturating concentration of the substrates were used (between 0.4–1.2 M). Some of the alcohols could not be used at 0.4–1.2-M concentration because of their limited solubility in the medium, and their final concentration is shown in the table. Other substances tested with no detectable activity included glucose, maltose, sucrose, pyruvate, lactate, citrate, and polyethylene glycol (results not shown).

D. Kinetic Studies. The kinetic properties for several different substrates for the testicular ADH were determined. Substrate concentrations were varied over a 10–100-fold range. The reaction rates were found to follow a linear relation when analyzed according to the method of Lineweaver and Burk,[17] suggesting that Michaelis constant (Km) for ethanol varied by one order of magnitude between the testicular and hepatic ADH's (TABLE 3). In contrast, the Km's for the other alcohols studied showed considerable differences between the two sources of enzyme. Both enzymes utilize NAD+ but not NADP+ and the Km for NAD+ using the testicular enzyme was 0.16 nM.

E. Inhibition by 4-Methylpyrazole. 4-Methylpyrazole, an inhibitor of the hepatic ADH, also inhibits the testicular enzyme. The inhibition was competitive in nature, as is evident by the double reciprocal plot shown in FIGURE 2. When the data were analyzed further according to Dixon[18] competitive inhibition was evident and the

TABLE 3. Comparison of Michaelis-Menten Constants (Km) of Rat Hepatic and Testicular Alcohol Dehydrogenase with Ethanol and Other Alcohol as Substrates[a]

Substrate	Km (mM)	
	Hepatic ADH	Testicular ADH
Ethanol	1.01	0.50
n-Propanol	0.47	0.088
n-Butanol	0.37	0.031
Isobutanol	0.64	0.059
Retinol	0.028	0.003

[a]Purified hepatic and partially purified testicular ADH were used as the source of enzymes. Kinetic constants were derived from the linear Lineweaver-Burk plots.

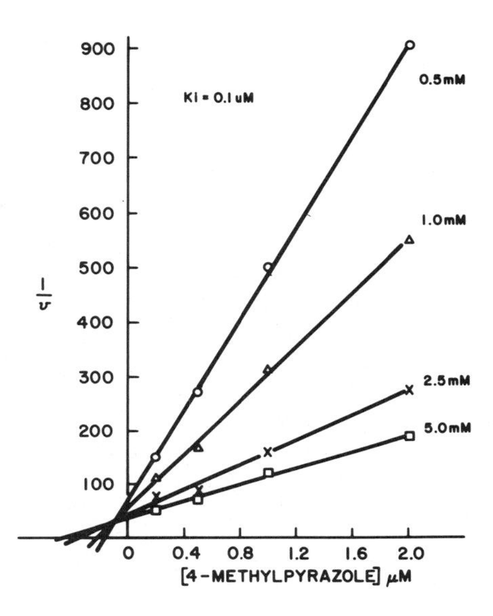

FIGURE 2. Competitive inhibition of rat-testicular ADH activity using ethanol as the substrate by 4-methylpyrazole analyzed according to the method of Dixon.[18]

inhibition constant (Ki) for 4-methylpyrazole using the testicular enzyme was 0.1 μM.

F. Inhibition of Retinol Oxidation by Testicular ADH. Retinol was found to have a very low Km (3 μM) for testicular ADH and therefore may serve as a substrate for the enzyme under physiological conditions. Ethanol inhibited the activity of testicular ADH for retinol oxidation and the inhibition was shown to be noncompetitive in nature, with a Ki of 1.9 mM (FIGURE 3).

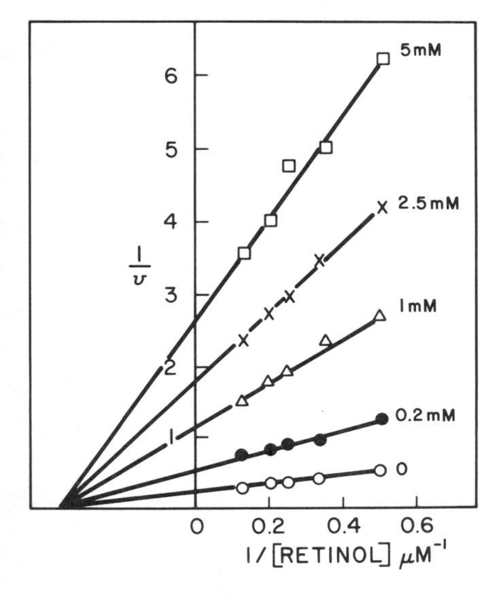

FIGURE 3. Inhibition of rat testicular ADH activity by ethanol using retinol as the substrate.

TABLE 4. Heat Stability of Rat Hepatic and Testicular Alcohol Dehydrogenases[a]

Temperature Treatment (°C)	$t_{1/2}$ (min)			
	Hepatic ADH		Testicular ADH	
	pH 8.5	pH 11.0	pH 8.5	pH 11.0
50°	114	125	49.0	55.0
60°	21	16	2.8	3.3

[a]Partially purified enzymes were prepared as described. Enzyme fractions (in 50 mM Tris-HCl, pH 7.4 buffer) were heated at either 50° C or 60° C for various times and the remaining activities were measured at both pH 8.5 and pH 11.0 as described in the Methods section. All studies were performed at a protein concentration of 10 mg/ml following the addition of BSA.

G. Heat Inactivation. The partially purified testicular ADH prepared as described above was shown to be relatively heat-labile at 50°C and to have a $t_{1/2}$ of about 3 min at 60°C (TABLE 4).

Testicular Diene Conjugate Content

Testicular mitochondrial lipids peroxidized *in vivo* exhibited an ultraviolet absorption spectra characteristic of conjugated dienes as can be seen in FIGURE 4. Alcohol-fed animals (0.455 ± 0.053 OD units) demonstrated a significant increase in their content of diene conjugates compared to that observed for the isocalorically fed control animals (0.382 ± 0.045; $p < 0.05$).

Testicular Mitochondrial-Free Fatty Acid Content

Further evidence of testicular peroxidation injury occurring as a consequence of alcohol administration was seen in the fatty acid composition of the testes. The increased production of diene conjugates observed in the alcohol-fed rats was accompanied by a decrease in the peroxidizable polyunsaturated fatty acids of testicular

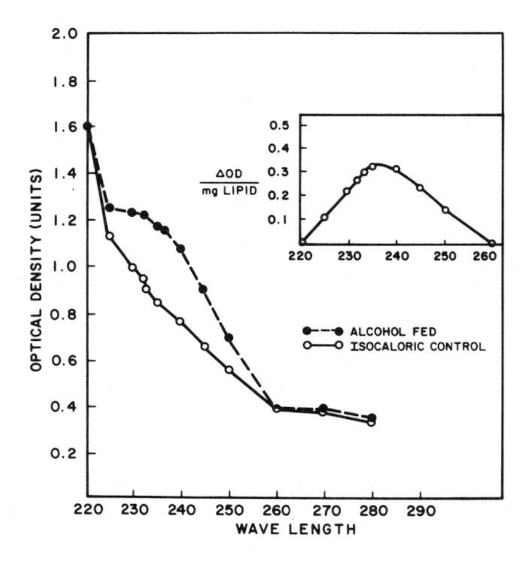

FIGURE 4. Rat testicular mitochondrial lipids obtained from alcohol fed and isocaloric controls and peroxidized *in vivo* (estimated by diene conjugation measured at 233 nm).

mitochondria. Similarly, the qualitatively different composition of the free fatty acids of testicular mitochondria is shown in FIGURE 5. Specifically, significant decreases in the major polyenes, arachidonic acid (20:4) (p 0.005) and docosopentanoic acid (22:5) ($p < 0.05$), were observed in the alcohol-treated animals as compared to isocaloric control-fed animals. In addition, a compensatory increase in palmitic (16:0), stearic (18:0), and oleic (18:1) acids was observed in the alcohol-fed animals but not in the isocalorically fed controls ($p < 0.05$). The other changes that were noted to be different between the two groups of animals were a decrease in the linoleic acid (18:2) and docosobutanoic acid (22:4) levels.

Testicular Mitochondrial Triglyceride Content

Additional evidence that alcohol interferes with testicular lipid metabolism was provided by an examination of the mitochondrial triglyceride content of the testes

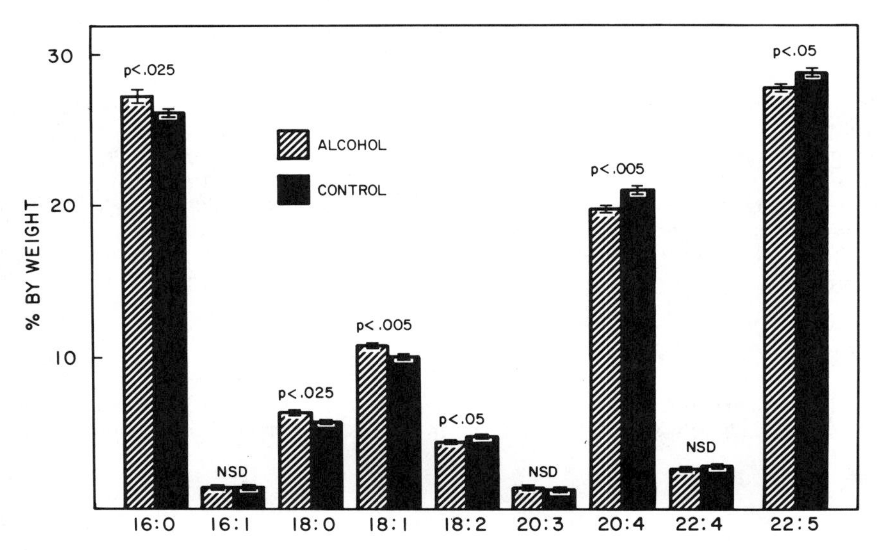

FIGURE 5. Rat testicular mitochondrial free fatty acid distribution in alcohol-fed and isocaloric control-fed animals. *Bars* represent mean values and the *brackets* S.E.M.

obtained from the two groups of animals studied (FIGURE 6). Not only did triglycerides accumulate in the testicular mitochondria of alcohol-fed animals, but changes occurred also in the triglyceride fatty acid composition. Specifically, as was the case with the free fatty acids, the docosopentanoic acid (22:5) levels in testicular mitochondrial triglycerides were reduced significantly ($p < 0.05$) in the alcohol-fed animals as compared to the isocalorically fed controls.

Testicular Malonaldehyde Formation

The effect of chronic alcohol administration on testicular mitochondrial lipids, as determined by the production of malonaldehyde *in vitro,* is shown in FIGURE 7.

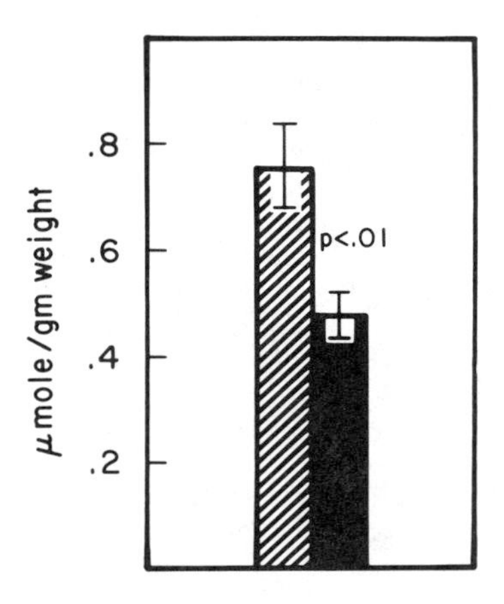

FIGURE 6. Rat testicular mitochondrial triglyceride content present in the testes of alcohol-fed and isocaloric control animals. The *cross-hatched bar* represents data obtained from alcohol-fed animals and the *solid bar* the data obtained from controls. The *brackets* represent the S.E.M.

Ingestion of alcohol markedly ($p < 0.05$) enhanced the ability of testicular mitochondria to produce malonaldehyde. Thus the ethanol-treated animals had approximately 1.2 times the rate of formation of malonaldehyde as compared to the isocalorically fed controls. Further, a negative correlation was found between the rate of malonaldehyde formation and the degree of testicular atrophy present in the alcohol-fed animals ($r = 0.923$; $p < 0.05$; $n = 10$ pairs). In order to determine whether peroxidation in mitochondria was representative of other testicular organelles, malonaldehyde formation was measured in microsomes. As with mitochondria, alcohol-fed animals demonstrated a significant ($p < 0.05$) increase in microsomal malonaldehyde formation as compared to the controls.

Testicular Reduced Glutathione Levels

The changes in testicular reduced glutathione levels that occurred as a result of chronic alcohol administration are shown in FIGURE 8. Animals fed alcohol demon-

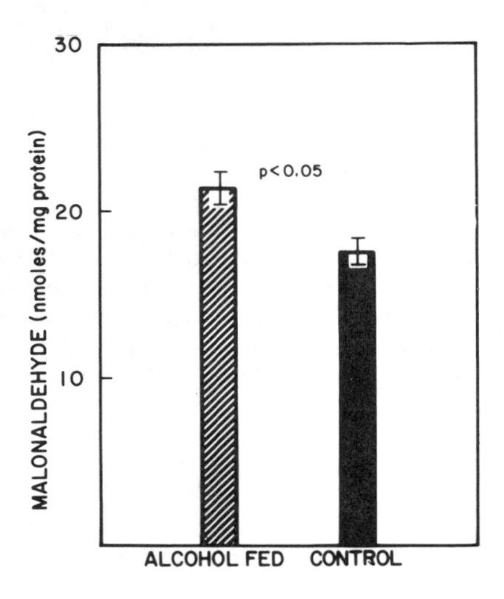

FIGURE 7. Malonaldehyde formation by rat testes mitochondria peroxidized *in vitro.* The *cross-hatched bar* represents the mean of the data obtained from alcohol-fed animals, while the *solid bar* represents the mean of the data obtained from isocaloric control-fed animals. The *brackets* represent the S.E.M.

strated a significant ($p < 0.05$) decrease in reduced glutathione levels as compared to the control-fed animals. Further, there was a direct positive correlation between the observed reduced glutathione levels and the degree of testicular atrophy present in the alcohol-fed animals ($r = 0.850$; $p < 0.05$; $n = 6$ pairs).

DISCUSSION

The present report confirms earlier results that have described testicular atrophy and reduced testosterone levels in alcohol-fed rats compared to isocaloric controls with both groups having similar LH concentrations, despite the apparent Leydig cell injury present in the alcohol-fed animals. However, FSH levels were increased ($p < 0.05$) in the alcohol-fed animals as compared to the isocaloric controls.

Compared to isocaloric controls, the Leydig cells of alcohol-fed animals were noted to be smaller ($p < 0.01$) and to have less cytoplasm ($p < 0.01$), a reduced absolute volume of SER (which was appropriate for the reduced cytoplasmic volume), and an enlarged mitochondrial volume ($p < 0.005$, which was even more apparent when corrected for the reduced cytoplasmic volume). These morphometric changes in Leydig cells induced by alcohol feeding are similar to those found in hepatocytes of animals fed alcohol and in human alcohol abusers.[19–25] Thus, the morphometric alterations produced in one organ, the liver, which is a widely accepted target for ethanol toxicity, are reproduced in yet another target of ethanol's toxicity, the Leydig cells.

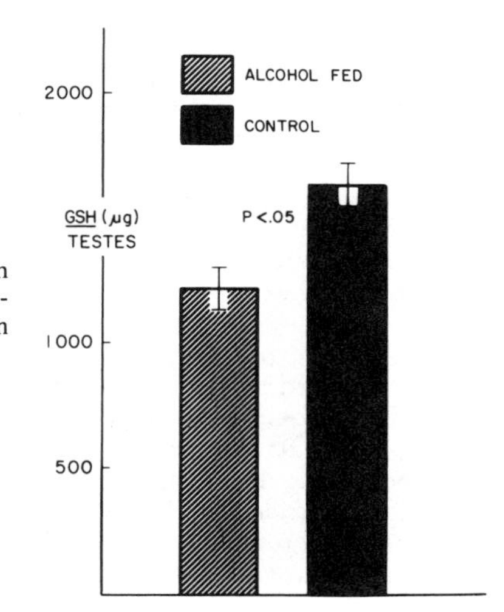

FIGURE 8. GSH content of rat testes obtained from alcohol-fed (*cross-hatched bar*) and isocaloric control animals (*solid bar*). The *bars* represent mean values and the brackets the S.E.M.

These data are consistent also with the biochemical data that suggest that the redox state of Leydig cells of alcohol-fed animals is reduced as is known to occur in the liver, and that microsomal enzyme activities of the testes of alcohol-fed animals are adversely affected as a result of ethanol feeding.[19–24] Thus, they provide circumstantial evidence in support of a common mechanism for ethanol-induced toxicity within the liver (as reported by others) and the testes (herein reported). These studies demonstrate also that an alcohol dehydrogenase is present in small yet readily detectable and identifiable quantities in rat testis. Moreover, this rat testicular enzyme has substrate specificities very similar to those obtained for liver ADH. Both enzymes have a Km for ethanol in the 0.5–1.0-mM range, use NAD+ but not NADP+ as a cofactor, and can use other aliphatic alcohols (*n*-butanol, isobutanol, etc.) as substrates at "physiologic" concentrations.

The most interesting putative substrate examined was retinol. The testicular ADH was found to have a very low Km for retinol (3 μM). In addition, ethanol was found to be a noncompetitive inhibitior of retinol oxidation by the enzyme. The calculated inhibition constant (Ki) was 1.9 mM.

Because retinal formed from retinol has been shown to be essential for spermatogenesis,[25] the observed inhibition of testicular retinol dehydrogenase activity by ethanol suggests that at blood ethanol levels as low as 8 mg/dl (well below the level of legal intoxication, *i.e.,* 100 mg/dl) retinal generation by the testes might be inhibited by as much as 50%. Such inhibition by ethanol of retinal generation within the testes might explain, at least in part, the phenomena of reduced fertility and testicular atrophy that are observed frequently in chronically alcoholic men and various animal models of alcoholism. It also allows for the possibility of free radical formation and peroxidation injury occurring as a consequence of testicular oxidation of ethanol and acetaldehyde. In this regard, the results herein presented strongly suggest that testicular lipid peroxidation is in fact a metabolic consequence of chronic alcohol administration to animals. Taken together, these data suggest that alcohol feeding modifies testicular metabolism and that mitochondrial membrane damage occurs as a result of lipid peroxidation manifested by reduced glutathione levels, an abnormal fatty acid distribution, malonaldehyde formation, and conjugated diene absorption.

The specific mechanisms by which lipoperoxidation produces its effect on testicular mitochondrial membranes in response to ethanol administration has not been determined by the present work. The present data only suggest, albeit quite strongly, that it occurs.

One might expect that depletion of GSH as it occurs with alcohol exposure could lead to enhanced peroxidative tissue damage in both the liver and testes. Consistent with such a sequence of events, the testicular content of GSH was observed to be reduced ($p < 0.05$) in alcohol-fed animals as compared to isocaloric controls. Moreover, there was a direct correlation between the gross testicular atrophy observed and the reduced testicular GSH levels present in the alcohol-fed animals. Whether the degeneration of testicular steroidogenesis and reduced spermatogenesis characteristic of alcohol-fed rats is directly related to the enhanced lipid peroxidation, however, remains to be proved. Nonetheless, the present study provides support for the concept that intracellular lipid peroxidation within the testes is a metabolic consequence of chronic alcohol ingestion.

REFERENCES

1. Bolender, R. P. 1978. Correlation of morphometry and sterology with biochemical analysis of cell fractions. Int. Rev. Cytol. **55:** 247–252.
2. Neischlag, E. & D. L. Loriaux. 1972. Radioimmunoassay for plasma testosterone. Z. Klin. Chem. u Klin. Biochem. **10:** 164–168.
3. Greenwood, R. C., W. M. Hunter & J. S. Glover. 1963. The preparation of 131 I-labelled human growth hormone of high specific activity. Biochem. J. **89:** 114–118.
4. Rodbard, D. & A. Chrambach. 1971. Estimation of molecular radius free motility and valence using polyacrylamide gel electrophoresis. Ann. Biochem. **40:** 95–105.
5. Sherins, R. J., J. Vaitukaitus & A. Chrambach. 1972. Physical characteristics of hFSH and its desialization products by isoelectric focusing and electrophoresis in polyactylamide gel. Endocrinology **92:** 1135–1140.

6. RODBARD, D. & J. E. LEWARD. 1970. Computer analysis of radio ligand assay and radioimmunoassay data. *In* Proceedings of the Second Symposium on Steroid Assay by Protein Binding. 75–103. Karolinska Institutet. Stockholm.
7. BONNICHSEN, R. K. & A. M. WASSEN. 1948. Crystalline alcohol dehydrogenase from horse liver. Arch. Biochem. Biophys. **18:** 361–363.
8. MEZEY, E. & P. R. HOLT. 1971. The inhibiting effect of ethanol on retinol oxidation by human liver and cattle retina. Exp. Mol. Pathol. **15:** 148–156.
9. OELTMAN, T. M. T. & E. C. HEATH. 1977. Binding of native glucosamine-labeled, luteinizing hormone to Leydig cells. Arch. Biochem. Biophys. **179:** 608–615.
10. LOWRY, O. H. *et al.* 1951. Protein measurement with the folin phenol reagent. J. Biol. Chem. **193:** 265–275.
11. BIERI, J. G. & A. A. ANDERSON. 1960. Peroxidation of lipids in tissue homogenates as related to vitamin E. Arch. Biochem. **90:** 105–110.
12. FOLCH, J., M. LEES & G. H. S. STANLEY. 1957. A simple method for the isolation and purification of total lipids from animal tissue. J. Biol. Chem. **226:** 497–506.
13. BLIGH, E. G. & W. J. DYER. 1959. A rapid method for total lipid extraction and purification. Can. J. Biochem. Physiol. **37:** 911–920.
14. HASHIMOTO, S & R. O. RECKNAGEL. 1968. No chemical evidence of hepatic lipid peroxidation in active ethanol toxicity. Exp. Mol. Pathol. **8:** 225–231.
15. HISSIN, R. & R. HIFF. 1976. A fluorometric method for determination of oxidized and reduced glutathione in tissues. Anal. Biochem. **74:** 214–220.
16. BRADFORD, M. 1976. A rapid and sensitive method for the quantitation of microgram quantities of protein utilizing the principle of protein-dye binding. Anal. Biochem. **72:** 248–255.
17. LINEWEAVER, H. & D. BURK. 1934. The determination of enzyme dissociation constants. J. Am. Chem. Soc. **56:** 658–666.
18. DIXON, M. 1953. The determination of enzyme inhibitor constants. Biochem. J. **55:** 170–171.
19. KIESSLING, K. H., L. PILSTROM & B. STRANDBERG. 1965. Ethanol and the human liver; correlation between mitochondrial size and degree of ethanol abuse. Acta Med. Scand. **178:** 533–535.
20. RUBIN, E. & C. S. LIEBER. 1968. Alcohol induced hepatic injury in nonalcoholic volunteers. N. Engl. J. Med. **278:** 869–876.
21. ISERI, O. A. & L. S. GOTTLIEB. 1971. Alcoholic hyalin and megamitochondria as separate and distinct entities in liver disease associated with alcoholism. Gastroenterology **60:**1027–1035.
22. RUBIN, E. 1973. The spectrum of alcoholic liver injury. Int. Acad. Pathol Monogr. **13:** 199–217.
23. HORVATH, E., K. KOVACS & R. C. ROSS. 1975. Alcoholic liver lesion. Frequency and diagnostic value of fine structural alterations in hepatocytes. Beitr. Pathol. **148:** 67–85.
24. BRUGUERA, M., A. BERTRAN, J. A. BOMBI & J. RODEC. 1974. Giant mitochondria in hepatocytes. Gastroenterology **73:** 1383–1387.
25. VAN THIEL, D. H., J. S. GAVALER & R. LESTER. 1974. Ethanol inhibition of vitamin A metabolism in the testes: possible mechanism for sterility in alcoholics. Science **186:** 941–942.

DISCUSSION OF THE PAPER

W. S. THAYER (*Hahnemann University School of Medicine, Philadelphia, PA*): Are you showing the susceptibility of membranes to lipid peroxidation rather than lipid peroxidation that has already occurred?

VAN THIEL: Yes. All we can say is that there is a potential for lipid peroxidation.

T. M. DONOHUE (*University of Nebraska College of Medicine, Omaha, NE*): Why is there an increase in estrogen levels in these chronic alcoholics?

VAN THIEL: It is not a consequence of alcohol-induced injury of endocrine cells. If you can't make androgens you can't possibly make estrogens, because androgens are the precursors of estrogens.

We measure androgens in nanomolar amounts, and estrogens in picomolar amounts, so a bit of androgen is a lot of estrogen. And you can demonstrate that. True androgen, that is testosterone and dihydrotestosterone, levels are suppressed as a consequence of alcohol-induced injury but proandrogens, namely androstenedione and dehydroepiandrosterone, probably of adrenal origin, are in fact increased. So you have an increased proandrogen concentration or weak androgen. In the periphery, primarily in fat stores and in skin, alcohol induces the enzyme aromatase, and an initial increase in FSH, which also induces aromatase activity. The result is an increased conversion of androgen to estrogens.

If you shift the conversion ratio from 3% to just 5%, you actually increase three orders of magnitude, as a result of the difference between nanograms and picograms. Normal estrogen levels can thus be maintained even in the face of androgen deficiency. Portal hypertension and interruption of enterohepatic circulation essentially eliminate the liver as a source of first pass removal of much of these proandrogens, which are distributed to the fat and skin. There they are converted to estrogens. In fact the slightly increased estrone levels and normal estradiol levels probably only represent the estrogen that refluxes out of the cells where it is converted from androgen to estrogen.

M. A. COLLINS (*Loyola University School of Medicine, Maywood, IL*): Is ADH in the testes similar to that in the brain?

VAN THIEL: Yes, it immunologically cross-reacts.

Y. ISRAEL (*University of Toronto, Toronto, Canada*): Unlike man, the rat has one major ADH in the liver. Have you compared your protein with that in the rat liver?

VAN THIEL: It's a very different protein.

K. ISSELBACHER (*Harvard Medical School, Boston, MA*): Does altering glutathione levels modify the testosterone effect?

VAN THIEL: It's theoretically possible that increasing the glutathione content of the testes would prevent peroxidation injury.

Hypermetabolic State, Hepatocyte Expansion, and Liver Blood Flow: An Interaction Triad in Alcoholic Liver Injury[a]

YEDY ISRAEL AND HECTOR ORREGO

Departments of Pharmacology and Medicine
University of Toronto
and
Addiction Research Foundation
Toronto, Canada M5S 2S1

In the present article we wish to pathogenically link some specific abnormalities that appear to determine the severity of alcoholic disease as expressed by their relationships to the risk of mortality. The abnormalities that are associated with a poor prognosis can be divided into two main groups, namely, (i) a reduction in the functional mass of the liver and (ii) the presence of portal hypertension and its consequences (ascites, encephalopathy, and renal abnormalities).

In relation to a reduced functional hepatic mass we will consider three elements, namely, (a) necrosis, (b) interposition of exchange barriers, and (c) the production of intra- and extrasystemic shunts. In relation to the pathogenesis of portal hypertension we will consider the possible role of (a) an increased hepatic resistance to portal blood flow and (b) an increased blood flow. While for clarity of presentation these are initially analyzed separately, we wish to postulate that liver disease indeed results from the interaction of these variables. We view fibrosis and cirrhosis as *one* of the elements interacting to produce hepatocellular dysfunction and a reduction in the functional hepatic mass. Excluding the *purely* hemodynamic consequences of portal hypertension, all other factors integrate to compromise hepatocyte function and eventually cell survival.

It is well accepted that alcoholics self-titrate themselves to high blood ethanol levels for many years before clinical liver disease ensues.[101,178,182] We have proposed that liver disease and, more specifically, the presence of necrosis occurs when a primary metabolic factor determined by alcohol interacts with a precipitating factor that occurs at random, and that is largely independent of alcohol—in essence a stochastic process.[66,67,71,137,141] Here we expand this view with the concept that the threshold required for the precipitating factor may in fact be altered by the exposure to ethanol. For example, it has been shown that an increase in liver oxygen consumption induced by ethanol results in cell necrosis when the liver is artificially exposed to conditions of reduced oxygen availability.[45,66,67,71,73,84,147,191] The threshold for the latter process may be, in fact, reduced by alcohol-related conditions, which may interfere with the delivery of oxygen to the liver or with its cellular exchange. Animal models have been classically considered to reproduce poorly the human disease picture as, for example, in relation to necrosis. This may be so precisely because the rigorous experimental design has excluded the precipitating factor.

[a]Supported by National Institute on Alcohol Abuse and Alcoholism Grant AA-06573 and Addiction Research Foundation Grant 5600.

REDUCTION IN FUNCTIONAL HEPATIC MASS

Necrosis

Necrosis is a hallmark of alcoholic hepatitis.[21,48,54,112,160] Recent studies indicate that the presence of necrosis is of major prognostic significance in alcoholic liver disease.[137,140] Individuals with alcoholic cirrhosis but without liver cell necrosis have a mortality that is not different from that of patients with only fatty liver.[137] Necrosis in alcoholic hepatitis occurs characteristically in the periphery of the hepatic acinus (Zone 3 or centrolobular area).[21,48,54,112,160,177] Any hypothesis on the mechanism of necrosis should account for this anatomical localization. Four general causes of necrosis have been postulated to lead to hepatocellular necrosis in alcoholic liver injury, namely hypoxic necrosis, immunological mechanisms, acetaldehyde-induced damage, and lipoperoxidation of essential membrane components.

Hypoxic necrosis is more likely a result in Zone 3 of the liver acinus (centrolobular area). Zone 3 is known to be hypoxic with respect to the periportal areas (Zone 1).[5,77,78,81,89,117,131,153,158] Therefore, Zone 3 is likely to be more susceptible to conditions where either the oxygen delivery to the liver is decreased or oxygen consumption increases without a concomitant compensation in supply.[102,103,175] On the other hand, until this time there has not been an *a priori* theoretical reason to expect a Zone-3 location of immunologically induced liver damage. The third factor, acetaldehyde, although highly reactive with macromolecules,[19,24,107] has not been actually shown to lead to cell necrosis at the concentration that can occur in the liver or in the circulation.[108,135] Acetaldehyde could, nevertheless, account for the centrolobular localization of cell necrosis, since acetaldehyde induced in the metabolism of ethanol along the sinusoid[199,200] exists in a higher concentration in blood leaving Zone 3 than in Zone 1.[107]

It can be argued, however, that should the immunological damage be associated with acetaldehyde, this combination could indeed be postulated as a cause of Zone-3 necrosis. In fact, as we will discuss, recent evidence suggests that acetaldehyde binds to proteins that can become neoantigens leading to the production of immunoglobulins,[65] which could result in cellular cytotoxicity. In addition, an acetaldehyde-stimulated, cell-mediated cytotoxicity has been postulated to play a role in cellular cytotoxicity.[9,65,82,188] Lipoperoxidation, which has the potential of producing cell necrosis,[159,183] has not been convincingly demonstrated to play a role in alcoholic liver disease, nor has it been shown to produce effects circumscribed to Zone 3 of the acinus.[191] We shall now expand on these possible mechanisms, which will be interrelated at the end.

HYPOXIC HYPOTHESIS

As indicated above, hypoxic conditions are well known to result in cell necrosis in a variety of tissues[92,167] including the liver.[102,103,136,175] A condition of cellular hypoxia should be seen as one in which the oxygen supply does not meet the demand of the tissue; thus, we should consider both supply and demand in relation to the production of hypoxic damage. The importance of the demand factor was demonstrated early for liver tissue by the studies of McIver and Winter[195] who showed that administration of thyroid hormones, which increase hepatic oxygen consumption, makes the liver more susceptible to conditions that reduce oxygen availability. It has been demonstrated that chronic consumption of alcohol leads to both an increased rate of oxygen consumption and an increased rate of alcohol metabolism in isolated organs and tissues, a condition that was termed a hypermetabolic state.[66,67,71,74,78,79,137,196,202,203] Such an effect has also been observed in humans[76,90] but not in baboons.[77] An increase in hepatic oxygen consumption has also been observed in animals, including baboons, following the acute administration of ethanol.[18,216,217,218,173,174,77,205,180] Although the mechanism for such an increased oxygen demand by the tissue has not been fully elucidated, it has been proposed that an increased oxygen demand by the mitochondria follows an increased availability of ADP to such organelles.[13,196,202,218] It is known that the rate of the mitochondrial respiratory chain and, thus, oxygen consumption is regulated by the availability of ADP and the ATP/ADP $\times$ Pi ratio.[26] The hypermetabolic state could be blocked by atractyloside, a compound that inhibits ADP translocation into the mitochondria.[196] Further, a greater production of ADP may be mediated by an increased activity of the sodium pump and $(Na^{+}K)$-ATPase.[13,50,71,74,163] Although this aspect has been controversial[218] other functional ATPases or processes that utilize ATP, yielding ADP, are likely to exist.[218] In fact, a number of studies have shown that the hepatic ATP/ADP=Pi rate is reduced following chronic alcohol consumption.[20,34,58,190]

An interesting feature of the hypermetabolic state, induced by both acute and chronic administration of ethanol, is the fact that thyroid hormone function appears to be *permissive* for such an effect to occur. While neither acute nor chronic ethanol administration to animals alters the circulation T4 or T3[71,75,193] levels, thyroidectomy or the administration of the antithyroid drug propylthiouracil (PTU) markedly suppresses or abolishes both acute and chronically induced hypermetabolic state.[67,74,195,216] Recent studies indicated that PTU administration to animals blocked the increases in ethanol metabolism induced by chronic ethanol consumption even though PTU increased alcohol dehydrogenase levels, thus showing that mitochondrial oxygen consumption constitutes an important regulatory step in the rate of ethanol metabolism in chronically treated animals.[20,154]

It should be noted that an increased metabolism of ethanol through the microsomal pathway would also increase the rate of oxygen consumption.[149] In fact, such a mechanism having the stoichiometry of a mixed function oxidase utilizes twice as much oxygen to transform one mole of ethanol than the ADH mitochondrial pathway.[71]

An increased oxygen demand, which exists in laboratory animals such as rats and baboons, is *per se* not condition enough to produce cell necrosis. This may be due to two factors: (a) the increased oxygen consumption is unable to exceed the passive safety factors of the biological system, and (b) there are active compensating mechanisms to increase the availability of oxygen to the liver. The oxygen reserve in blood leaving the liver constitutes a safety factor. Normal oxygen tensions of 30–35 mmHg[76,81,131] indicate a saturation of hemoglobin of the order of 50–60%. However, whether such a level of oxygen tension actually constitutes a reserve safety factor can be challenged.[77,78,117,153,173] First, it should be remembered that such a value constitutes an average O_2 tension resulting from a mixture of blood flowing through sinusoids of different lengths such that, as shown by Kessler *et al.*[89] and Sato *et al.*,[173] oxygen tensions as low as 2 mmHg (approx. 0.1 mM) can also be observed. It is not clear what the oxygen tension is that is required to support normal hepatocyte function. The Km for cytochrome oxidase is of the order of 0.2 mM.[25] However, there are diffusion barriers that might markedly reduce the availability of oxygen. In fact, studies by Quistorff and Chance,[153] Thurman *et al.*,[195] and Jauhonen *et al.*,[77] who have measured the redox potential of liver cells, have suggested that Zone 3 of the liver is normally in a partial state of relative hypoxia. Both the permeability barriers and the range of oxygen tensions at the end of different sinusoidal lengths could combine to finally determine a *focal* hypoxic state in conditions of an increased oxygen demand. The existence of different degrees or areas of hypoxia in different periacinar cells within Zone 3 of the acinus might explain the focal nature of the typical necrotic lesion in the alcoholic.[21,48,51,160]

In addition to a passive safety factor the acute administration of alcohol triggers a compensatory mechanism by which portal liver blood flow is markedly increased by 50–60%.[1,18,120,180,205] Such a mechanism is dependent on ethanol metabolism *per se*, as it can be blocked by 4-methylpyrazole,[120] an alcohol dehydrogenase inhibitor.[105] It has been further shown that the increase in portal blood flow is independent of the concentration of blood ethanol when these are above those that saturate the alcohol dehydrogenase system.[120] Such an efficient compensatory mechanism can play an important role in protecting the liver against hypoxic necrosis.

As indicated earlier, the hypoxic theory of liver necrosis is based on the concept that an increased oxygen consumption should be associated with a precipitating factor reducing oxygen availability to the liver, the combination of which results in pathological oxygen deprivation to Zone 3 hepatocytes and in cell damage. These are discussed below.

Precipitating Factors in Hypoxic Liver Injury

a. Causes Leading to a Reduced Oxygenated Hemoglobin Supply

A reduction in oxygen availability based on either a reduced hemoglobin level or a reduced hemoglobin saturation constitutes an important precipitating factor. In animal models, chronic administration of ethanol resulted in hepatocellular necrosis when associated with conditions of low atmospheric oxygen tension[45,73,147] and experimental anemia.[66,137,191] These experiments demonstrated that the rat, an animal that does not show hepatocellular necrosis, can produce it when alcohol administration is combined with a precipitating factor, the latter occurring in human alcoholics.[66] In the alcoholic a number of conditions can result in a reduced oxygen availability that can act as precipitating factors.[66] Among these is anemia of different causes, of common occurrence in the alcoholic with or without liver disease.[54,57,112,121,126] Anemia could be due to nutritional problems or upper gastrointestinal bleeding.[42,59] We have shown that in alcoholics with liver disease a strong relationship exists between anemia and mortality.[137] Anemia is also prevalent in women due to menstrual blood loss. In the U.S.A., 20% of women present anemia with Hb values of less than 12 g percent.[160] It is of interest that women are markedly more susceptible to the development of alcoholic liver disease and to presenting a worse prognosis than men.[2,29,49,96,128,179,198,210] Such an enhanced susceptibility is reduced, or may even be reversed, after menopause.[179,210]

Alcoholics are also at a greater risk of presenting reduced oxygen situations. Respiratory dysfunction occurs in 52% of alcoholic men and in 77% of alcoholic women.[187,214] Also, pneumonias[187] and sleep apnea[27] are known to occur with a greater frequency in alcoholics. Heavy smoking, in which a significant percentage of hemoglobin can be found in the form of carboxyhemoglobin, is frequently associated with alcoholism.[157,203]

b. Interference with a Compensatory Increase in Blood Flow

There are very few studies on this important aspect of alcohol and the liver. As mentioned above, the acutely or the chronically ethanol-induced increase in liver oxygen consumption is normally compensated for by an increase in portal blood flow. This compensation occurs when alcohol is present[18,77] but is not seen in early withdrawal.[89,76] It is likely that alterations in this mechanism might play a role in the production of hypoxic liver damage in certain circumstances. It is known that the compensatory increase in splanchnic blood flow can be markedly inhibited and even completely suppressed under certain specific circumstances such as the administration of intra-arterial fructose and if the animal has received anaesthetic agents (ketamine, thiopental, or fentanyl).[23] Although the mechanism of the inhibition induced by these agents has not been defined, it is clear that there is the possibility of interference with this compensatory mechanism. Further, the compensatory increase in portal blood flow can be seriously impaired in patients presenting portal hypertension with porto-systemic shunting. In these patients, the lack of a compensatory increase in portal blood flow might result in further liver damage. This factor could be another factor explaining the poor prognosis associated with the presence of porto-systemic shunts.[192]

c. *Sinusoidal Compression and Decreased Blood Flow*

Hepatomegaly is a constant finding in alcoholic liver disease.[140] Liver enlargement is also observed in animals fed alcohol chronically.[69,72,138] In alcoholics with varying degrees of liver dysfunction, liver size, as measured by ultrasonography, has been shown to correlate positively with the severity of liver disease.[104] Both in animals and in humans this is a rapidly reversible condition upon discontinuation of alcohol ingestion.[15,104]

Studies by Baraona *et al.*[7] and in our laboratory[15,69,70,72] have shown that liver enlargement is due to an increase in the size of the hepatocytes rather than to an increase in their number. While fat accumulation accounts for 20–25% of the increase in hepatocyte size, most of the increase (50–60%) is due is an accumulation of intracellular water, which is accounted for by an increase in total intracellular K^+, the main intracellular cation.[70,72] While an increase in protein accumulation has also been proposed to account for the accumulation of water,[7] other investigators have disputed this mechanism.[70,72]

Studies both in chronically alcohol-fed animals and in alcoholics have shown that an increase in hepatocyte volume results in marked compression of the extracellular, plasma volume associated with compression of sinusoids.[69,72,204] In alcoholics, mean sinusoidal caliber (in formalin-fixed biopsies) has been found to be reduced to only 15–20% of that found in nonalcoholic individuals with normal liver biopsies or in patients with nonalcoholic liver disease.[204] Such a striking change is expected to lead to profound alterations in liver microcirculation. It has now become evident that the normal sinusoidal caliber does not allow for more than one erythrocyte per cross section at a time. Such a feature results in a stirring action of the red cells on the space of Disse (endothelial massage).[212] Thus, it is not surprising that an increase in hepatocyte volume compressing the sinusoid caliber would result in an interference in red cell circulation. Recent *in-vivo* studies[35,56] have confirmed that both animals fed alcohol chronically and alcoholics with liver disease present a significant reduction in hepatic hemoglobin, and thus in the maximal oxygen delivery capacity of hemoglobin at a single time.

The combination of a hypermetabolic state due to the continued alcohol consumption with a decrease in hepatocyte perfusion might in itself trigger hepatic necrosis or might significantly lower the threshold for other precipitating factors reducing oxygen availability.

Another element in relation to hepatocyte enlargement that could potentially contribute to the production of a functional hypoxia by a reduction of the mitochondrial to sinusoidal oxygen gradient is the fact that in enlarged hepatocytes a reduced surface/volume relationship should result in a potential reduction in oxygen exchange between the cell and its milieu.

d. Alterations in Endothelial Fenestrae, Hepatocyte Microvillae and Collagenization of the Space of Disse, as Mechanisms Preventing Diffusional Exchanges

Recent studies have demonstrated that the endothelial fenestrae, doughnut-like structures through which by diffusion different molecules gain access to the hepatocyte microvillae, are now known to be active structures with variable diameters, apparently influenced by contractile proteins.[164] In baboons fed alcohol chronically the number of open fenestrae is significantly decreased.[115] The same phenomena have been observed in humans with cirrhosis.[61,63]

Among different epithelia the hepatocyte, along with the absorptive, small intestinal cells and the nephron, are the cells endowed with the most prominent system of microvillae. Recent studies have demonstrated that microvillae are markedly decreased, or are even nonexistent, when combined with an accumulation of collagen in the space of Disse.[61,138,142,148,162] The latter space, which separates the hepatocytes from the endothelial cell, is often found in the alcoholics to be filled with bundles of collagen fibres, the extreme case being the formation of a continuous basement membrane leading to the functional capillarization of the sinusoid.[15,41,61,62,63,138,142,148,162,176]

Although the three types of abnormalities have not been clearly demonstrated to rate limit the exchange of oxygen between the sinusoidal blood and the hepatocyte intracellular space, we have included them, since they are general barriers for the diffusion of larger molecules. For example, collagenization and capillarization of the space of Disse have been shown to significantly reduce the diffusion of albumin into the space of Disse in cirrhosis.[62,63]

ACETALDEHYDE AS A MECHANISM OF HEPATOCELLULAR DAMAGE

A number of studies have demonstrated that acetaldehyde, a highly reactive metabolite,[107] can bind to a number of molecules of biological importance, including proteins such as hemoglobin,[184] tubulin,[116] and albumin,[38,39,47,113,189] DNA,[166] phospholipids,[88] and serotonin, dopamine, and norepinephrine, yielding pharmacologically active compounds.[107,155] It is not clear, however, whether these interactions are related to cell necrosis,[9] nor has an animal model been produced in which acetaldehyde has been involved as the direct cause of hepatotoxicity.

A number of investigators have proposed that acetaldehyde-mediated cell death might have an immunological component.[9,82,138] Two immune mechanisms are well established to lead to cell necrosis in a variety of tissues: (a) cellular cytotoxicity mediated by lymphokins occurs when specific T-lymphocytes recognize antigens in the cell surface; and (b) humoral cytotoxicity occurs upon the combination of circulating immunoglobulins with cellular surface antigens. Cell damage is mediated by complement activation, and by killer cell and neutrophil linkage.[146]

Both mechanisms have been proposed to play a role in acetaldehyde-mediated cell lysis, which essentially depends on the existence of either T-cell or immunoglobulins that can recognize acetaldehyde.[9,82,188] Early studies by Leevy and co-workers[82,188] and by Actis and co-workers[1b] indicated that lymphocytes of patients with alcoholic necrosis can be activated when incubated with acetaldehyde. Lymphocytes of alcoholics presenting only fatty liver or inactive cirrhosis were not activated by acetaldehyde. It is unlikely that lymphocytes recognize a small molecule like acetaldehyde *per se*. On the basis of recent findings[65] it is probable that when lymphocytes are exposed to acetaldehyde, condensation products occur in the lymphocyte cell surface that trigger the immune response.

Recently, it was demonstrated that proteins modified by acetaldehyde can become active immunogens leading to the production of immunoglobulins that can specifically recognize the acetaldehyde-containing epitope.[65] For example, immunization of mice with acetaldehyde conjugated to keyhole limpet hemocyanin (KLH), a nonmammalian protein, leads to the production of circulating antibodies that can recognize mammalian proteins containing the acetaldehyde moiety, but not the unmodified proteins *per se*. Similar findings were obtained by immunizing with mammalian proteins unrelated to the ones against which the antibody is tested. These studies suggest that small determinants in a variety of proteins can trigger an immune response to acetaldehyde-containing epitope(s) independently of the carrier protein. These immunoglobulins are likely to recognize acetaldehyde adducts in hepatocyte plasma membranes, which could lead to complement-mediated lysis. That circulating immunoglobulins to acetaldehyde-modified proteins normally exist in blood is suggested by recent studies indicating that normal human sera incubated with liver plasma membrane vesicles conjugated to acetaldehyde activates complement.[9,10] Such activation is not observed in vesicles that were not previously exposed to acetaldehyde. This may occur because the general nonalcoholic population is exposed to low levels of alcohol either by consumption or by endogenous production of ethanol and especially acetaldehyde by the intestinal bacterial flora.[6,16,80,91,97,123] The existence of low titers of antibodies, and acetaldehyde-modified proteins in normal populations has been observed by us (Niemelä, Orrego, and Israel).

The presence of circulating antibodies recognizing hepatocytes altered by ethanol or by its metabolites is also suggested by studies indicating that sera of alcoholics can induce cytotoxicity to hepatocytes isolated from animals pretreated with ethanol. Such an effect was not observed in sera from control individuals.[132] Furthermore, this cytotoxicity was inhibited if the animals were pretreated with 4-methylpyrazole, an inhibitor of alcoholic dehydrogenase, and enhanced when the animals were treated with disulfiram, an inhibitor of acetaldehyde dehydrogenase. These data were interpreted as an indication that ethanol metabolism is required for the expression of the ethanol-related determinant and suggest that an impaired ability to metabolize acetaldehyde could lead to the development of immunological reactions to new liver membrane antigens.[33]

These results are not necessarily in conflict with the above studies suggesting the presence of antibodies against acetaldehyde bound to hepatocyte membrane vesicles in the sera of normal individuals, since the relative dilutions and thus the titers of antibodies in the two studies are not comparable. It should be noted, however, that in the latter study the cytotoxicity of the sera was not dependent on the presence or absence of hepatocellular necrosis in the donor.[132] Thus, the studies may be interpreted as an indication that chronic alcohol consumption increases the levels of immunoglobins that can react with hepatocyte membrane proteins,[22,33,60,122,172,185,186,194,211] likely following modification by acetaldehyde, but that actual necrosis will depend *not only* on the existence of immunoglobulins, but also on the number of antigenic sites exposed in surface of the hepatocytes.

Also, the possibility of Mallory hyalin being involved in an immunological mechanism has been proposed. The cytotoxicity of lymphocytes in patients with alcoholic liver disease has been related to the presence of Mallory's bodies in liver biopsies[30,44,145,219] and antibodies against Mallory's hyalin have been detected in serum of patients with alcoholic hepatitis.[85,221] Mallory's material has been found to stimulate the production of fibrogenic factor,[99] migration inhibitory factor,[220] and transfer factor[86] from lymphocytes. There is some controversy, though, as to the purity of

Mallory-body antigens,[30] such that other molecules present in the preparation might have been responsible for the production of the antibodies.

A fundamental question relates to the mechanism by which an intracellular antigen is recognized by the extracellular immune system in order to cause cell death. A possible explanation for the interrelation between the existence of antibodies against Mallory bodies and cell necrosis in the alcoholic could be given by the recent findings that acetaldehyde bound to proteins can be recognized by antibodies independently of the nature of the protein.[65] If both Mallory bodies and cell membranes contained sites modified by acetaldehyde, circulating antibodies would reorganize the acetaldehyde-containing sites in both proteins but would act only on cell membranes to produce cytotoxicity. The same concepts may apply to the role of a "liver specific protein" in immunologically induced liver damage.[11,12,28,43]

Should liver cell necrosis in the alcoholic be of immunological nature[43,52,124,152,220] it would be expected that women should be more susceptible to develop the disease, as is the case for many immunitary diseases[96,114,125,215] As indicated earlier women are indeed more susceptible than men to the development of alcoholic liver disease.

PATHOGENESIS OF PORTAL HYPERTENSION

Portal hypertension is one of the most serious consequences of alcoholic liver disease and is pathogenetically related to a number of conditions that constitute the major causes of death, such as esophageal varices and upper-gastrointestinal hemorrhage, ascites, porto-systemic encephalopathy, and the hepatorenal syndrome.[32]

From the hemodynamic point of view, portal hypertension can result from: (a) an increase in liver resistance to blood flow, (b) an increase in portal blood flow, or (c) a combination of both factors.

Some investigators propose that portal hypertension is primarily the result of an increase in liver resistance,[53,64,118,209] while others suggest that the "forward" (increase in portal blood flow)[206,207,213] hypotheses would explain the mechanism of portal hypertension. Since splanchnic blood flow corresponds to about 20–25% of the cardiac output,[165] it is difficult to conceive that the very large elevations in portal pressure seen in alcoholic liver disease (frequently in the range of 20–30 mmHg or more), can result solely from an increase in portal blood flow. However, increases in blood flow would potentiate the effects on portal pressure of an increase in hepatic resistance.[222] Recently it has been shown that in cirrhosis approximately 60% of portal pressure is the result of an increase in liver resistance, and 40% can be attributed to an increase in portal blood flow.[4] The mechanisms by which acute alcohol administration can lead to an increased portal flow have been indicated above.

In alcoholic liver disease the increase in resistance to blood flow has been attributed to either a sinusoidal or a postsinusoidal compression by: (a) expanding regenerative nodules, (b) postsinusoidal venous outflow blockade resulting from sclerosis or occlusive lesions of the hepatic veins (central vein sclerosis), (c) compression of sinusoidal caliber by widening of the space of Disse due to collagen accumulation, and (d) enlargement of the hepatocytes encroaching on the sinusoids.

a. "Regenerative" Nodules

One of the hallmarks of cirrhosis is the presence of nodules in which parenchymal cells are surrounded by fibrous septae and in which the normal vasculature of the liver is not discernible.[150] It is believed that these are "regenerating" nodules expanding and compressing the vascular space within the liver capsule. Initially it was thought that the increase in resistance was of postsinusoidal origin and located at the level of the venous return vessels, which are more easily compressible than the portal veins contained in the more rigid portal spaces.[3,27,87] Nevertheless, most of the recent evidence suggests that in cirrhosis the resistance is located at the level of the sinusoids.[181] The role of the "regenerative" nodules in the production of portal hypertension in alcoholic liver disease remains controversial for the following four reasons. (i) The expansive character of the nodules remains undemonstrated. (ii) Portal hypertension can be found in patients without cirrhosis. We have observed that 34% of the patients with alcoholic liver disease and portal hypertension do not present cirrhosis on liver biopsy.[15] (iii) A good correlation between reliable measurements of nodularity and portal pressure has never been established. (iv) A common observation is that portal hypertension decreases soon after withdrawal from alcohol, while the nodularity is likely to remain unchanged.[100,161]

b. *Terminal Hepatic Vein Sclerosis*

This lesion comprises a process of fibrogenesis progressively obstructing the terminal hepatic vein (central veins).[63] Terminal hepatic vein sclerosis can occur at a rather early, precirrhotic state of alcoholic liver disease. It has been claimed that this lesion has a predictive value in determining progression to cirrhosis in both humans[129] and baboons.[201] It should be noted, nevertheless, that other authors have shown no relationship between terminal hepatic vein sclerosis and the clinical or histological severity of human liver disease.[130] In the baboon, terminal hepatic vein sclerosis correlates with mild elevations of portal pressure in the precirrhotic stage.[127] It should be noted that terminal hepatic vein sclerosis would result in a postsinusoidal rather than in a sinusodial site of resistance, the latter being the case in human alcoholic liver disease.[181]

c. *Collagen in the Space of Disse*

Accumulation of collagen in the space of Disse has been associated with the presence of portal hypertension in several types of liver disease including idiopathic portal hypertension.[93,138,169,171] We have found a highly significant correlation between the amount of collagen, including capillarization, in the space of Disse and the height of portal pressure ($r = 0.84$, $p < 10^{-6}$, $n = 70$)[15,138] in patients with alcoholic liver disease with or without cirrhosis. Although this close correlation indicates an interrelation between the two factors, it does not permit a conclusion as to the cause-effect relationship. A strong causal role, however, is not certain, because collagen accumulation did not significantly change in patients in whom portal pressure decreased in a six-month period of observation.[15]

d. *Hepatocyte Enlargement*

We have proposed that the increase in hepatocyte size compresses the sinusoids,[15,69,70,72,138] the site that most authors believe is the origin of the intrahepatic resistance observed in alcoholic liver disease. This increase in resistance to portal blood flow occurs after a threshold in cell size is exceeded, resulting in an increase in intrahepatic and portal pressures.[69] We found a strong correlation between cell size and portal pressure in 129 patients with alcoholic cirrhosis ($r = 0.66$, $p < 10^{-4}$) and in fifty patients with noncirrhotic alcoholic liver disease ($r = 0.80$, $p < 10^{-6}$). The correlation was independent of the histological diagnosis of fatty liver, alcoholic hepatitis, or cirrhosis. While cirrhotics, as a group, have higher portal pressures than noncirrhotics, this could be explained, on the basis of the hepatocyte expansion hypothesis, by the fact that cirrhotics have significantly larger hepatocytes than noncirrhotics ($769 \pm 26\ \mu m^2$ and $564 \pm 21\ \mu m^2$, respectively), *rather than* by the presence of cirrhosis *per se.*[15] We have further observed that in patients in whom liver biopsies and portal pressure measurements were performed simultaneously on at least two occasions within 6 months, the direction of the change in portal pressure in 91% of the cases was the same as the direction of the change in cell size ($p < 0.001$).[15]

Since the existence of good correlations between hepatocyte size and portal hypertension described above is not enough to establish causality, we have further studied the relationship between hepatocyte enlargement and portal pressure by applying the general concept that animal cells act as osmometers in a model of perfused rat liver. In this model, graded hepatocyte enlargement was induced by perfusing the liver with hypotonic medium. In line with the hepatocyte enlargement hypothesis this procedure resulted in: swelling of the hepatocytes, increase in liver weight, compression of the sinusoids, reduction in the extracellular spaces, and graded increases in portal pressure.[31]

Conditions leading to increases of resistance of postsinusoidal origin should increase or leave unaltered the sinusoidal diameter. On the other hand, conditions increasing resistance at the sinusoidal level should result in sinusoidal compression. Patients with alcoholic liver disease show a marked reduction ($p < 0.001$) in relative sinusoidal area ($995 \pm 135\ \mu m^2$, $n = 19$) when compared to nonalcoholic patients with normal liver histology ($5{,}100 \pm 389\ \mu m^2$, $n = 19$). In addition, there is a significant inverse correlation between hepatocyte size and sinusoidal area ($r = 0.66$, $p < 10^{-6}$, $n = 44$) indicating that larger hepatocytes were related to sinusoidal compression.[204] In the alcoholic patients, portal pressure correlated inversely ($r = -0.77$, $p < 0.0001$) with sinusoidal areas only after the sinusoidal area was reduced to values below 20% of normal. Such a threshold was never reached in patients with nonalcoholic liver disease, in whom no correlation of portal hypertension with either sinusoidal area or hepatocyte size was observed.[204] We have also reported a significant decrease of the liver vascular space of rats treated chronically with ethanol and presenting enlarged hepatocytes.[69,204] Both of these findings, in rats and in patients with alcoholic liver disease, have recently been confirmed by Hayashi *et al.*[98] using organ reflectance spectrophotometry, which estimates the amount of hemoglobin in the liver *in vivo.*[55]

The dramatic reduction in sinusoidal area that occurs in alcoholic liver disease and its relationship to hepatocyte size is in line with the concept of a sinusoidal compression determined by the enlarged hepatocytes as a mechanism for portal hypertension in this condition. It should be noted that while the hypothesis of hepatocyte enlargement as a cause of portal hypertension appears most attractive, this postulate does not exclude the possible contribution of the other factors mentioned above.

INTERACTIVE MECHANISMS

An investigator analyzing the mechanisms that lead to death in a given individual with alcoholic liver disease will inescapably be confronted with a multitude of factors that ultimately lead to the demise of the individual. In fact, many of the causal factors leading to death result from concatenated abnormalities of extrahepatic origin. In conditions of less severe disease, a lesser number of factors concatenate, and therefore the analysis is simplified. In essence *the severity of the disease is a function of an expansive process of continuous recruitment of interactive mechanisms*. We feel that it may be naive to consider a disease that leads to *death* as resulting from only a single, noninteractive, pathogenic cause. Alcoholic liver disease is unlikely an exception. Therefore, we will analyze some interactive systems that constitute vicious cycles *per se* and that when combined with each other further potentiate themselves. It is impossible to be exhaustive and many more cycles conceivably play a role. Mortality due to liver damage, as indicated earlier, can be seen as due to two main causes, (i) hepatocellular dysfunction, of which cell necrosis is one extreme, and (ii) portal hypertension. These interact in several ways. Portal hypertension results in (a) upper gastrointestinal bleeding, (b) collateral circulation bypassing the liver, (c) reduced functional blood perfusion, and (d) hyperesplenism and secondary anemia. All of these reduce oxygen availability to the liver and therefore, in the presence of continued alcohol intake, they increase the risk of hypoxic hepatocellular necrosis. All necrosis, on the other side, triggers collagen deposition, accompanied by a reduction in hepatocyte microvillae and in the number of fenestra, which (a) interferes with the optimum exchange of oxygen and nutrients, and (b) may increase the rigidity of the space of Disse, interfering with the stirred layer "massaging" action of red cells along the hepatocyte brush border. Cell expansion *per se,* which appears to be the primary mechanism of portal hypertension, can reduce sinusoidal perfusion through functional pathways. An increased intra- and extrahepatic blood shunting results in a reduction in the effectiveness of the compensatory mechanisms increasing portal blood flow in the presence of ethanol.

All of these cycles decrease oxygen availability to the liver cell. Other cycles may exist at lesser degrees of severity. Although in normal persons the blood levels of acetaldehyde following ethanol ingestion are extremely low,[108,223] acetaldehyde levels increase in alcoholics, who metabolize alcohol at higher rates.[94,95,108,109,111,133,134,135,143,144] The presence of liver disease in which acetaldehyde dehydrogenase is reduced will also diminish the removal of acetaldehyde.[133,144] The latter may create an additional vicious cycle through the formation of adducts, immune complexes, and immune liver injury.

From the above, it is clear that in the presence of interactive systems the disease is self-fuelled, requiring ever lessening degrees of both external precipitating causes and amounts of alcohol.

REFERENCES

1. Abrams, M. A. & C. Cooper. 1976. Mechanism of increased hepatic uptake of unsterified fatty acid from serum of ethanol-treated rats. Biochem. J. **156:** 47–54.

1b. Actis, G. C., A. Ponzetto, M. Rizzetto & G. Verme. 1978. Cell-mediated immunity to acetaldehyde in alcoholic liver disease demonstrated by leukocyte migration test. Am. J. Dig. Dis. **23:** 883–886.

2. Audigier, J. C., H. Coppere, & C. Barthelemy. 1984. Consommation d'alcool, cirrhose: aspects epidemiologiques. Gastroenterol. Clin. Biol. **8:** 925–933.

3. Baggenstoss, A. H. 1961. Post-necrotic cirrhosis: morphology, etiology and pathogenesis. *In* Progress in Liver Diseases. H. Popper & F. Schaffner, Eds. Vol. 1: 14–25. Grune & Stratton. New York, NY.

4. Benoit, J. N., W. A. Womack, L. Hernandez & D. N. Granger. 1985. "Forward" and "backward" flow mechanisms of portal hypertension. Relative contributions in the rat model of portal vein stenosis. Gastroenterology **89:** 1092–1096.

5. Baraona, E., P. Janhonen, H. Miyakawa & C. S. Lieber. 1983. Zonal redox changes as a cause of selective perivenular hepatotoxicity of alcohol. Pharmacol. Biochem. Behav. **18:** 449–454.

6. BARAONA, E., R. JULKUNEN, L. TANNENBAUM & C. S. LIEBER. 1986. Role of intestinal bacterial overgrowth in ethanol production and metabolism in rats. Gastroenterology **90:** 103–110.
7. BARAONA, E., M. A. LEO, S. A. BOROWSKY & C. S. LIEBER. 1975. Alcoholic hepatomegaly, accumulation of protein in the liver. Science **190:** 794–795.
8. BARAONA, E., Y. MATSUDA, P. PIKKARAINEN, F. FINKELMAN & C. S. LIEBER. 1981. Effects of ethanol on hepatic protein secretion and microtubules. Possible mediation by acetaldehyde. *In* Currents in Alcoholism. M. Galanter, Ed. Vol. 7: 421–434. Grune & Stratton. New York, NY.
9. BARRY, R. E. & J. D. MCGIVAN. 1985. Acetaldehyde alone may initiate hepatocellular damage in acute alcoholic liver disease. Gut **26:** 1065–1069.
10. BARRY, R. E., J. D. MCGIVAN & M. HAYES. 1984. Acetaldehyde binds to liver cell membranes without affecting membrane function. Gut **25:** 412–416.
11. BEHRENS, U. J. & F. PARONETTO. 1979. Studies on "liver-specific" antigens. I. Evaluation of the liver specificity of "LSP and LP-2." Gastroenterology **77:** 1045–1052.
12. BEHRENS, U. J., S. VERNANCE & F. PARONETTO. 1976. Studies on "liver specific" antigens. II. Detection of serum antibodies to liver and kidney cell membrane antigens in patients with chronic liver disease. Gastroenterology **77:** 1053–1061.
13. BERNSTEIN, J., L. VIDELA & Y. ISRAEL. 1973. Metabolic alterations produced in the liver by chronic ethanol administration. II. Changes related to energetic parameters of the cell. Biochem. J. **134:** 515–521.
14. BHUYAN, U. N., N. C. NAYAK, M. G. KEO & V. RAMALINGASWAMI. 1965. Effect of dietary protein on carbontetrachloride-induced hepatic fibrogenesis in albino rats. Lab. Invest. **14:** 184–190.
15. BLENDIS, L. M., H. ORREGO, I. R. CROSSLEY, J. E. BLAKE, A. MEDLINE & Y. ISRAEL. 1982. The role of hepatocyte enlargement in hepatic pressure in cirrhotic and noncirrhotic alcoholic liver disease. Hepatology **2:** 539–546.
16. BODE, J. C., S. RUST & C. BODE. 1984 The effect of cimetidine treatment on ethanol formation in human stomach. Scand. J. Gastroent. **19:** 853–856.
17. BOVERIS, A., C. G. FRAGA, A. I. VARSAVSKY & O. R. KOCH. 1983. Increased chemiluminescence and superoxide production in the liver of chronically ethanol-treated rats. Arch. Biochem. Biophys. **227:** 534–541.
18. BREDFELDT, J. E., E. M. RILEY & R. J. GROSZMANN. 1985. Compensatory mechanisms in response to an elevated hepatic oxygen consumption in chronically ethanolfed rats. Am. J. Physiol. **248:** 507–511.
19. BRIEN, J. F. & C. W. LOOMIS. 1983. Pharmacology of acetaldehyde. Can. J. Physiol. Pharmac. **61:** 1–22.
20. BRITTON, R. S. & Y. ISRAEL. 1980. Effect of 6-n-propyl-2-thiouracil on the rate of ethanol metabolism in rats treated chronically with ethanol. Biochem. Pharmacol. **29:** 2951–2955.
21. BRUNT, P. W., M. C. KEW, P. J. SCHEUER & S. SHERLOCK. 1974. Studies in alcoholic liver disease in Britain. I. Clinical and pathological patterns related to natural history. Gut **15:** 52–58.
22. BURT A. D., R. S. ANTHONY, W. S. HISLOP, I. A. D. BOUCHIER & R. N. M. MACSWEEN. 1982. Liver membrane antibodies in alcoholic liver disease. I. Prevalence and immunoglobulin class. Gut **23:** 221–225.
23. CARMICHAEL, F. J., J. P. MCKAIGNEY, V. SALDIVIA & H. ORREGO. 1985. Effect of anesthetic agents on ethanol-induced increase in splanchnic blood flow in the rat. Hepatology **5:** 978(A).
24. CEDERBAUM, A. I. & E. RUBIN. 1976. Mechanism of the protective action of cysteine and penicillamine against acetaldehyde-induced mitochondrial injury. Biochem. Pharmacol. **25:** 2179–2185.
25. CHANCE, B. 1977. Molecular basis of O_2 affinity for cytochrome oxidase. *In* Oxygen and Physiological Function. F. F. Jobsis, Ed. 14–21. Professional Information Library. Dallas, TX.
26. CHANCE, B. & P. K. MIATRA. 1963. Determination of the intracellular phosphate potential of ascites cells by reversed electron transfer. *In* Control Mechanisms in Respiration and Fermentation. B. Wright, Ed. 307–312. Rouald Press. New York, NY.

27. CHERNIAK, N. S. 1984. Sleep apnea and its causes. J. Clin. Invest. **73:** 1501–1506.
28. CHISARI, F. V. 1980. Liver-specific protein in perspective. Gastroenterology **78:** 168–170.
29. COATES, R. A., M. L. HALLIDAY, J. G. RANKIN, S. V. FEINMAN & M. M. FISHER. 1986. Risk of fatty infiltration of cirrhosis of the liver in relation to ethanol consumption: a case-control study. Clin. Invest. Med. **9:** 26–32.
30. COCHRANE, A. M. G., A. MOUSSOUROS, B. PORTMANN, I. G. MCFARLANE, A. D. THOMSON, A. L. W. F. EDDLESTON & R. WILLIAMS. 1977. Lymphocyte cytotoxicity for isolated hepatocytes in alcoholic liver disease. Gastroenterology **72:** 918–923.
31. COLMAN, J. C., R. S. BRITTON, H. ORREGO, V. SALDIVIA, A. MEDLINE & Y. ISRAEL. 1983. Relation between osmotically-induced hepatocyte enlargement and portal hypertension. Am. J. Physiol. **245:** 383–387.
32. CONN, H. O. & R. J. GROSZMANN. 1982. The pathophysiology of portal hypertension. *In* The Liver: Biology and Pathobiology. I. M. Arias, H. Popper, D. Schacter & D. A. Shafritz, Eds. 821–848. Raven Press. New York, NY.
33. CROSSLEY, I. R., J. NEUBERGER, M. DAVIS, R. WILLIAMS & A. L. W. F. EDDLESTON. 1986. Ethanol metabolism in the generation of new antigenic determinants on liver cells. Gut **27:** 186–189.
34. CUNNINGHAM, C. C., G. SINTHUSEK, P. I. SPACH & C. LEATHERS. 1981. Effect of dietary ethanol and cholesterol on metabolic functions of hepatic mitochondria and microsomes from the monkey, macaca nemestrina. Alcoholism: Clin. Exp. Res. **5:** 410–416.
35. DAWSON, A. G. 1983. Ethanol oxidation in systems containing soluble and mitochondrial fractions of rat liver. Regulation by acetaldehyde. Biochem. Pharmacol. **32:** 2157–2165.
36. DEGOTT, C., B. DEVERGIE, B. RUEFF & F. POTET. 1981. La necrose hyaline dans l'hepatite alcoolique aigue: valeur prognostique d'une étude histologique semiquantitative. Gastroenterol. Clin. Biol. **5:** 161–167.
37. DHINAGRA, R., N. KANAGASUNDARAM & C. M. LEEVY. 1980. Mechanism of intrahepatic polymorphonuclear leukocyte (PML) accumulation in alcoholic hepatitis. Gastroenterology **79:** 1013(A).
38. DONOHUE, T. M., D. J. TUMA & M. F. SORRELL. 1983. Acetaldehyde adducts with proteins binding of ^{14}C acetaldehyde to serum albumin. Arch. Biochem. Biophys. **220:** 239–246.
39. DONOHUE, T. M., D. J. TUMA & M. F. SORRELL. 1983. Binding of metabolically derived acetaldehyde to hepatic proteins in vitro. Lab. Invest. **49:** 226–229.
40. EDMONDSON, H. A. 1980. Pathology of alcoholism. Am. J. Clin. Pathol. **74:** 725–742.
41. EGUCHI, T., N. KIEJIRI & M. KAWAGUCHI. 1975. Scanning and transmission electron microscopic observation of human and rat liver sinusoid in liver cirrhosis. J. Clin. Electron. Microsc. **8:** 5–6.
42. EICHNER, E. R. 1973. The hematologic disorders of alcoholism. Am. J. Med. **54:** 621–630.
43. FEIGHERY, C. & D. G. WEIR. 1980. How specific is liver-specific protein? Gastroenterology **79:** 179 (correspondence).
44. FOX, R. A. 1977. Immune mechanisms in alcoholic liver disease. *In* Alcohol and the Liver. M. M. Fisher & G. J. Rankin, Eds. 309–320. Plenum Press. New York, NY.
45. FRENCH, S. W., N. C. BENSON & P. S. SUN. 1984. Centrilobular liver necrosis induced by hypoxia in chronic ethanol-fed rats. Hepatology **4:** 912–917.
46. FRENCH, S. W., B. H. RUEBNER, E. MEZEY, T. TAMURA & C. H. HALSTED. 1983. Effect of chronic ethanol feeding on hepatic mitochondria in the monkey. Hepatology **3:** 34–40.
47. GAINES, K. C., J. M. SALHANY, D. J. TUMA & M. F. SORRELL. 1977. Reactions of acetaldehyde with human erythrocyte membrane proteins. FEBS Lett. **75:** 115–119.
48. GALAMBOS, J. T. 1974. Alcoholic hepatitis. *In* The Liver and Its Diseases. F. Schaffner, S. Sherlock & C. M. Leevy, Eds. 255–272. Intercontinental Medical Book Corp. New York, NY.
49. GAVALER, J. S. 1982. Sex-related differences in ethanol-induced liver disease: artifactual or real? Alcoholism: Clin. Exp. Res. **6:** 186–196.
50. GONZALEZ-CALVIN, J. L., J. B. SAUNDERS, I. R. CROSSLEY, C. J. DICKENSON, H. M.

SMITH, J. M. TREDGER & R. WILLIAMS. 1985. Effects of ethanol administration on rat liver plasma membrane-bound enzymes. Biochem. Pharmacol. **34:** 2685–2689

51. GREEN, J., S. MISTILIS & L. SCHIFF. 1963. Acute alcoholic hepatitis. A clinical study of fifty cases. Arch. Intern. Med. **112:** 113–124.
52. HADZYJKANNIS, S., T. FEIZI & P. J. SCHJEUER. 1969. Immunoglobulin-containing cells in the liver. Clin. Exp. Immunol. **5:** 499–514.
53. HALES, M.R., J. S. ALLAN & E. M. HALL. 1959. Injection-corrosion studies of normal and cirrhotic livers. Am. J. Pathol. **35:** 909–927.
54. HARINASUTA, U. & H. ZIMMERMAN. 1971. Alcoholic steato-necrosis. I. Relationship between severity of hepatic disease and presence of Mallory bodies in the liver. Gastroenterology **60:** 1036–1046.
55. HAYASHI, N., A. KASAHARA, K. KUROSAWA, H. YOSHIHARA, Y. SASAKI, H. FUSAMOTO & N. SATO. 1985. Hepatic hemodynamics in alcoholic liver injuries assessed by reflectance spectrophotometry. Alcohol **2:** 453–456.
56. HAYASHI, N., A. KASAHARA, K. JUROSAWA, Y. SASAKI, H. FUSAMOTO, N. SATO & T. KAMADA. 1985. Oxygen supply to the liver in patients with alcoholic liver disease assessed by organ-reflectance spectrophotometry. Gastroenterology **88:** 881–885.
57. HERBERT, V. & G. TISMAN. 1975. Hematologic effects of ethanol. 1975. Ann. N.Y. Acad. Sci. **252:** 307–315.
58. HERNANDEZ-MUNOZ, R., A. SANTAMARIA, GARCIA-SAINZE, E. PINA & V. CHAGOYA. 1978. On the mechanism of ethanol-induced fatty liver and its reversibility by adenosine. Arch. Biochem. Biophys. **190:** 155–162
59. HILLMAN, R. S. 1975. Alcohol and hematopoiesis. Ann. N.Y. Acad. Sci. **252:** 297–315.
60. HOPF, U., K. H. MEYER ZUM BUSCHENFELD & J. FREUDENBER. 1974. Liver-specific antigens of different species. II. Localization of a membrane antigen at cell surfaces of isolated hepatocytes. Clin. Exp. Immunol. 16: 117–124
61. HORN, T., J. JUNGE & P. CHRISTOFFERSSEN. 1985. Early alcoholic liver injury: changes of the Disse space in acinar Zone 3. Liver **5:** 301–310.
62. HUET, P. M., C. A. GORESKY, J. P. VILLENEUVE, D. MARLEAU & J. O. LOUGH. 1982. Assessment of liver microcirculation in human cirrhosis. J. Clin. Invest. **70:** 1234–1244.
63. HUET, P. M., J. P. VILLENEUVE, G. POMIER-LAYRARGUES & D. MARLEAU. 1985. Hepatic circulation in cirrhosis. Clinics in Gastroenterology **14:** 155–168.
64. INOMATA, T., G. A. RAO & H. TSUKAMOTO. 1986. Lipid peroxidation may not be important in an early stage of alcohol-induced liver injury. Fed. Proc. **45:** 567(A).
65. ISRAEL, Y., E. HORWITZ, O. NIEMELÄ & R. ARNON. 1986. Specific antibodies against acetaldehyde containing epitopes in acetaldehyde-protein condensates. Proc. Natl. Acad. Sci. USA **83:** 7923–7929.
66. ISAREL, Y., H. KALANT, H. ORREGO, J. M. KHANNA, M. J. PHILLIPS & D. J. STEWART. 1979. Hypermetabolic state: oxygen availability and alcohol-induced liver damage. *In* Biochemistry and Pharmacology of Ethanol. E. Majchrowicz & E. P. Noble, Eds. Vol. 1: 433–444. Plenum Press. New York, NY.
67. ISRAEL, Y., H. KALANT, H. ORREGO, J. M. KHANNA, L. VIDELA & M. J. PHILLIPS. 1975. Experimental alcohol-induced hepatic necrosis: suppression by propylthiouracil. Proc. Natl. Acad. Sci. USA **72:** 1137–1141.
68. ISRAEL, Y., J. M. KHANNA & J. LIN. 1970. Effect of 2-4-denitrophenol on the rate of ethanol elimination in the rat in vivo. Biochem. J. **120:** 447–448.
69. ISRAEL, Y., J. M. KHANNA, H. ORREGO, G. RACHAMIN, S. WAHID, R. BRITTON, A. MACDONALD & H. KALANT. 1979. Studies on metabolic tolerance to alcohol, hepatomegaly and alcoholic liver disease. Drug Alcohol Dep. **4:** 109–118.
70. ISRAEL, Y. & H. ORREGO. 1983. On the characteristics of alcohol-induced liver enlargement and its possible hemodynamic consequences. Pharmacol. Biochem. Behav. **18**(S.1):433–437.
71. ISRAEL, Y. & H. ORREGO. 1984. Hypermetabolic state hypoxic liver damage. *In* Recent Developments in Alcoholism. M. Galanter, Ed. 119–133. Plenum Press. New York, NY.
72. ISRAEL, Y., H. ORREGO, J. C. COLMAN & R. S. BRITTON. 1982. Alcohol-induced

hepatomegaly: pathogenesis and role in the production of portal hypertension. Fed. Proc. **41:** 2472–2477.

73. ISRAEL, Y., H. ORREGO, J. M. KHANNA, D. J. STEWART M. J. PHILLIPS & H. KALANT. 1977. Alcohol-induced susceptibility to hypoxic liver damage: possible role in the pathogenesis of alcoholic liver disease. *In* Alcohol and the Liver. M. M. Fisher & J. G. Rankin, Eds. 323–345. Plenum Publishing Co. New York, NY.
74. ISRAEL, Y., L. VIDELA & J. BERNSTEIN. 1975. Hypermetabolic state after chronic ethanol consumption. Hormonal interrelations and pathogenic implications. Fed. Proc. **34:** 2052–2059.
75. ISRAEL, Y., P. G. WALFISH, H. ORREGO, J. BLAKE & H. KALANT. 1979. Thyroid hormones in alcoholic liver disease. Effect of treatment with 6-n-propylthioracil. Gastroenterology **76:** 116–122.
76. ITURRIAGA, H., G. UGARTE & Y. ISRAEL. 1980. Hepatic vein oxygenation, liver blood flow and the rate of ethanol metabolism in recently abstinent patients. Eur. J. Clin. Invest. **10:** 211–218.
77. JAUHONEN, P., E. BARAONA, H. MIYAKAWA & C. S. LIEBER. 1982. Mechanism for selective perivenular hepatotoxicity of ethanol. Alcoholism: Clin. Exp. Res. **6:** 350–357.
78. JI, S., J. J. LEMASTERS, V. CHRISTENSON & R. G. THURMAN. 1982. Periportal and pericentral pyridine nucleotide fluorescence from the surface of the perfused liver: evaluation of the hypothesis that chronic treatment with ethanol produces pericentral hypotoxia. Proc. Natl. Acad. Sci. USA **79:** 5415–5419.
79. JI, S. J. J. LEMASTERS, V. CHRISTENSON, R. G. THURMAN. 1983. Selective increase in percentral oxygen gradient in perfused rat liver following ethanol treatment. Pharmacol. Biochem. Behav. **18:** 439–442.
80. JULKUNAN, R. J. K., C. DI PADOVA & C. S. LIEBER. 1985. First pass metabolism of ethanol a gastrointestinal barrier against systemic toxicity of ethanol. Life Sci. **37:** 567–573.
81. JUNGERMANN, K. & N. KATZ. 1982. Functional hepatocellular heterogeneity. Hepatology **2:** 385–395.
82. KAKUMU, S. & C. M. LEEVY. 1977. Lymphocyte cytotoxicity in alcoholic hepatitis. Gastroenterology **72:** 594–597.
83. KHANNA, J. M. & Y. ISRAEL. 1980. Ethanol metabolism. *In* Liver and Biliary Tract Physiology. N. B. Javitt, Ed. Vol. 1:275–296. University Park Press. Baltimore, MD.
84. KALANT, H., Y. ISRAEL, M. J. PHILLIPS, N. WOO, J. M. KHANNA & H. ORREGO. 1975. Necrosis produced by hepatic arterial ligation in alcohol-fed rats. Fed. Proc. **34:** 719(A).
85. KANAGASUNDARAM, N., T. CHEN, & C. M. LEEVY. 1977. Autoimmunity in alcoholic hepatitis. Gastroenterology **73:** 1227(A).
86. KANAGASUNDARAM, N. & C. M. LEEVY. 1975. Transfer factor in alcoholic hepatitis: its occurrence and significance. Gastroenterology **69:** 833(A).
87. KELTY, R. H., A. H. BAGGENSTOSS & H. R. BUTT. 1950. The relation of the regenerated hepatic nodule to the vascular bed in cirrhosis. Proc. Mayo Clin. **25:** 17–26.
88. KENNEY, W. C. Acetaldehyde adducts of phospholipids. 1982. Alcoholism: Clin. Exp. Res. **6:** 412–416.
89. KESSLER, M., L. GORNANDT & H. LANG. 1973. Correlation between oxygen tension in tissue and hemoglobin dissociation curve. *In* Oxygen Supply. Theoretical and Practical Aspects of Oxygen Supply and Microcirculation of Tissue. M. Kessler, Ed. 156–159. University Park Press. Baltimore, MD.
90. KESSLER, B. J., J. B. LIEBLER, G. J. BRONFIN & M. SASS. 1954. The hepatic blood flow and splanchnic oxygen consumption in alcoholic fatty liver. J. Clin. Invest. **33:** 1338–1345.
91. KLIPSTEIN, F. A., L. V. HOLDEMAN, J. J. CORCINO & W. E. C. MOORE. 1973. Enterotoxigenic intestinal bacteria in tropical sprue. Ann. Intern. Med. **79:** 632–634.
92. KLONER, R. A., C. E. GANOTE, D. A. WHALEN & R. B. JENNINGS. 1974. Effect of a transient period of ischemia on myocardial cells. II. Fine structure during the first few minutes in reflow. Am. J. Pathol. **74:** 399–422.
93. KLUGE, T., H. SOMMERSCHILD & A. FLATMARK. 1970. Sinusoidal portal hypertension. Surgery **68:** 294–300.

94. KOIVULA, T. & K. O. LINDROS. 1975. Effects of long term ethanol treatment on aldehyde and alcohol dehydrogenase activities in rat liver. Biochem. Pharmacol. **24:** 1937–1940.
95. KORSTEN, M. A., S. MATSUZAKI, L. FEINMAN & C. S. LIEBER. 1975. High blood acetaldehyde levels after ethanol administration. Difference between alcoholic and non-alcoholic subjects. N. Engl. J. Med. **292:** 386–389.
96. KRASNER, N., M. DAVIS, B. PORTMANN & R. WILLIAMS. 1977. Changing pattern of alcoholic liver disease in Great Britain: relation to sex and signs of autoimmunity. Br. J. Med. **1:** 1497–1500.
97. KREBS, H. A. & J. R. PERKINS. 1970. The physiological role of liver alcohol dehydrogenase. Biochem. J. **118:** 635–644.
98. LEE, B. M., M. M. WINTROBE & H. F. BUNN. 1980. Iron-deficiency anaemia and the sideroblastic anemias. *In* Harrison's Principles of Internal Medicine, 9th Ed. K. J. Isselbacher, R. D. Adams, E. Braunwald, R. G. Peterdorf & J. D. Wilson, Eds. 1514–1518. McGraw-Hill Book Company. Toronto, Canada.
99. LEEVY, C. M., T. CHEN, A. LUISADA-OPPER, N. KANAGASUNDARAM & R. ZETTERMAN. 1976. Liver disease of the alcoholic: role of immunologic abnormalities in pathogenesis, recognition and treatment. *In* Progress in Liver Diseases. H. Popper & F. Schaffner, Eds. 516–530. Grune & Stratton. New York, NY.
100. LEEVY, C. M., M. ZINKE, J. BABER & W. Y. CHEY. 1958. Observations on the influence of medical therapy on portal hypertension in hepatic cirrhosis. Ann. Intern. Med. **49:** 837–851.
101. LELBACH, W. L. 1974. Organic pathology related to volume and pattern of alcohol use. *In* Research Advances in Alcohol and Drug Problems. R. J. Gibbins, Y. Israel, H. Kalant, R. E. Popham, W. Schmidt & R. G. Smart, Eds. Vol. 1: 93–198. John Wiley. Toronto, Canada.
102. LEMASTERS, J. J., S. JI, C. J. STEMKOWSKI & R. G. THURMAN. 1983. Hypoxic hepatocellular injury. Pharmacol. Biochem. Behav. **18:** 455–459.
103. LEMASTERS, J. J., S. JI & R. G. THURMAN. 1981. Centrilobular injury following hypoxia in isolated, perfused rat liver. Science **213:** 661-663.
104. LEUNG, N. W. Y., P. FARRANT & T. J. PETERS. 1986. Liver volume measurement by ultrasonography in normal subjects and alcoholic patients. J. Hepatol. **2:** 157–164.
105. LI, T. K. & H. THEORELL. 1969. Human liver alcohol dehydrogenase: inhibition by pyrazole and pyrazole analogs. Acta Chem. Scand. **23:** 892–902.
106. LIEBER, C. S. & L. M. DE CARLI. 1976. Animal models of ethanol dependence and liver injury in rats and baboons. Fed. Proc. **35:** 1232–1236.
107. LINDROS, K. O. 1978. Acetaldehyde—its metabolism and role in the actions of alcohol. *In* Research Advances in Alcohol and Drug Problems. Y. Israel, F. B. Glaser, H. Kalant, R. E. Popham, W. Schmidt & R. G. Smart, Eds. Vol. 1:111–176. Plenum Publishing Corporation. New York, NY.
108. LINDROS, K. O. 1982. Human blood acetaldehyde levels: with improved methods, a clearer picture emerges. Alcoholism: Clin. Exp. Res. **6:** 70–75.
109. LINDROS, K. O., L. PENNANEN & T. KOIVULA. 1979. Enzymatic and metabolic modification of hepatic ethanol and acetaldehyde oxidation by the dietary protein level. Biochem. Pharmacol. **28:** 2313–2320.
110. LINDROS, K. O. & A. STOWELL. 1982. Effects of ethanol-derived acetaldehyde on the phosphorylation potential and on the intramitochondrial redox state in intact rat liver. Arch. Biochem. Biophys. **218:** 429–437.
111. LINDROS, K. O., A. STOWELL, P. PIKKARAINEN & M. SALASPURO. 1980. Elevated blood acetaldehyde in alcoholics with accelerated ethanol elimination. Pharm. Biochem. Behav. **13:** 119–124.
112. LISCHNER, M. W., J. F. ALEXANDER & J. T. GALAMBOS. 1971. Natural history of alcoholic hepatitis. I. The acute disease. Am. J. Dig. Dis. **16:** 481–494.
113. LUMENG, L. & P. J. DURANT. 1985. Regulation of the formation of stable adducts between acetaldehyde and blood proteins. Alcohol **2:** 397–400.
114. MACSWEEN, R. N. M. & P. A. BERG. 1976. Autoimmune diseases of the liver. *In* Immunological Aspects of the Liver and Gastrointestinal Tract. A. Ferguson & R.N.M. MACSWEEN, Eds. 345–386. University Park Press. Baltimore, MD.
115. MAK, K. M. & C. S. LIEBER. 1984. Alterations in endothelial fenestrations in liver

sinusoids of baboons fed alcohol: a scanning electronmicroscope study. Hepatology **4:** 386–391.

116. MATSUDA, Y., E. BARAONA, M. SALASPURO & C. S. LIEBER. 1979. Effects of ethanol on liver microtubles and Golgi apparatus. Possible role in altered hepatic secretion of plasma proteins. Lab. Invest. **41:** 455–463.
117. MATSUMARA, T. & R. THURMAN. 1983. Measuring rates of O_2 uptake in periportal and pericentral regions of liver lobule: stop-flow experiments with perfused liver. Am. J. Physiol. **244:** G656–G659.
118. MCINDOE, A. H. 1929. Vascular lesions of portal cirrhosis. Arch. Pathol. **5:** 23–42.
119. MCIVER, M. A. & E. A. WINTER. 1943. Deleterious effects of anoxia on the liver of the hyperthyroid animal. Arch. Surg. **46:** 171–185.
120. MCKAIGNEY, J. P., F. J. CARMICHAEL, V. SALDIVIA, Y. ISRAEL & H. ORREGO. 1986. Role of ethanol metabolism in the ethanol-induced increase in splanchnic circulation. Am. J. Physiol. **250:** G518–523.
121. MCMARTIN, K. E. 1984. Increased urinary folate excretion and decreased plasma folate levels in the rat after acute ethanol treatment. Alcoholism: Clin. Exp. Res. **8:** 172–178.
122. MEYER ZUM BUSCHENFELDE, K. H. & P. A. MIESCHER. 1972. Liver-specific antigens. Purification and characterization. Clin. Exp. Immunol. **10:** 89–102.
123. MEZEY, E., A. L. IMBEMBO, J. J. POTTER, K. C. RENT, R. LOMBARDO & P. R. HOLT. 1975. Endogenous ethanol production and hepatic disease following jejunoileal bypass for morbid obesity. Am. J. Clin. Nutr. **28:** 1277–1283.
124. MIHAS, A. A., D. M. BULL & C. S. DAVISON. 1975. Cell-mediated immunity to liver in patients with alcoholic hepatitis. Lancet **1:** 951–953.
125. MISTILIS, S. P. 1968. Liver disease in pregnancy with particular emphasis on the cholestatic syndromes. Aust. Ann. Med. **17:** 248–260.
126. MISTILIS, S. P. & G. D. BARR. 1980. Alcohol and the liver. Acute alcoholic hepatitis. Med. J. Aust. **2:** 182–188.
127. MIYAKAWA, H., S. ILDA, M. A. LEO, R. J. GREENSTEIN, D. S. ZIMMON & C. S. LIEBER. 1985. Pathogenesis of precirrhotic portal hypertension in alcohol-fed baboons. Gastroenterology **88:** 143–150.
128. MORGAN, M. Y. & S. SHERLOCK. 1977. Sex-related differences among 100 patients with alcoholic liver disease. Br. Med. J. **1:** 939–941.
129. NAKANO, M., T. M. WORMER & C. S. LIEBER. 1982. Perivenular fibrosis in alcoholic liver injury: ultrastructure and histologic progression. Gastroenterology **83:** 777–785.
130. NASRALLAH, S. M., V. H. NASSAR & J. T. GALAMBOS. 1980. Importance of terminal hepatic verule thickening. Arch. Pathol. Lab. Med. **104:** 84–86.
131. NAUCK, M., D. WOLFLE, N. KATZ & K. JUNGERMANN. 1981. Modulation of the glucagon-dependent induction of phosphoenolpyruvate carboxykinase and tyrosine aminotransferase by arterial and venous oxygen concentrations in hepatocyte cultures. Eur. J. Biochem. **119:** 657–661.
132. NEUBERGER, J., I. R. CROSSLEY, J. B. SAUNDERS, M. DAVIS, B. PORTMANN, A. L. W. F. EDDLESTON & R. WILLIAMS. 1984. Antibodies to alcohol-altered liver cell determinants in patients with alcoholic liver disease. Gut **25:** 300–304.
133. NILIUS, R., B. ZIPPRICH & S. KRABBE. 1983. Aldehyde dehydrogenase (E.C.1.2.1.3) in chronic alcoholic liver disease. Hepato-gastroenterology **30:** 134–136.
134. NUUTINEN, H., K. O. LINDROS & M. SALASPURO. 1983. Determinants of blood acetaldehyde levels during ethanol oxidation in chronic alcoholics. Alcoholism: Clin. Exp. Res. **7:** 163–168.
135. NUUTINEN, H. U., M. P. SALASPURO, M. VALLE & K. O. LINDROS. 1984. Blood acetaldehyde concentration gradient between hepatic and antecubital venous blood in ethanol-intoxicated alcoholics and controls. Eur. J. Clin. Invest. **14:** 306–311.
136. OKUNO, F., H. ORREGO & Y. ISRAEL. 1983. Calcium requirement for anoxic liver cell injury. Res. Comm. Chem. Pathol. Pharmacol. **39:** 437–444.
137. ORREGO, H., J. E. BLAKE, A. MEDLINE & Y. ISREAL. 1985. Interrelation of the hypermetabolic state, necrosis, anemia and cell enlargement as determinants of severity in alcoholic liver disease. Acta. Med. Scand. **218**(S.703): 81–95.
138. ORREGO, H., L. M. BLENDIS, I. R. CROSSLEY, A. MEDLINE, A. MACDONALD, S. RITCHIE

& Y. Israel. 1981. Correlation of intrahepatic pressure with collagen in the Disse space and hepatomegaly in humans and in the rat. Gastroenterology **80:** 5546–5567.

139. Orrego, H. & Y. Israel. Biochemical, morphological and clinical correlates of alcoholic liver disease. *In* Alcohol and Aldehyde Metabolizing Systems IV. R. G. Thurman, Ed. 497–508. Plenum Press. New York, NY.
140. Orrego, H., Y. Israel, J. E. Blake & A. Medline. 1983. Assessment of prognostic factors in alcoholic liver disease; toward a global quantitative expression of severity. Hepatology **3:** 896–905.
141. Orrego, H., Y. Israel & L. M. Blendis. 1981. Alcoholic liver disease: information in search of knowledge? Hepatology **1:** 267–283.
142. Orrego, H., A. Medline, L. M. Blendis, R. J. Rankin & D. A. Kreaden. 1979. Collagenisation of the Disse space in alcoholic liver disease. Gut **20:** 673–679.
143. Palmer, K. R. & W. J. Jenkins. 1982. Impaired acetaldehyde oxidation in alcoholics. Gut **23:** 729–733.
144. Palmer, K. R. & W. J. Jenkins. 1985. Aldehyde dehydrogenase in alcoholic subjects. Hepatology **5:** 260–263.
145. Paronetto, F. & S. Vernace. 1975. Immunological studies in patients with chronic active hepatitis. Cytotoxic activity to autochthonous liver cells grown in tissue culture. Clin. Exp. Immunol. **19:** 99–104.
146. Paul, W. E. 1984. Fundamental Immunology. Raven Press. New York, NY.
147. Perrissoud, D., M. F. Maignan & J. H. Dumont. 1985. Antinecrotic effect of 3-palmitoyl(+)-catechin against liver damage induced by galactosamine or ethanol in the rat. Liver **5:** 55–63.
148. Phillips, J. M. & J. W. Steiner. 1966. Electron microscopy of cirrhotic nodules. Lab. Invest. **15:** 801–817.
149. Pirola, R. C. & C. S. Lieber. 1976. Hypothesis: energy wastage in alcoholism and drug abuse; possible role of hepatic microsomal enzymes. Amer. J. Clin. Nutr. **29:** 90–93.
150. Popper, H. 1977. Pathologic aspects of cirrhosis. Am. J. Path. **87:** 228–264.
151. Popper, H. & C. S. Lieber. 1980. Histogenesis of alcoholic fibrosis and cirrhosis in the baboon. Amer. J. Path. **98:** 695–710.
152. Popper, H. & F. Paronetto. 1984. Problems in the immunology of hepatic diseases. Hepato-gastroenterology **31:** 1–5.
153. Quistorff, B., B. Chance & H. Takeda. 1978. Two- and three-dimensional redox heterogeneity of rat liver. Effects of anoxia and alcohol on the lobular redox pattern. *In* Frontiers of Biological Energetics. Vol. 2: 1487–1497. Academy Press. New York, NY.
154. Rachamin, G., F. Okuno & Y. Israel. 1985. Inhibitory effect of propylthiouracil on the development of metabolic tolerance to ethanol. Biochem. Pharmacol. **34:** 2377–2383.
155. Rahwan, R. G. 1975. Toxicology of ethanol: possible role of acetaldehyde, tetrahydroisoquinolines and tetrahydro-beta-carbonies. Toxicol. Appl. Pharmacol. **34:** 3–29.
156. Rankin, J. G. D., H. Orrego, J. Deschenes, A. Medline, J. E. Findlay & A. I. M. Armstrong. 1978. Alcoholic liver disease: the problem of diagnosis. Alcoholism: Clin. Exp. Res. **2:** 327–338.
157. Rankin, J. G. & P. Wilkinson. 1971. Alcohol and tobacco smoking. *In* The Health of a Metropolis. J. Krupinski & A. Stoller, Eds. 61–67. Heineman Educational Australia. Australia.
158. Rappaport, A. M. 1973. The microcirculatory hepatic unit. Microvasc. Res. **6:** 212–228.
159. Recknagel, R. O. 1967. Carbon tetrachloride hepatotoxicity. Pharmacol. Rev. **19:** 145–208.
160. Review by an International Group. 1981. Alcoholic liver disease; morphological manifestations. Lancet **1:** 707.
161. Reynolds, T. B., H. M. Geller, O. I. T. Kuzma & A. G. Redeker. 1960. Spontaneous decrease in portal pressure with clinical improvement in cirrhosis. N. Engl. J. Med. **263:** 734–739.
162. Reynolds, T. B., R. Hidemura, H. Michel & R. Peters. 1969. Portal hypertension without cirrhosis in alcoholic liver disease. Ann. Int. Med. **70:** 497–506.
163. Ricci, R. L., S. S. Crawford & P. B. Miner. 1981. The effect of ethanol on hepatic

sodium plus potassium activated adenosine triphosphatase activity in the rat. Gastroenterology **80:** 1445–1450.
164. Rice, J., Z. Gatmaitan, R. Mikkelson & I. M. Arias. 1985. On the structure and function of the fenestra of hepatic endothelial cells. Hepatology **5:** 1049(A).
165. Richardson, P. D. & P. G. Withrington. 1981. Liver blood flow. I. Intrinsic and nervous control of liver blood flow. Gastroenterology **81:** 159–173.
166. Ristow, H. & G. Obe. 1978. Acetaldehyde induces cross-links in DNA and causes sister chromatid exchanges in human cells. Mutat. Res. **58:** 115–119.
167. Robinson, J. W. L., V. Mirkovitch, B. Winistorfer & F. Saegesser. 1981. Response of the intestinal mucosa to ischaemia. Gut **22:** 512–527.
168. Rogers, A. E., J. C. Fox & L. S. Gottlieb. 1981. Effects of ethanol and malnutrition on non-human primate liver. *In* Frontiers of Liver Disease. P. D. Berk & T. C. Chalmers, Eds. 167–175. Thieme-Stratton, Inc. New York, NY.
169. Russell, R. M., S. A. Bagheri, J. L. Boyer & Z. Hruban. 1970. Hepatic injury and chronic hypervitaminosis A. N. Engl. J. Med. **291:** 435–440.
170. Ryle, P. R., J. Chakrabory & A. D. Thomson. 1985. The roles of the hepatocellular redox state and the hepatic acetaldehyde concentration in determining the ethanol elimination rate in fasted rats. Biochem. Pharmacol. **34:** 3577–3583.
171. Sama, S. K., S. Bhargava, N. Gopi Nath, J. R. Talwar, N. C. Nayak, B. N. Tandon & K. L. Wig. 1971. Non-cirrhotic portal fibrosis. Amer. J. Med. **51:** 160–169.
172. J. Sanchez-Tapias, H. C. Thomas & S. Sherlock. 1977. Lymphocyte populations in liver biopsy specimens from patients with chronic liver disease. Gut **18:** 472–475.
173. Sato, N., T. Kamada, S. Kawano, N. Hayashi, Y. Kishida, H. Mercer, H. Yoshihara & H. Abe. 1983. Effect of acute and chronic ethanol consumption on hepatic tissue oxygen tension in rats. Pharmacol. Biochem. Behav. **18:** 443–447.
174. Sato, N., T. Kamada, T. Schichiri, T. Matsumura & H. Abe. 1980. Effect of ethanol hemoperfusion and oxygen sufficiency in livers in situ. Adv. Ex. Med. Biol. **132:** 355–362.
175. Schaffner, F. 1970. Oxygen supply and the hepatocyte. Ann. N.Y. Acad. Sci. **170:** 67–74.
176. Schaffner, F. & H. Popper. 1963. Capillarization of hepatic sinusoids. Gastroenterology **44:** 239–242.
177. Schaffner, F. & H. Popper. 1970. Alcoholic hepatitis in the spectrum of ethanol-induced liver injury. Scand. J. Gastroenterol. **5**(S.7): 69–78.
178. Schmidt, W. 1977. The epidemiology of cirrhosis of the liver; a statistical analysis of mortality data with special reference to Canada. *In* Alcohol and the Liver. M. M. Fisher & J. G. Rankin, Eds. 1–26. Plenum Press. New York, NY.
179. Schmidt, W. & R. E. Popham. 1980. Sex differences in mortality. A comparison of male and female alcoholics. *In* Alcohol and Drug Problems in Women. O. J. Kalant, Ed. 365–384. Plenum Publ. Corp. New York, NY.
180. Shaw, S., E. A. Heller, H. S. Friedman, E. Baraona & C. S. Lieber. 1977. Increased hepatic oxygenation following ethanol administration in the baboon. Proc. Soc. Exp. Biol. Med. **156:** 509–513.
181. Shibayama, Y. & K. Nakata. 1985. Localization of increased hepatic vascular resistance in liver cirrhosis. Hepatology **5:** 643–648.
182. Skog, O. J. 1985. The wetness of drinking cultures; a key variable in epidemiology of alcoholic liver cirrhosis. Acta Med. Scand. **218**(5)(S.703): 157–184.
183. Slater, T. F. 1975. The role of lipid peroxidation in liver injury. *In* Pathogenesis and Mechanisms of Liver Cell Necrosis. D. Keppler, Ed. 209–223. University Park Press. Baltimore, MD.
184. Stevens, V. J., W. J. Fantl, C. B. Newman, R. V. Sims, A. Cerami, C. M. Peterson. 1981. Acetaldehyde adducts with hemoglobin. J. Clin. Invest. **67:** 361–369.
185. Strickland, R. G., E. Diaz-Jovanen & R. C. Williams. 1975. Application of lymphocyte markers as a probe in various disease states. *In* Lymphocytes and Their Interactions. 133–155. Raven Press. New York, NY.
186. Smith, M. G., A. L. W. F. Eddleston & R. Williams. 1975. Immunological factors in the evolution of active chronic hepatitis and other autoimmune liver disease. Clin. Gastroenterol. **4:** 297–313.

187. SMITH F. & D. L. PALMER. 1976. Alcoholism, infection and altered host defences: a review of clinical and experimental observations. J. Chron. Dis. **29:** 35–49.
188. SORRELL, M. F. & C. M. LEEVY. 1972. Lymphocyte transformation and alcoholic liver injury. Gastroenterology **63:** 1020–1025.
189. SORRELL, M. F. & D. J. TUMA. 1985. Hypothesis: alcoholic liver injury and the covalent binding of acetaldehyde. Alcoholism: Clin. Exp. Res. **9:** 306–309.
190. SPACH, P. I., J. W. PARCE & C. C. CUNNINGHAM. 1979. Effect of chronic ethanol administration on energy metabolism and phospholipase A2 activity in rat liver. Biochem. J. **178:** 23–33.
191. SPEISKY, H., D. BUNOUT, H. ORREGO, H. G. GILES, A. GUNASEKARA & Y. ISRAEL. 1985. Lack of changes in diene conjugate levels following ethanol-induced glutathione depletion or hepatic necrosis. Res. Comm. Chem. Pathol. Pharmacol. **48:** 77–90.
192. SYROTA, A., A. PARAF, C. GAUDEBOUT & A. DESGREZ. 1981. Significance of intra- and extrahepatic portasystemic shunting in survival of cirrhotic patients. Dig. Dis. Sci. **26:** 878–885.
193. TESCHKE, R., F. MORENO, E. HEINEN, J. HERRMANN, H. L. KRUSKEMPER & G. STROHMEYER. 1983. Hepatic thyroid hormone levels following chronic alcohol consumption: direct experimental evidence in rats against the existence of a hyperthyroid hepatic state. Hepatology **3:** 469–474.
194. THOMSON, A. D., M. A. G. COCHRANE & I. G. MCFARLANE. 1974. Lymphocyte cytotoxicity to isolated hepatocytes in chronic active hepatitis. Nature **252:** 721–722.
195. THURMAN, R. G., S. JI, T. MATSUMURA & J. J. LEMASTERS. 1984. Is hypoxia involved in the mechanism of alcohol-induced liver injury? Fundam. Appl. Toxicol. **4:** 125–133.
196. THURMAN, R. G., W. R. MCKENNA & T. B. MCCAFFREY. 1976. Pathways responsible for the adaptive increase in ethanol utilization following chronic treatment with ethanol: inhibition studies with hemoglobin-free perfused rat liver. Mol. Pharmacol. **12:** 156–166.
197. TUYNS, A. J. & G. PEQUIGNOT. 1984. Greater risk of ascitic cirrhosis in females in relation to alcohol consumption. Intl. J. Epidemiol. **13:** 53–57.
198. TYGSTRUP, N., P. K. ANDERSEN & B. L. R. THOMSEN. 1985. Prognostic evaluation in alcoholic cirrhosis. Acta Med. Scand. **218**(S.703): 81–95.
199. VAANANEN, H. & K. O. LINDROS. 1985. Comparison of ethanol metabolism in isolated periportal or perivenous hepatocytes: effects of chronic ethanol treatment. Alcoholism: Clin. Exp. Res. **9:** 315–321.
200. VAANANEN, H., M. SALASPURO & K. LINDROS. 1984. The effect of chronic ethanol ingestion on ethanol metabolizing enzymes in isolated periportal and perivenous rat hepatocytes. Hepatology **4:** 862–866.
201. VAN WAES, L. & C. S. LIEBER. 1977. Early perivascular sclerosis in alcoholic fatty liver: an index of progressive liver injury. Gastroenterology **73:** 646–650.
202. VIDELA, L., J. BERNSTEIN & Y. ISRAEL. 1973. Metabolic alterations produced in the liver by chronic ethanol administration: increased oxidative capacity. Biochem. J. **134:** 507–514.
203. VIDELA, L. & Y. ISRAEL. 1970. Factors that modify the mechanism of ethanol in rat liver and adaptative changes produced by its chronic administration. Biochem. J. **118:** 275–281.
204. VIDINS, E. I., R. S. BRITTON, A. MEDLINE, L. M. BLENDIS, Y. ISRAEL & H. ORREGO. 1985. Sinusoidal caliber in alcoholic and non-alcoholic liver disease: diagnostic and pathogenic implications. Hepatology **5:** 408–414.
205. VILLENEUVE, J. P., G. POMIER & P. M. HUET. 1981. Effect of ethanol on hepatic blood flow in unanesthetized dogs with chronic portal and hepatic vein catheterization. Can. J. Physiol. Pharmacol. **59:** 598–603.
206. VOROBIOFF, J., J. E. BREDFELDT & R. J. GROSZMANN. 1983. Hyperdynamic circulation in portal-hypertensive rat model: a primary factor for maintenance of chronic portal hypertension. Am. J. Physiol. **244:** G52–57.
207. VOROBIOFF, J., J. E. BREDFELDT & R. J. GROSZMANN. 1984. Increased blood flow through the portal system in cirrhotic rats. Gastroenterology **87:** 1120–1126.
208. WALD, N., S. HOWARD, P. G. SMITH & A. BAILEY. 1975. Use of carboxy-hemoglobin

levels to predict the development of diseases associated with cigarette smoking. Thorax **30:** 133–139.
209. WHIPPLE, A. O. 1945. The problem of portal hypertension in relation to the hepatosplenopathies. Ann. Surg. **122:** 449–475.
210. WILKINSON, P. 1980. Sex differences in morbidity of alcoholics. *In* Alcohol and Drug Problems in Women. O. J. Kalant, Ed. 331–364. Plenum Publishing Corp. New York, NY.
211. WILLIAMS, R. & M. DAVIS. 1977. Nutrition: effects of alcohol. Alcoholic liver injury. Proc. R. Soc. London, Ser. B **70:** 333–336.
212. WISSE, E., R. B. DE ZANGER, K. CHARELS, P. VAN DER SMISSEN & R. S. MCCUSKEY. 1985. The liver sieve: considerations concerning the structure and function of endothelial fenestra, the sinusoidal wall and the space of Disse. Hepatology **5:** 683–692.
213. WITTE, C. L. & M. H. WITTE. 1983. Splanchnic circulatory and tissue fluid dynamics in portal hypertension. Fed. Proc. **42:** 1685–1689.
214. WOLFE, J. D., D. P. TASHKIN, F. E. HOLLY, M. B. BRACHMAN & M. G. GENOVESI. 1977. Hypoxemia of cirrhosis. Detection of abnormal small pulmonary vascular channels by a quantitative radionuclide method. Amer. J. Med. **63:** 746–754.
215. WRIGHT, R. 1977. Chronic hepatitis. *In* Immunology of Gastrointestinal and Liver Disease. Current Topics in Immunology. J. Turk, Ed. 93–109. E. Arnold Publishers, Ltd., London.
216. YUKI, T., Y. ISRAEL & R. G. THURMAN. 1982. The swift increase in alcohol metabolism: inhibition of propylthiouracil. Biochem. Pharmacol. **31:** 2403–2407.
217. YUKI, T. & R. G. THURMAN. 1979. Mechanism of the swift increase in alcohol metabolism "SIAM" in the rat. *In* Alcohol and Aldehyde Metabolizing Systems, Vol. IV. R. G. Thurman, Ed. Vol. 4: 689–696. Plenum Press. New York. NY.
218. YUKI, T. & R. G. THURMAN. 1980. The swift increase in alcohol metabolism: time course for the increase in hepatic oxygen uptake and the involvement of glycolysis. Biochem. J. **186:** 119–126.
219. ZETTERMAN, R. K. & C. M. LEEVY. 1975. Immunologic reactivity and alcoholic liver disease. Bull. N.Y. Acad. Sci. **15:** 533–544.
220. ZETTERMAN, R. K., A. LUISADA-OPPER & C. M. LEEVY. 1976. Alcoholic hepatitis: cell-mediated immunological response to alcoholic hyalin. Gastroenterology **70:** 382–384.
221. ZINNEMANN, H. H. 1975. Autoimmune phenomena in alcoholic cirrhosis. Am. J. Dig. Dis. **20:** 337–345.
222. ZIMMON, D. S. & R. E. KESSLER. 1980. Effect of portal venous blood flow diversion on portal pressure. J. Clin. Invest. **65:** 1388–1397.

DISCUSSION OF THE PAPER

M. A. ROTHSCHILD (*New York University Medical Center, New York, NY*): Your data suggest that as the blood flow decreases, the hemoglobin content appears to decrease. This suggests that the intrahepatic hematocrit is depressed. Where does this "skimming" occur, outside the liver or sequestered in some other place?

ISRAEL: It stays outside the liver, because there is a 30% reduction in total hemoglobin in the liver, along with a compression of the sinusoidal space.

E. RUBIN (*Thomas Jefferson University School of Medicine, Philadelphia, PA*): Whatever the mechanism the antibody to acetaldehyde adducts, it might be used as a marker, similar to the glycosylated hemoglobin in diabetes.

ISRAEL: We're actively looking in this area. We have a small clinical sample, and the data look very good.

The Effect of Ethanol on Superoxide Production in Alveolar Macrophages

RAYMOND J. DORIO,[a] JAN B. HOEK,[b] HENRY J. FORMAN,[a] AND EMANUEL RUBIN[b]

[a]*Department of Pediatrics*
Children's Hospital of Los Angeles
Los Angeles, California 90027
and
[b]*Department of Pathology*
Thomas Jefferson University
Philadelphia, Pennsylvania 19103

Human alcoholics are predisposed to develop pneumonia.[1] Resident macrophages are normal alveolar constituents that produce superoxide and other free radicals in response to bacterial products. Superoxide production in these cells is elicited by factors that stimulate either of two interrelated pathways,[2] one involving calcium mobilization and the other protein kinase C activation. Superoxide production is initiated when agonists bind cell surface receptors. However, these cell surface receptors can be bypassed by increasing intracellular calcium with ionophores or by activating protein kinase C directly with phorbol esters. We have found that ethanol, at physiologic levels, has several effects on superoxide production by alveolar macrophages. These effects may be important in modulating the response of these cells to physiologic stimuli.

Phorbol 12-myristate 13-acetate (PMA) is a potent agonist for superoxide production (1–2 nmoles/min/10^6 cells at maximally effective PMA concentrations). Ethanol pretreatment results in two effects—synergism between ethanol and PMA at low ethanol concentrations (10 and 25 mM) and inhibition of PMA-induced superoxide production at higher ethanol concentrations (>75 mM) (FIG. 1). The inhibition of PMA-induced superoxide production is immediate with 500 mM ethanol; however, at lower ethanol concentrations, inhibition increases with time, up to 30 min after ethanol addition (data not shown).

Ethanol itself is a weak agonist for superoxide production (FIG. 2). Superoxide production is directly related to the concentration of ethanol (10.6 ± 0.5 pmol/min/10^6 cells at 75 mM ethanol and 33.9 ± 1.7 pmol/min/10^6 cells at 500 mM ethanol). The level of superoxide produced in response to ethanol is at least 100-fold less than that produced in response to PMA.

We can only speculate on the mechanisms responsible for these effects. Ethanol, at concentrations that stimulate superoxide production and inhibit PMA-induced superoxide production, perturbs membrane lipids. Perturbation of the membrane lipid structure after ethanol addition may affect membrane proteins that depend on the lipid environment for activity. The superoxide oxidase and protein kinase C are both dependent on lipids for activity.[3,4] An alternative explanation may involve mobilization of intracellular calcium. Hoek has shown that ethanol induces phosphatidylinositide breakdown and intracellular calcium mobilization in hepatocytes.[5] If ethanol also mobilizes intracellular calcium in macrophages this finding might provide a mechanism for two of the effects—stimulation of superoxide production and synergism between PMA and ethanol. The calcium ionophore A23187 is, for instance, a

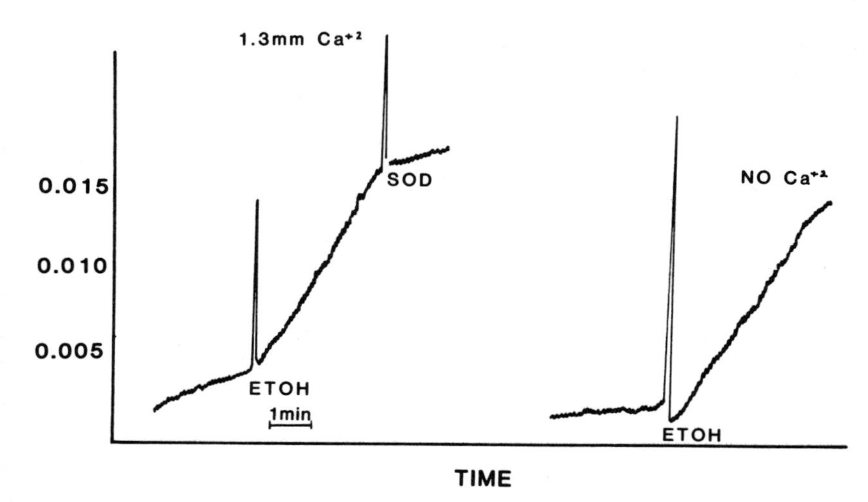

FIGURE 1. Ethanol induces superoxide production in alveolar macrophages. Superoxide production in rat alveolar macrophages is measured after addition of 163 mM ethanol and by following reduction of ferricytochrome C at $\triangle A = 550–540$ nm. Superoxide dismutase inhibits the reaction.

low-potency agonist for superoxide production, and also acts synergistically with PMA to produce superoxide in these cells. Although direct stimulation of superoxide production of ethanol is small, ethanol may play a more significant role in desensitizing the superoxide generating system to other, stronger agonists.

In summary, we find that ethanol, at physiologic levels, has several effects of

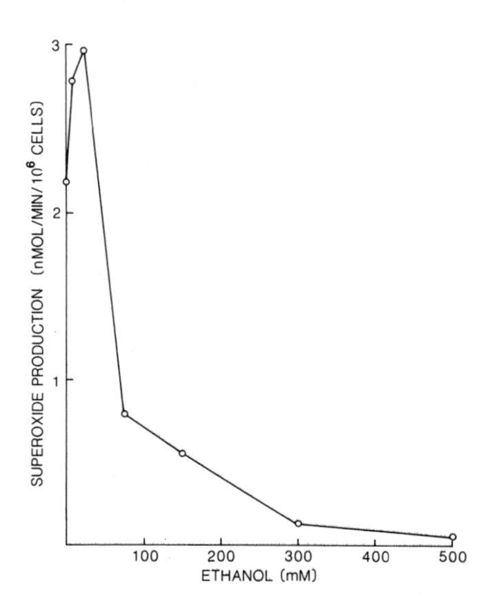

FIGURE 2. Synergistic and inhibitory actions of various concentrations of ethanol on phorbol-ester-induced superoxide production. Macrophages are preincubated with varying concentrations of ethanol for 15 min at 37° C. PMA (10 ng/ml) is then added and the rate of superoxide generation measured.

superoxide production by alveolar macrophages. These effects may be important in modulating the response of these cells to physiologic stimuli.

REFERENCES

1. SCHMIDT, W. & J. DE LINT. 1972. J. Stud. Alcohol **33:** 171–85.
2. MCPHAIL, L.C. & R. SNYDERMAN. 1984. Contemp. Top. Immunobiol. **14:** 247–281.
3. BALBIOR, B.M. & W.A. PETERS. 1981. J. Biol. Chem. **256:** 2321–2323.
4. ASHENDEL, C.L. 1985. Biochim. Biophys. Acta **822:** 219–242.
5. HOEK, J.B. 1987. Ann. N.Y. Acad. Sci. This volume.

Ethanol Decreases Corticotropin-Releasing Factor Binding, Adenylate Cyclase Activity, Pro-Opiomelanocortin Biosynthesis, and β-Endorphin Release in Cultured Pituitary Cells

JITENDRA R. DAVE[a] AND ROBERT L. ESKAY

Laboratory of Clinical Studies
Division of Intramural Clinical and Biological Research
National Institute on Alcohol Abuse and Alcoholism
Bethesda, Maryland, 20892

Although ethanol is commonly considered to be anxiolytic, its administration produces a variety of physiological changes that mimic a stress response. In both humans and rodents, acute or chronic administration of ethanol is reported to stimulate adrenal corticosteroid release. In a recent study, we reported that animals exposed continuously for 14 days to ethanol vapor in an inhalation chamber exhibited lower corticotropin-releasing factor (CRF) binding and adenylate cyclase activity in anterior (AL) and neurointermediate lobe (NIL) membranes of the pituitary gland, as compared to controls not treated with ethanol.[1] This treatment also resulted in decreased circulating immunoreactive β-endorphin (BE) levels and caused a reduction in pro-opiomelanocortin (POMC) mRNA levels in both AL and NIL of the pituitary gland.[1] In a recent study,[2] pretreatment of cultured anterior pituitary cells with 0.2% ethanol for 24 h, but not acute exposure to ethanol, was reported to result in decreased basal and CRF-stimulated ACTH secretion. This finding suggests that chronic ethanol administration reduces pituitary responsiveness. Since ethanol-induced alterations of pituitary gland function are insufficiently characterized at the present time, this study was undertaken to determine the effect of chronic ethanol exposure of various cellular events associated with the synthesis and secretion of POMC-derived peptides from cultured rat anterior pituitary (AP) cells and from a mouse-derived anterior pituitary tumor-cell line (AtT-20 cells). The specific objectives were to determine if chronic ethanol treatment could i) modify the binding of CRF and membrane-associated adenylate cyclase activity, ii) alter POMC biosynthesis, and iii) modify BE secretion from AP and AtT-20 cells.

RESULTS AND DISCUSSION

The present study demonstrates that chronic ethanol treatment of both AP and AtT-20 cells produce a dose-related decrease in ^{125}I-labeled CRF binding and basal

[a]Address correspondence to: Dr. J. R. Dave, LPPS, NIAAA, 12501 Washington Ave., Rockville, MD 20852.

TABLE 1. Effect of Chronic Ethanol Treatment on CRF Binding, Adenylate Cyclase Activity, β-Endorphin Release, and Pro-opiomelanocortin mRNA Levels in Cultured Rat Anterior Pituitary Cells[a]

Treatment	^{125}I-rCRF Specific Binding (% of Control)	Adenylate Cyclase Activity		β-Endorphin Release (% of Control)	Pro-opiomelanocortin mRNA (% of Control)
		Basal (% of Control)	CRF (0.4 nM) (% of Basal)		
Control	100 ± 8	100 ± 4	250 ± 10	100 ± 4	100 ± 5
Ethanol 0.1%	95 ± 5	97 ± 2	200 ± 11	100 ± 4	80 ± 10
0.2%	80 ± 6	85 ± 3	150 ± 8	100 ± 4	60 ± 10
0.4%	60 ± 4	78 ± 4	90 ± 8	75 ± 5	40 ± 10
0.6%	61 ± 6	71 ± 4	70 ± 4	60 ± 6	40 ± 10

[a] AP cells were cultured in DMEM media and used on day 5 after the dispersion. Prior to the experimental manipulations, culture media was removed and the cells were incubated in the presence of various concentrations of ethanol at 37° C (10% CO_2) in DMEM containing 2% fetal calf serum for 24 hr. Other details of cell culture are described in an earlier study.[3] ^{125}I-rat CRF binding and AC activity were determined as described earlier.[1] For peptide release studies, a medium containing the indicated concentrations of ethanol was added to each well. Following the 1-hr experimental incubation of triplicate wells, the medium was removed and centrifuged for 20 sec at 8,000 × g in a Beckman Microfuge to remove detached cells. The supernatant fluid was removed for the determination of BE by radioimmunoassay.[4] For POMC mRNA studies, the medium was removed at the end of the experimental period (24 hr), cells were washed twice with 1 ml of PBS and incubated with 0.5 ml of SET-PK (1% SDS, 10 mM Tris-HCl, pH 7.8, 5 mM EDTA containing 100 μg/ml proteinase K) solution for 90–120 min at 42° C, and total nucleic acids were extracted.[1]

TABLE 2. Effect of Chronic Ethanol Treatment on CRF Binding, Adenylate Cyclase Activity, β-Endorphin Release, and Pro-opiomelanocortin mRNA Levels in Cultured Mouse-Derived AtT-20's[a]

Treatment	^{125}I-rCRF Specific Binding (% of Control)	Adenylate Cyclase Activity: Basal (% of Control)	Adenylate Cyclase Activity: CRF (0.4 nM) (% of Basal)	β-Endorphin Release (% of Control)	Pro-opiomelanocortin mRNA (% of Control)
Control	100 ± 5	100 ± 5	320 ± 20	100 ± 3	100 ± 10
Ethanol 0.1%	90 ± 3	90 ± 2	310 ± 15	100 ± 3	100 ± 5
0.2%	83 ± 6	82 ± 3	180 ± 14	100 ± 4	100 ± 10
0.4%	55 ± 4	65 ± 5	100 ± 12	70 ± 6	40 ± 10
0.6%	50 ± 6	45 ± 10	80 ± 7	55 ± 7	20 ± 10

[a]AtT-20 cells were cultured in 90% DMEM/10% fetal calf serum in humidified 90% air/10% CO_2 at 37° C. Cells were plated in 22.6-mm-diameter culture wells (12 wells per plate, Costar) at a density of 80,000 cells per well and grown to 60–80% confluency (1 million cells per well at 5–6 days after plating). Prior to experimental procedures, the culture medium was removed and the cells were incubated in the presence of various concentrations of ethanol at 37° C (10% CO_2) in DMEM containing 2% fetal calf serum. The other procedures were identical to those described in the footnote to TABLE 1.

and CRF-stimulated adenylate cyclase activity (TABLES 1 & 2). This treatment also produced a decrease in BE secretion and POMC mRNA levels. However, in AP cells, chronic ethanol treatment did not alter prolactin (PRL) specific mRNA levels (data not presented). Furthermore, chronic ethanol treatment of AP and AtT-20 cells for 24 hr did not affect intracellular BE levels. Taken together, these observations suggest that the observed decrease of BE secretion in this study may be due to a direct action of ethanol on the plasma membrane. The results of the present *in-vitro* study are consistent with our previous *in-vivo* findings and suggest that ethanol-mediated alterations of pituitary membranes, combined at least in part with a direct effect of ethanol on the POMC gene transcription, may result in decreased POMC mRNA levels. Further studies on the effect of ethanol on the transcription rate of the POMC gene should provide additional insight into the mechanism of action of ethanol on the hypothalamic-pituitary-adrenal axis.

REFERENCES

1. DAVE, J. R., L. E. EIDEN, J. W. KARANIAN & R. L. ESKAY. 1986. Ethanol exposure decreases pituitary corticotropin-releasing factor binding, adenylate cyclase activity, proopiomelanocortin biosynthesis, and plasma beta-endorphin levels in the rat. Endocrinology **118:** 280–286.
2. RIVIER, C., T. BRUHN & W. VALE. 1984. Effect of ethanol on the hypothalamic-pituitary-adrenal axis in the rat: role of corticotropin-releasing factor (CRF). J. Pharmacol. Exp. Ther. **229:** 127–135.
3. DAVE, J. R., L. E. EIDEN & R. L. ESKAY. 1987. Elevation of intracellular cyclic AMP by corticotropin-releasing factor links secretion of beta-endorphin and biosynthesis of proopiomelanocortin in cultured anterior pituitary and AtT-20 cells. 3rd Biological Colloquium, N.Y. Acad. Sci. In press.
4. DAVE, J. R., N. RUBINSTEIN & R. L. ESKAY. 1985. Evidence that beta-endorphin binds to specific receptors in rat peripheral tissue and stimulates the adenylate cyclase/cyclic AMP system. Endocrinology **117:** 1389–1396.

Implications of Alcohol-Induced Alterations in the Prostaglandin Profile of the Vascular System

JOHN W. KARANIAN, JAMES YERGEY,
AND NORMAN SALEM, JR.

Laboratory of Clinical Studies
Division of Intramural Clinical and Biological Research
National Institute on Alcohol Abuse and Alcoholism
Bethesda, Maryland 20892

Epidemiologic studies have suggested that low rates of alcohol consumption may lower the incidence of atherosclerosis-related cardiovascular pathology, whereas high exposure correlates with a high incidence of vasospastic disorders.[1] We have examined the proposition that the prostaglandin (PG) profile in rat aorta, cerebral microvasculature and plasma is modulated by alcohol dose and the duration of exposure and thereby contributes to changes in the cardiovascular state. Isolated rat aorta were either exposed to alcohol *in vitro* (FIGURE 1) or by an inhalation technique using our automated apparatus for precise control of alcohol vapor and blood alcohol concentrations (BAC) (FIGURE 2).[2] A wide range of mean BAC (50–450 mg%) were obtained based upon multiple blood ethanol determinations for each rat throughout the exposure period. An *acute* exposure to moderate alcohol levels *in vitro* decreases the amplitude of contraction to a thromboxane-mimic, whereas higher levels do not effect the response. The observed depression in aortic contractility is associated with a

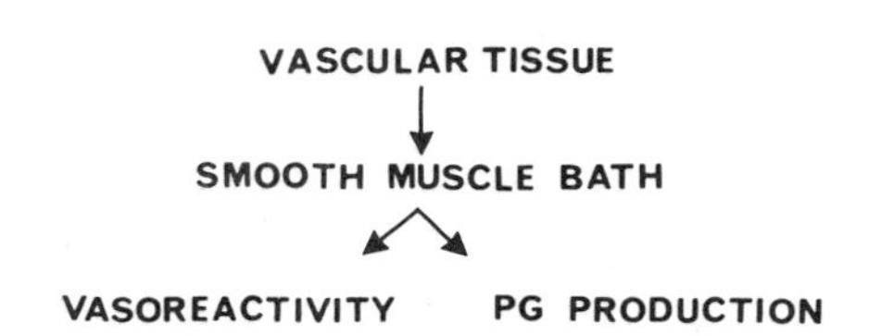

FIGURE 1. *In-vitro* exposure. Rat aorta, placed in oxygenated Krebs Ringer Bicarbonate, were exposed to graded concentrations of ethanol. Vasoreactivity (tension) and prostaglandin (PG) production (RIA) were determined at various time points.

significant increase in the PGI_2 (vasodilator, anti-aggregator)/TXA_2(vasoconstrictor, aggregator) ratio. At higher levels, alcohol acts as a spasmogen, increasing intracellular CA^{++}, and in this way may counteract the depressive effect of PGs, such as PGI_2, on vascular contractility and platelet aggregability.[3] *Chronic* alcohol inhalation significantly depressed aortic PG levels (27% decrease by moderate BAC; 45% decrease by high BAC) and induced a mild hyperreactivity to the TXA_2-mimic. This depression in PG levels may be due in part to alcohol-induced depletion of the precursor fatty acids dihomogammalinolenic acid (20:3w6) and arachidonic acid (20:4w6); their content was inversely related to the duration of exposure and decreased by as much as 50%. Similarly, long-term inhalation exposure to a high BAC induced a marked hypersensitivity to the pressor effects of noradrenaline and the TXA_2-mimic (22% and 44% increase, respectively; monitored via tail artery cannulation) and depressed plasma PGI_2 and TXA_2 levels (21% decrease by moderate BAC; 37%

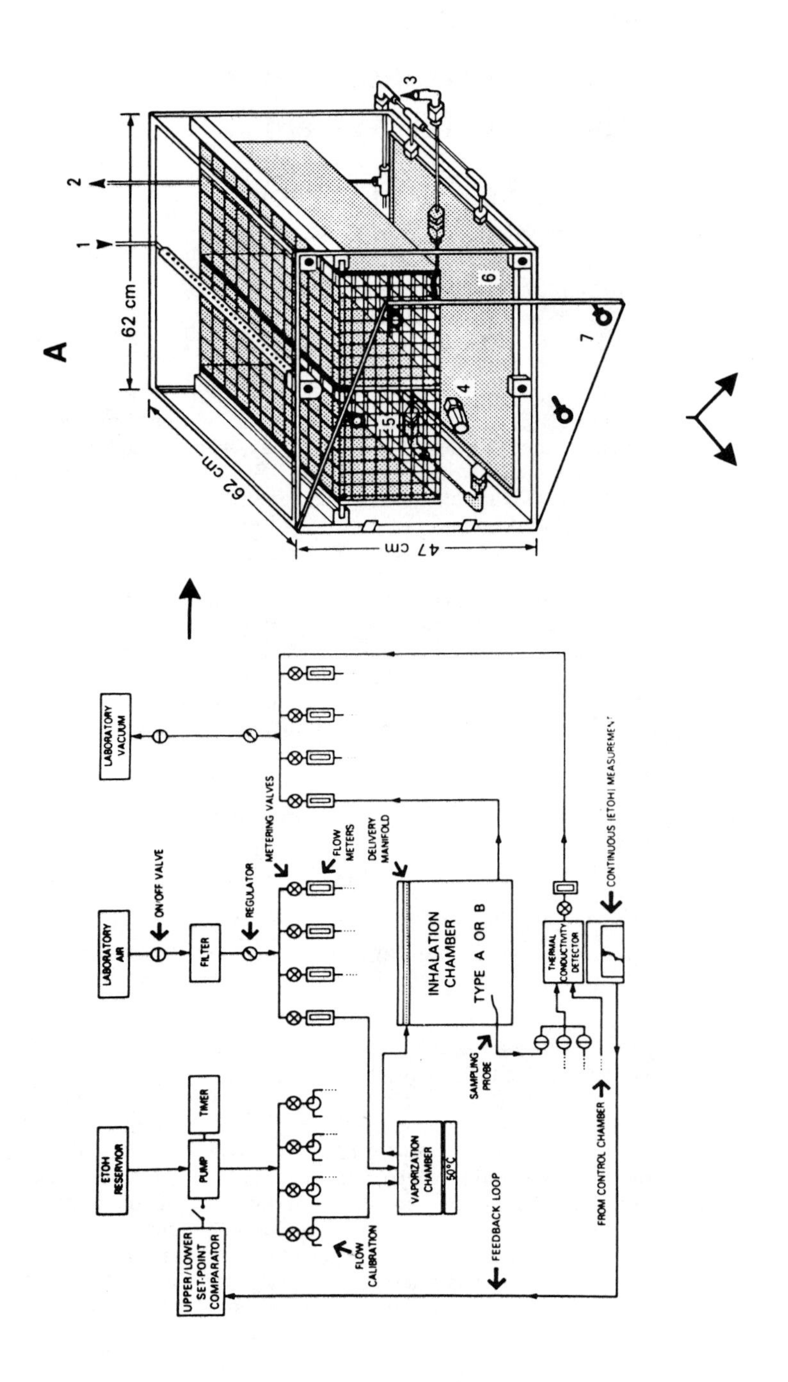
A
62 cm
62 cm
47 cm
1
2
3
4
5
6
7
ETOH RESERVIOR
TIMER
PUMP
UPPER/LOWER SET-POINT COMPARATOR
FLOW CALIBRATION
VAPORIZATION CHAMBER
50°C
FEEDBACK LOOP
LABORATORY AIR
ON/OFF VALVE
FILTER
REGULATOR
METERING VALVES
FLOW METERS
DELIVERY MANIFOLD
INHALATION CHAMBER
TYPE A OR B
SAMPLING PROBE
THERMAL CONDUCTIVITY DETECTOR
FROM CONTROL CHAMBER
LABORATORY VACUUM

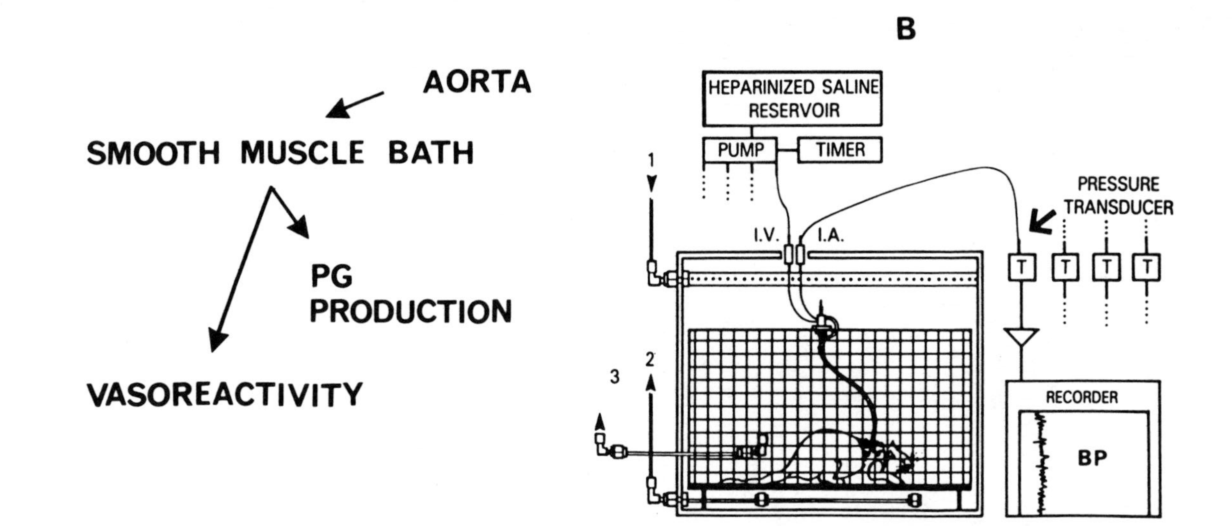

FIGURE 2. Inhalation technique: schematic representation of ethanol metering and vaporization apparatus and inhalation chamber vapor concentration sensing device and control system. Inhalation chamber type **A** for acute and chronic ethanol exposure. Vasoreactivity and PG production were subsequently determined in isolated rat aorta. Inhalation chamber type **B** for harnessed rats with an indwelling catheter. Withdrawal of blood samples for plasma PG determination and monitoring of blood pressure and its reactivity in unanesthetized rats was possible during acute or chronic ethanol exposure. **1**) Stainless steel inlet manifold (2 cm × 580 cm with 0.75-mm holes drilled at 120 and 240° from top) from laboratory air and ethanol metering apparatus and vaporization flask; **2**) Chamber exhaust via 4 ports (1/4″ copper tubing) to vacuum metering apparatus; **3**) Tube (1/8″ stainless steel) to thermal conductivity detector; **4**) Sampling port; **5**) Two-compartment, tracked sheet metal cage with mesh bottom and separate mesh top; **6**) Sheet metal pan for droppings; **7**) 1/2″-thick hinged Plexiglass door with airtight gasket seal and quick release mechanism.

decrease by high BAC). As expected, nonsteroidal anti-steroidal anti-inflammatory drug treatment of normal rats induced a comparable increase (48%) in vascular reactivity to the TXA_2-mimic. Pretreatment of alcoholic rats with aspirin (45 mg/kg) induced a mild (nonsignificant) increase in this response. These data are consistent with the hypothesis that chronic alcohol exposure has an aspirinlike effect on the prostanoid system as reflected by altered contractility. A cerebral microvasculature fraction has shown qualitatively similar effects with respect to PG levels; acute alcohol inhalation increased the PGI_2/TXA_2 ratio (fourfold) and chronic exposure decreased the PGI_2/TXA_2 ratio (32%). Hyperreactivity and the depression in total PG levels associated with chronic alcohol abuse could contribute to the hypertensive and vasospastic disorders found in alcoholics. In contrast, hyporeactivity and an increase in total PG levels associated with moderate alcohol use may be one mechanism contributing to the lower incidence of atherosclerotic heart disease in this group. Thus, modification of the vascular state, at least in part, will be determined by the summation of the biological effects of various prostaglandins.[4]

REFERENCES

1. DOLL, R. 1983. Prospects for prevention. Br. J. Med. **286:** 445–449.
2. KARANIAN, J. W., J. YERGEY, R. LISTER, N. D'SOUZA, M. LINNOILA & N. SALEM, JR. 1986. Characterization of an automated apparatus for precise control of inhalation chamber ethanol vapor and blood ethanol concentrations. Alcoholism: Clin. Exp. Res. **10**(3): 19–25.
3. KARANIAN, J. W., & N. SALEM, JR. 1986. The effects of acute and chronic ethanol exposure on the response of rat aorta to a thromboxane-mimic U46619, *in vitro*. Alcoholism: Clin. Exp. Res. **10**(1): 1–6.
4. KARANIAN, J. W., M. STOJANOV & N. SALEM, JR. 1985. The effect of ethanol on prostacyclin and thromboxane A_2 synthesis in rat aortic rings, *in vitro*. Prostanglandins, Leukotrienes and Medicine **20:** 175–186.

The Increased Vulnerability of Kidneys of Chronic Alcoholic Rats to the Insult of a Nephrotoxin[a]

MASAAKI ISHIGAMI, S. TSUYOSHI OHNISHI, TEIRYO MAEDA, AND SO YABUKI

Membrane Research Institute
University City Science Center
Philadelphia, Pennsylvania 19104

Kanto Rosai Hospital
Kawasaki, Japan

Toho University School of Medicine
Tokyo, Japan

The effects of chronic alcohol ingestion on the liver, heart, and brain has been extensively studied, and it is known that subcellular components such as mitochondria and Na-K-ATPase enzyme are impaired by alcoholism.[1-4] However, such effects on the kidney are less well defined. Growing evidence suggests that renal failure occurs with higher frequency in cirrhotic alcoholics. This suggests that there may be a link between liver damage and kidney function, and that the effects of chronic alcoholism on the kidney may also be clinically significant. The condition in which this link is expressed is called the hepatorenal syndrome, but the precise mechanism has not been elucidated.[5] Thus, we undertook this work to study the effects of chronic alcohol ingestion on renal function. Our goal was to study the susceptibility of the alcoholic kidney to trauma. We report here on our use of a nephrotoxin insult as a trauma model (ischemic trauma in the kidney[6] was also shown to have an effect similar to that of a nephrotoxin).

At the Department of Pathology, Hahnemann University School of Medicine, male Sprague-Dawley rats (weight ranges between 250 and 350g) were fed a liquid diet containing ethanol (ethanol provided 36% of total calories) for 5 weeks to induce chronic alcoholism. Sodium thiobarbital (Inactin, 100 mg/Kg; i.p.) was used for anesthesia. An endotracheal tube was inserted and a polyethylene catheter was placed into an external jugular vein for infusion containing both I-125-iothalamate (7,000 cpm/ml) and I-131-iodohyppurate (7,000 cpm/ml) for the determination of glomerular filtration rate (GFR) and renal blood flow (RBF). The left ureter was cannulated with a polyethylene tube for urine collection. Six hours later after mercuric chloride injection (1 mg/kg; i.v.), urine and blood samples were collected, and the renal functions were measured. In order to assess the statistical significance of the data, the paired t-test was used. Results with a p value smaller than 0.05 were considered to be significant.

As shown in FIGURE 1, renal functions of both alcoholic rats and nonalcoholic rats were the same. Five weeks of alcohol ingestion did not significantly alter either GFR (403 ± 27 μl/min/100g body weight (BW) vs 380 ± 30 μl/min/100g BW in

[a]Supported in part by National Institutes of Health Grant AM36216.

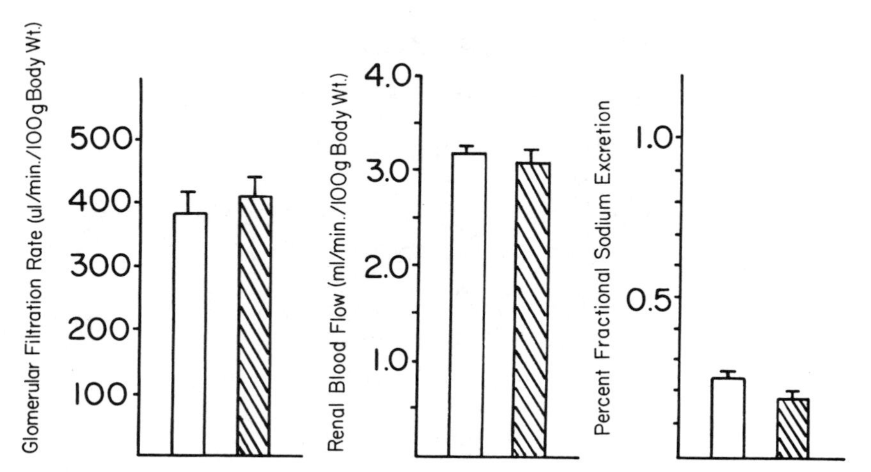

FIGURE 1. Comparison of renal functions between control rats (*open columns*, N = 10) and 5-week alcoholic rats (*cross-hatched columns*, N = 10).

nonalcoholic rats) or RBF (3.1 ± 0.1 m/min/100g BW vs 3.1 ± 0.2 ml/min/100g BW in nonalcoholic rats).

The percent fractional sodium excretion of these two groups was also not significantly different. However, the difference between these two types of rats became apparent when animals were subjected to mercuric chloride, a nephrotoxin; it caused a significantly larger impairment of kidney functions in alcoholic rats than in normal rats. Six hours after mercuric chloride (1 mg/kg) was administered (i.v.), GFR and RBF in nonalcoholic rats were 241 ± 20.5 μl/min/100g BW (37% decrease; the decrease from the original value) and 1.8 ± 0.08 ml/min/100g BW (42% decrease), respectively, whereas they were 76 ± 23.5 ml/min/100g BW (68% decrease) and 1.0 ± 0.25 ml/min/100g BW (81% decrease) in alcoholic rats (FIG. 2). It was also found that the sodium excretion was better preserved in nonalcoholic rats (data not shown).

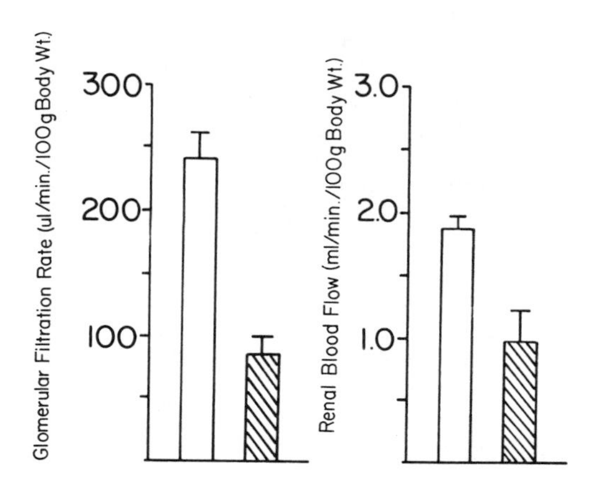

FIGURE 2. Decrease of renal functions of control rats (*open columns*, N = 5) and 5-week alcoholic rats (*cross-hatched columns*, N = 5) 6 hr after mercuric chloride injection (1 mg/kg; i.v.).

A direct alcoholic effect on liver cells and cardiac cells, such as damage to mitochondrial respiratory activity,[1,2] has been suggested. It was also reported that acetaldehyde, a metabolite of ethanol, impairs cardiac mitochondrial function as well as the ATPase activity of cardiac muscle.[3,4] Several mechanisms of acute renal failure have been suggested, which include vascular effects (determined by functional studies) and tubular effects (determined by morphological studies).[7,8] However, the mechanism of alcoholic effects in the kidney has not been characterized. It is possible that ethanol and/or acetaldhyde may cause some damage to the membrane structure and function of renal cells. That damage is not recognized under normal conditions. However, as our observation clearly demonstrated, when the kidney is exposed to the insult of a nephrotoxin (or ischemia; data not shown), the kidney of the chronic alcoholic is more vulnerable to such an insult. This model may be useful for exploring the molecular mechanism of reduced tolerance of biological membranes in chronic alcoholism.

ACKNOWLEDGMENTS

The authors thank Dr. E. Rubin and B. C. Ponnappa of Hahnemann University School of Medicine for providing us with alcoholic and control rats.

REFERENCES

1. LIEBER, C. S. & E. RUBIN. 1968. Alcoholic fatty liver in man on high protein and low fat diet. Am. J. Med. **44:** 200–206.
2. HASUMURA, Y., R. TESCHKE & C. S. LIEBER. 1974. Acetaldehyde oxidation by hepatic mitochondria; Its decrease after chronic ethanol consumption. Gastroenterology **66:** 415–422.
3. SARMA, J. S. M., S. IKEDA, R. FISCHER, Y. MARUYAMA, R. WEISHAAR & R. J. BING. 1976. Biochemical and contractile properties of heart muscle after prolonged alcohol administration. J. Mol. Cell. Cardiol. **8:** 951–972.
4. PACHINGER, O. M., H. TILLMANS, J. C. MAO, J. M. FAUVEL & R. J. BING. 1973. The effect of prolonged administration of ethanol on cardiac metabolism and performance on the dog. J. Clin. Invest. **52:** 2690–2696.
5. BRENNER, B. M. & J. M. LAZARUS. 1983. *In* Acute Renal Failure. 56–65. W. B. Saunders Publ. Company. Philadelphia, PA.
6. BRENNER, B. M. & J. M. LAZARUS. 1983. *In* Acute Renal Failure. 483–491. W. B. Saunders Publ. Company. Philadelphia, PA.
7. HSU, C. H., T. W. KURTZ, J. ROSENZWEIG & J. M. WELLER. 1977. Renal hemodynamics in $HgCl_2$-induced acute renal failure. Nephron **18:** 326–332.
8. BIBER, T. U. L., M. MYLLE, A. D. BAINES, C. W. GOTTSCHALK, J. R. OLIVER & M. C. MACDOWELL. 1968. A study of micropuncture and microdissection of acute renal damage in rats. Am. J. Med. **44:** 664–705.

Functional Analysis of T Lymphocyte Subsets in Chronic Experimental Alcoholism in Rats

OMAR BAGASRA, ALY HOWEEDY, AND ANDRE KAJDACSY-BALLA

Department of Pathology and Laboratory Medicine
Hahnemann University
Philadelphia, Pennsylvania 19102

Human alcoholics have a constellation of alcohol-associated conditions, which may contribute to the alcoholic's increased susceptibility to infections. These conditions include liver cirrhosis, malnutrition, cancer, and immunosuppression. However, whether an altered immune responsiveness is due to the effects of ethanol and its metabolites or to the other frequently associated complications of chronic alcohol

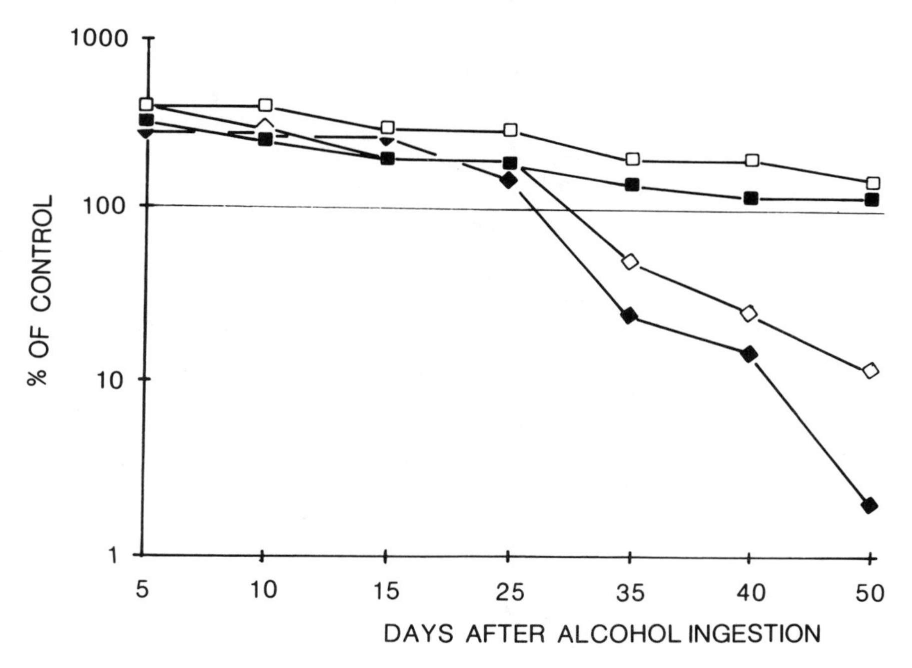

FIGURE 1. Primary splenic PFC (♦—♦) and serum antibody (◊—◊) responses of chronically alcoholic rats to SRBC. Also primary splenic PFC (■—■) and serum antibody (□—□) responses to SIII antigen. Alcoholic rats and their littermate controls were immunized with SRBCs (3×10^8 cells/animal) or SIII (0.5 μg/animal) i.p. at various intervals after ethanol ingestion. Primary splenic PFC and serum antibody responses to the antigens were measured 5 days postimmunization. Mean IgM PFC control counts are shown on a PFC/10^6 cell basis: serum antibody titers are those of representative experiments. Mean responses are the direct results obtained from 3 animal pairs per experimental variables.

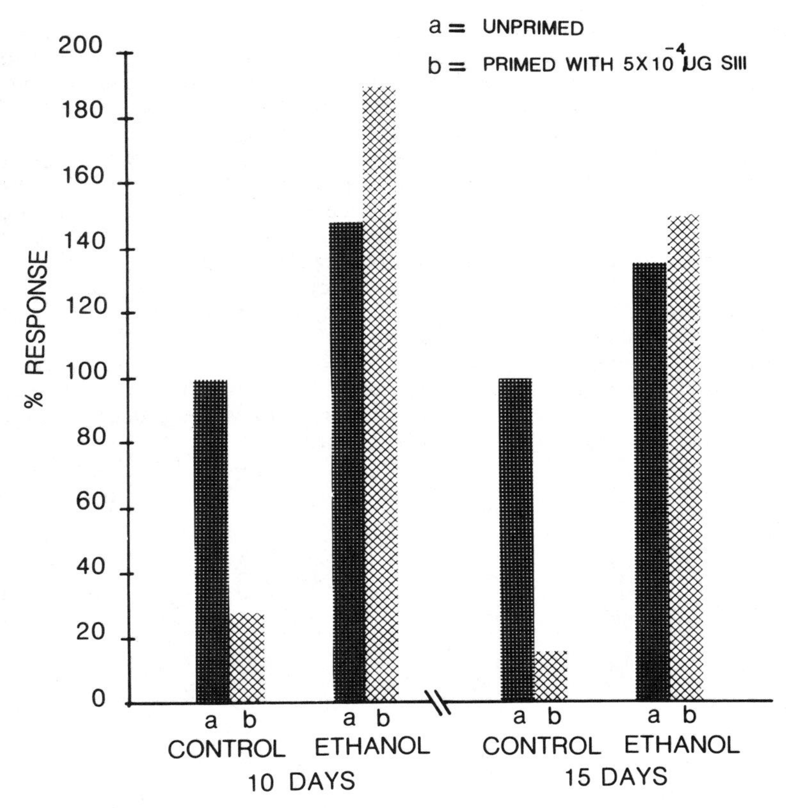

FIGURE 2. Failure to suppress the primary splenic PFC responses of alcoholic rats to SIII after low-dose priming. Alcoholic rats from day 10 and day 15 and their littermate controls were primed with 0.0005 μg SIII/animal. i.p. 3 days later these animals were immunized with an optimal dose (0.5 μg) of SIII by the same route. Primary splenic PFC were measured 5 days later. Results recorded as PFC/10^6 spleen cells and compared to the responses of unprimed rats and expressed as the percentage of control responses. Results of a representative experiment.

consumption has not been determined satisfactorily. In the present study we have used a well established rat model of experimental alcoholism, originally described by Lieber *et al.,* in which malnutrition, infection, and underlying hepatic dysfunctions are well controlled.

In order to obtain a better understanding of immune system function in chronic alcoholism, we have assessed primary B cell responses to helper T cell independent (TI) and dependent (TD) antigens in chronic alcoholic Sprague-Dawley male rats fed totally liquid diet containing ethanol. Pair-fed littermates received the same diet except that ethanol was isocalorically replaced by carbohydrate, which accounted for 36% of total calories. Induction of a B cell response to TI antigens like SIII (pneumococcal polysaccharide type III) is not dependent upon helper T cells or macrophages, but the magnitude of the developing response is normally regulated by suppressor T cells. Elimination of T suppressor cells or functional defect in this subset will result in enhanced humoral immune response as compared to littermate controls.

Therefore, SIII could be used to assess B cell as well as suppressor T cell function *in vivo,* whereas, SRBC is TD antigen and could be used to assess helper T cell function. The ability of alcoholic animals to mount primary *in-vivo* splenic plaque-forming cell (PFC) responses to SIII was elevated throughout 50 days of observations when compared to pair-fed controls; serum antibody responses to SIII paralleled to enhanced PFC responses. Primary *in-vivo* B cell responses to TD antigen (SRBC) were initially elevated but were found to be significantly suppressed 30 days after chronic ethanol consumption. The degree of immunosuppression increased with length of chronic ethanol consumption (FIGURE 1). The elevated primary splenic PFC responses observed during chronic ethanol consumption to TI (SIII) antigen may be attributed to loss of T suppressor cell control, since alcoholic rats' splenic cells did not respond to low-dose priming with SIII (FIGURE 2). We suggest that either loss of function and/or actual depletion of accessory and regulatory cells (T suppressor and T helper) may be responsible for irregularities in B cell function observed during chronic alcoholism. Currently we are examining the degree of susceptibility of alcoholic animals to various microorganisms, dividing them into two categories: a) where primary defense is humoral and b) where primary defense is cell-mediated.

Electrophoretic Variants of Arylsulfatase A in Alcoholic Patients and Controls[a]

PAUL MANOWITZ, LUANN VILLANO FINE,
RENA NORA, KEVIN PERNICANO, PETER NATHAN,
AND SUDHANSU CHOKROVERTY

University of Medicine and Dentistry of New Jersey-Robert Wood Johnson Medical School Piscataway, New Jersey 08854

Veterans Administration Medical Center Lyons, New Jersey 07939

Rutgers, The State University of New Jersey New Brunswick, New Jersey 08903

The purpose of this study is to determine if electrophoretic variant forms of arylsulfatase A are found more frequently in hospitalized alcoholic patients than in controls, and, if so, to determine which forms are most closely associated with alcoholic patients. Arylsulfatase A (EC 3.1.6.1) is a lysosomal enzyme that degrades sulphatides, a constituent of myelin. Sulphatides have been implicated as a component of the opiate receptor,[1] the GABA receptor,[2] and Na-K ATPase.[3] The enzyme is widely distributed throughout the body including in the brain, liver, kidney, spleen, and blood.

Previous work has demonstrated that electrophoretic variants of arylsulfatase A can be detected in lysates of leucocytes and platelets.[4] The lysates for this study were obtained from 20-ml blood samples to which Dextran T500 had been added to sediment the red blood cells differentially. The leucocytes and platelets were washed in saline solutions and then lysed by freeze-thawing. After dialysis, the lysates were analyzed for arylsulfatase-A-specific activity and subjected to discontinuous electrophoresis. To date, several electrophoretic variants of the enzyme have been studied in leucocytes and platelets from 79 alcoholic patients, 166 psychiatric controls without alcoholism, and 125 normal controls among whom were subjects studied previously.[5]

The variant forms of arylsulfatase A have been found in twenty percent of alcoholic patients, but only two to four percent of the psychiatric and normal control groups as shown in TABLE 1. The patterns of the enzyme were labeled according to the number of bands present. For example, the IV_A pattern has four bands. The subscripts designate different patterns. Five variant patterns have been found in addition to the IV_A pattern, which was found in the majority of patients and normal controls.

"Normality" for these controls in this study was defined as a self report of no current medical problem and no history of a psychiatric problem. Neuropsychological and neurological testing as well as a psychiatric interview were given to three of the five normal subjects with a variant enzyme. This has been particularly helpful in deciding whether or not a particular variant was associated with alcoholism. For example, the normal control subject with the III_A variant has severe deficits. His scores on two of the tests, the Benton visual retention test and the trailmaking test, were in the brain-

[a]Supported by a grant from the UNICO Foundation, Inc. and by National Institute on Alcohol Abuse and Alcoholism Grant AA06564.

damaged range. By the criteria of self report of the study, he was normal but by more rigorous criteria, he would not be considered normal.

The main finding of this study is that certain variants of arylsulfatase A are clearly associated with patients with alcoholism. For example, the III_A variant is found in 5 out of 79 (6%) of patients with alcoholism, none of the 117 schizophrenic patients without alcoholism, and 1 of the 125 normal controls (the one who, as just discussed, has abnormalities on neuropsychological testing). On the other hand, no convincing data to date has been obtained to associate the eight-band variant with alcoholism. Hence, certain variants of arylsulfatase A appear to be found predominantly in alcoholic patients and others do not.

Recently, a seven-band variant of arylsulfatase A has been identified. It was possible to identify this new variant by using purified platelets, which gave a clearer, more distinct electrophoretic banding pattern of arylsulfatase A than the leucocytes-plus-platelet preparations routinely used. The new variant has been found in one schizophrenic patient without alcoholism and one normal control. However, since the purified platelet isolation procedure has not been used routinely, it remains to be seen if this variant will be associated with mental illness.

TABLE 1. Distribution of Forms of Arylsulfatase A among Alcoholic Patients and Controls

Forms of Arylsulfatase A	Alcoholic Patients	Schizophrenics without Alcoholism	Other Psychiatric Patients	Normal Controls
Normal enzyme, IV_A	63	114	48	120[a]
Variant IV_B	9	2	1	1[a]
Variant III_A	5	0	0	1[b]
Variant III_B	1	0	0	0
Variant VIII	1	0	0	2[c]
Variant VII	0	1	0	1[a]
Total	79	117	49	125

[a]Normality was defined in terms of self report. No neuropsychological tests were administered to these subjects.

[b]This control has severe cognitive deficits in the brain-damaged range on neuropsychological tests.

[c]One of these two subjects showed specific deficits in performance on spatial relationship tasks as well as neurological deficits. The other subject showed no abnormalities.

Our hypothesis is that certain variants of arylsulfatase A have a reduced ability to degrade sulphatides. This abnormality may be exacerbated by excessive ethanol intake as is suggested by the fact that animals fed ethanol chronically have elevated brain sulphatides.[6] The accumulation of abnormally high levels of sulphatides may lead eventually to demyelination and generalized cerebral atrophy, analogous to the situation in metachromatic leukodystrophy, a genetic disease associated with very low levels of arylsulfatase A and abnormally high levels of sulphatides.

Note: Since this paper was submitted for publication, a five band (V) variant of arylsulfatase A has been identified. Some of the subjects identified incorrectly as having the IV_B variant in this paper have the V variant of arylsulfatase A. The identifications of the other variants are correct.

REFERENCES

1. CRAVES, F. B., B. ZALC, L. LEYBIN, N. BAUMANN & H. H. LOH. 1980. Antibodies to cerebroside sulfate inhibit the effects of morphine and β-endorphin. Science **207:** 75–76.
2. EBADI, M. & A. CHWEH. 1980. Inhibition of arylsulphatase A of Na-independent (^{3}H)-GABA and (^{3}H)-muscimol binding to bovine cerebellar synaptic membranes. Neuropharmacology **19:** 1105–1111.
3. KARLSSON, K. A. 1977. Aspects on structure and function of sphingolipids in cell surface membranes. *In* Structure of Biological Membranes. S. Abrahamsson and I. Pascher, Eds. 245–274. Plenum Press. New York, N.Y.
4. MANOWITZ, P., L. GOLDSTEIN & R. NORA. 1981. An arylsulfatase A variant in schizophrenic patients: Preliminary report. Biol. Psychiatry **16:** 1107–1113.
5. HULYALKAR, A. R., R. NORA & P. MANOWITZ. 1984. Arylsulfatase A variants in patients with alcoholism. Alcoholism **8:** 337–341.
6. RAWAT, A. K. 1974. Lipid metabolism in brains from mice chronically-fed ethanol. Res. Commun. Chem. Pathol. Pharmacol. **8:** 461–469.

Ethanol-Induced Cytoskeletal Dysgenesis in Developing Skeletal and Cardiac Myocytes[a]

E. D. ADICKES AND T. J. MOLLNER

University of Nebraska Medical Center
and
Veterans Administration Medical Center
Omaha, Nebraska 68105

Ethanol induces embryopathic cytoskeletal dysgenesis in fetal alcohol syndrome (FAS) in humans and fetal alcohol effects (FAE) in animals.[1,2] Cytoskeletal maldevelopment occurring in myocytes during organogenesis contributes to cellular dysmaturity with resultant hypotonia.

Parallel *in-vivo* and *in-vitro* studies support these observations. Female Sprague Dawley rats were fed the Leiber-DeCarli liquid diets of 10% (LP-E) and 25% (HP-E) protein-derived calories with 36% of the calories derived from ethanol. In control animals, maltose dextrin was substituted for ethanol (low protein, LP; high protein, HP rat chow, C). Ultrastructural studies of muscle biopsies from pups of ethanol-fed dams on days 1, 5, 10, and 15 showed dysmaturity and myofibrillar dysgenesis. Myofilament damage ranged from focal sarcomeric dysplasia with disorganization of actin-Z-band assembly to total disruption with sarcoplasm reduced to masses of disorganized filaments (FIG. 1a). Myocyte dysmaturity persisted to 15 days in pups from both ethanol-fed groups characterized by retention of central nuclei, voids of sarcomeric development, and cytoskeletal dysgenesis (FIG. 2). Muscle from control pups did not show dysgenesis, and central nuclei did not persist past day five.

To determine if ethanol is the teratogenic agent, cardiac and skeletal myocyte cultures from newborn Sprague Dawley rat pups were grown in Dulbecco's modified Eagle's medium containing 10% fetal calf serum. After 24 hours of growth, 5- and 10-mM concentrations of ethanol were added to the medium in separate cultures with daily fresh media-ethanol changes.

Morphologic studies of sequentially sacrificed monolayer cultures on days 5, 10, and 15 showed disorganized cytoskeletal development compared to controls. Antiactin immunofluorescence revealed less concentrated, nonlinear, and nonparallel actin-Z-band assembly in both 5- and 10-mM cultures. Ultrastructural confirmation of dysgenesis included sarcomeric dysplasia and actin filament disorganization in both skeletal and cardiac myocytes exposed to ethanol (FIG. 1b). The dysgenesis was similar to the *in-vivo* studies.

When developing myocytes are exposed to ethanol during embryogenesis-organogenesis, *in vivo,* and growth and development, *in vitro,* cytoskeletal dysgenesis is evident. This dysgenesis contributes to cellular dysfunction resulting in embryopathic damage of the fetal alcohol syndrome and fetal alcohol effects.

[a]Supported by the Veterans Administration Research Advisory Group (EDA).

FIGURE 1. Comparison of myofibrillar dysgenesis in rat model (**a**) and myocardial cell culture (**b**). Both show dysgenic sarcomeric development with abnormal actin-Z-band assembly. (a) 7,000X; (b) 20,000X.

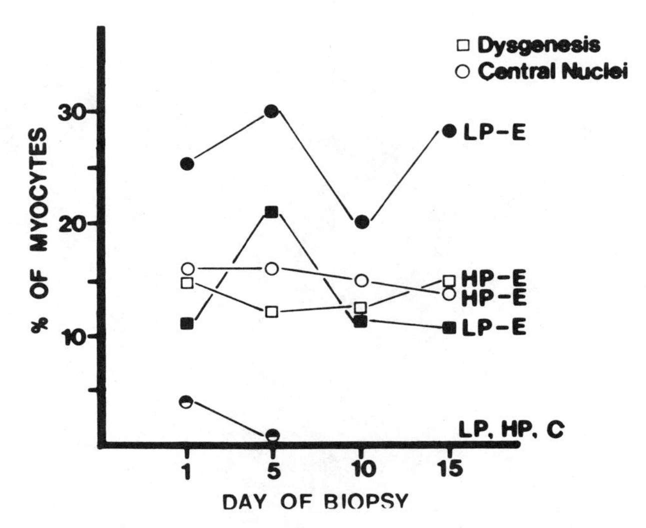

FIGURE 2. Presence of central nuclei and dysgenic myocytes in high and low protein ethanol animals (HP-E and LP-E) and controls (LP, HP, and C).

REFERENCES

1. ADICKES, E. D. & R. M. SHUMAN. 1983. Fetal alcohol myopathy. J. Ped. Path. **1:** 369–384.
2. ADICKES, E. D. & T. J. MOLLNER. 1986. Ethanol induced cytoskeletal dysgenesis with dietary protein manipulations. Alcohol and Alcoholism **21:** 347–355.

Mechanism for Ethanol-Induced Increase in Urinary Folate Excretion: Plasma Binding and Renal Clearance Studies[a]

KENNETH E. McMARTIN, BERNARD H. EISENGA, AND WILLIAM R. BATES

Department of Pharmacology
Section of Toxicology
Louisiana State University Medical Center
Shreveport, Louisiana 71130

Acute administration of ethanol to rats markedly increases urinary folate excretion and subsequently decreases plasma folate levels.[1] When ethanol treatment is repeated daily for four days, the excess urinary folate loss is cumulative,[2] indicating a possible role in the development of the folate deficiency of chronic alcoholism.[3] Mechanisms for the increase in urinary folate excretion include increased delivery of filterable folate to the glomerulus due to reduced plasma folate binding, or a change in the kidney's processing of folate. This study evaluates the effects of ethanol on urinary clearance of folate and inulin and also on plasma folate binding.

Plasma folate binding was studied in male Sprague-Dawley rats dosed with ethanol (1 g/kg each at 0, 1, 2, and 3 hr) or with isocaloric glucose. The total and unsaturated folate binding capacities (TFBC and UFBC) in plasma collected at 2.5, 4, and 6 hr were measured by the method of Colman and Herbert.[4] UFBC represents the direct binding of ^{3}H-folic acid by a specific plasma protein, and TFBC represents the total capacity for binding after endogenous ligands have been stripped from the protein. Urinary clearance was studied after rats were anesthetized with pentobarbital and implanted with tracheal, jugular, carotid, and ureter catheters. Loading doses of mannitol (250 mg/kg), inulin (50 mg/kg), ^{3}H-folic acid (3, 7.5, and 15 μmol/kg), and ethanol (1 g/kg) were given i.v., followed by maintenance doses at 1% of the loading doses per min. Following equilibration, urine and plasma samples were collected at 15-min intervals through 1 hr. Inulin was assayed by colorimetric analysis,[5] ^{3}H-folic acid by liquid scintillation counting,[6] and ethanol by GLC.[7]

Initial studies of the plasma FBC showed that 16 hr of food deprivation significantly ($p < 0.05$) increased the TFBC (158 ± 29 pg fol bound/ml serum in fasted rats vs 58 ± 5 in fed rats) and the UFBC (56 ± 3, fasted, vs 15 ± 3, fed). This increase in plasma folate binding could be one means by which the urinary folate excretion is decreased by fasting.[2] Ethanol administration appeared to decrease the plasma FBC in fasted rats compared to controls (FIG. 1), but none of these differences was significant ($p > 0.05$).

In control rats, the urinary clearance of folate with respect to that of inulin was constant over the last three 15-min collection periods. The ratios of folate/inulin clearance were in the range from 1.2–1.8, indicating net secretion of folate into the urine by the rat. The ratios were not different among the 3 dose levels of folate (FIG. 2), indicating that the folate doses were high and clearance had reached its maximum. In

[a]Supported in part by United States Public Health Service Grant AA05308.

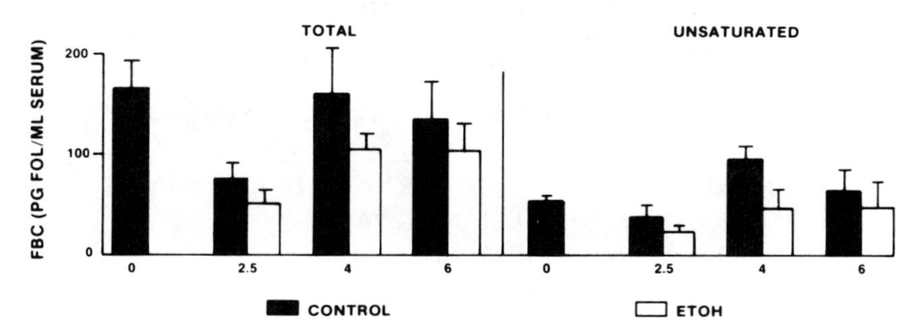

FIGURE 1. Effect of acute ethanol treatment on total and unsaturated folate binding capacities in rat plasma. Ethanol was given in a total dose of 4 g/kg and isocaloric glucose was given to controls. *Numbers beneath the bars* represent hours after initiation of treatment. *Bars* represent the mean value + S.E.M. (n = 4).

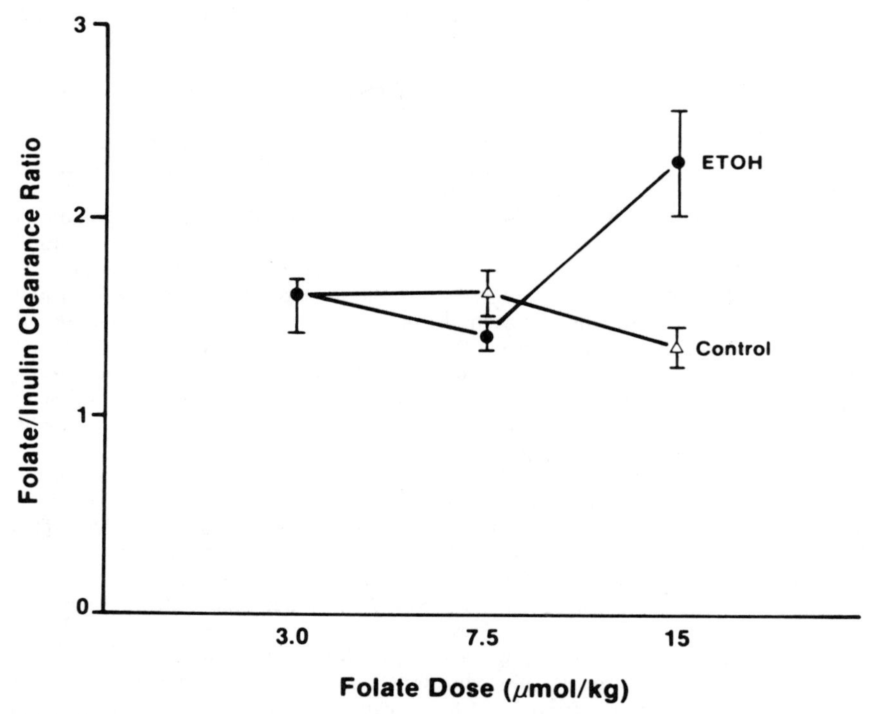

FIGURE 2. Effect of ethanol treatment of rats on renal clearance of folate with respect to that of inulin. The folate/inulin clearance ratio for each animal was averaged over the last 45 min of the collection period. Plotted values represent the mean for the rats in each group + S.E.M. (n = 24).

the ethanol-treated rats, ratios of folate/inulin clearance varied from 1.4 to 2.6. Plasma ethanol levels remained constant throughout the 60-min collection period. The mean values were 123 ± 7, 89 ± 5, and 105 ± 3 mg/dl at 3, 7.5, and 15 μmol folate/kg, respectively. Ethanol did not alter folate clearance with respect to inulin at the lower 2 doses of folate, but did increase it at 15 μmol/kg (FIG. 2). The lack of marked effects at these ethanol levels could be explained by previous studies, in which higher ethanol levels were needed to increase urinary folate excretion.[8] These studies suggest that high doses of ethanol may increase the delivery of folate to the kidney, but that moderate levels do not alter the renal handling of folate.

REFERENCES

1. McMARTIN, K. E. 1984. Increased urinary folate excretion and decreased plasma folate levels in the rat after acute ethanol treatment. Alcoholism: Clin. Exp. Res. **8:** 172–178.
2. McMARTIN, K. E., T. D. COLLINS & L. BAIRNSFATHER. 1986. Cumulative excess urinary excretion of folate in rats after repeated ethanol treatment. J. Nutr. **116:** 1316–1325.
3. WEIR, D. G., P. G. McGING & J. M. SCOTT. 1985. Folate metabolism, the enterohepatic circulation and alcohol. Biochem. Pharmacol. **34:** 1–7.
4. COLMAN, N. & V. HERBERT. 1976. Total folate binding capacity of normal human plasma, and variations in uremia, cirrhosis, and pregnancy. Blood **48:** 911–921.
5. HEYROVSKY, A. 1956. A new method for the determination of inulin in plasma and urine. Clin. Chim. Acta **1:** 470–474.
6. McMARTIN, K. E., V. VIRAYOTHA & T. R. TEPHLY. 1981. High-pressure liquid chromatography separation and determination of rat liver folates. Arch. Biochem. Biophys. **209:** 127–136.
7. BAKER, R. N., A. L. ALENTY & J. F. ZACK, JR. 1969. Simultaneous determination of lower alcohols, acetone, and acetaldehyde in blood by gas chromatography. J. Chromatogr. Sci. **7:** 312–314.
8. McMARTIN, K. E., T. D. COLLINS, C. Q. SHIAO, L. VIDRINE & H. M. REDETZKI. 1986. Study of dose-dependence and urinary folate excretion produced by ethanol in humans and rats. Alcoholism: Clin. Exp. Res. **10:** 419–424.

PART VI. EFFECTS OF ETHANOL ON CELLS OF THE NERVOUS SYSTEM

Electrophysiology of Ethanol on Central Neurons[a]

G.R. SIGGINS,[b] F.E. BLOOM,[b] E.D. FRENCH,[c]
S.G. MADAMBA,[b] J. MANCILLAS,[d]
Q.J. PITTMAN,[e] AND J. ROGERS[f]

[b]*Alcohol Research Center*
Scripps Clinic and Research Foundation
La Jolla, California

[c]*Maryland Psychiatric Research Center*
Baltimore, Maryland

[d]*Laboratory of Molecular Biology*
Medical Research Council Centre
Hills Road
Cambridge, England

[e]*Faculty of Medicine*
University of Calgary
Alberta, Canada

[f]*Institute for Biogerontology Research*
Sun City, Arizona

INTRODUCTION

The neuronal mechanisms of ethanol intoxication still remain essentially unknown despite considerable past research. Diverse actions have been ascribed to ethanol (see REFS. 1–5 for reviews) probably because past studies have used a variety of invertebrate and vertebrate model neuronal systems. Furthermore, a wide range of concentrations of ethanol have been studied after various routes of administration; as a result many reports of ethanol action occur with tissue drug exposure that are often higher than the blood levels associated with moderate intoxication. Many electrophysiological studies have used peripheral vertebrate or isolated invertebrate preparations for ease of intracellular recording. These studies have revealed dose-dependent and cell specific mechanisms of ethanol action on neuronal excitability that generally fall into one of three categories:[1,5] 1) alterations in the voltage-dependent ionic conductances underlying action potentials; 2) changes in passive membrane conductances and therefore resting membrane potentials; 3) pre- or postsynaptic changes in synaptic transmission, including alteration in transmitter release or in the sensitivity of the postjunctional membrane receptors or transmitter-activated conductances.

Although most studies describe a depressant effect of ethanol on neuronal excitability (see *e.g.*, REFS. 6–8), in some instances ethanol enhances excitability.[9–13]

[a]Supported by National Institute on Alcohol Abuse and Alcoholism Grants AA-06420 and AA-07456, National Institute on Drug Abuse Grant DA-03665, the Alexander-von-Humboldt-Stiftung, and the Canadian Medical Research Council.

Many of these effects may also be time-dependent, even displaying an early effect followed by a later effect of opposite direction.[9,11] Data from several invertebrate and vertebrate preparations suggest that junctional or synaptic transmission may be more sensitive to low concentrations of ethanol than are either spike electrogenesis or passive membrane properties.[1-5,10] Although depression of synaptic potentials is the usual effect, recent studies of some skeletal muscle, spinal cord, and cerebral cortical neurons have shown a facilitatory action of ethanol on both excitatory and inhibitory synapses.[10,12,14-18]

In our laboratory we have adopted a "multimodel" approach for the examination of alcohol effects, which includes: 1) *in-vivo* awake animals, 2) *in-vivo* anesthetized animals, 3) *in-vitro* brain slices, and 4) *in-vitro* cultures. The awake animal model has the advantage of normal physiology without the presence of anesthetics, but suffers from the inability to determine mechanisms of drug action by intracellular recording. The *in-vivo* anesthetized model has fairly normal physiology and neuronal connectivity, where it is possible to make a fast survey of alcohol actions; but an anesthetic is present, the concentrations of drugs applied to neurons are unclear, and the mechanisms underlying actions are very difficult to probe.

The use of *in-vitro* brain slices has the advantages that known drug concentrations can be tested and mechanisms of transmitter and drug action can much more easily be defined. However, there is still the question of a cellular physiology altered by the methods of preparation and the bathing situation, and many neuronal connections are disrupted. *In-vitro* culture systems have many of the advantages of the slice preparations, and in addition are accessible to single-channel analysis, but again, altered physiology and connections are a concern. In addition, cells in most cultures are studied in an immature and/or undifferentiated state.

This multimodel approach is helpful for minimizing the disadvantages inherent in each model alone, by comparing the results derived from each model. The following account describes several recent studies from three of these models that may provide insight into the electrophysiological effects of intoxicating doses of ethanol on nerve cells. The results show interactions between ethanol and anesthetics and between ethanol and certain putative neurotransmitters, and indicate that synaptic transmission may be the step most sensitive to low concentrations of ethanol.

IN-VIVO STUDIES: INFERIOR OLIVARY NEURONS

The Purkinje cell of the rat cerebellar cortex *in vivo* recently has been a useful model for study of ethanol effects.[1-5,8,13] Previous work in our laboratory has shown a significant time- and dose-dependent increase in climbing fiber-mediated bursts of action potentials in rat cerebellar Purkinje cells after acute ethanol, with little change in the discharge of single (intrinsic) spikes.[13] Recently, Sinclair *et al.*[19,20] reported an exactly opposite effect: after acute ethanol, a significant decline in climbing fiber-mediated Purkinje cell bursts was observed. However, different anesthetics and routes of ethanol administration were employed in the two studies.

The climbing fibers arise from neurons in the inferior olivary nucleus that send axons to the cerebellar cortex. These axons provide a powerful excitatory input to their Purkinje neuron targets via generation of a pronounced excitatory postsynaptic potential (EPSP), resulting in a burst of spikes in Purkinje cells.[21] Therefore, we reasoned that a more direct assessment of the effects of ethanol in this system would require recording of the olivary source cells for cerebellar climbing fibers. We used several means of animal immobilization, including anesthesia with chloral hydrate, halothane vapor, or urethane or tubocurare plus local (Xylocaine) anesthesia to determine if the anesthetic might influence responses to ethanol.[22]

The extracellular electrical activity of single neurons of the inferior olivary nucleus was recorded by standard methods,[22] and the location of the neurons was verified histologically and physiologically. Firing frequency and pattern were quantified by ratemeter and computer, using software that generates interspike interval histograms. Single inferior olive neurons were isolated at least 10 min before ethanol was administered and changes in firing frequency and pattern were observed for 1–2 hr. The overall effects of ethanol were analyzed by one-way analysis of variance for each anesthetic condition, using the baseline and 10–60 min postethanol recording data (baseline, 10, 20, 40, 80, and > 80 min epochs).

As shown in the averaged data of FIGURE 1, at 10–60 min postethanol (2 g/kg) there is a significant increase in olivary unit activity if the rat is under chloral hydrate ($F = 8.32$, $p < 0.05$, 6 rats), halothane vapor ($F = 12.97$, $p < 0.05$, 12 rats), or local ($F = 5.35$, $p < 0.05$, 4 rats) anesthesia. The results for chloral hydrate and halothane

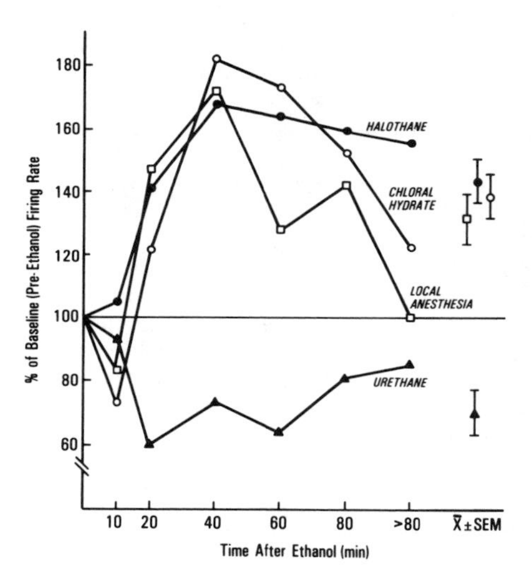

FIGURE 1. Effects of 2 g/kg intraperitoneal ethanol on inferior olive neurons under the various anesthetic conditions. The data are expressed as percent of the pre-ethanol baseline firing rate, averaged over all cells studied at the given time epochs. Data points with error bars at *far right* of the graph are overall means ± S.E.M. for firing rate changes after ethanol. Local anesthesia does not differ significantly from chloral hydrate or halothane ($p \ll 0.05$), but does from urethane ($F = 25.04$, $p < 0.05$). (From Rogers *et al.*[13] Reprinted by permission from *Brain Research.*)

do not differ significantly ($p \gg 0.05$) from those for local anesthesia and d-tubocurare immobilization. In these conditions, olivary activation peaks around 40 min after ethanol administration at levels 70–80% over baseline. A similar time span for cerebellar climbing fiber activation is seen after 2 g/kg acute intraperitoneal ethanol.[5,13] Also as in the cerebellum, ethanol-induced changes in olivary activity show evidence of recovery approximately 80 min or more after the substance is injected (FIG. 1).

As for individual cells, two distinct changes in firing pattern were observed. Some olivary units simply increased their overall rate within 10–20 min of ethanol injection. Many other units, however, responded with rhythmic flurries of spikes (FIG. 2). This pattern is paralleled by the response of many cerebellar Purkinje cells, which, after acute ethanol, begin to fire rhythmic and abnormally lengthy bursts of climbing fiber-mediated spikes.[13] These effects of ethanol are remarkably similar to the actions of harmaline[23] (see below).

The overall ethanol-induced activation of olivary units was the major consistent

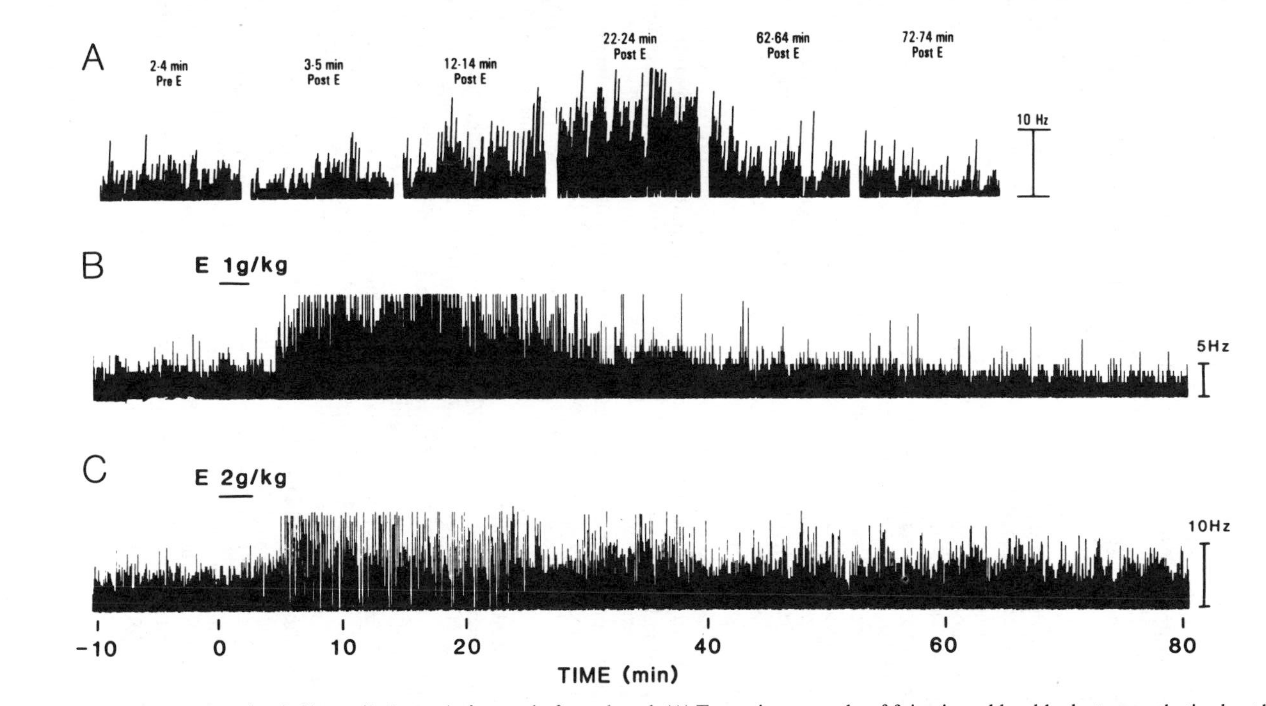

FIGURE 2. Ratemeter records of olivary discharge before and after ethanol. (**A**) Two-minute epochs of firing in a chloral hydrate-anesthetized rat before and after ethanol. (**B**) Longer time-base record of the excitatory effect of 1 g/kg ethanol given to a halothane-anesthetized rat. (**C**) Same time-base in another halothane-anesthetized rat receiving 2 g/kg ethanol. Note that in all three cases there is a generally abrupt transition from slow, somewhat regular firing to periods of fast bursts separated by relative (*panel A*) or complete (*panels B and C*) quiescence. (From Rogers *et al.*[13] Reprinted by permission from *Brain Research*.)

response under halothane, chloral hydrate, or local anesthesia. However, a brief depression of firing occurred from 1–10 min after ethanol administration (FIG. 1) in the majority of cells recorded under chloral hydrate or local anesthesia and in at least one-third of cells recorded under halothane.

In contrast to the other anesthetics and local anesthesia, ethanol injected under urethane anesthesia significantly decreases olivary firing ($F = 4.38$, $p < 0.05$, 8 rats; see FIG. 1). For example, 40 min after ethanol, average olivary firing in urethane anesthetized rats falls 25% below baseline, whereas it has risen 75% above baseline in locally anesthetized animals (FIG. 1). Urethane-ethanol depression of inferior olive activity is even more striking when ethanol is given intravenously (FIG. 3), the method of administration employed by Sinclair *et al.*[19,20] The interaction of urethane and ethanol to decrease olivary firing also has a clear parallel in the cerebellum, since this drug combination is known to depress climbing fiber-mediated burst activity of Purkinje neurons.[19,20] In some experiments, 1-g/kg ethanol doses were tested in combination with halothane or urethane anesthesia. These studies gave results qualitatively similar to those with 2-g/kg doses: clear excitation of olivary neurons with halothane anesthesia (5 rats) and depression of olivary firing under urethane (2 rats).

In summary, under all the anesthetic conditions tested there is an exact parallel between the effects of ethanol on the olive and the effects of ethanol on the cerebellum. When urethane is used, depression occurs.[19,20] When other anesthetics or local anesthesia are used, activation occurs. These results thus extend our findings that acute ethanol increases climbing fiber-mediated bursts in cerebellar Purkinje cells[5,13] by showing that this cerebellar activation is very likely secondary to inferior olive activation.

The present data may also help resolve differences between our previous work[5,13] and that of Sinclair *et al.*[19,20] Since the decreased olivocerebellar activity with urethane anesthesia is opposite to that seen in all the other anesthetic conditions tested, including local anesthesia, we conclude that results based on urethane anesthesia may not reflect the natural, physiologic effects of ethanol. However, this should not be taken to mean that chloral hydrate and halothane completely lack interactive effects with ethanol. Rather, it can only be concluded that at the doses employed, the effects of ethanol under chloral hydrate or halothane anesthesia are very similar to those observed in locally anesthetized animals.

Although the mechanism by which ethanol causes olivocerebellar activation remains unclear, several possibilities have been considered.[22] The indirect mediation of this effect by some metabolite of ethanol, such as a harmaline-like β-carboline (formed by condensation of acetaldehyde and serotonin) is a possibility that would explain the contrasting effects of systemically versus locally administered ethanol on several cells, including olivocerebellar neurons. Thus, systemic ethanol elevates cerebellar Purkinje cell[5,13] and inferior olivary neuron activity,[22] whereas locally applied (*e.g.*, by electroosmosis or pressure ejection) ethanol has a depressant effect on unit activity.[5,8]

Other hypotheses are tenable and also should be explored. In particular, the inferior olive receives input from several sources, including the locus coeruleus, cerebral cortex, and spinal cord, as well as the raphe (see Rogers *et al.*[22] for references). Ethanol-induced changes in the activity of these inputs could mediate changes in the olive. An action of ethanol or its metabolites directly on the membrane of olivary neurons also remains an important possibility.

IN-VIVO STUDIES: HIPPOCAMPAL PYRAMIDAL CELLS

The hippocampal pyramidal cell (HPC) recently has provided much information on ethanol effects. Extracellular recordings *in vivo* indicate that systematically

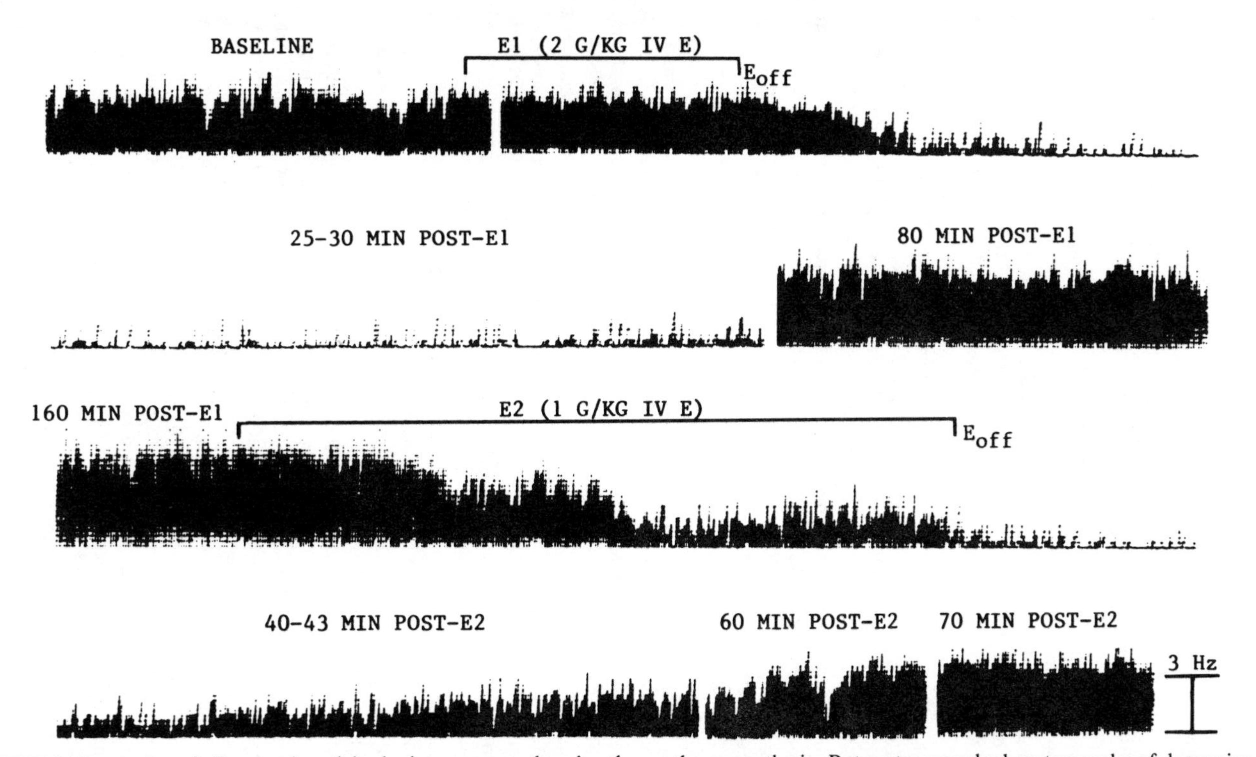

FIGURE 3. Depression of olivary unit activity by intravenous ethanol under urethane anesthesia. Ratemeter records show two cycles of depression and recovery demonstrable in the same cell over a 4-hr period. The first ethanol administration, 2 g/kg, was given continuously over a 10-min period (note break in record). The second administration, 1 g/kg, was given over a 5-min period as shown. Calibration: 10 Hz/30 sec. Time relative to last ethanol administration is shown above each ratemeter trace. (From Rogers *et al.*[13] Reprinted by permission from *Brain Research.*)

administered ethanol can alter hippocampal EEGs and multi- and single unit firing rates, as well as field potential and single unit responses evoked by stimulation of afferent inputs.[1–7,11,12] Although most of these studies found depressant effects (usually at high supra-intoxicating doses of ethanol), we have noted[12] that systemic ethanol also can facilitate both excitatory and inhibitory responses to afferent stimulation. Local application to HPCs of ethanol by microelectroosmosis or pressure can produce an early excitatory response, sometimes followed by a depression at higher doses, although depression alone is also seen.[5,9]

To determine the transmitters involved in the enhancement by ethanol of the excitatory and inhibitory responses of pyramidal cells to stimulation of afferent pathways, we[24] tested the effect of ethanol on the responses of identified pyramidal cells in the halothane-anesthetized rat to iontophoretically-applied transmitters (see REF. 24 for details). Ethanol (0.75 or 1.5 g/kg) was injected intraperitoneally over a 8–10-min period. The ethanol doses were chosen to yield blood ethanol levels of about 80 or 150 mg% in this size rat.[25] Drug effects were analyzed on line by a computer that generated peri-drug response interval histograms for subsequent quantification. Pyramidal cells were identified[24] on the basis of bursting discharge behavior (FIG. 4) and their stereotaxic coordinates. Their control responses to periodically pulsed, alternating high and low doses of a neurotransmitter were recorded. Ethanol was then injected i.p. and the magnitude of the responses to the given transmitter re-evaluated periodically, usually at 5, 15, 20, 30, 45, 60, 90, and 120 min postethanol. Changes in ACh responses were evaluated using the matched pairs sign test.[24]

Systemic ethanol (0.75 or 1 g/kg) markedly enhanced excitatory responses to iontophoresis of acetylcholine (ACh) (FIG. 4) in ten rats, decreased them in two, and had no clear statistically significant effects in one. Such ethanol-induced increases in ACh responses were observed in all six CA1 cells. In CA3, an increase in ACh responses was seen in four cells, a decrease in two, and no change in one. Significant ACh-facilitatory effects were evident by 15 min after injection of ethanol (FIG. 4b) and reached a peak at about 30 min. Recovery from the actions of ethanol usually occurred about 60 min after injection, although in some cases recovery was slow and not complete until after 2 hr (FIG. 4d). The enhancement of ACh excitations by ethanol appears to be selective as no comparable effect was observed on glutamate-induced excitation in three cells (20 trials) tested alternatively with both transmitters (FIG. 4). This effect of ethanol appears not to be due to the specific experimental conditions, complications of iontophoresis, or anesthesia, since in preliminary studies in the hippocampal slice preparation, defined low-ethanol concentrations also potentiated responses to superfused ACh.[26]

Systemic ethanol also significantly increased the amplitude and duration of inhibitory responses to iontophoretically applied somatostatin-14 (SS-14) (24 tests in six cells, five in CA3 and one in CA1). This enhancement was clearly evident in all cells at 10–15 min after ethanol injection, with recovery at 60–80 min. Ethanol had no statistically significant effect on inhibitory responses to serotonin or norepinephrine (21 tests in three cells for each transmitter, two in CA3 and one in CA1). Because of reports[14] that GABA effects are enhanced by very low doses of ethanol, we also examined GABA actions in some detail. Ethanol seemed to cause a small (average: 25%) but consistent potentiation of inhibitory responses to GABA in 10 rats (two in CA3 and eight in CA1), a reduction in one (CA3), and no clear change in four (two in CA3 and two in CA1). Most of these "potentiations," do not recover up to 3 hr after ethanol administration. Furthermore, a similar increased responsiveness to GABA was also observed in 5 of 6 saline-injected control subjects in which individual hippocampal neurons were tested with GABA repeatedly for comparable periods of observation. These and other control experiments[24] suggest that the apparent potentiation of GABA was an artifact of the repetitive iontophoretic application under these experimental conditions rather than a true pharmacological interaction. In contrast, changes in responses to ACh and SS-14 were only observed after ethanol injections, and all recovered within 1–2 hr.

These findings parallel those obtained from vertebrate neuromuscular junctions, where ACh is known to be the transmitter and alcohol facilitates excitatory junctional transmission.[15–18] In frog sympathetic ganglia, ethanol has a dual effect, with low doses facilitating and high doses depressing ACh-mediated synaptic transmission.[27] Ours is,

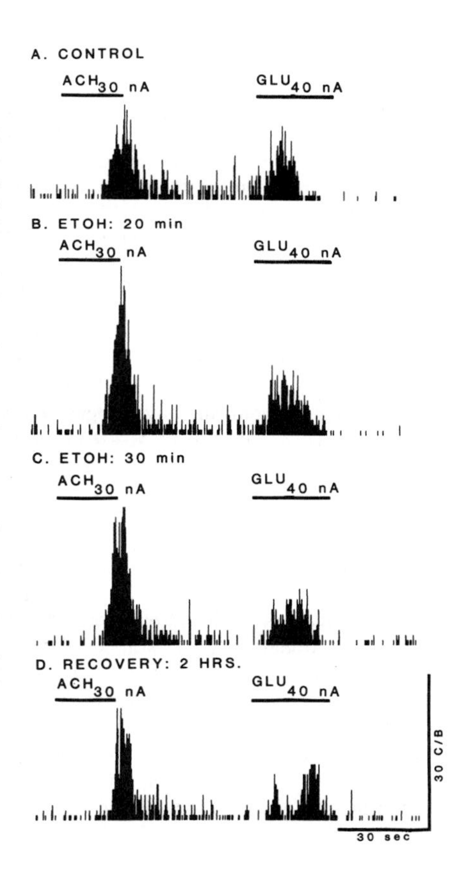

FIGURE 4. Ethanol selectively enhances excitatory responses to acetylcholine. Peri-drug interval histograms, all composed of 2 sweeps of 2 drug pulses each. (**A**) Control responses to brief iontophoretic pulses of acetylcholine (ACH) and glutamate (GLU). The *subscripts* (30 and 40 nA, respectively) indicate the current used to eject the drug; the *bar* indicates the period of application. ACh causes a 763% increase in firing rate over baseline, and glutamate, an increase of 800%. (**B**) Twenty min after i.p. ethanol (1.5 g/kg), ACh causes a 1044% increase in firing above baseline (*i.e.*, a 37% larger response than the control), whereas responses to glutamate are not significantly different (836% = 4% larger than the control response). (**C**) Thirty min after ethanol, responses to ACh average 13% larger than control (863% increase above baseline), while those of glutamate are 17% smaller than control (662% above baseline). (**D**) Recovery from the effects of ethanol is complete after two hr. ACh responses are 8% smaller than control (702% above baseline), whereas those of glutamate are 27% smaller (582%). The effect of ethanol on ACh excitations was statistically significant (matched pairs sign test, $p < 0.05$). (Calibration: 30 sec, 30 counts per bin.) (From Mancillas *et al.*[24] Reprinted by permission from *Science*.)

to our knowledge, the first report[24] of an effect of systemic ethanol on responses to somatostatin. However, we have not been able to confirm the previously reported enhancement of presumed GABAergic transmission seen in rat cortical neurons.[14] That enhancement may reflect the co-localization (and presumed co-release) of SS and GABA detected immunocytochemically in interneurons.[28,29] Inhibitory effects of such co-released SS, rather than GABA, might then have been potentiated, as we observed in the hippocampus.[24]

It is perhaps relevant that we have observed SS-14 potentiation of responses to ACh in the hippocampus (Mancillas *et al.*, in preparation). Thus it is possible that ethanol-induced enhancement of responses to ACh may be secondary to an enhancement of the effects of endogenously released somatostatin, which in turn enhances postsynaptic responses to iontophoretically-applied ACh. The ethanol-induced alteration of ACh and SS-14 responses seen in this study may help to explain the enhancement by ethanol of inhibitory and excitatory synaptic transmission previously described in hippocampus.[12]

IN-VITRO STUDIES: HIPPOCAMPUS SLICE

The *in-vitro* hippocampal slice preparation has special advantages that allow: 1) activation of intrinsic synaptic connections between the cells, 2) administration of drugs in known concentrations, and 3) long-term stable intracellular recording. Studies in other laboratories using this preparation have shown acute ethanol effects, usually inhibitory, on the evoked population field potentials (see REFS. 6, 7). Recently, with intracellular recording of HPCs in slices, others[6] have reported a variety of effects of acute ethanol, including hyperpolarization of CA1 pyramidal cells and enlargement of IPSPs and postburst afterhyperpolarizations.[6] However, our preliminary studies[30] did not yield comparable results. Therefore, we[31] performed additional intracellular studies of the effects of ethanol on CA1 pyramidal cells *in vitro* and, in addition, gathered new data on the responses of CA3 pyramidal neurons to ethanol.

Transverse hippocampal slices 350–400 μm thick were prepared from rats by standard methods[31,32] and placed in a chamber where the slice was completely submerged and continuously superfused with carbogenated artificial spinal fluid (ACSF) at constant rate (1.5–4.0 ml/min), as previously described.[32] Intracellular recording was performed with glass micropipettes filled with either 3M KCl, a 1:10 mixture of KCl (1.0 M) and K-citrate (1.6 M), or K-acetate (2 M), having tip resistances of 60–200 MΩ. CA1 and CA3 pyramidal neurons were penetrated, identified, and recorded by conventional means,[32] with a bridge circuit to inject current pulses. Ethanol solutions of known concentrations (10–350 mM) were introduced into the slice chamber without disrupting the flow of the superfusate, by means of a multiple valve/reservoir system. The 34 CA1 and 16 CA3 pyramidal cells included in our analysis all had input resistances above 30 MΩ, large overshooting action potentials (60–105 mV), and resting membrane potentials of 59.2 $\pm$ 11.7 mV (CA1; mean $\pm$ S.D.) and 55.3 $\pm$ 7.3 mV (CA3).

Resting Membrane Properties. Superfusion of ethanol was performed in 46 complete tests on the 34 CA1 neurons, and in 25 tests on the 16 CA3 neurons. In CA3, the predominant response (seen in 50% of tests) to low ethanol concentrations (50 mM) was weak depolarization (2–4 mV). At high ethanol concentrations (200 mM), the depolarization was followed by a small hyperpolarization in 67% of tests. Hyperpolarization alone was seen in only 9% of all tests, and even then in only 2 tests of the highest concentration (200 mM). In CA3, these effects most often were accompanied by no change in input resistance.

In CA1, membrane potential was unchanged in 43% of all ethanol tests, whereas 33% revealed weak (range = 2–7 mV; mean = 3.2 mV) depolarizing actions. Hyperpolarizations (2–5 mV; mean = 2.9 mV) were seen in only 21% of tests and usually at higher concentrations (43 mM or more). Ethanol concentrations below 20 mM most often had no measurable effect on membrane potential. There was no clear relationship in either CA1 or CA3 between membrane potential changes and input resistance measures. Thus, in about one-half of both CA1 and CA3 tests showing a change in membrane potential, changes in input resistance could not be detected. Furthermore, in those responding cells that did show a change in input resistance, it was as likely to increase as to decrease. Many CA1 cells seemed quite resistant to ethanol even at high concentrations (86 mM or more), showing no change in input resistance, size of EPSPs or IPSPs, or membrane potential.

Evoked Synaptic Potentials. Stimulation of the hilar region of the slice to activate the mossy fiber pathway to CA3 pyramidal cells or of the SR input to CA1 usually evoked, at relatively low stimulus strengths, a small graded depolarizing potential (probably an EPSP) followed by a longer graded hyperpolarizing potential likely to be an inhibitory postsynaptic potential (IPSP; FIG. 5). As stimulus strengths are increased, EPSPs become larger and eventually trigger action potentials. These synaptic responses were used to probe ethanol actions.

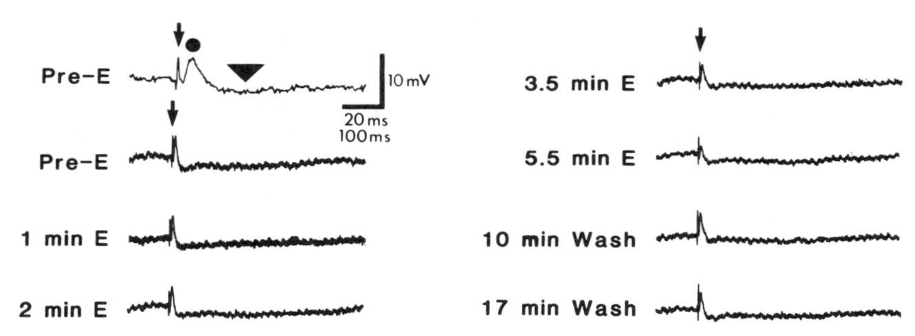

FIGURE 5. An intermediate concentration of ethanol (*E*; 50 mM) reduces IPSPs first, followed by a reduction of EPSPs in a CA1 pyramidal cell. This neuron did not fire spontaneously and had a relatively large resting potential (−78 mV), and therefore the evoked EPSPs were relatively large and the evoked IPSPs relatively small. *Top left panel* shows expanded trace for demonstration of the EPSP. *Arrows* indicate point of SR stimulation, which is followed rapidly by an EPSP (*closed circle*), IPSP (*closed triangle*) sequence. *Pre-E* = control; *Wash* = washout of ethanol with ACSF. Note that the hyperpolarizing IPSP (recorded with a K-acetate electrode) is diminished at 1–2 min after ethanol superfusion, whereas EPSPs are not diminished until about 2–3.5 min after ethanol. Also, the IPSP shows only partial recovery at 17 min, whereas the EPSP nearly reaches control amplitude even before ethanol perfusion is terminated.

In the preponderance of tests (78–100%) of CA3 pyramidal cells, ethanol superfusion at all concentrations tested reduced both the evoked EPSPs (by 17–70%) and the IPSPs (by 10–90%) (TABLE 1). In only 17% of tests were IPSPs increased (by 10–40%) by ethanol. In CA1 neurons, SR-evoked synaptic potentials were also quite sensitive to ethanol (TABLE 1; FIG. 5). Thus, in 41% of CA1 tests EPSPs were reduced and in 60% of tests IPSPs were reduced by ethanol superfusion (TABLE 1). In CA1 only a small percentage of tests showed EPSPs (24%) and IPSPs (7%) increased in size. These synaptic effects were most often observed at ethanol concentrations greater than 20 mM. Concentrations of 10–15 mM ethanol usually had no effect on the PSPs. Ethanol 20–25 mM reduced them in about 50% of tests. IPSPs of both CA3 and CA1 neurons sometimes were diminished at earlier ethanol perfusion times and at lower ethanol concentrations than were the EPSPs (FIG. 5). In the majority of tests in both CA1 and CA3, synaptic potentials changed without changes in resting membrane properties (FIG. 5), or even with changes in membrane properties that would be antagonistic (via driving force or resistance) to the observed alteration in synaptic potentials.

Spontaneous Activity, Spike Size, and Spike Threshold. Many cells in CA3 fired spontaneous action potentials, often in bursts. In these cells, ethanol superfusion (50–200 mM) measurably increased the rate of firing in 32% of 25 tests, decreased it in 24% (6 tests), biphasically increased and then decreased firing in another 32% (8 tests), and was ineffective in 12% (3 tests). Often the excitations appeared to arise from a reduction in the size and duration of the postburst afterhyperpolarization (AHPs) (see below), allowing subsequent spikes to discharge at an earlier time.

CA1 pyramidal neurons did not usually discharge frequent spontaneous action potentials and rarely diplayed bursts. However, in those that did fire spontaneously, ethanol superfusion (10–350 mM) elevated firing frequency in 9% (3 of 34 tests), slowed it in 50% (17 tests), biphasically increased and then decreased firing in 12% (4 tests), and had no effect in 29% (10 tests). The increase in rate and the biphasic responses were more apparent at low (10–43 mM) ethanol concentrations. The

TABLE 1. Effect of Ethanol on Postsynaptic Potentials Evoked by Stimulation of Hilum (CA3) or Stratum Radiatum (CA1)[a]

		EPSPs			IPSPs		
	EtoH	↑	↓	NC	↑	↓	NC
CA3	50	0	6	0	1	9	0
	100–150	UD	UD	UD	0	1	1
	200	0	5	0	2	4	0
	Totals	0	11	0	3	14	1
CA1	10–25	0	4	4	0	6	7
	43–50	2	2	3	0	6	1
	86–100	1	4	2	0	6	2
	150–200	2	1	1	2	0	0
	350	2	1	0	UD	UD	UD
	Totals	7	12	10	2	18	10

[a]Ethanol concentration (EtoH) in mM. Other numbers refer to number of tests with each concentration. ↑ = increased; ↓ = decreased; NC = no change; UD = undetermined.

spontaneous rate change was not always logically correlated with a change in membrane potential (*e.g.*,depolarization might be accompanied by decreased firing). Spike amplitude was usually not measurably affected by ethanol concentrations in the 10–150 mM range. Although a concentration of 200 mM occasionally reduced spike size, in many other tests at these concentrations no change in spike size was seen. However, 350 mM consistently reduced the spike amplitude. Thresholds for generating action potentials via activation of SR (CA1) or hylum (CA3) were usually (18 of 25 tests) increased (by 5–80%) by all concentrations of ethanol tested above 10–15 mM.

Postburst Afterhyperpolarizations. The AHPs that follow spontaneous bursts of spikes in CA3 neurons, or bursts evoked by depolarizing current injection in CA1 neurons, are probably generated by activation of a Ca^{++}-dependent K^+ conductance,[33-35] triggered by the Ca^{++} that enters the cell during the action potential bursts. In CA3, 50 mM ethanol either produced no change or reduced the magnitude of the spontaneous AHP by 10–24%, an effect that often appeared to cause an increase in the spontaneous firing rate. In CA1, ethanol usually had no effect on current-evoked AHPs (56% of 23 tests); however, 22% of tests showed slightly increased and 22% of tests showed decreased AHP amplitudes. The decreases in the AHP amplitudes in CA1 ranged from 5–20% and occurred at ethanol concentrations as low as 15–22 mM, whereas AHP increases usually were seen at higher concentrations. In CA1, the number of spikes evoked by the depolarizing constant current pulse used to generate AHPs (a measure of accomodation and also dependent upon the Ca^{++}-dependent K^+ conductance (cf. REF. 33)) was usually not affected by ethanol superfusion, although in 20% of tests the number of spikes decreased during ethanol. No cells showed an increase in spike number.

GABA Responses. We examined the effects of ethanol superfusion on the responses of CA1 pyramidal cells to GABA in an effort to confirm previous reports (see REF. 14) that ethanol potentiated GABAergic transmission and to determine the pre- or postsynaptic locus of action for our finding that ethanol reduces IPSP size. GABA was applied locally by micropipette and uniform responses were obtained in 5 cells by repetitive application of pulses of constant duration and pressure at constant interpulse intervals. In all cells, GABA elicited hyperpolarizations of 5–10 mV, accompanied by cessation of action potential discharge in spontaneously active cells. In all 5 cells studied, superfusion of 22–80 mM ethanol had little effect on the responses to GABA. One cell showed a barely detectable prolongation (by about 15%) of the GABA response, although even in this case the magnitude of the GABA-induced hyperpolarization was not increased by ethanol. In another cell ethanol slightly reduced (by about 25%) the magnitude and duration of the response to GABA.

Interpreting the Intracellular Data. Our primary goal in this study was to monitor transmembrane and synaptic properties of mammalian brain neurons while presenting ethanol to these neurons at drug levels comparable to those obtained in blood during intoxication in order to identify possible membrane mechanisms that could underlie intoxication. An interesting aspect of our results is that concentrations of ethanol as large as 150–200 mM (equivalent to about 690–920 mg%, or fatal levels *in vivo*) had no detectable effect on action potential size, membrane potential, or resistance in many pyramidal neurons, although membrane potential and resistance in several other pyramidal neurons were sensitive to ethanol concentrations as low as 10–25 mM. This latter range would be equivalent to blood levels (45–115 mg%) considered minimally to clearly intoxicating in rats or humans. The variability in responsiveness of resting membrane properties to ethanol was seen in both CA1 and CA3 neurons, and is also reinforced by the relative weakness and wide range of response direction observed. The general lack of ethanol effects on membrane potential or resistance in a high percentage of cells would seem to indicate that the direct alteration of resting membrane properties in significant numbers of hippocampal pyramidal cells is not a necessary condition of behavioral intoxication. Thus, in our model, low intoxicating doses of ethanol do not appear to have reproducible effects on action potential conductances (category one of the Introduction) or on passive membrane conductances (category two).

In contrast, ethanol had more consistent effects on synaptic potentials. Thus, in the majority of pyramidal cells, both EPSPs and IPSPs were depressed, even at low (10–50 mM) ethanol concentrations (equivalent to 46–230 mg%). Hence, it is possible that an alteration of synaptic activity in hippocampal pyramidal cells is a contributing factor in the cellular basis of some aspect of intoxication, based simply on comparison of ethanol doses required.

There are several inconsistencies between our data and the intracellular data also obtained from the rat hippocampal slice preparation by Carlen and co-workers,[6] who found that ethanol elicited marked hyperpolarizations of CA1 pyramidal neurons and augmented the SR-evoked IPSPs and current-evoked AHPs. In general our data is in disagreement with these findings. We have seen few ethanol-evoked hyperpolarizations, and IPSPs appear to be more often reduced than enhanced, in accord with biochemical studies showing ethanol-induced reduction of GABA release from brain tissue.[36–38] In our studies, the AHPs in most CA1 and CA3 neurons were not enhanced following ethanol exposure. Indeed, many CA3 cells exhibited a reduction of spontaneous postburst AHPs. These discrepancies could arise from several methodological differences, including the use by Carlen *et al.*[6] of "microdrops" of ethanol applied to an exposed slice surface versus our use of total slice immersion with continuous superfusion. The microdrop method suffers from the possibility that the medium used in the microdrop pipette could differ markedly from that in the extracellular space. Furthermore, it has been unusual to see evidence of recovery using the microdrop method, whereas in our studies all data are from experiments in which recovery was eventually realized after long washout periods.

Our results also diverge from the *in-vivo* extracellular findings of Newlin *et al.*,[12] who observed increases in responses to both excitatory and inhibitory afferents to hippocampal pyramidal cells after systemic injection of 3 g/kg ethanol. However, in this *in-vivo* study, measurements were not taken until 20 min after ethanol injection, a duration of ethanol exposure not often reached in our *in-vitro* studies, and the afferents activated *in vivo* (commissural inputs) were different from those in our study. Nonetheless, this discrepancy between results may indicate differences in the effects of ethanol applied to neurons directly (locally) versus the systemic route, perhaps because of the involvement of a metabolic mediator such as a tetrahydroisoquinoline (see REF. 9) in the *in-vivo* situation.

Our results also do not support the conclusion that ethanol enhances GABAergic IPSPs, as suggested by the extracellular findings of Nestaros[14] that systemic ethanol prolongs the inhibitory period produced by surface stimulation or iontophoretic GABA in cortical neurons. In our *in-vitro* study, ethanol superfusion increased the size or duration of IPSPs in only 5 of 47 tests, and had little effect on the hyperpolarizing, inhibitory responses to GABA applied locally to pyramidal neurons, in accord with our *in-vivo* hippocampal studies (see above).

Regarding interactions between ethanol and other neurotransmitters, in recent studies we (Siggins and Madamba, in preparation) have sought *in-vitro* correlates of the potentiation of ACh responses seen *in vivo* with systemic ethanol (described above). Our preliminary intracellular recordings in the hippocampal slice reveal that low concentrations of ACh (1–10 μM) or the pure muscarinic agonist muscarine (0.2–2 μM) usually exert, like ethanol, only weak effects on membrane potential (depolarizing or occasionally hyperpolarizing) but cause pronounced depressions of postburst AHPs and synaptic potentials (especially IPSPs). Thus, the ethanol-induced potentiations of ACh seen *in vivo* may arise as a summation of like effects, with the increases in firing rate resulting from loss of the inhibitions derived from synaptic input and afterhyperpolarizations. However, in at least one pyramidal cell showing a pronounced (8 mV) depolarization to muscarine (200 nM), superfusion of muscarine plus a concentration of ethanol (11 mM) having no direct effect by itself resulted in a near doubling of the depolarization (to 15 mV). This suggests that in some cells ethanol may truly potentiate ACh effects by augmenting ACh-induced depolarizations. Further studies will be required to verify this suggestion and to determine the mechanism of the effect.

SUMMARY AND CONCLUSIONS

With respect to the theme of this volume, the results of our recent studies on three neuronal model systems point to several relevant conclusions: 1) ethanol may interact electrophysiologically with certain anesthetics such as urethane; 2) ethanol can selectively enhance responses to certain neurotransmitters; 3) resting membrane properties of individual neurons show a wide range of sensitivities to ethanol and are generally fairly insensitive; 4) the synapse—independent of specific transmitters—seems most sensitive to ethanol.

As regards the first point, it has long been known that ethanol and anesthetics have features in common, including the ability to alter the lipid components of biological membranes (see R. A. Harris *et al.*, L. L. M. van Deenen *et al.*, M. J. Hudspith *et al.*, E. Rubin *et al.*, and C. C. Cunningham & P. I. Spach in this volume), so interactions between the two are not unexpected. However, our electrophysiological findings suggest great caution and appropriate controls be used in *in-vivo* studies of anesthetized animals, as the interactions derived may actually reverse the usual effect of ethanol. The enhancement of responses to ACh and SS (second point) might be assumed to arise postsynaptically in the target cells recorded and are seen with low, intoxicating doses of ethanol. Whether this potentiation involves enhancement of specific agonist binding to the receptor or facilitation of the function of the ionic channel linked to the receptor remains to be determined. It is not hard to imagine that ethanol could perturb membrane properties near receptors, to alter their conformation and ligand binding, or perhaps even uncover hidden receptors. The relative insensitivity of the resting membrane properties (third point) may suggest that membrane channels responsible for these functions (*e.g.*, 'leak' channels for Na^+ and K^+ ions) do not usually interact with the lipid components affected by ethanol, at least at low, 'intoxicating' ethanol concentrations.

Finally, the reduction of synaptic potentials by ethanol may indicate a presynaptic locus of action, as the response to the transmitter for at least one of these synaptic potentials (GABA) was not altered. These data would seem to indicate that synaptic release of the transmitter is reduced by ethanol, at least in the hippocampal slice. The high sensitivity of this presynaptic element for ethanol could indicate that the machinery for synaptic release, such as conductances for calcium entry (see REF. 39) or the action of second messenger systems (*e.g.*, those leading to synapsin phosphorylation) are particularly sensitive to ethanol. Studies on other presynaptic model systems such as the squid giant synapse could help resolve some of these questions more directly.

ACKNOWLEDGMENT

We thank Nancy Callahan for manuscript typing.

REFERENCES

1. BERRY, M. S. & V. W. PENTREATH. 1980. The neurophysiology of alcohol. *In* Psychopharmacology of Alcohol. M. Sandler, Ed., 43–72. Raven Press. New York, NY.
2. FABER, D. S. & M. R. KLEE. 1977. Actions of ethanol on neuronal membrane properties and synaptic transmission. *In* Alcohol and Opiates: Neurochemical and Behavioral Mechanisms. K. Blum, Ed. 41–63. Academic Press. New York, NY.

3. GRENELL, R. G. 1972. Effects of alcohol on the neuron. *In* Biology of Alcoholism. B. Kissin & H. Begleiter, Eds. Vol. 1: 1–19. Plenum Press. New York, NY.
4. KALANT, H. 1974. Ethanol and the nervous system. Experimental neurophysiological aspects. Int. J. Neurol. **9:** 111–124.
5. SIGGINS, G. R. & F. E. BLOOM. 1981. Alcohol-related electrophysiology. Pharmacol. Biochem. Behav. **13:** 203–211.
6. CARLEN, P. L., N. GUREVICH & D. DURAND. 1982. Ethanol in low doses augments calcium-mediated mechanisms measured intracellularly in hippocampal neurons. Science **215:** 306–309.
7. DURAND, D., W. A. CORRIGAL, P. KUJTAN & P. L. CARLEN. 1981. Effects of low concentrations of ethanol on CA1 hippocampal neurons in vitro. Can. J. Physiol. Pharmacol. **59:** 979–984.
8. SIGGINS, G. R. & E. FRENCH. 1979. Central neurons are depressed by iontophoretic and micro-pressure applications of ethanol and tetrahydropapaveroline. Drug Alcohol Depend. **4:** 239–243.
9. BERGER, T., E. D. FRENCH, G. R. SIGGINS & F. E. BLOOM. 1982. Ethanol and some tetrahydroisoquinolines alter the discharge of rat hippocampal neurons in vivo when applied by microelectroosmosis or pressure: Relationship to opiate action. Pharmacol. Biochem. Behav. **17:** 813–821.
10. GROUL, D. L. 1982. Ethanol alters synaptic activity in cultured spinal cord neurons. Brain Res. **243:** 25–33.
11. GRUPP, L. A. 1980. Biphasic action of ethanol on single units of the dorsal hippocampus and the relationship to the cortical EEG. Psychopharmacology **70:** 95–103.
12. NEWLIN, S. A., J. MANCILLAS-TREVINO & F. E. BLOOM. 1981. Ethanol causes increases in excitation and inhibition in area CA3 of the dorsal hippocampus. Brain Res. **209:** 113–128.
13. ROGERS, J., G. R. SIGGINS, J. A. SCHULMAN & F. E. BLOOM. 1979. Physiological correlates of ethanol intoxication, tolerance, and dependence in rat cerebellar Purkinje cells. Brain Res. **196:** 183–198.
14. NESTEROS, N. S. 1980. Ethanol specifically potentiates GABA-mediated neurotransmission in feline cerebral cortex. Science **209:** 708–710.
15. BRADLEY, R. J., K. PEPER & R. STERZ. 1980. Nature (Lond.) **284:** 60–62.
16. GAGE, P. W. 1965. J. Pharmacol. Exp. Ther. **150:** 236–243.
17. QUASTEL, D. M., J. T. HACKETT & K. OKAMOTO. 1972. Canad. J. Physiol. Pharmacol. **50:** 279–284.
18. OKADA, K. 1967. Effects of alcohols and acetone on the neuromuscular junction of frog. Jap. J. Physiol. **17:** 245–261.
19. SINCLAIR, J. G. & G. F. LO. 1981. The effects of ethanol on cerebellar Purkinje cell discharge pattern and inhibition evoked by local surface stimulation. Brain Res. **204:** 465–471.
20. SINCLAIR, J. G. & G. F. LO & A. F. TIEM. 1980. The effects of ethanol on cerebellar Purkinje cells in naive and alcohol-dependent rats. Can. J. Physiol. Pharmacol. **58:** 429–432.
21. ECCLES, J. C., M. ITO & J. SZENTAGOTHAI. 1967. The Cerebellum as a Neuronal Machine. Springer-Verlag. New York, NY.
22. ROGERS, J., S. G. MADAMBA, D. A. STAUNTON & G. R. SIGGINS. 1986. Ethanol increases single unit activity in the inferior olivary nucleus. Brain Res. **385:** 253–262.
23. ECCLES, J. C., R. LLINAS & K. SASAKI. 1966. The excitatory synaptic action of climbing fibers on the Purkinje cells of the cerebellum. J. Physiol. **182:** 268–296.
24. MANCILLAS, J., G. R. SIGGINS & F. E. BLOOM. 1986. Systemic ethanol: selective enhancement of responses to acetylcholine and somatostatin in the rat hippocampus. Science **231:** 161–163.
25. BLOOM, F. E., P. LAD, Q. PITTMAN & J. ROGERS. 1982. Blood alcohol levels in rats: non-uniform yields from intraperitoneal doses based on body weight. Brit. J. Pharmacol. **75:** 251–254.
26. SIGGINS, G. R. & S. MADAMBA. 1985. Acetylcholine (ACh) and muscarine reduce CA1 field potentials in the hippocampal slice: potentiation by ethanol. Soc. Neurosci. Abstr. **10:** 300.

27. MONTOYA, G. A., W. K. RIKER & N. J. RUSSELL. 1977. J. Pharmacol. Exp. Ther. **200:** 320–327.
28. HENDRY, S. H. C., E. G. JONES, J. DEFELIPE, D. SCHMECHEL, C. BRANDON & P. C. EMSON. 1984. Proc. Natl. Acad. Sci. USA **81:** 6526–6530.
29. SOMOGYI, A. J., A. HODGSON, M. DAVID SMITH, M. GRACIA NUNZI, A. GORIO & WU JANG-YEN. 1984. J. Neurosci. **4:** 2590–2603.
30. PITTMAN, Q. J. & G. R. SIGGINS. 1980. Ethanol has multiple actions on electrophysiological properties of hippocampal (HPC) pyramidal neurons in vitro. Soc. Neurosci. Abstr. **6:** 771.
31. SIGGINS, G. R., Q. J. PITTMAN & E. D. FRENCH. 1986. Effects of ethanol on CA1 and CA3 pyramidal cells in the hippocampal slice preparation: an intracellular study. Brain Res. In press.
32. SIGGINS, G. R. & W. ZIEGLGÄNSBERGER. 1981. Morphine and opioid peptides reduce inhibitory synaptic potentials in hippocampal pyramidal cells in vitro without alteration of membrane potential. Proc. Natl. Acad. Sci. USA **78:** 5235–5239.
33. ALGER, B. E. & R. A. NICOLL. 1980. Epileptiform burst afterhyperpolarization: calcium-dependent potassium potential in hippocampal CA1 cells. Science **210:** 1122–1124.
34. HOTSON, J. R. & D. A. PRINCE. 1980. A calcium-activated hyperpolarization follows repetitive firing in hippocampal neurons. J. Neurophysiol. **43:** 409–429.
35. SCHWARTZKROIN, P. A. & C. E. STAFSTROM. 1980. Effects of EGTA on the calcium-activated afterhyperpolarization in hippocampal CA3 pyramidal cells. Science **210:** 1125–1126.
36. CARMICHAEL, F. J. & Y. ISRAEL. 1975. Effects of ethanol on neurotransmitter release by rat brain cortical slices. J. Pharmacol. Exp. Ther. **193:** 824–834.
37. HOWERTON, T. C. & A. C. COLLINS. 1984. Ethanol-induced inhibition of GABA release from LS and SS mouse brain slices. Alcohol **1:** 471–477.
38. STRONG, R. & W. G. WOOD. 1984. Membrane properties and aging: in vivo and in vitro effects of ethanol on synaptosomal gamma-aminobutyric acid (GABA) release. J. Pharmac. Exp. Ther. **299:** 726–730.
39. OAKES, S. G. & R. S. POZOS. 1982. Electrophysiologic effects of acute ethanol exposure. II. Alterations in the calcium component of action potentials from sensory neurons in dissociated culture. Develop. Brain Res. **5:** 251–255.

DISCUSSION OF THE PAPER

B. TABAKOFF (*National Institute on Alcohol Abuse and Alcoholism, Bethesda, MD*): In terms of the cholinergic channels, or the channels that are linked to the muscarinic cholinergic receptors, do you know how the receptor is linked to those channels?

SIGGINS: In hippocampus it is not clear. In some of the simpler systems, sympathetic ganglia and so forth, these events are just being worked out. The interesting thing about acetylcholine effects in the hippocampus is that they may involve a closing rather than the opening of a channel. The depolarizing responses are accompanied by a decrease in conductance, a decrease in permeability. Potassium channels are probably closed.

J. M. LITTLETON (*King's College, London, England*): Is it possible that there are presynaptic muscarinic receptors on the GABAergic neurons?

SIGGINS: I think that's what's happening. One of the ways that acetylcholine probably excites cells in the hippocampus is by blocking inhibitory mechanisms, probably presynaptically on the GABAergic neurons, by preventing release of GABA.

Acetylcholine and muscarine also block release of excitatory transmittors. If you apply acetylcholine to the dendrites, where the excitatory terminals are, you block those EPSPs. We see similar things with superfusion of muscarine.

LITTLETON: You said that you got more or less the same kinds of effects with ethanol as with muscarinic receptor activation. Did you get a potentiation of those muscarinic effects on the IPSPs with ethanol as well?

SIGGINS: Yes. In several cases we've seen merely summation, but in about half of the cells we've seen a true potentiation, in which some threshold doses of ethanol potentiate the muscarinic effect on the presynaptic elements. There seems to be a muscarinic effect, both pre- and postsynaptically, and ethanol potentiates those effects pre- and postsynaptically.

F. WEIGHT: We've done some similar experiments in the hippocampus and our data agree, on CA3 neurons. In voltage clamp experiments with steps there is an increase in a slow outward current, and that current may be similar to the one that muscarine inhibits. It is not clear whether the action of ethanol is on the muscarinic mechanism per se or on a voltage-activated current that muscarine inhibits.

On the other side of the coin, in other experiments in PC-12 cells (a theochromacytoma cell line that has electrical excitability), muscarine activation causes a release of a transmitter, and that is inhibited by ethanol. So in different cell types you may find different types of phenomena.

SIGGINS: Yes, we see effects of ethanol that vary from cell to cell even within the same type of identified cell in hippocampus. In CA1 or CA3, some cells are very sensitive to ethanol or to muscarine, while others are not. We see variability not only to ethanol but also to transmitters.

Adaptation to Ethanol in Cultured Neural Cells and Human Lymphocytes From Alcoholics[a]

ADRIENNE S. GORDON,[b,c,e] BRADLEY WRUBEL,[b]
KATHLEEN COLLIER,[b] WILLIAM ESTRIN,[b,c,f]
AND IVAN DIAMOND[b,c,d,e]

[b]*Ernest Gallo Clinic and Research Center*
Departments of [c]*Neurology,*
[d]*Pediatrics, and* [e]*Pharmacology*
[f]*Occupational Health Center*
University of California at San Francisco
San Francisco General Hospital
San Francisco, California 94110

INTRODUCTION

The neurological and behavioral complications of alcoholism are major problems in medicine and in society. And yet, despite the widespread incidence of alcoholism, the molecular events accounting for intoxication, tolerance, and physical dependence after alcohol abuse are poorly understood. Current evidence suggests that ethanol intercalates into neural cell membranes and induces adaptive changes in neuronal function.[1-4] Acute and chronic ethanol-induced changes in membrane order and membrane constituents have been reported (*e.g.* cholesterol/phospholipid ratios[5-8] and fatty acids[9-11]), but experiments using different animals, different strains of the same animals, or the same strains in different laboratories have yielded conflicting results.[12] This may be due in part to unknown genetic factors, the use of disrupted tissue preparations, and the mode of ethanol administration. A most serious problem derives from the heterogeneity of brain. Biochemical changes restricted to selected brain regions, cell types, or specific membrane components may be difficult to detect in crude brain preparations. Also, it is nearly impossible to distinguish primary biochemical effects of ethanol in the nervous system from secondary or tertiary responses related to a series of synaptic events or systemic, metabolic, and hormonal influences.

ETHANOL INTERACTION WITH INHIBITORY SYNAPTIC MECHANISMS

Unlike most psychoactive agents that interact at nanomolar concentrations with specific receptors in the nervous system, ethanol does not produce clinical effects until millimolar concentrations are achieved in the brain. Since neural receptors are localized in specialized membranes, it is possible that ethanol produces biochemical and functional changes in synaptic activity by altering the physical and molecular

[a]Supported in part by grants from the National Institute on Alcohol Abuse and Alcoholism and from the Alcoholic Beverage Medical Research Foundation.

milieu of certain receptors and other membrane proteins required for signal transduction.

Ethanol produces inhibitory responses in brain, and after withdrawal alcoholic patients and animals exhibit intense neuronal hyperactivity. Therefore, ethanol might acutely activate inhibitory synaptic mechanisms in the brain, while an adaptive response to the continued presence of ethanol would be to diminish the activity of the same inhibitory synaptic mechanism. During withdrawal, neural hyperexcitability might be related to decreased inhibition in the nervous system. These predictions, illustrated in FIGURE 1, were tested in a model cell culture system[13–15] using the neural cell line NG108-15 to study the interaction of ethanol with intact neural cells and to determine the adaptive responses of these cells to acute and chronic ethanol exposure under controlled conditions.

IN-VITRO STUDIES WITH NG108-15 CELLS: ACUTE ETHANOL STIMULATION OF ADENOSINE RECEPTOR-DEPENDENT cAMP LEVELS

Adenosine is an inhibitory neurotransmitter, which appears to mediate some of the acute effects of ethanol in the brain.[16,17] The adenosine receptor in NG108-15 cells is coupled positively to adenylate cyclase and appears to be identical to the A_2 receptor in

ETHANOL INTERACTION WITH INHIBITORY SYNAPSES

CONDITION	INHIBITORY ACTIVITY
NORMAL	+++
ACUTE ETHANOL STIMULATION	++++++
CHRONIC ADAPTATION TO ETHANOL	+++
ACUTE ETHANOL WITHDRAWAL	+

FIGURE 1. Hypothetical effects of ethanol at inhibitory synapses.

brain.[18] We have found that the adenosine agonist phenylisopropyladenosine (PIA) increases the intracellular levels of cAMP in these cells in a concentration-dependent manner.[15] Since the brain is a major target for ethanol, it is possible that neuronal sensitivity in brain may be related to the fact that neurons in the central nervous system are not dividing. Therefore, we examined the effects of ethanol both on rapidly growing cells in complete serum-free medium and on cells growing slowly in medium that was depleted of growth factors. FIGURE 2 shows that ethanol added acutely stimulates adenosine receptor-dependent cAMP production in NG108-15 cells. In this experiment, NG108-15 cells were treated with a phosphodiesterase inhibitor to block cAMP hydrolysis and incubated with PIA and varying concentrations of ethanol. Rapidly growing cells showed increases in cAMP levels at ethanol concentrations greater than 100 mM. However, when cell division was reduced in the absence of growth factors, acute sensitivity to ethanol was increased. Concentrations of ethanol as low as 50 mM caused a 20–40% increase in PIA-stimulated cAMP levels. Under these conditions of study, ethanol had no significant effect on basal cAMP levels in the absence of PIA.

The increase in adenosine receptor-dependent cAMP levels produced by ethanol might be related to changes in osmolality of the medium or other nonspecific factors. On the other hand, if the response of the cells was due to a specific effect of ethanol, then the increase in cAMP levels should be related to the solubility of ethanol in cellular membranes. Therefore, it would be anticipated that alcohols of increasing

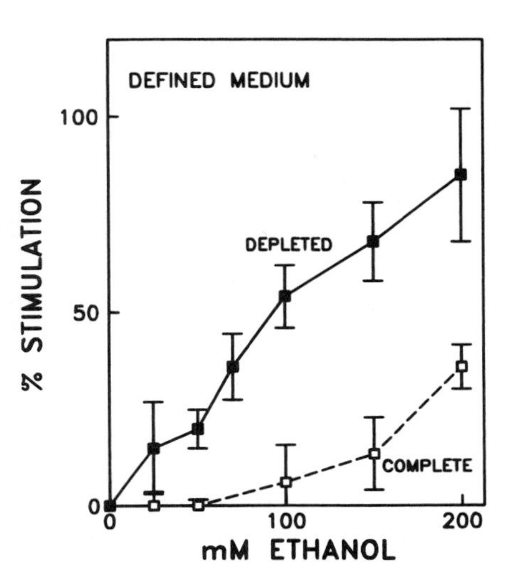

FIGURE 2. Acute ethanol stimulation of cAMP levels in NG108-15 cells in the presence of 0.1 mM phenylisopropyladenosine (PIA). Cells were grown in complete defined medium (□) or defined medium depleted of oleic acid, insulin, and transferrin (■), preincubated for 60 min with the phosphodiesterase inhibitor Ro 20-1724 and incubated with PIA in the presence or absence of ethanol for 30 min. cAMP was assayed as described in REFERENCE 15.

chain length should show increasing stimulation of adenosine receptor-dependent cAMP levels. The results of such an experiment are shown in FIGURE 3. Stimulation of adenosine receptor-dependent cAMP levels increased with increases in the chain length of the added alcohol.

ADAPTIVE RESPONSES OF NG108-15 CELLS TO CHRONIC ETHANOL

As organisms become tolerant to ethanol, the acute effects of ethanol should be diminished or counterbalanced by an opposing physiologic mechanism (FIG. 1). In this model cellular system, NG108-15 cells adapted to the long-term presence of ethanol by decreasing adenosine receptor-stimulated cAMP levels. FIGURE 4 shows relative PIA-stimulated cAMP levels in NG108-15 cells grown with or without ethanol for

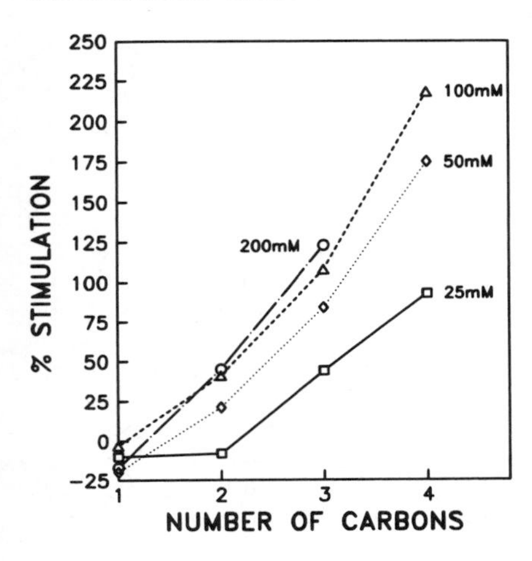

FIGURE 3. Acute alcohol stimulation of cAMP levels in NG108-15 cells in the presence of 0.1 mM PIA. Assay as described in FIGURE 2.

several days. Assays were carried out either with or without ethanol present during PIA stimulation. There was a 35–48% reduction in PIA-stimulated cAMP levels after 48 hr in ethanol, when measured in the absence of ethanol. If this decrease in receptor-dependent cAMP accumulation is a compensatory response related to chronic ethanol exposure (*i.e* a form of cellular tolerance to ethanol), then ethanol should be required for the chronically-treated cells to exhibit normal adenosine receptor-activated cAMP levels. Therefore, cells grown in complete defined medium for 2 days in 200 mM ethanol were washed and rechallenged with 200 mM ethanol. FIGURE 4 also shows that the chronically-treated cells still responded to acute ethanol exposure by increasing PIA-stimulated cAMP levels. Most important, in the presence of ethanol, the chronically treated cells showed PIA-stimulated cAMP levels that were the same as those in untreated control cells. These findings suggest that NG108-15 cells undergo compensatory membrane changes that appear to result from cellular tolerance to ethanol; receptor-dependent cAMP levels in chronically treated cells were normal in the presence of ethanol. These cells were also "dependent" on ethanol, since

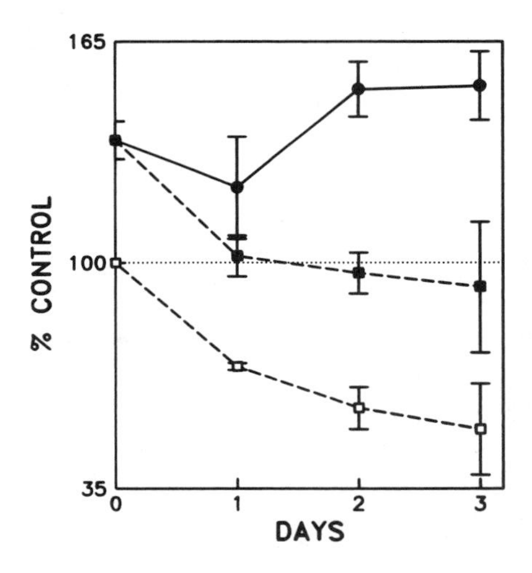

FIGURE 4. Cellular adaptation to ethanol. cAMP levels were measured in the presence of 0.1 mM PIA in NG108-15 cells grown without (●) or with (□,■) 200 mM EtOH for the indicated number of days in complete defined medium. Assays were carried out as described in FIGURE 2 with (●,■) or without (□) 200 mM ethanol present during the PIA incubation. (From Gordon *et al.*[15] Reprinted by permission from the National Academy of Sciences USA.)

they showed reduced PIA-stimulated cAMP levels when ethanol was removed. Moreover, the observed effects of chronic ethanol treatment were reversible; reduced adenosine receptor-dependent cAMP levels in ethanol-tolerant cells returned to control values 48 hr after ethanol was removed (FIG. 5).

Our results, which are schematically presented in FIGURE 6, suggest that NG108-15 cells in culture can be used as a model to study "intoxication," "tolerance," and "physical dependence" at a cellular level. The results also suggest that inhibitory synaptic mechanisms may be important targets for ethanol in the brain. The major findings are: (a) acute ethanol exposure increased adenosine receptor-stimulated cAMP levels above normal; (b) chronic ethanol exposure decreased adenosine receptor-stimulated cAMP levels below normal; (c) acute ethanol restored chronically treated cellular cAMP to normal levels; and (d) recovery occurs after withdrawal. These observations suggest that adaptation to ethanol occurs in intact cells and that depressed adenosine receptor-stimulated cAMP levels in cells might be a biochemical change of pathophysiologic significance in chronic alcoholism. For example, it is

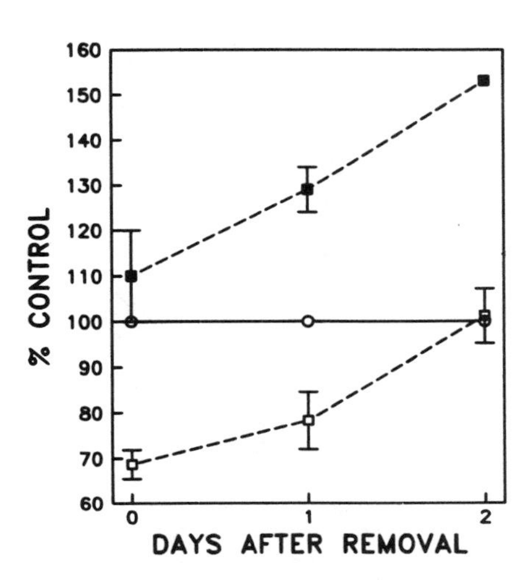

FIGURE 5. Withdrawal from chronic ethanol. cAMP levels were measured in the presence of 0.1 mM PIA at various times after removal of ethanol. The cells had been grown for 48 hr in depleted defined medium with 100 mM ethanol. Assays were carried out either with (■) or without (□) 100 mM ethanol present during the PIA incubation. (From Gordon *et al.*[15] Reprinted by permission from the National Academy of Sciences USA.)

possible that depressed adenosine receptor-stimulated cAMP levels occur in alcoholic patients and contribute to the hyperactive syndrome encountered immediately after ethanol withdrawal.

STUDIES WITH HUMAN LYMPHOCYTES

Human lymphocytes have the same adenosine A_2 receptor as NG108-15 cells.[19] Therefore, we undertook an age- and sex-matched controlled study of basal and adenosine receptor-stimulated cAMP levels in lymphocytes from chronic alcoholics,

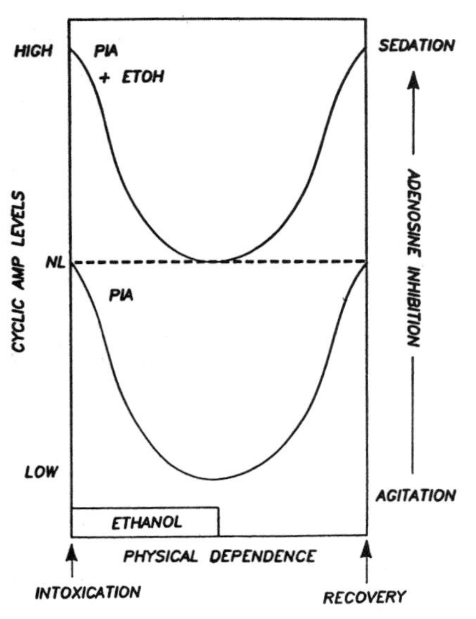

FIGURE 6. Cellular model for acute and chronic effects of ethanol on cAMP levels and physiological responses.

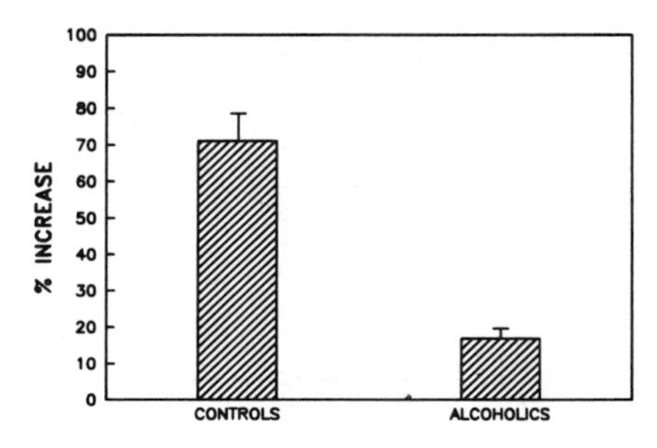

FIGURE 7. Ethanol stimulation of cAMP levels measured in the presence of 0.1 mM PIA in lymphocytes from alcoholics and controls. 5 × 10^5 cells were preincubated for 5 min with Ro 20-1724 and then incubated for 60 min with 80 μM PIA in the presence or absence of 80 mM ethanol. Data presented as % increase in the presence of ethanol as compared to values in the absence of ethanol.

normal subjects, and patients with nonalcoholic liver disease. There were 10 patients in each group. The alcoholics and patients with liver disease exhibited abnormalities of some liver enzymes, but there were no statistically significant differences between the alcoholic and the matched liver disease group. Extensive clinical laboratory values in the alcoholic patients were virtually indistinguishable from patients with liver disease and were similar to normal controls except for significantly higher SGOT and globulin. There was no evidence of malnutrition in the alcoholic patient population. We found that normal human lymphocytes showed increases in intracellular cAMP levels when exposed to increasing concentrations of PIA. Moreover, ethanol added *in vitro* to the isolated lymphocytes further increased PIA-stimulated cAMP without changing basal cAMP levels. This suggested that lymphocytes from alcoholics might show an adaptive response to chronic ethanol abuse, which could involve a reduction in adenosine receptor-dependent cAMP levels. We found that lymphocytes from alcoholics showed a nearly fourfold reduction of both basal and PIA-stimulated cAMP levels when compared to normal subjects or patients with nonalcoholic liver disease (TABLE 1). Moreover, there was no difference between lymphocytes from control subjects and those from patients with liver disease.

Our findings suggest that lymphocytes from alcoholics may have adapted to chronic alcohol abuse by reducing intracellular levels of cAMP. We then examined the effect of acute alcohol on PIA-stimulated cAMP levels. We found that lymphocytes from alcoholics were strikingly resistant to the acute stimulating effect of ethanol added *in vitro*. FIGURE 7 shows that there was a fourfold reduction in the percent of ethanol stimulation of PIA-dependent cAMP accumulation in lymphocytes from alcoholics compared to lymphocytes from controls.

TABLE 1. Lymphocyte Studies in Controls, Alcoholics, and Patients with Nonalcoholic, Noninfectious Liver Disease[a]

	Controls (Mean ± S.E.M.)	Alcoholics (Mean ± S.E.M.)	Liver Disease (Mean ± S.E.M.)
Basal cAMP (pmol/10^6 cells)	9.55 ± 1.65	2.30 ± 0.34[b] p = 0.0004	8.33 ± 1.29[c] p = 0.0005
PIA-Stimulated cAMP (pmol/10^6 cells)	15.81 ± 2.52	3.72 ± 0.53[b] p = 0.0002	14.04 ± 1.93[c] p = 0.0005

[a]p values calculated by Wilcoxon Rank sum test comparing controls to alcoholics and alcoholics to patients with liver disease.

[b]Compared to control group.

[c]Compared to alcoholic group.

Taken together, these results suggest that lymphocytes from alcoholics show changes in receptor-mediated signal transduction that resemble some of the changes found in neural cells in culture. Cells from alcoholics exhibited (a) reduced basal cAMP levels, (b) reduced PIA-stimulated cAMP levels, and (c) increased resistance to acute ethanol stimulation of adenosine receptor-dependent cAMP levels. Similar findings were also observed in T-cells isolated from 6 alcoholics when compared to T-cells from 6 nonmatched normal subjects. T-cells from alcoholics showed 27% and 37% of the basal and adenosine receptor-stimulated cAMP levels, respectively, when compared to levels measured in control T-cells ($p < 0.02$).

There were two striking differences between the results in human lymphocytes and those with neural cells in culture. First, the unstimulated intracellular levels of cAMP were reduced in lymphocytes from alcoholics but unchanged in neural cells in culture. Second, and perhaps most important, the acute response to ethanol was markedly reduced in lymphocytes from alcoholics, while neural cells that adapted to ethanol *in vitro* showed no change in the magnitude of their acute response when rechallenged with ethanol. These differences may be due to such variables as ethanol concentration, time of exposure, and differences in the cellular preparations.

FUTURE STUDIES

It is possible that measurements of basal and receptor-stimulated cAMP levels in lymphocytes may distinguish alcoholic and nonalcoholic human populations. The reduced levels of cAMP in lymphocytes from alcoholics may reflect an acquired membrane abnormality caused by chronic alcohol abuse. On the other hand, our findings could be related to a genetically determined difference in the membrane response of cells of alcoholics. These possibilities can be tested by examining the interaction of ethanol with human lymphocytes obtained from alcoholics and controls and maintained in culture over long periods of time. Such studies may also help to identify specific nutritional requirements that are necessary for cells to exhibit an adaptive response to ethanol. Most important, ethanol studies with intact lymphocytes *in vitro* may help to explain the molecular basis of the apparent membrane defect in signal transduction that develops in alcoholism, and might even be useful in helping to identify individuals at risk to develop alcoholism.

ACKNOWLEDGMENTS

We thank Mr. William Meraz and Dr. David Feigel for their contributions to this work, and Ms. R. C. Webb for manuscript preparation.

REFERENCES

1. GOLDSTEIN, D. B. 1983. Pharmacology of Alcohol. Oxford University Press. New York, NY.
2. CHIN, J. H. & D. B. GOLDSTEIN. 1981. Mol. Pharm. **19:** 425–431.
3. FRANKS, N. P. & W. R. LIEB. 1982. Nature **300:** 487–493.
4. ROTTENBERG, H., A. WARING & E. RUBIN. 1981. Science **213:** 583–584.
5. JOHNSON, D. A., N. M. LEE, R. COOKE & H. H. LOH. 1979. Mol. Pharmacol. **15:** 739–746.

6. CHIN, J. H., L. M. PARSONS & D. B. GOLDSTEIN. 1978. Biochim. Biophys. Acta **513:** 358–363.
7. PARSONS, L. M., E. H. GALLAHER & D. B. GOLDSTEIN. 1982. J. Pharmacol. Exp. Ther. **223:** 472–476.
8. SMITH, T. L. & M. H. GERHART. 1982. Life Sci. **31:** 1419–1425.
9. WARING, A. J., H. ROTTENBERG, T. OHNISHKI & E. RUBIN. 1981. Proc. Natl. Acad. Sci. USA **78:** 2582–2586.
10. LITTLETON, J. M. & G. R. JOHN. 1977. J. Pharm. Pharmacol. **29:** 579–580.
11. REITZ, R. C., L. WANG, R. J. SCHILLING, G. H. STARICH, J. D. BERGSTROM & J. A. THOMPSON. 1982. Prog. Lipid Res. **20:** 209–213.
12. LYON, R. C. & D. B. GOLDSTEIN. 1983. Mol. Pharmacol. **23:** 86–91.
13. CHARNESS, M. E., A. S. GORDON & I. DIAMOND. 1983. Science **222:** 1246–1248.
14. CHARNESS, M. E., L. A. QUERIMIT & I. DIAMOND. 1986. J. Biol. Chem. **261:** 3164–3169.
15. GORDON, A. S., K. COLLIER & I. DIAMOND. 1986. Proc. Natl. Acad. Sci. USA **83:** 2105–2108.
16. PROCTOR, W. R. & T. V. DUNWIDDIE. 1984. Science **224:** 519–521.
17. DAR, M. S., S. J. MUSTAFA & W. R. WOLLES. 1983. Life Sci. **33:** 1363–1374.
18. SNYDER, S. H. 1985. Ann. Rev. Neurosci. **8:** 103–124.
19. HYNIE, S., F. LANEFELT & B. B. FREDHOLM. 1980. Acta Pharmacol. Toxicol. **47:** 58–65.

The Influence of Catecholamine Systems and Thyroid Function on the Actions of Ethanol in Long-Sleep (LS) and Short-Sleep (SS) Mice[a]

NORMAN WEINER,[b] JEFFREY K. DISBROW,[c]
THOMAS A. FRENCH,[c] AND JOSEPH M. MASSERANO[c]

[b]*Abbott Laboratories*
Abbott Park, Illinois 60064
and
[c]*Department of Pharmacology*
University of Colorado School of Medicine
Denver, Colorado 80262

During the past two decades, a considerable number of investigations have been conducted on the effects of ethanol on catecholamine levels and turnover in both the central and peripheral nervous systems. Whereas the bulk of the investigations have failed to demonstrate an effect of acute ethanol administration on levels of brain norepinephrine or dopamine,[1-4] many investigators have reported that ethanol may affect biogenic amine turnover in the central nervous system. However, conflicting results have been reported on the nature of the effects elicited. Corrodi and co-workers[5] examined norepinephrine turnover by monitoring the decline in brain norepinephrine content after administration to rats of alpha-methyl-p-tyrosine, an inhibitor of tyrosine hydroxylase. Brain norepinephrine levels fell more rapidly shortly after rats were treated with 2 g/kg of ethanol, suggesting that an enhancement of brain norepinephrine turnover occurs following ethanol administration. These workers reported no effect of ethanol on dopamine turnover shortly after ethanol. At later times following ethanol administration, the turnover rates of both norepinephine and dopamine were reduced.

Utilizing the incorporation of radiolabeled tyrosine into brain catecholamines as an index of catecholamine synthesis and turnover, Carlsson *et al.*[2] observed that the oral administration of 7 g/kg of ethanol was associated with an increased rate of incorporation of ^{3}H-tyrosine into dopamine and, to a lesser extent, norepinephrine. These authors claimed that there was a direct correlation between the intensity of the behavioral effects produced by ethanol and the extent of the enhancement of the incorporation of tyrosine into brain dopamine and norepinephrine in these rats. Carlsson and Lindqvist[6] demonstrated that the intraperitoneal administration of ethanol is associated with a significant increase in the accumulation of L-3,4-dihydroxyphenylalanine (dopa) in the brain following inhibition of L-aromatic amino acid decarboxylase. These effects were observed in regions of the brain that contain both norepinephrine and dopamine neurons. The authors concluded that ethanol stimulates the synthesis of both catecholamines. Ahtee and Sarstrom-Fraser[7] reported that chronic ethanol administration fails to alter turnover rates of either dopamine or

[a]Supported by Alcohol, Drug Abuse, and Mental Health Administration Grant AA03527.

norepinephrine in brains of rats. However, an increase in the turnover of these amines was observed during the withdrawal period. Pohorecky and colleagues[3,8,9] showed that small doses of ethanol are associated with a transient reduction in norepinephrine metabolites in the brain, followed by an increase in these substances. Larger doses of ethanol, which produce loss of righting reflex and unconsciousness, depressed the levels of norepinephrine metabolites. These results suggest that small doses of ethanol may increase norepinephrine turnover in the brain, whereas larger doses tend to depress the turnover of this catecholamine. Pohorecky and Jaffee[9] reported that, during chronic administration of ethanol, there is increased turnover of norepinephrine in the brain. Withdrawal following chronic ethanol administration was also associated with enhanced norepinephrine turnover. Liljequist and Carlsson[10] reported that the administration of ethanol markedly reduces the formation in brain of the 0-methylated dopamine metabolite, 3-methoxytyramine, whereas the accumulation of normetanephrine, the 0-methylated norepinephrine metabolite, is unaffected. In contrast, Bacapoulos and co-workers[11] observed that the administration of ethanol, 2 g/kg, i.p., to rats failed to alter norepinephrine turnover in most brain regions, whereas this dose of ethanol reduced the turnover of dopamine in the substantia nigra and caudate nucleus.

Bustos and Roth[12] were unable to demonstrate an effect of ethanol on the *in-vivo* accumulation of dopa in the neostriatum following inhibition of L-aromatic amino acid decarboxylase. However, ethanol administration was associated with an increase in the accumulation of dihydroxyphenylacetic acid (dopac) in a dose-dependent manner. Bustos and Roth concluded that ethanol may enhance the release of dopamine from nerve terminals, but the dopamine present in the synaptic cleft may, by stimulation of presynaptic autoreceptors, mediate inhibition of tyrosine hydroxylase activity in the nigro-striatal dopaminergic nerve terminals. In contrast, Fadda *et al.*[13] showed that acute administration of ethanol to rats was associated with increased levels of dopac and enhanced dopa formation in the caudate nucleus. However, similar effects were not observed in the substantia nigra or frontal cortex.

Eriksson and Carlsson[14] reported that ethanol administration to rats is associated with considerable increases in concentrations of brain neutral amino acids such as tyrosine, tryptophan, 5-hydroxytryptophan, and dopa. They attributed these changes to an increase in the transfer of the aromatic neutral amino acids across the blood brain barrier as a consequence of reduced levels of nonaromatic neutral amino acids in the plasma. The latter are known to compete with aromatic amino acids for uptake into the brain. The increase in levels of aromatic amino acids as a consequence of ethanol administration may contribute significantly to the enhanced synthesis and turnover of catecholamines and 5-hydroxytryptamine, which have been reported in some studies. However, Pohorecky *et al.*[15] were unable to demonstrate elevations in brain or plasma tyrosine after acute ethanol administration to rats, although plasma, but not brain, tyrosine was elevated after chronic ethanol administration.

In summary, the bulk of the studies reported indicate that acute administration of ethanol over a wide range of doses produces an initial increase in norepinephrine turnover. The initial increase in norepinephrine turnover is followed by a later decrease in turnover rate. Similarly, although the results are somewhat conflicting, the bulk of the evidence suggests that dopamine turnover in the neostriatum is increased following acute administration of ethanol. Much of the controversy in these turnover studies may be attributed to differences in experimental design, since many of the investigators elected to employ different doses of ethanol, different routes of administration, and different times following ethanol administration at which turnover is measured. The differences in species of animal and strain of rodent employed also undoubtedly contribute to the variability in the results that have been obtained. The level of excitation or stress imposed, either deliberately or inadvertently, on the animals during ethanol administration also may confound the results obtained. Brick and Pohorecky[16] have shown that ethanol tends to elicit a stressful response in a tranquil rodent, but alcohol reduces stress in a previously stressed animal, as revealed by elevated plasma levels of glucocorticoids, nonesterified fatty acids, and other generally recognized indices of stress.

In order to investigate more definitively the direct effects of ethanol on catecholamine turnover, we employed two lines of mice that were bred selectively to differ markedly in their sensitivity to ethanol with respect to duration of loss of righting reflex.[17,18] When administered the same dose of ethanol intraperitoneally (approximately 4 g/kg), the short-sleep (SS) line of mice exhibit a loss of righting reflex for approximately 10 to 15 min, whereas the long-sleep (LS) line of mice are incapable of righting themselves for at least 110 to 140 minutes.[19]

Catecholamine levels in different brain regions were compared in the two lines of mice.[19] Brain regions that were examined included cortex, striatum, hypothalamus, locus coeruleus, hippocampus, substantia nigra, and cerebellum. The norepinephrine content of the cerebellum was approximately 30% less in the LS mice than in the SS mice. No significant differences in norepinephrine content were noted in the other brain regions. Neither were any differences seen in dopamine content of LS and SS mice in either cortex, striatum, hypothalamus, or substantia nigra.

The lower norepinephrine content in the cerebellum of LS mice is consistent with the observation that dopa accumulation following administration of an L-aromatic amino acid decarboxylase inhibitor is also approximately 30% less in the cerebellum of LS mice than in SS mice. The latter results indicate that basal cerebellar norepinephrine turnover is significantly less in LS than in SS mice.[19]

The lower level of norepinephrine and the lower norepinephrine turnover rate in the cerebellum of the LS mice as compared with SS mice is of particular interest, since the major behavioral difference noted in these animals following ethanol administration is a much more prolonged loss of righting reflex in the LS line. The cerebellum is known to be intimately involved in the ability of animals to maintain normal posture. Sorensen *et al.*[20] observed a profoundly lower sensitivity to the depressant effects of ethanol of cerebellar Purkinje neurons of SS as compared with LS mice.

Ethanol administration is associated with a reduction in norepinephrine turnover in the cerebellum and hippocampus of both LS and SS mice. However, in hypothalamus, locus coeruleus, and frontal cortex, a reduction in norepinephrine synthesis, as measured by the accumulation of dopa following inhibition of L-aromatic amino acid decarboxylase, was apparent only in the LS line. In contrast, in both lines of mice, dopamine turnover in the striatum was enhanced following ethanol administration.[19]

We also examined possible differences in the metabolism of catecholamines in the adrenal glands of LS and SS mice.[21] Basal levels of tyrosine hydroxylase activity and epinephrine and norepinephrine content were significantly higher in the adrenal glands of LS mice. Ten minutes after the intraperitoneal administration of ethanol, tyrosine hydroxylase activity in the adrenal gland of LS mice is decreased significantly, whereas SS mouse adrenal tyrosine hydroxylase activity is considerably increased. At 25 and 125 min after ethanol, SS mouse adrenal gland tyrosine hydroxylase activity has returned to control levels, but the decrease in tyrosine hydroxylase activity in adrenals of LS mice persists throughout this period. At 125 min, the catecholamine content in the adrenal gland of LS mice is also significantly depressed. Kinetic analysis of the tyrosine hydroxylase prepared from adrenal glands of these animals shortly after ethanol administration reveals that the Km for the pterin cofactor, 6-methyltetrahydropterin, is decreased in the adrenal glands of SS mice and increased in the adrenal glands of LS mice. The Vmax is unaffected by acute ethanol treatment in either line.

The kinetic changes in the enzyme could explain the *in-situ* alterations in tyrosine hydroxylase activity that were observed following ethanol administration, as determined by dopa accumulation in the adrenal glands of rats challenged with an inhibitor of L-aromatic amino acid decarboxylase. The activating affect of ethanol on tyrosine hydroxylase in SS mice appears to be transynaptically mediated, since the administration of chlorisondamine, a ganglionic blocking agent, is able to antagonize this effect. In contrast, chlorisondamine treatment fails to modify the depressant effects of ethanol on adrenal tyrosine hydroxylase activity in LS mice. Pentobarbital, in doses adequate to eliminate the righting reflex in both LS and SS mice, fails to affect adrenal gland tyrosine hydroxylase in either line. These results suggest that ethanol administration is associated with a specific, transient stimulation of adrenal medullary secretion in SS, but not LS, mice and that this may be mediated via the central nervous system. It is possible that the selective stimulation of the adrenal medulla and release of catecholamines in SS mice may contribute to the difference in ethanol sensitivity observed between the two lines following ethanol administration. This notion is supported by the observation that chlorisondamine administration increases the duration of the loss of righting reflex in the SS mice that is induced by ethanol but fails to affect the duration of the loss of righting reflex induced by ethanol in the LS line of mice. Interestingly, chlorisondamine prolongs the duration of the loss of righting reflex induced by pentobarbital in both lines of mice to a comparable degree.[21]

The centrally mediated increase in tyrosine hydroxylase activity in the adrenal medulla of SS mice following ethanol administration and the decrease in the activity of this enzyme in the adrenal gland of LS mice may provide insight into the possible neurochemical basis for differences in sensitivity of the two lines of mice to ethanol. It is possible that the resistance of the SS mice to the depressant effects of ethanol is related to an adaptive increase in central and peripheral noradrenergic and adrenergic neural activities that mediate arousal behavior and may therefore counter the depressant effects of the ethanol. In contrast, the adaptive response of LS mice to the administration of ethanol is either minimal or absent, and, as a consequence, they are more sensitive to the depressant effects of this agent. Consistent with this notion are the observations that intraventricular administration of catecholamine agonists to LS mice results in a reduction in the duration of ethanol-induced loss of righting reflex, whereas in SS mice ethanol-induced sleep time is increased following intraventricular administration of norepinephrine, dopamine, epinephrine, or isoproterenol.[22] These results suggest that there is an inadequate central catecholamine response to the depressant effects of ethanol in LS mice. In contrast, the neurochemical adaptation seen in SS mice may be optimal; additional stimulation of catecholamine receptors in the central nervous system may fail to produce any additional response to neutralize the depressant actions of ethanol. In fact, excessive stimulation of central catecholamine receptors may interfere with this adaptive response.

Additional support for this notion is derived from the fact that either (a) shortly after administration of alpha-methyl-p-tyrosine,[23] (b) 10 days following administration of reserpine,[24] or (c) administration of 6-hydroxydopamine during development,[24] each of which reduces brain norepinephrine levels in nerve terminal regions of adult rats, there is a considerable increase in the duration of the loss of righting reflex in SS mice following ethanol (TABLE 1). In contrast, acute administration of alpha-methyl-p-tyrosine or perinatal administration of 6-hydroxydopamine reduces brain catecholamine levels in adult LS mice, but does not affect the duration of the loss of righting reflex in this line.[24] These results support the notion that central catecholamine systems are not contributing significantly to counter the depressant actions of ethanol in LS mice. However, ten days following reserpine administration, when brain catecholamine levels are still modestly decreased, but tyrosine hydroxylase activity is

TABLE 1. The Effect of 6-Hydroxydopamine, Reserpine, and Alpha-methyl-p-tyrosine on Ethanol-Induced Sleep Times in LS and SS Mice

	Sleep Time (Minutes)	
	LS	SS
Control	154 ± 10	11 ± 3
6-Hydroxydopamine[a]	131 ± 7	32 ± 5[d]
Control	134 ± 8	18 ± 2
Alpha-methyl-p-tyrosine[b]	119 ± 5	28 ± 4[d]
Control	100 ± 4	22 ± 5
Reserpine[c]	55 ± 4[d]	49 ± 7[d]

[a]LS and SS mice were administered 6-hydroxydopamine, 100 mg/kg, s.c., at 5, 7, and 9 days of age and ethanol-induced (4 mg/g, i.p.) loss of righting reflex (sleep time) was measured at 60 days.

[b]LS and SS mice were administered alpha-methyl-p-tyrosine, methyl ester, 60 mg/kg, i.p., at 0, 4, and 8 hr. At 12 hr ethanol (4 mg/g, i.p.) was administered and loss of righting reflex (sleep time) was measured.

[c]Reserpine (5 mg/kg, i.p.) was administered 10 days prior to administration of ethanol and measurement of sleep time. The animals were maintained in a 28° environment for the 4 days following reserpine administration in order to minimize the toxic effects of reserpine.

[d]$p < 0.05$ vs. corresponding untreated group.

enhanced, administration of ethanol results in a reduced duration of the loss of the righting reflex in LS mice (TABLE 1). It is generally believed that newly synthesized neurotransmitter is the functionally important component of the norepinephrine pool in the central nervous system. The elevated tyrosine hydroxylase activity may indicate that catecholamine adaptive responses are enhanced ten days following reserpine.[22,24]

The significantly greater basal levels of catecholamines and tyrosine hydroxylase activity in the adrenal gland of LS mice, as compared with the SS line,[21] are reminiscent of the biochemical findings of Lau and Slotkin,[25] who examined the effects of the hypothyroid state on peripheral catecholamine systems. These workers observed that adult rats that had been rendered hypothyroid by administration of propylthiouracil perinatally, exhibited elevated levels of catecholamines and catecholamine biosynthetic enzymes in the adrenal gland. LS mice differ from SS mice in several other

TABLE 2. Postnatal Age and Sensitivity of LS and SS Mice to Ethanol[a]

Postnatal Age (Days)	Ethanol mg/g	Sleep Time (Minutes)	
		LS	SS
9	1.0	62 ± 4	52 ± 6
10	1.0	65 ± 7	25 ± 3[b,c]
11	1.0	60 ± 7	0[d]
12	1.0	42 ± 6	0[d]
14	4.1	260 ± 20	105 ± 5[b]
20	4.1	144 ± 5	20 ± 4[b]
65	4.3	168 ± 5	17 ± 3[b]

[a]Ethanol was administered i.p. to mice and duration of loss of righting reflex (sleep time) was determined.

[b]$p < 0.01$ vs. corresponding LS sleep time.

[c]$p < 0.05$ vs. day 9 or 10 LS mice or day 9 SS mice.

[d]Animals failed to lose righting reflex.

characteristics that suggest that the latter may enjoy a more abundant degree of thyroid function.[26] SS mice are leaner, more aggressive, and more active than the more docile LS mice. The cortical electroencephalogram of SS mice is more active than that of LS mice.[27] The density of beta adrenergic receptors in the cerebral cortex has been reported to be greater in SS than in LS mice,[28] and SS mice manifest a more rapid basal heart rate than that of LS mice.[27]

It is well known that the normal development of the brain is dependent upon the level of thyroid hormones. This may be particularly critical during the first ten days of postnatal life in rodents, a period that coincides with the important phase in the morphological and functional development of the central nervous system in these animals.[29,30] As discussed above, many of the differences between LS and SS mice suggest that these lines of mice may differ in thyroid status, at least during the developmental period, and this may affect some of the characteristics of the adult lines of mice, including ethanol sensitivity. In order to evaluate this, we have examined the thyroid status and ethanol sensitivity of LS and SS mice during development and at maturity.[26] Both LS and SS mice are more sensitive to the depressant effects of ethanol at 9 to 11 days of age than at later stages of development or at maturity (TABLES 2 and 3). Ethanol, 1 g/kg, i.p., is sufficient to induce a loss of righting reflex of greater than 50 min in both LS and SS mice at 9 days of age. Interestingly, there is no significant difference in the sensitivity of the LS and SS lines of mice to the depressant effects of ethanol at 9 days of age. Differential sensitivity to ethanol in the two lines of mice first becomes apparent at 10 days of age and increases progressively from 10 to 20 days of age, at which time the animals exhibit a degree of sensitivity to ethanol that is comparable to the respective mature animals. Serum levels of thyroxine in LS and SS mice are not significantly different at 6 and 8 days of age postnatally. However, by 10 days of age, the level of serum thyroxine is significantly higher in SS than in LS mice. This difference is accentuated at days 10 and 12 and persists until adulthood (TABLE 4).

It is of interest that the greatest difference in thyroxine levels in LS and SS mice occurs between days 11 and 14 postpartum, which includes the critical period during which thyroid hormones are presumed to affect brain development. The weights of the parathyroid-thyroid glands from adult SS mice are approximately 50% greater than those from LS mice. Histological appearance of the gland from SS mice also suggests

TABLE 3. Waking Serum Ethanol Levels in LS and SS Mice of Different Postnatal Ages[a]

Postnatal Age (Days)	Ethanol mg/g	Waking Serum Ethanol mg/100 ml	
		LS	SS
9	1.0	160 ± 8	165 ± 7
10	1.0	162 ± 8	186 ± 8[b,c]
11	1.0	166 ± 7	—[d]
12	1.0	178 ± 8	—[d]
14	4.1	214 ± 9	391 ± 14[b]
20	4.1	259 ± 7	474 ± 18[b]
65	4.3	270 ± 10	496 ± 14[b]

[a]For experimental details, see TABLE 2.
[b]$p < 0.01$ vs. corresponding LS mice.
[c]$p < 0.05$ vs. 9 day SS value.
[d]Animals failed to lose righting reflex (sleep time = 0).

TABLE 4. Serum Thyroxine Levels in LS and SS Mice of Different Postnatal Ages

Postnatal Age (Days)	Serum Thyroxine[a] μg/100 ml	
	LS	SS
6	3.9 ± 0.4	4.6 ± 0.3
8	6.2 ± 0.4	6.0 ± 0.2
10	7.6 ± 0.6	10.2 ± 0.8[b]
12	7.9 ± 0.7	11.7 ± 0.8[b]
14	8.1 ± 0.5	12.1 ± 0.7[b]
16	7.7 ± 0.7	10.2 ± 0.6[b]
20	6.3 ± 0.5	8.2 ± 0.3[b]
60	4.7 ± 0.3	6.8 ± 0.3[b]

[a]Serum thyroxine was measured by radioimmunoassay.
[b]$p < 0.01$ vs. corresponding LS value.

that the gland is more active in these animals as compared with the LS mice. These studies indicate that the LS mice may be in a hypothyroid state, relative to the thyroid status of SS mice, both during development and in adult life.[26] The differences in thyroid function during development and at maturity may contribute significantly to the differential sensitivity to the depressant effects of ethanol of the two lines of mice.

REFERENCES

1. HAGGENDAL, J. & M. LINDQVIST. 1961. Ineffectiveness of ethanol on noradrenaline, dopamine or 5-hydroxytryptamine levels in brain. Acta Pharmacol. Toxicol. **18:** 278–280.
2. CARLSSON, A., T. MAGNUSSON, T. H. SVENSSON & B. WALDECK. 1973. Effect of ethanol on the metabolism of brain catecholamines. Psychopharmacology **30:** 27–36.
3. POHORECKY, L. A. 1974. The effects of ethanol on central and peripheral noradrenergic neurons. J. Pharmacol. Exp. Ther. **189:** 380–391.
4. HUNT, W. A. & E. MAJCHROWICZ. 1974. Alterations in the turnover of brain norepinephrine and dopamine in alcohol-dependent rats. J. Neurochem. **23:** 549–552.
5. CORRODI, H., K. FUXE & T. HOKFELT. 1966. The effect of ethanol on the activity of central catecholamine neurons in rat brain. J. Pharm. Pharmacol. **18:** 821–823.
6. CARLSSON, A. & M. LINDQVIST. 1973. Effect of ethanol on the hydroxylation of tyrosine and tryptophan in rat brain *in vivo.* J. Pharm. Pharmacol. **25:** 437–440.
7. AHTEE, L. & M. SVARSTROM-FRASER. 1975. Effects of ethanol dependence and withdrawal on the catecholamines in rat brain and heart. Acta Pharmacol. Toxicol. **36:** 289–298.
8. POHORECKY, L. A., L. S. JAFFEE & H. A. BERKELEY. 1974. Ethanol withdrawal in the rat. Involvement of noradrenergic neurons. Life Sci. **15:** 427–437.
9. POHORECKY, L. A. & L. S. JAFFEE. 1975. Noradrenergic involvement in the acute effects of ethanol. Res. Com. Chem. Path. Pharmacol. **12:** 433–447.
10. LILJEQUIST, S. & A. CARLSSON. 1978. Alteration of central catecholamine metabolism following acute administration of ethanol. J. Pharm. Pharmacol. **30:** 728–730.
11. BACOPOULOS, N. G., R. K. BHATNAGAR & L. S. VAN ORDEN, III. 1978. The effects of subhypnotic doses of ethanol on regional catecholamine turnover. J. Pharmacol. Exp. Ther. **204:** 1–10.
12. BUSTOS, G. & R. H. ROTH. 1976. Effect of acute ethanol treatment on transmitter synthesis and metabolism in central dopaminergic neurons. J. Pharm. Pharmacol. **28:** 580–581.
13. FADDA, F., A. ARGIOLAS, M. R. MELIS, G. SERRA & G. L. GESSA. 1980. Differential effect of acute and chronic ethanol on dopamine metabolism in frontal cortex, caudate nucleus and substantia nigra. Life Sci. **27:** 979–986.

14. ERIKSSON, T. & A. CARLSSON. 1980. Ethanol-induced increase in brain concentrations of administered neutral amino acids. Naunyn-Schmeideberg's Arch. Pharmacol. **314:** 47–50.
15. POHORECKY, L. A., B. NEWMAN, J. SUN & W. H. BAILEY. 1978. Acute and chronic ethanol ingestion and serotonin metabolism in rat brain. J. Pharmacol. Exp. Ther. **204:** 424–432.
16. BRICK, J. & L. A. POHORECKY. 1983. The neuroendocrine response to stress and the effect of ethanol. *In* Stress and Alcohol Use. J. Brick & L. A. Pohorecky, Eds. 389–400. Elsevier Science Publ. Co. New York, NY.
17. MCCLEARN, G. E. & R. KAKIHANA. 1981. Selective breeding for ethanol sensitivity: short-sleep and long-sleep mice. *In* Development of Animal Models as Pharmacogenetic Tools. G. E. McClearn, R. A. Deitrich & V. G. Erwin, Eds. 147–159. U.S. Government Printing Office. Washington, DC.
18. HESTON, W. D. W., V. G. ERWIN, S. M. ANDERSON & H. ROBBINS. 1974. A comparison of the effects of alcohol on mice selectively bred for differences in ethanol sleep time. Life Sci. **14:** 365–370.
19. FRENCH, T. A. & N. WEINER. 1984. Effect of ethanol on tyrosine hydroxylase in brain regions of long and short sleep mice. Alcohol **1:** 247–252.
20. SORENSEN, S., M. PALMER, T. DUNWIDDIE & B. HOFFER. 1980. Electrophysiological correlates of ethanol-induced sedation in differentially sensitive lines of mice. Science **210:** 1143–1145.
21. FRENCH, T. A., J. M. MASSERANO & N. WEINER. 1985. Ethanol-induced changes in tyrosine hydroxylase activity in adrenal glands of mice selectively bred for differences in sensitivity to ethanol. J. Pharmacol. Exp. Ther. **232:** 315–321.
22. MASSERANO, J. M., & N. WEINER. 1982. Investigations into the neurochemical mechanisms mediating differences in ethanol sensitivity in two lines of mice. J. Pharmacol. Exp. Ther. **221:** 404–409.
23. FRENCH, T. A., K. L. CLAY, R. C. MURPHY & N. WEINER. 1985. Alpha-methyl-paratyrosine effects in mice selectively bred for differences in sensitivity to ethanol. Biochem. Pharmacol. **34:** 3811–3821.
24. FRENCH, T. A., J. M. MASSERANO & N. WEINER. 1986. Further studies on the neurochemical mechanisms mediating differences in ethanol sensitivity in LS and SS mice. Alcoholism: Clin. Exp. Res. In Press.
25. LAU, C. & T. A. SLOTKIN. 1982. Maturation of sympathetic neurotransmission in rat heart. VIII. Slowed development of noradrenergic synapses resulting from hypothyroidism. J. Pharmacol. Exp. Ther. **220:** 629–636.
26. DISBROW, J. K., J. M. MASSERANO & N. WEINER. 1986. The thyroid status during postnatal maturation of the brain in mice genetically bred for differences in ethanol sensitivity. J. Pharmacol. Exp. Ther. In Press.
27. RYAN, L. J., J. E. BAN, B. SANDERS & S. K. SHARPLESS. 1979. Electrophysiological responses to ethanol, pentobarbital and nicotine in mice genetically selected for differential sensitivity to ethanol. J. Comp. Physiol. Psychol. **93:** 1035–1052.
28. DIBNER, M. D., N. R. ZAHNISER, B. B. WOLFE, R. A. RABIN & P. B. MOLINOFF. 1980. Brain neurotransmitter receptor systems in mice genetically selected for differences in sensitivity to ethanol. Pharmacol. Biochem. Behav. **12:** 509–513.
29. FORD, D. H. 1968. Central nervous system-thyroid interrelationship. Brain Res. **7:** 329–349.
30. DEMEMES, D., C. DECHESNE, C. LEGRAND & A. SANS. 1986. Effects of hypothyroidism on postnatal development in the peripheral vestibular system. Dev. Brain Res. **25:** 147–152.

DISCUSSION OF THE PAPER

D. H. VAN THIEL (*University of Pittsburgh School of Medicine, Pittsburgh, PA*): What are the TSH levels of short-sleep versus long-sleep animals?

WEINER: Using human TSH radio-immuno assay, not the rat assay, we found in respect to TSH, the radio iodine uptake was greater in the SS than in the LS animals.

VAN THIEL: Without knowing the thyroid binding globulin activity, the fact that they have more uptake just suggests that you may have a defect in organification.

WEINER: We haven't looked at that, but we have looked at the serum binding to thyroid binding protein. There is somewhat less in the SS animals, so the percent of free thyroxin is higher in the SS. The free thyroxin index, which presumably is a measure of peripheral thyroid target hormone availability, was even higher in the SS than in the LS, than the serum thyroxin levels would indicate.

E. RUBIN (*Thomas Jefferson University School of Medicine, Philadelphia, PA*): Have you done any experiments on adrenalectomized animals?

WEINER: We haven't, primarily because that's very stressful, particularly in the developing animals. We would also get rid of our adrenal medulla as well.

E. NOBLE (*University of California, Los Angeles, CA*): With regard to Dr. Rubin's question, Galgiano when she was in my laboratory did some adrenelectomy studies on long-sleep versus short-sleep and it appears to be that one of the main things is the difference in the brain sensitivity.

The second comment I have is again related to the thyroid. It's known that thyroidtropin releasing hormone has some amathistic properties. That raises the immediate question, do the short sleepers put out more TRH, which makes them sober quicker than the other strain?

WEINER: We have administered TRH during that developmental period without any marked differential effect on the two strains. TRH administered to the adult animal will shorten the sleep time of both LS and SS to a comparable degree.

B. TABAKOFF (*National Institute on Alcohol Abuse and Alcoholism, Bethesda, MD*): The cerebellar development in these animals takes place over the same time course as the differential sensitivity development. That correlation could explain the developmental differences. Is there a good correlation between cerebellar development and the changes that one sees in sensitivity?

WEINER: The studies that have been done in terms of the electrophysiological sensitivity suggest a similar developmental pattern.

False Neurotransmitters and the Effects of Ethanol on the Brain

BRIAN R. SMITH AND ZALMAN AMIT

Centre for Studies in Behavioral Neurobiology
Department of Psychology
Concordia University
1455 de Maisonneuve Boulevard
Montreal, Quebec H3G 1M8, Canada

Alcohol abuse has been a severe problem for society for many years, and the search for effective prevention and treatment programs has led to an extensive examination of this problem. Despite these enormous research efforts, there is, as yet, no concensus regarding the nature of the precise biological and pharmacological mechanisms that subserve alcohol dependence. Over the years, several hypotheses were offered in an attempt to account for this phenomenon. One recent hypothesis that attracted a great deal of attention in both the scientific and medical community is what has been referred to as the "Multi-Metabolite Theory."[1,2] This notion suggests that the primary pharmacological mechanism that underlies alcohol dependence involves amine-aldehyde condensation products that act as false neurotransmitters in the brain and thereby mediate voluntary alcohol consumption.

The purpose of the present paper is to critically examine this notion within its theoretical context and in light of experimental evidence. More importantly, we shall attempt to determine its viability in light of existing data on alcohol dependence and, more specifically, alcohol self-administration.

Amine-aldehyde condensation products can occur spontaneously via a chemical reaction known as the Pictet-Spengler reaction. Holmstedt[3] described the historical background of the research on these compounds and indicated that these isoqinoline derivatives were in fact synthesized as early as 1909. However, it was not until 1970 that it was first reported, on the basis of *in-vitro* experiments, that catecholamines and the alcohol metabolite, acetaldehyde, could condense to form tetrahydroisoquinoline (TIQ) alkaloids and that this reaction could, in principle, occur in biological tissues.[4,5,6] Cohen and Collins[4] demonstrated that TIQ alkaloids could be detected in cow adrenal glands following their perfusion with acetaldehyde. Davis and Walsh[5,6] observed similar alkaloid formation in perfused rat brain homogenates.

Cohen and his colleagues then reported the results of a series of studies examining the neurochemical properties of these TIQ compounds. They observed that these substances were endowed with all the necessary properties enabling them to act as false neurotransmitters (for review see REFS. 7, 8). It was observed that they were taken up and stored by catecholamine neurons[9,10] and as a result, the reuptake of naturally occurring catecholamines was inhibited.[11,12] Furthermore, TIQ alkaloids have been shown, through electron microscopy studies, to be stored in catecholamine synaptic vesicles[13] as well as released into the synapse by electrical or chemical stimulation.[14,15] They were also shown to activate catecholamine receptors.[16] In addition to these transmitterlike properties, TIQ alkaloids were shown to bind competitively to enzyme systems that synthesize or limit the actions of catecholamines.[17–19] Given these neurochemical properties, it was suggested that TIQ alkaloids are in fact false neurotransmitters and their formation may in some fashion contribute to the regulation of the neurophysiological and neuropharmacological effects of alcohol consumption.[7,8]

As early as 1972, the notion that amine-aldehyde condensation products may play a role in alcohol dependence was beginning to receive increased attention. Amit[20] proposed a model suggesting a mediational role for TIQ alkaloids in the regulation of ethanol intake. It was suggested that these condensation products may promote alcohol intake, possibly through some acquired homeostatic mechanism that would have the ability to regulate alcohol consumption. Following the later findings of Cohen,[7,8] others also proposed a role for TIQ alkaloids in alcohol addictions.[1,8,21] These notions which still constitute the basis of the current formulation of this concept suggest that TIQ alkaloids, through their "false neurotransmitter" properties within the catecholamine system, produce some alteration in the functioning of the adrenergic system that subsequently contributes to the development of "physical dependence" upon alcohol (for a detailed description see REF. 2). There is no question that this notion was, and in fact remains, an intriguing one. Also, it seems clear that this interest is in some part due to the fact that tetrahydropapaveroline (THP), one of the TIQ alkaloids most commonly studied, is a biological precursor to an opiate alkaloid found in the opium poppy.[22] As well, the observation concerning the possible presence of endogenous levels of salsolinol (dopamine-acetaldehyde condensation product) in both nonalcoholic and alcoholic subjects contributed to the strong attraction that this theory held for researchers.[23,24]

The first step in determining the viability of this notion rested with the attempts to measure the presence of meaningful *in-vivo* quantities of TIQ alkaloids in subjects exposed to alcohol. Early attempts were successful, however, only when subjects received additional manipulations that retarded alcohol and catecholamine metabolism or increased catecholamine levels.[25-27] More recently, however, several studies demonstrated the presence of TIQ alkaloids following alcohol exposure in the absence of pharmacological manipulations.[23,28-32] However, it has been demonstrated that levels of TIQ excretion may be modulated by variables unrelated to alcohol consumption, one of which may be diet. This in turn suggested that interpretations with regard to TIQ formed following alcohol ingestion should be made cautiously.[33] It should also be mentioned that much of this work focused on the identification of salsolinol or a metabolite of this substance. To date, there have been no reports of the synthesis and presence of THP, *in vivo,* under normal drinking conditions. The absence of these data is particularly unfortunate, since it is this particular compound that has been favored for further investigation.[5] In addition, it was this compound upon which much of the later psychopharmacological research concerning TIQ alkaloids and alcohol intake was based.[2]

The above data would seem to support the notion that TIQ alkaloids do exist *in vivo,* although their precise relationship to alcohol ingestion remains unclear. However, it is the reports of their possible psychopharmacological effects that have perhaps supplied the main impetus for the "Multi-Metabolite Theory" of alcohol dependence. In a series of studies, Myers and co-workers reported that the intracerebroventricular infusion of minute quantities of TIQ alkaloids would result in a dramatic and long-lasting increase in alcohol consumption (for review see REF. 34). It was on the basis of these results that the notion of a TIQ-induced alteration of central nervous system activity leading to dependence was founded.

A great deal of controversy enveloped the attempts to evaluate the significance of these findings. The controversy centered particularly on variations in experimental methodology used by different investigators.[35,36] Several reports have been published that have addressed this question directly,[35,36,38] though a detailed re-examination of this issue is beyond the scope of this paper. However, it is clear that the effects of TIQ alkaloids on alcohol consumption is a phenomenon that is extremely difficult to reproduce.[35–42] Duncan and Deitrich,[36] while reporting some modest effect of TIQ infusions on ethanol drinking, were also unable to replicate completely the original findings. In fact they found that under a certain dose range used in the original investigation, they obtained results that contradicted the original report. In a study conducted in the original author's laboratory, Sinclair failed to observe any change in ethanol consumption following intraventricular infusions of TIQ alkaloid.[39] It has been suggested that the difficulty in reliably reproducing this phenomenon lies in the precise experimental procedures used and the necessity of obtaining the exact dose of TIQ for each individual animal.[2] It would appear, therefore, that the observation of the effect is based more on the nature of the experimental paradigm per se than on the manipulation itself. This would strongly suggest that even if the phenomenon is real, it is nevertheless a fickle and nonrobust one, particularly, when a simple change in screening procedures for ethanol intake can interfere with the observation of the effect. This suggestion is in strong contrast to the consistent findings that alcohol consumption is a very robust, if heterogeneous phenomenon. It was reliably shown in numerous studies that 30–60% of a randomly selected group of rats will readily consume relatively large amounts of ethanol.[43–45] It has also been shown that naive animals will learn, without any form of prior training, a complex task in order to receive ethanol.[46–48] Alterations in response contingencies for the self-administration of ethanol can also be made, and animals will continue to self-administer.[49] Given this information, it would appear unlikely that a behavior as strong and as resistant to disruption as voluntary consumption of alcohol would be mediated and regulated by a mechanism that itself is so easily disrupted that minute alterations in experimental methodology will render it so difficult to reproduce.

While the preceding discussion evaluted the experimental findings, there remain serious difficulties with the proposed theory of action of amine-aldehyde condensation products. Much of the supporting data has concentrated on the effects of dopamine-based TIQ alkaloids, and in fact it is the proposed formation of THP that is most often suggested as regulating the possible alterations reported in ethanol intake.[2] However, much of the existing research examining the involvement of the catecholamines has implicated norepinephrine (NE) and not dopamine (DA) in the regulation of ethanol self-administration. Central depletions of NE by the dopamine-beta-hyroxylase (DBH) inhibitors FLA-63 and FLA-57 were observed to attenuate voluntary ethanol intake.[50–51] In addition, Davis and co-workers[52,53] demonstrated that the administration of FLA-57 and the DBH inhibitor U14,624 blocked intragastric ethanol self-administration, while the DA receptor blocker, haloperidol, was shown to be ineffective.[52] Despite the fact that DA neurons were not manipulated, presumably leaving the formation of THP and/or salsolinol unaffected, animals were observed to reduce their ethanol intake. Central DA neurons have also been chemically lesioned following the administration of the neurotoxin 6-hydroxydopamine, yet only when NE neurons were lesioned did ethanol intake change.[54–58] Perhaps the most significant experiment in this series involved the neurochemical lesioning of DA pathways.[55] In this study the authors reported that this manipulation failed to alter ethanol drinking behavior. It is presumed that formation of these DA-based TIQ alkaloids should be disrupted by these manipulations, yet the drinking behavior persisted.

It has been suggested that the primary action of TIQ alkaloids is to produce an alteration in the functioning of the adrenergic system in such a way as to create an imbalance in this system.[2] This hypothetical imbalance is then thought to create a "craving" or an "uncontrollable urge to ingest alcohol," in fact, a state of physical dependence on alcohol.[2] It is the onset of this "need" state that presumably leads to the dependence on alcohol. However, it has been argued that the phenomenon of physical dependence on the one hand and voluntary consumption of alcohol on the other may be unrelated.[49,59,60] In fact, it has been frequently demonstrated that animals made dependent do not show a preference for ethanol in a subsequent free-choice test despite the presence of withdrawal symptoms (*e.g.* REFS. 61, 62). There has been growing acceptance of the notion that alcohol self-administration in humans is a distinct set of responses governed primarily by its reinforcing properties.[63–68] In similar fashion, it has been demonstrated that ethanol drinking in animals is also an operant response that can be modified by the same manipulations as any other operant response.[46,47,69–72] It would appear from these findings that the self-administration of alcohol in both animals and humans is controlled by its reinforcing properties and that the production of dependence may not play a significant role in further consumption. The notion of dependence leading to the consumption of alcohol as proposed by the TIQ-alkaloid hypothesis would seem to be in direct contrast to much of the data in this area.

Since the first observations of the formation of amine-aldehyde condensation products were made, a great deal of research has been directed towards an attempt to determine what if any role these compounds play in the actions of alcohol. Early notions focussed on their possible involvement in alcohol intake and the subsequent development of dependence. The idea seems to possess much face validity, and it is probably due to this that it remains a popular notion today. Unfortunately, there is little in the way of solid, consistent data to support this theory. While the answers to alcohol dependence are still elusive, there remains an extensive literature that any attempt at theorizing must account for. The discussion above details some of the major difficulties with the "Multi-Metabolite Theory," and it would appear that it is not viable in light of current information. This is not meant to suggest that these condensation compounds are not involved in some fashion in some alcohol effects. The consequences of alcohol abuse are extensive and while related to intake, all of them are not necessarily related to the regulation of consumption. It is possible that some action of the amine-aldehyde condensation products may yet be determined that would account for some of these actions of alcohol. One intriguing possibility suggested by Collins[73] is that the oxidative metabolites of TIQ alkaloids may be toxic agents responsible in part for the neuropathology of chronic alcoholism. As in any scientific endeavour, theories in psychopharmacology are put forward to describe phenomena, and these provide the impetus and direction for further investigation. Yet, as with all theories, there comes a time when existing data goes beyond the initial formulation, and they must be discarded and new ones formed. The arguments presented here suggest that the amine-aldehyde condensation theory does not appear to be a viable explanation of the mechanisms regulating alcohol consumption and that further research in this area may be more fruitfully directed to other avenues of investigation.

REFERENCES

1. MYERS, R. D. 1980. Pharmacological effects of amine-aldehyde condensation products. *In* Alcohol Tolerance and Dependence. H. Rigter & J. C. Crabbe, Eds. 337–370. Elsevier Biomedical Press. Amsterdam.
2. MYERS, R. D. 1985. Multiple metabolite theory, alcohol drinking and the alcogene. *In* Aldehyde Adducts in Alcoholism. M. A. Collins, Ed. 201–220. Alan R. Liss. New York, NY.
3. HOLMSTEDT, B. 1982. Betacarbolines and tetrahydroisoquinolines: historical and ethnopharmacological background. *In* Beta-carbolines and Tetrahydroisoquinolines. F. Bloom, J. Barchas, M. Sandler & E. Usdin, Eds. 3–13. Alan R. Liss. New York, NY.
4. COHEN, G. & M. A. COLLINS. 1970. Science **167:** 1749–1751.
5. DAVIS, V. E. & M. J. WALSH. 1970. Science **167:** 1005–1007.

6. WALSH, M. J., V. E. DAVIS & Y. YAMANAKA. 1970. J. Pharmacol. Exp. Ther. **174:** 388–400.
7. COHEN, G. 1976. Biochem. Pharmacol. **25:** 1123–1128.
8. COHEN, G. 1979. Interaction of catecholamines with acetaldehyde to form tetrahydroisoquinoline neurotransmitters. *In* Membrane Mechanisms of Drugs of Abuse. C. W. Sharp & L. Abood, Eds. 73–90. Alan R. Liss. New York, NY.
9. COHEN, G., C. MYTILINEOU & R. E. BARRETT. 1972. Science **175:** 1269–1272.
10. LOCKE, S., G. COHEN & D. DEMBIEC. 1973. J. Pharmacol. Exp. Ther. **187:** 56–67.
11. HEIKKILA, R., G. COHEN & D. DEMBIEC. 1971. J. Pharmacol. Exp. Ther. **179:** 250–258.
12. TUOMISTO, L. & J. TUOMISTO. 1973. Arch. Pharmacol. Exp. Pathol. **279:** 371–380.
13. TENNYSON, R. M., G. COHEN, C. MYTILINEOU & R. HEIKKILA. 1973. Brain Res. **51:** 161–169.
14. GREENBERG, R. S. & G. COHEN. 1973. J. Pharmacol. Exp. Ther. **184:** 119–128.
15. RAHWAN, R. G., P. J. O'NEILL & D. D. MILLER. 1974. Life Sci. **14:** 1927–1938.
16. MYTILINEOU, C., G. COHEN & R. BARRETT. 1974. Eur. J. Pharmacol. **25:** 390–401.
17. COHEN, G. & S. KATZ. 1975. J. Neurochem. **25:** 719–722.
18. COLLINS, M. A., J. L. CASHAW & V. E. DAVIS. 1973. Biochem. Pharmacol. **22:** 2337–2348.
19. GIOVINE, A., M. RENIS & A. BERTOLINO. 1976. Pharmacology **14:** 86–94.
20. AMIT, Z. & M. H. STERN. 1972. Electrochemical interactions in the medial forebrain bundle and alcohol dependence in the laboratory rat. *In* Biological Aspects of Alcohol Consumption. O. Forsander & K. Eriksson, Eds. 225–231. Finnish Foundation for Alcohol Studies. Helsinki.
21. MYERS, R. D. & C. L. MELCHIOR. 1977. Science **196:** 554–556.
22. KIRBY, G. W. 1967. Science **155:** 170–173.
23. SJOQUIST, B., S. BORG & H. KVANDE. 1980. Drug Alcohol Depend. **6:** 73–74.
24. SJOQUIST, B., S. BORG & H. KVANDE. Subst. Alcohol Actions Misuse **2:** 73–77.
25. COLLINS, M. A. & M. G. BIGDELI. 1975. Life Sci. **16:** 585–602.
26. SANDLER, M., S. G. CARTER, R. HUNTER & G. M. STERN. 1973. Nature **241:** 439–443.
27. TURNER, A. J., K. M. BAKER, S. ALGERI, A. FRIGENIO & S. GARATTINI. 1974. Life Sci. **14:** 2247–2257.
28. HAMILTON, M. G., M. HIRST & K. BLUM. 1979. Life Sci. **25:** 2205–2210.
29. COLLINS, M. A., W. P. NIJM, G. F. BORGE, G. TEAS & C. GOLDFARB. 1979. Science **206:** 1184–1186.
30. COLLINS, M. A. 1980. Neuroamine condensations in human subjects. *In* Biological Effects of Alcohol. H. Begleiter, Ed. 87–102. Plenum Press. New York, NY.
31. SJOQUIST, B., S. BORGE & H. KVANDE. 1981. Subst. Alcohol Actions Misuse **2:** 63–72.
32. SJOQUIST, B., S. LILJEQUIST & J. ENGEL. 1982. J. Neurochem. **39:** 259–262.
33. HIRST, M., D. R. EVANS, C. W. GOWDEY & M. A. ADAMS. 1985. Pharmacol. Biochem. Behav. **22:** 993–1000.
34. MYERS, R. D. 1978. Alcoholism: Clin. Exp. Res. **2:** 145–154.
35. AMIT, Z., B. R. SMITH, Z. W. BROWN & R. L. WILLIAMS. 1982. An examination of the role of TIQ alkaloids in alcohol intake: reinforcers, satiety agents or artifacts. *In* Betacarbolines and Tetrahydroisoquinolines. F. Bloom, J. Barchas, M. Sandler & E. Usdin, Eds. 345–364. Alan R. Liss. New York, NY.
36. DUNCAN, C. & R. A. DEITRICH. 1980. Pharmacol. Biochem. Behav. **13:** 265–281.
37. SMITH, B. R., Z. W. BROWN & Z. AMIT. 1980. Subst. Alcohol Actions Misuse **1:** 209–221.
38. BROWN, Z. W., Z. AMIT & B. SMITH. 1980. Examination of the role of tetrahydroisoquinoline alkaloids in the mediation of ethanol consumption in rats. *In* Biological Effects of Alcohol. H. Begleiter, Ed. 103–120. Plenum Press. New York, NY.
39. SINCLAIR, J. D. & R. D. MYERS. 1982. Subst. Alcohol Actions Misuse **3:** 5–24.
40. DEUTCH, A. Unpublished observations.
41. BOLAND, F. Unpublished observations.
42. KOOB, G. Unpublished observations.
43. KAHN, M. & G. STELLAR. 1960. J. Comp. Physiol. Psychol. **53:** 571–575.
44. RICHTER, C. P. & K. H. CAMPBELL. 1940. Science **91:** 507–508.
45. WILSON, C. W. M. 1972. The limiting factors in alcohol consumption. *In* Biological Aspects

of Alcoholism. O. Forsander & K. Eriksson, Eds. 207–215. Finnish Foundation for Alcohol Studies. Helsinki.
46. AMIT, Z. & M. H. STERN. 1969. Psychonom. Sci. **15:** 162–163.
47. SINCLAIR, J. D. 1974. Nature **244:** 590–592.
48. WERNER, T. E., S. G. SMITH & W. M. DAVIS. 1977. Physiol. Psychol. **5:** 453–454.
49. MEISCH, R. A. 1980. Ethanol as a reinforcer for rats, monkeys and humans. *In* Animal Models in Alcohol Research. K. Eriksson, J. D. Sinclair & K. Kiianmaa, Eds. 153–158. Academic Press. London.
50. AMIT, Z., D. E. LEVITAN & K. O. LINDROS. 1976. Arch. Inter. Pharmacodyn. Ther. **223:** 114–119.
51. AMIT, Z., Z. W. BROWN, D. E. LEVITAN & S.-O. OGREN. 1977. Arch. Inter. Pharmacodyn. Ther. **230:** 65–75.
52. DAVIS, W. M., S. G. SMITH & T. E. WERNER. 1978. Pharmacol. Biochem. Behav. **9:** 369–374.
53. DAVIS, W. M., T. E. WERNER & S. G. SMITH. 1979. Pharmacol. Biochem. Behav. **11:** 545–548.
54. BROWN, Z. W. & Z. AMIT. 1977. Neurosci. Let. **5:** 333–336.
55. KIIANMAA, K., K. ANDERSSON & K. FUXE. 1979. Pharmacol. Biochem. Behav. **10:** 603–608.
56. KIIANMAA, K., K. FUXE, G. JONSSON & L. AHTEE. 1975. Neurosci. Lett. **1:** 41–45.
57. KIIANMAA, K. 1980. Eur. J. Pharmacol. **64:** 9–19.
58. MASON, S. T., M. E. CORCORAN & H. C. FIBIGER. 1979. Neurosci. Lett. **12:** 137–142.
59. AMIT, Z., E. A. SUTHERLAND & N. WHITE. 1976. Drug Alcohol Depend. **1:** 435–440.
60. SINCLAIR, J. D. 1980. Comparison of the factors which influence voluntary drinking in humans and animals. *In* Animal Models in Alcohol Research. K. Eriksson, J. D. Sinclair & K. Kiianmaa, Eds. 119–137. Academic Press. London.
61. FREUND, G. 1969. Arch. Neurol. **21:** 315–320.
62. HUNTER, B. E., D. W. WALKER & J. N. RILEY. 1974. Pharmacol. Biochem. Behav. **2:** 523–529.
63. AMIT, Z. & E. A. SUTHERLAND. 1976. Drug Alcohol Depend. **1:** 1–13.
64. BIGELOW, G. & I. LIEBSON. 1972. Psychol. Rec. **22:** 305–314.
65. FUNDERBUNK, F. R. & R. P. ALLEN. 1977. J. Stud. Alcohol **38:** 410–425.
66. LUDWIG, A. M., A. WIKLER & L. H. STARK. 1974. Arch. Gen. Psychiat. **30:** 539–547.
67. MELLO, N. K. & J. H. MENDELSON. 1965. Nature **206:** 43–46.
68. SANDERS, R. M., P. E. NATHAN & J. S. O'BRIEN. 1976. Brit. J. Addict. **71:** 307–319.
69. ANDERSON, W. W. & T. THOMPSON. 1974. Pharmacol. Biochem. Behav. **2:** 367–377.
70. MEISCH, R. A. 1977. Ethanol self-administration: infrahuman studies. *In* Advances in Behavioral Pharmacology. T. Thompson & P. B. Dews, Eds. Vol. 1:35–84. Academic Press. London.
71. MEISCH, R. A. & P. BEARDSLEY. 1975. Psychopharmacology **43:** 19–23.
72. PENN, D. E., W. S. MCBRIDE, L. LUMENG, T. M. GAFF & T. K. LI. 1978. Pharmacol. Biochem. Behav. **8:** 475–481.
73. COLLINS, M. A. 1982. Trends Pharmacol. Sci. **3:** 373–375.

Ethanol Differentially Up-Regulates Functional Receptors in Cultured Neural Cells[a]

MICHAEL E. CHARNESS AND LISA A. QUERIMIT

Department of Neurology
and
The Ernest Gallo Clinic and Research Center
University of California, San Francisco
San Francisco General Hospital
San Francisco, California 94110

Neural adaptation to ethanol may include cellular responses that limit ethanol's entry into the plasma membrane[1] or compensate for its effects within the membrane.[2] We have developed a cellular model to study this adaptive response using the neuroblastoma x glioma hybrid cell line, NG108-15.[3,4]

Ethanol inhibits opioid peptide binding to the delta-opioid receptor.[3,5] When NG108-15 hybrid cells are grown with 25–200 mM ethanol, opioid receptor density increases up to twofold without a change in receptor affinity.[3] Increased receptor expression does not appear to be a compensatory response to diminished binding of endogenous opioid peptides.[4] The opiate antagonist, naloxone, also increases opioid receptor number, but produces a smaller effect than ethanol with greater fractional inhibition of binding. Furthermore, ethanol does not inhibit receptor down-regulation by etorphine, a potent opiate agonist. Ethanol increases opioid receptor expression in NG108-15 cells treated with actinomycin D, but not cycloheximide; hence, normal protein synthesis, but not DNA transcription, may be required for this response. The opioid receptors induced in ethanol-treated cells are subject to normal up-regulation by naloxone, down-regulation by etorphine, and acute inhibition of agonist binding by Na_+. These receptors therefore appear to occupy a membrane domain wherein they are subject to normal regulatory mechanisms.

NG108-15 cells express alpha^2-adrenergic and muscarinic receptors, which like the delta-opioid receptor are coupled to adenylate cyclase by the inhibitory GTP binding protein, N^i.[6] Acute ethanol exposure inhibited tritiated-antagonist binding to these receptors in the order alpha^2-adrenergic > delta-opioid > muscarinic. Exposing neuroblastoma x glioma NG108-15 hybrid cells to 200 mM ethanol for 48 hr increased the expression of delta-opioid and alpha^2-adrenergic receptors nearly twofold and muscarinic cholinergic receptors less than 50%. Thus, although acute inhibition of receptor binding by ethanol is not solely responsible for subsequent receptor up-regulation in the continued presence of ethanol,[4] the sensitivity of each receptor to acute inhibition appears to predict the magnitude of the chronic response.

To determine the functional consequences of these receptor alterations, we exposed NG108-15 cells to 200mM ethanol for 48 hr and inhibited phenylisopropyladenosine-

[a]Supported in part by National Institute on Alcohol Abuse and Alcoholism Research Scientist Development Award AA00083 and Research Grant AA06662, a grant from the Alcoholic Beverage Medical Research Foundation, and a Basil O'Connor Starter Research Grant from the March of Dimes (all to M.E.C.).

stimulated cAMP synthesis with etorphine, oxymetazoline, and carbachol, agonists for the opioid, adrenergic, and muscarinic receptors, respectively. All cells were pretreated with Ro 20-1724 to prevent cAMP degradation by phosphodiesterase. Chronic ethanol treatment increased the potency of etorphine 3.5-fold (IC^{50} 0.77 vs 2.72 nM), oxymetazoline threefold (IC^{50} 1.78 vs 5.36 μM), and carbachol just 53% (IC^{50} 1.05 vs 1.61 μM). Thus, ethanol differentially increases the expression of functional neurotransmitter receptors in NG108-15 cells, causing a corresponding increase in cellular sensitivity to opiate, alpha^{2}-adrenergic, and muscarinic cholinergic agonists. Neural cells may adapt to ethanol, in part, by changing the expression of neurotransmitter receptors that regulate cellular levels of cAMP.

REFERENCES

1. Rottenberg, H., A. Waring & E. Rubin. 1981. Science **213:**583–585.
2. Hill, M. W. & A. D. Bangham. 1975. Adv. Exp. Med. Biol. **59:**1–9.
3. Charness, M. E., A. S. Gordon & I. Diamond. 1983. Science **222:**1246–1248.
4. Charness, M. E., L. A. Querimit & I. Diamond. 1986. J. Biol. Chem. **261:**3164–3169.
5. Hiller, J. M., L. M. Angel & E. J. Simon. 1981. Science **214:**468–469.
6. Kurose, H., T. Katada, T. Amano & M. Ui. 1983. J. Biol. Chem. **258:**4870–4875.

Effects of Alcohols on Voltage-Dependent Conductances in *Aplysia* Neurons

STEVEN N. TREISTMAN, PATRICIA CAMACHO-NASI, AND ANDREW WILSON

Worcester Foundation for Experimental Biology
222 Maple Avenue
Shrewsbury, Massachusetts 01545

INTRODUCTION

The question of how ethanol (EtOH) exerts its effects in the nervous system can be approached on many levels. In our studies, we use the simple nervous system from the marine mollusc, *Aplysia*, to examine the actions of alcohols on individual nerve cells. It is probable that some subset of neurons within complex nervous systems is most sensitive to EtOH and dictates the response of the entire system. We have approached this question in *Aplysia* and find that in one group of identified neurons, the early potassium current, I_A, shows a greatly increased inactivation time constant when exposed to EtOH, whereas a second group is insensitive at similar concentrations (Treistman and Wilson, in preparation). At another level, we can ask which sites in nerve cells are most sensitive to alcohols. Here, we describe the relative sensitivity of four voltage-dependent membrane currents. Finally, we can define the molecular target for anesthetic molecules. Either or both the channel proteins and the lipid matrix in which they are imbedded are possibilities. We describe experiments to examine the effects of channel activity on the subsequent response to alcohols in two voltage-dependent currents. A dependency on prior activity is suggestive of a direct interaction with channels, rather than a nonspecific lipid effect.

RESULTS

Our experiments used standard two-electrode voltage clamp techniques in axotomized cell bodies. Pharmacological blocking agents and appropriate ionic media were used to isolate individual currents for study.

The four currents examined were inward sodium (I_{Na}), delayed rectifier potassium (I_K), early transient potassium (I_A), and inward calcium (I_{Ca}). In all of the cells studied, I_{Ca} was the most sensitive of the four. Significant reductions of I_{Ca} amplitude occurred in 50 mM EtOH concentrations. Effects on the other currents were not apparent until concentrations 2–4 times higher were reached. The effects of EtOH were reversible. Since I_{Ca} is of critical importance to many neural processes including the activation of other membrane currents and the release of transmitters from terminals, the sensitivity of this current is likely to be significant. FIGURE 1 shows the sensitivity of the four currents.

The effects of channel history on sensitivity to EtOH were examined for I_A and I_{Ca} channel populations. FIGURE 2 shows the effect of prepulse protocol (channel history) on the ethanol sensitivity of I_A inactivation. After a period at -70 mV, during which I_A inactivation was removed, the cell was stepped to -40 mV for increasing periods of time. At −40 mV, I_A inactivated in a time-dependent manner, without any apparent opening of I_A channels. The amplitude of I_A was measured during steps to −20 mV, which immediately followed each −40 mV prepulse. An increase in inactivation is manifested as a greater reduction in I_A following a prepulse. The plot of the data shows that in the presence of ethanol, a shorter prepulse was necessary to produce a given amount of inactivation than in the control. This is in contrast to the prolongation of I_A

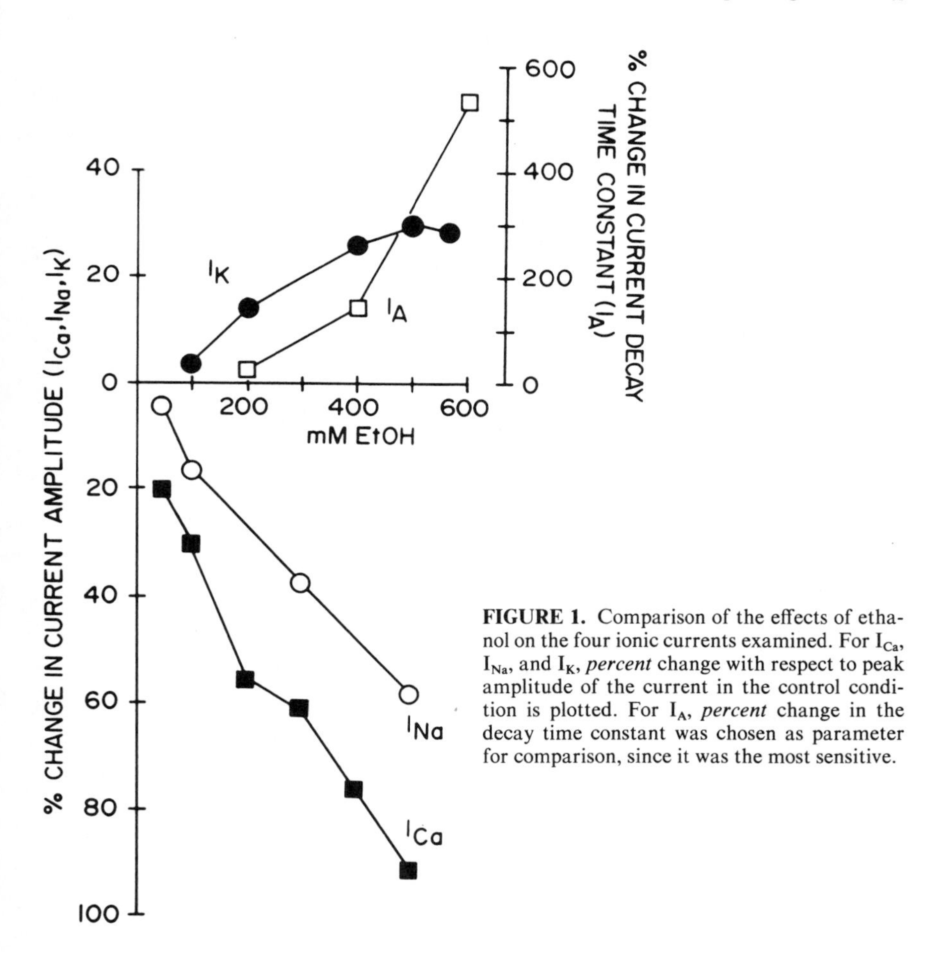

FIGURE 1. Comparison of the effects of ethanol on the four ionic currents examined. For I_{Ca}, I_{Na}, and I_K, *percent* change with respect to peak amplitude of the current in the control condition is plotted. For I_A, *percent* change in the decay time constant was chosen as parameter for comparison, since it was the most sensitive.

inactivation time constant observed when inactivation occurred in an open channel population at −20 mV.

To examine the influence of calcium channel history on EtOH's effects, we used both calcium and barium as charge carriers. Barium blocks inactivation of I_{Ca}, and comparison of the results obtained with the two carriers allowed dissociation of inactivation effects from other effects. A number of protocols were used to examine the influence of channel activity. When barium was charge carrier, the effects of EtOH were unaffected by any of the protocols. When calcium was carrier, the cumulative decrement of I_{Ca} attributed to inactivation processes during successive clamp steps was augmented by EtOH.

Fluorescence studies that determine the lateral mobility of lipid analogues give some indication of membrane "state." This knowledge can be useful in formulating hypotheses relating the lipid environment to changes in neuronal excitability. We have performed such studies with EtOH and butanol (BuOH), observing the diffusibility of two different fluorescent lipid analogues within the plasma membranes of intact *Aplysia* neurons. Neither EtOH nor BuOH affected the diffusibility of NBD-labeled phosphatidylcholine. Lipid analogues may selectively partition into specific domains within the plasma membrane and, therefore, report not on the bulk membrane fluidity

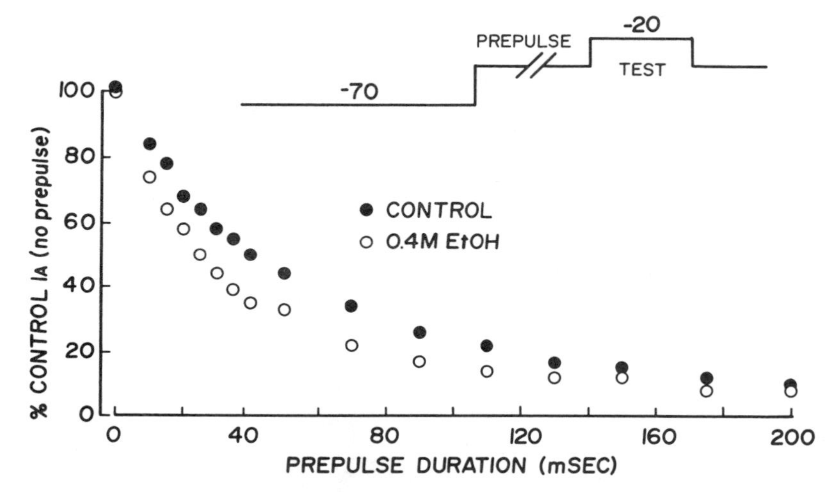

FIGURE 2. The effect of prepulse protocol (channel history) on the ethanol sensitivity of I_A inactivation. After a period at -70 mV, during which I_A inactivation was removed, the cell was stepped to -40 mV for increasing periods of time. At -40 mV, I_A inactivates in a time-dependent manner without any apparent opening of I_A channels. The amplitude of I_A was measured during step to -20 mV, which immediately followed each -40 mV prepulse. The plot of the data shows that in the presence of ethanol, a shorter prepulse was necessary to produce a given amount of inactivation than in the control. That is, inactivation developed faster in the presence of ethanol. This is in contrast to the slowing of I_A decay produced by ethanol when inactivation occurs without a prepulse, in which case some of the I_A channels are open prior to inactivation.

but on the fluidity of their local environment. To address this issue, we considered the effects of the same alcohols on a different lipid analogue, rhodamine-tagged PE. In contrast to their lack of effect on NBD-PC, both EtOH and BuOH increase the diffusion rate of Rh-PE. These results suggest a complexity in the lipid matrix that may be important in the actions of ethanol.

CONCLUSIONS

1) The four voltage-dependent currents showed variable sensitivities to ethanol. I_{Ca} is the most sensitive. 2) For I_A, inactivation is affected by ethanol very differently,

depending on whether the voltage clamp protocol measures decay of current after channels have opened or not. This suggests that a simple perturbation of a bulk lipid is not sufficient to explain ethanol's actions on this current. For barium current carried through Ca channels, channel history has little effect on the channel block produced by ethanol, consistent with a lipid site of action. However, when calcium is the charge carrier, the history of the channel has an influence, presumably because ethanol is affecting the inactivation of current. This may reflect a direct interaction with the channel. 3) Fluorescence photobleaching data indicate that the membrane lipid is not homogeneous, and this finding has relevance for the selective actions of ethanol on different channel populations. Such selectivity would be expected if different channels inhabited different domains, which were selectively perturbed by particular alcohols.

Selective Effects of Ethanol on Neurotransmitter Receptor-Effector Coupling Systems in Different Brain Areas[a]

PAULA L. HOFFMAN,[b] TOSHIKAZU SAITO,[c] AND BORIS TABAKOFF[b]

[b]*National Institute on Alcohol Abuse and Alcoholism*
Division of Intramural Clinical and Biological Research
Bethesda, Maryland 20892
and
[c]*Department of Psychiatry*
Sapporo Medical College
Sapporo 060, Japan

Ethanol *in vitro* enhances basal and neurotransmitter-stimulated adenylate cyclase (AC) activity in membranes from cerebral cortex and striatum of C57BL mice. In the presence of guanine nucleotides, low, physiologically attainable concentrations of ethanol (50–75 mM) significantly increase AC activity in both brain areas.[1,2] However, ethanol has specific and differential sites of action in each brain region. In cerebral cortical tissue, ethanol reduced the EC_{50} for guanine nucleotide (Gpp(NH)p) stimulation of AC activity, and also reduced the lag time for this activation (TABLE 1). The effect on lag time was similar to that produced by the beta-adrenergic agonist,

TABLE 1. Effect of Ethanol on Stimulation of Adenylate Cyclase Activity in Brain Membranes from C57BL Mice[a]

	Cerebral Cortex			Striatum		
	Basal	+ISO	+Ethanol	Basal	+DA	+Ethanol
Gpp (NH)p EC_{50} (μM)	1.9	n.d.[b]	1.0[c]	1.0[d]	n.d.	1.0[d]
Half-time for Gpp (NH)p activation (min)	6.4	4.6[c]	4.6[c]	4.8	2.5[c]	4.8
Mg^{++} EC_{50} (mM)	1.9	0.8[c]	1.4[c]	23[e]	14[e]	23[e]

[a]Adenylate cyclase activity was assayed in cortical or striatal membrane preparations as described.[1,2] The concentration of isoproterenol (ISO) was 1 μM, that of dopamine (DA) 10 μM, and that of ethanol 250 mM or 500 mM (for Gpp(NH)p EC_{50}). Half-time for activation by Gpp(NH)p was determined at 30°C. Values represent the means from at least three experiments.
[b]n.d. = not determined.
[c]$P < 0.05$ compared to basal (Student's t-test).
[d]Taken from REFERENCE 1.
[e]Estimated from REFERENCE 4.

[a]Supported in part by the Banbury Foundation.

TABLE 2. Effect of Ethanol on Inhibition of Adenylate Cyclase Activity in Brain Membranes from C57BL Mice[a]

Additions to Assay	Per Cent Inhibition by 300 μM Oxotremorine			
	Cerebral Cortex		Striatum	
	1.0 mM Mg^{++}	2.0 mM Mg^{++}	1.0 mM Mg^{++}	2.0 mM Mg^{++}
None	24.3 ± 5.5	21.0 ± 3.5	36.3 ± 4.0	26.3 ± 4.2
250 mM Ethanol	19.4 ± 1.3	15.3 ± 2.1	30.0 ± 7.0	29.6 ± 7.6

[a]Adenylate cyclase activity was assayed in cortical or striatal membranes as previously described[1,2] in the presence of 100 mM Nacl, 100 μM GTP, and 10 μM forskolin. The Mg^{++} concentrations indicated are *total* Mg^{++}. Values represent mean ± S.E.M. from 3–5 experiments.

isoproterenol (TABLE 1), suggesting that ethanol, like a neurotransmitter, promoted the activation of the guanine nucleotide-binding protein, N_s. This interpretation is supported by the finding that ethanol, like isoproterenol, also decreased the EC_{50} for Mg^{++} stimulation of cerebral cortical AC (TABLE 1). Mg^{++} plays a key role in N_s activation by stimulatory agonists,[3] and the proposed activation of N_s by ethanol may also be mediated by the change in affinity for Mg^{++}. In contrast, in striatal membranes, ethanol did not affect the EC_{50} for Gpp(NH)p or the lag time for activation of AC by Gpp(NH)p (TABLE 1). Ethanol also had no effect on the EC_{50} for Mg^{++} activation of striatal AC.[4] These results suggest that ethanol's effects differ with respect to activation of cortical N_s and striatal N_s. Another difference was that ethanol increased the activity of digitonin-solubilized AC from cerebral cortex,[1] but not from striatum. Thus, ethanol stimulated cerebral cortical AC activity at N_s and the catalytic unit of AC (as well as at the β-adrenergic receptor[1]), but did not affect striatal AC at these sites. In both brain areas, however, ethanol increased the activity of AC that had been preactivated by incubation with Gpp(NH)p, suggesting that ethanol enhanced the interaction of activated N_s with AC.[1,2] The differential effects of ethanol may result from differences in membrane lipid microenvironments of striatal and cortical AC systems, or from different characteristics of the lipid-protein interactions in the neuronal membranes derived from these two brain areas.

Ethanol *in vitro* did not significantly alter oxotremorine-induced inhibition of forskolin-stimulated AC activity in membranes from striatum or cortex at any Mg^{++} concentration tested (TABLE 2). Although Mg^{++} also plays a key role in inhibition of AC, the concentration of free Mg^{++} required for N_i activation is in the micromolar range.[5] Ethanol appears to alter the properties and function of N_s selectively at concentrations of Mg^{++} normally found intracellularly, and not to affect N_i. These selective effects of ethanol on regulation of AC activity could contribute to the specific effects on synaptic transmission produced by ethanol.

REFERENCES

1. SAITO, T., J. M. LEE & B. TABAKOFF. 1985. J. Neurochem. **44:** 1037–1044.
2. LUTHIN, G. R. & B. TABAKOFF. 1984. J. Pharmacol. Exp. Ther. **228:** 579–587.
3. IYENGAR, R. & L. BIRNBAUMER. 1981. J. Biol. Chem. **256:** 11036–11041.
4. RABIN, R. A. & P. B. MOLINOFF. 1983. J. Pharmacol. Exp. Ther. **227:** 551–556.
5. BOCKAERT, J., B. CANTAU & M. SEBBEN-PEREZ. 1984. Mol. Pharmacol. **26:** 180–186.

Genetic Variation in Ethanol Sensitivity in C57BL/6 and DBA/2 Mice: A Further Investigation of the Differences in Brain Catalase Activity

C. M. G. ARAGON AND Z. AMIT

Concordia University
Center for Studies in Behavioral Neurobiology
Department of Psychology
1455 de Maisonneuve Blvd. W.
Montreal, Quebec H3G 1M8, Canada

The large differences in voluntary ethanol consumption observed in different inbred strains of mice are well documented. Studies of this phenomenon have particularly focussed on two inbred strains, the C57BL/6 and DBA/2, the former a high-drinking strain and the latter an ethanol-avoiding strain. These two strains differ not only in their level of voluntary ethanol consumption but also in their sensitivity to the effects of ethanol seen in a wide variety of behavioral paradigm.[1] The mechanisms underlying these differences were variously attributed to differential neurosensitivity, differential acute and chronic tolerance, differences in the metabolism of ethanol and acetaldehyde in the liver, and differences in membrane fatty acid composition among others. However, despite the large body of accumulated data, there has been less than general concensus concerning the above observations of possible mechanisms. These disagreements as to the nature of the differences in ethanol sensitivity and their putative mechanism between the two strains tends to strengthen the notion that yet another mechanism may be mediating at least in part the differences between the two strains. During the past several years we presented data suggesting that brain more than liver acetaldehyde may be mediating many of the psychopharmacological effects of ethanol. We have provided data suggesting that acetaldehyde may be formed in brain via the peroxidatic activity of catalase and that this centrally formed acetaldehyde may be mediating many but not all the behavioral effects of ethanol.[2,3] We therefore considered the possibility that differential catalase activity in the brain may be one of the mechanisms underlying the differential sensitivity to ethanol. Male mice of the DBA/2J and C57BL/6J strains were sacrificed by decapitation, and whole brains were removed and assayed for catalase activity.[2,3] Mean catalase activity in the brains of DBA/2 mice was $8.18 \pm 0.4 \times 10^{-2}$ units/mg protein/min ($n = 8$). Mean catalase activity in brains of C57BL/6 was $5.46 \pm 0.4 \times 10^{-2}$ units/mg protein/min ($n = 8$). A two-tailed student t-test yielded significant results ($t = -4.73$, $p < 0.001$). Catalase activity in brains of C57BL/6 was found to be approximately 35% less than that found in brains of DBA/2 mice. The present study reveals a large and significant difference in brain catalase activity between the two mice strains. These results are in line with earlier studies reporting similar differences between these two strains in liver catalase activity.[4,5] The mouse strain C57BL/6 has a catalitically altered catalase as compared with DBA/2 in liver and brain tissues. An examination of the differential behavioral reaction to ethanol in these two strains reveals a remarkable similarity to the behavioral response of rats whose brain catalase was inhibited by 3-amino-1,2,4-

TABLE 1. Similarity Between Inbred Strain Differences in Neurosensitivity to Ethanol in Mice (C57BL/6, DBA/2) and Differences in Behavioral Responses to Ethanol in Rats Pretreated with Aminotriasol (AT) or Saline (C)

Measure of Neurosensitivity	Dose (g/kg)	Rank Order in Sensitivity	References
Conditioned aversion to saccharin induced by alcohol injections	1.0–4.0 1.2	DBA > C57 C > AT	Horowitz & Whitney, 1975 Aragon *et al.*, 1985
Changes in rectal temperature following i.p. alcohol dose	3.0 3.0	C57 > DBA AT > C	Moore & Kakihana, 1978 Aragon & Amit, 1985
Regaining of the righting reflex	3.5–5.0 3.0	DBA > C57 C > AT	Tabakoff & Kiianmaa, 1982 Kiianmaa *et al.*, 1983 P. La Droitte *et al.*, 1984 Aragon *et al.*, 1986
Ethanol induced locomotor activity changes	1.35 2.0	DBA > C57 C > AT	Tabakoff *et al.*, 1982 Kiianmaa *et al.*, 1983 Aragon *et al.*, 1985
Corticosterone response	0.8–1.6 1.0–4.0	DBA > C57 C > AT	Kakihana *et al.*, 1968 Aragon *et al.*, 1986

triazole as compared to controls (see TABLE 1). This comparison between the behavioral response to ethanol observed in C57 and DBA mice and its striking similarity to the observations in rats whose brain catalase was inhibited support the notion that the differences between the two strains may be mediated by different levels of brain catalase activity.

In a second study we examined the effect of pretreatment with a catalase inhibitor, 3-amino-1,2,4-triazole (AT) (1 g/kg, i.p.) for its capacity to modify the magnitude of the corticosterone response to an acute dose of ethanol (1.5 g/kg, i.p.) in these two inbred strains of mice. Mice were pretreated with AT or saline four hours before injection of ethanol or saline. One hour after ethanol administration mice were sacrificed and trunk blood was collected for plasma corticosterone and ethanol determination. Brains were removed for catalase activity assays. The plasma corticosterone levels are shown in TABLE 2. Ethanol produced a significant release of corticosterone in both strains independent of the pretreatment regimen. C57BL/6 mice pretreated with AT had a 23% decrease in the plasma levels of corticosterone as compared with saline pretreated animals; DBA/2 AT-pretreated mice had a decrease of 21% as compared with controls. No significant differences in mean blood ethanol were observed between ethanol-injected animals pretreated with AT or saline.

TABLE 2. Effect of Pretreatment with 3-Amino-1,2,4-Triasole in Plasma Levels of Corticosterone in C57BL/6 and DBA/2 in Response to an Acute Dose of Ethanol[a]

Mouse Strain	C57BL/6	DBA/2
Pretreatment—Treatment	Corticosterone (mg/ml)	
Saline—Saline	10.20 ± 1.60	9.50 ± 0.86
AT—Saline	12.20 ± 1.30	11.36 ± 1.04
Saline—Ethanol	21.04 ± 2.25	25.88 ± 1.9
AT—Ethanol	16.24 ± 1.2	20.34 ± 2.0

[a]Corticosterone was determined fluorometrically.

AT-pretreated animals were observed to have a 50% decrease in catalase activity in both strains as compared with controls (S-C57: 8.74 ± 0.5 μmol O_2/min/mg protein; AT-C57: 4.61 ± 0.3; S-DBA: 10.88 ± 0.6; AT-DBA: 5.61 ± 0.6). Inhibition of brain catalase by AT results in an attenuation of the corticosterone response to an acute dose of ethanol in both strains of mice. These results agree with previous studies in rats where inhibition of brain catalase by AT resulted in an attenuation of ethanol-induced behaviors such as conditioned taste aversion,[3] motor depression, duration of righting reflex, and corticosterone response. These results provide support for the notion of ethanol oxidation in mouse brain *in vivo* via catalase. They suggest that acetaldehyde is produced directly in the brain and may mediate at least in part some of the psychopharmacological properties of ethanol in rodents.

REFERENCES

1. BELKNAP, J. K. 1980. Genetic factors in the effects of ethanol neurosensitivity, functional tolerance and physical dependence. *In* Alcohol Tolerance and Dependence. H. Rigter & J. C. Crabbe, Jr., Eds. 157–180. Elsevier/North. Holland.
2. ARAGON, C. M. G., G. STERNKLAR & Z. AMIT. 1985. Alcohol **2:** 353–356.
3. ARAGON, C. M. G., K. SPIVAK & Z. AMIT. 1985. Life Sci. **37:** 2077–2084.
4. GANSCHOW, R. E. & R. T. SHIMKE. 1969. J. Biol. Chem. **244:** 4649–4652.
5. GANSCHOW, R. E. & R. T. SHIMKE. 1970. Genetics **4:** 157–167.

Alteration of Na,K-ATPase Regulation by Ethanol and Noradrenergic Manipulations *In Vivo*[a]

ALAN C. SWANN

Department of Psychiatry
University of Texas Medical School
and
Mental Sciences Institute
Houston, Texas 77025

Chronic treatment with ethanol leads to adaptations that reduce behavioral and biochemical sensitivity to ethanol. One apparent reflection of ethanol tolerance is reduced sensitivity of Na,K-ATPase to ethanol.[1,2]

Specific phases of the Na,K-ATPase reaction cycle are sensitive to the amount of order in the environment. The transition from E1, the Na^+-dependent protein kinase conformation, and E2, the K^+-dependent phosphatase conformation, has a large negative enthalpy and entropy, so the equilibrium between them, and the antagonism between ATP and K^+ binding, are strongly influenced by the amount of order in the membrane.[3,4] Ethanol, polyunsaturated fatty acids, or high temperature favor ATP binding and the El conformation.[5] The effect of ethanol is potentiated by norepinephrine, apparently through a similar effect on enzyme conformation.[6]

We examined the effects of noradrenergic stimulation (by repeated treatment with yohimbine) or depletion *in vivo,* combined with chronic ethanol treatment using a liquid diet, on phases of Na,K-ATPase regulation that are sensitive to membrane order, including 1) sensitivity to ethanol *in vitro,* 2) competition between ATP and K^+, and 3) thermodynamic constants for cation binding and for the E1-E2 transition.

Chronic ethanol reduced sensitivity to ethanol *in vitro,* while noradrenergic stimulation had the opposite effect. Noradrenergic stimulation reduced the effect of ethanol. Depletion of norepinephrine with the relatively selective toxin DSP4 reduced sensitivity to ethanol *in vitro;* this effect was additive to that of chronic ethanol treatment.

Chronic ethanol treatment reduced the apparent affinity of ATP as expressed by competition with allosteric K^+ binding. Noradrenergic stimulation increased the effectiveness of ATP.

In control rats, either ethanol or norepinephrine *in vitro* reduced apparent affinity for K^+. The combination was more effective than either agent alone. These effects were reduced by chronic ethanol. Noradrenergic stimulation reduced the effect of norepinephrine *in vitro* but increased the effect of ethanol.

Chronic ethanol increased, and noradrenergic stimulation reduced, the transition temperatures for Arrhenius plots of phosphatase activity. The negative enthalpy and entropy changes associated with K^+-activation of phosphatase activity or with the

[a]Supported by United States Public Health Service Grants AA05785, MH37141, and MH00415.

E1-E2 transition were increased by noradrenergic stimulation and decreased by chronic ethanol.

The effects of chronic ethanol were opposite to those of ethanol *in vitro* and were consistent with reduced membrane fluidity. Norepinephrine, on the other hand, had effects consistent with increased membrane fluidity *in vitro* and after chronic treatment. These data suggest that adaptation to membrane effects of ethanol occurs *in vivo* and is expressed in regulation of Na,K-ATPase activity. Noradrenergic stimulation has effects resembling those of increased membrane fluidity.

REFERENCES

1. LEVENTAL, M. & B. TABAKOFF, 1980. J. Pharmacol. Exp. Ther. **212:** 315–319.
2. BEAUGE', F., C. FLEURET-BALTER, J. NORDMANN & R. NORDMANN. 1984. Alcoholism: Clin. Exp. Res. **8:** 167–171.
3. SWANN, A. C. & R. W. ALBERS. 1979. J. Biol. Chem. **254:** 4540–4544.
4. SWANN, A. C. 1983. Arch. Biochem. Biophys. **221:** 148–157.
5. SWANN, A. C. 1983. J. Biol. Chem. **258:** 11780–11786.
6. KALANT, H. & N. RANGARAJ. 1981. Eur. J. Pharmacol. **70:** 157–166.

Ethanol Stimulates Protein Phosphorylation and Taurine Release from Astroglial Cells

W. SHAIN, V. MADELIAN, D.L. MARTIN, AND S. SILLIMAN

Wadsworth Center for Laboratories and Research
New York State Department of Health
Albany, New York 12201

Ethanol, like beta-adrenergic agonists,[1-3] can stimulate the release of taurine from astroglial cells. Taurine is an inhibitory amino acid causing increases in Cl^- permeability in neurons,[4] is an anticonvulsant,[5] and prolongs ethanol-induced sleep times.[6] Astroglia in cell culture concentrate taurine to intracellular concentrations $\geqslant 50$ mM[7] and spontaneously release taurine very slowly.[2] Stimulation of beta-adrenergic receptors results in cyclic AMP-dependent taurine release.[3,4] In this report we demonstrate that ethanol and neurotransmitters stimulate taurine release from astroglial cells by similar cellular processes.

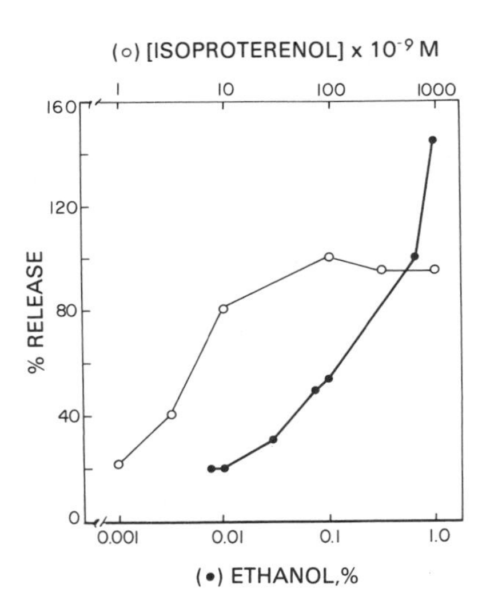

FIGURE 1. Ethanol-stimulated taurine release is dose dependent. Dose-dependent release of taurine release from LRM55 astroglial cells by isoproterenol (○—○) and ethanol (●—●). Release data were calculated by determining the total amount of taurine release following a 3-min application of agonist at each concentration.

LRM55 astroglial cells were used for measuring ^3H-taurine release using previously described procedures.[2]

Three principal findings were made by these studies. (A) Ethanol-stimulated taurine release was dose-dependent (FIGURE 1) but did not saturate at the highest concentrations tested (1.0%). Comparison of the dose-response curves for ethanol and the beta-adrenergic agonist isoproterenol (IPR) indicate that at concentrations associated with intoxication, *i.e.* 0.1%, ethanol-stimulated release was similar in magnitude to that observed with the EC_{50} concentration of IPR. (B) Ethanol-stimulated release is similar to neurotransmitter stimulated release as indicated by two observations. First, a 30-min application of ethanol, like IPR, resulted in three phases of taurine release—activation, inactivation, and an elevated steady state that was maintained as long as ethanol was present. Second, ethanol (0.1%) treatment resulted in the phosphorylation of three proteins that have been temporally associated with neurotransmitter-stimulated taurine release (data not shown). (C) Ethanol-stimulated taurine release was not mediated via cAMP, but rather may result from activation of another second messenger system. Three observations are consistent with this conclusion. First, ethanol did not stimulate cAMP formation in LRM55 astroglial cells. Second, simultaneous application of ethanol (0.33%) and 100 nM IPR, a concentration producing a maximal response, resulted in an additive response (5694 cpm observed vs 5211 cpm expected, FIG. 2). Third, pre-exposure of cells to either ethanol or IPR did not result in desensitization of a response to the other secretogogue.

These results indicate that ethanol-stimulated taurine release from astroglial cells is dose dependent, occurs at pharmacologically pertinent doses, and may be regulated by cellular mechanisms similar to those observed for neurotransmitter-stimulated taurine release. This *in-vitro* system provides a unique opportunity to study acute effects of ethanol and a model system for studying the basic mechanism underlying ethanol's actions in the central nervous system.

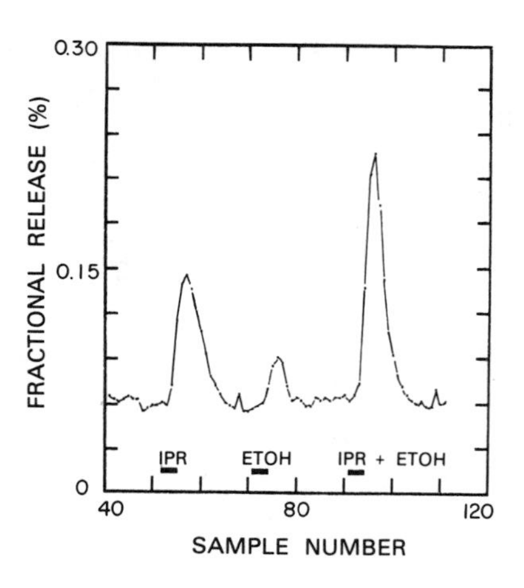

FIGURE 2. Simultaneous application of ethanol and isoproterenol results in additive release. LRM55 astroglial cells were exposed to isoproterenol (IPR; 100 nM), ethanol (ETOH; 0.33%), and isoproterenol and ethanol (IPR + ETOH) at the indicated times.

REFERENCES

1. SHAIN, W. & D. L. MARTIN. 1984. Cell Molec. Neurobiol. **4:** 191–196.
2. MADELIAN, V., D. L. MARTIN, R. LEPORE, M. PERRONE & W. SHAIN. 1985. J. Neurosci. **5:** 3154–3162.
3. SHAIN, W., V. MADELIAN, D. L. MARTIN, H. K. KIMELBERG, M. PERRONE & R. LEPORE. 1986. J. Neurochem. **46:** 1298–1303.
4. OKAMOTO, K., H. KIMURA & Y. SAKAI. 1983. Brain Res. **260:** 261–269.
5. VAN GELDER, N. M. 1978. Glutamic acid and epilepsy: the action of taurine. *In* Taurine and Neurological Disorders. A. Barbeau and R. J. Huxtable, Eds. 387–402. Raven Press. New York, NY.
6. MATTUCCI-SCHIAVONE, L. & A. P. FERKO. 1985. Eur. J. Pharmacol. **113:** 275–278.
7. MARTIN, D. L. & W. SHAIN. 1979. J. Biol. Chem. **254:** 7076–7084.

Tissue Formation and Enzyme Inhibitory Effects of Dopamine-Derived 3,4-Dihydroisoquinolines (DIQs): Possible Roles in Chronic Alcoholism

MICHAEL A. COLLINS, BHI Y. CHENG,
AND THOMAS C. ORIGITANO

Department of Biochemistry and Biophysics
Loyola University School of Medicine
Maywood, Illinois 60153

Two tetrahydroisoquinoline condensation products of dopamine, namely, salsolinol (SAL), the primary adduct with acetaldehyde, and SAL-1-carboxylic acid (SAL-1-CA), the adduct with pyruvate, appear to be increased in the brain of humans and rats during ethanol exposure.[1-3] SAL-1-CA is of particular interest because its (oxidative) decarboxylation produces the DIQ, 1,2-dehydroSAL (DSAL), which, as suggested earlier,[4] conceivably could be a neurotoxic agent in alcohol-induced brain damage (FIG. 1). This report summarizes portions of our recent *in-vitro* studies on the mechanisms of formation of DSAL from SAL-1-CA in liver, brain, and kidney, and on the potent inhibition by DSAL and other DIQs of catechol-O-methyltransferase (COMT). Our results show that SAL-1-CA produces DSAL nonenzymatically in liver and brain *via* a O_2^--mediated process, and in kidney by an apparent enzyme(?)-regulated process as well.

Methods. SAL-1-CA, DSAL, and O-methyl products were synthesized as described.[5] Incubations of SAL-1-CA in buffer or rat tissue (9000xg supernatants of homogenates; 4.5 mg protein in 1 ml vol.) were followed by isocratic HPLC with electrochemical detection. For COMT inhibition studies, rat liver supernatant with SAL-1-CA substrate and S-adenosyl methionine was incubated 5 min with DSAL or other known inhibitors; O-methylated substrate production was assessed by HPLC.[6]

Results and Discussion. DSAL, formed linearly with time from SAL-1-CA in pH 8 Na_2HPO_4 buffer, was suppressed by EDTA and, furthermore, by removal of trace metal ions with Chelex-100. There were no differences between the production of DSAL in buffer and in boiled and unboiled preparations of liver or brain. However, unboiled kidney preparations demonstrated, in a saturable fashion, DSAL formation that was threefold greater than boiled kidney, implicating some form of (heat-labile) catalysis. Nonenzymatic DSAL formation in buffer or boiled tissues was greatly suppressed by added superoxide dismutase, but less so by catalase, and negligibly by DMSO, a $HO^·$ trap. The results of COMT inhibition experiments showed that DSAL was a potent uncompetitive inhibitor of this enzyme. In terms of K_i's, DSAL was 30–40 times more effective than the known COMT inhibitors, tropolone, or SAL. No O-methylated products of DSAL were detectable *in vitro* or *in vivo*.[6]

Based on our results, a plausible mechanism for the O_2^--dependent oxidative decarboxylation of SAL-1-CA is metal-ion generation of O_2^- followed by oxidation of the catechol to the o-quinone, and expulsion of the 1-carboxyl group to produce the

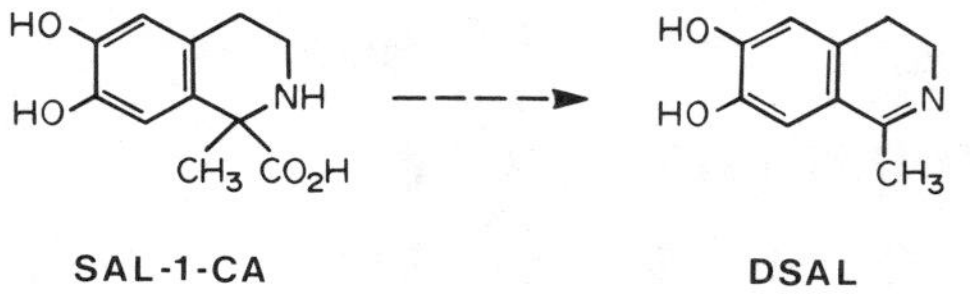

SAL-1-CA ----> DSAL

FIGURE 1. Salsolinol-1-carboxylic acid (SAL-1-CA) through its (oxidative) decarboxylation produces 1,2-dehydrosalsolinol (DSAL).

quinoidamine of DSAL. It is likely that DSAL is an endogenous product of the accumulated SAL-1-CA in chronic alcoholics, forming nonenzymatically in most tissues including brain, and possibly enzymatically in kidney. While it is an extremely effective COMT inhibitor, whether its levels become high enough during alcoholism to inhibit this and other neuronal enzymes is not yet known.

REFERENCES

1. Myers, W. D., L. Mackenzie, K. T. Ng, G. Singer, G. A. Smythe & M. Duncan. 1985. Life Sci. **36:** 309–314.
2. Sjoquist, B. 1985. *In Aldehyde Adducts in Alcoholism.* M. A. Collins, Ed. 115–124. Alan R. Liss Publ. Co. New York, N.Y.
3. Sjoquist, B. & S. Ljungquist. 1985. J. Chrom. **343:** 1–8.
4. Collins, M. A. 1982. Trends Pharmacol. Sci. **3:** 373–375.
5. Collins, M. A. & T. C. Origitano. 1983. J. Neurochem. **41:** 1569–1575.
6. Cheng, B. T., T. C. Origitano & M. A. Collins. 1987. J. Neurochem. **48:** 779–786.

Morphometric Evidence for Modification of Purkinje Cell Response to Ethanol During Aging[a]

ROBERTA J. PENTNEY AND PATRICIA QUIGLEY

Department of Anatomical Sciences
State University of New York at Buffalo
Buffalo, New York 14214

Cerebellar Purkinje neurons undergo structural changes in their dendritic networks during normal aging.[1–6] The current experiment sought for morphometric evidence of further modification of neuron structure following chronic ethanol treatment during the time these age-related dendritic changes were expressed.

A liquid (Sustacal) diet with 35% ethanol-derived calories (EDC) was fed to Fischer 344 rats for 6 months beginning at 10 (N = 8) and 20 (N = 6) months of age. Equal numbers of age-matched pair-fed rats received this diet with isocaloric replacement of ethanol by sucrose (0% EDC). All rats were then returned gradually to a solid laboratory chow diet (LCD) for a 6-week recovery period. Mean blood alcohol levels (BAL) after 6 weeks of dietary ethanol were 90.3 ± 33.8 mg/dl and 90.1 ± 27.7 mg/dl, respectively. Rats fed only the LCD diet, aged 10 (N = 8), 18 (N = 8), and 28 (N = 8) months, were used to assess network changes due to aging alone.

From coded Golgi-Cox stained parasagittal sections of the cerebellar vermis (120 μm), undamaged well-impregnated Purkinje cells, selected randomly from all folia, were drawn at a final magnification of 2000X through a microscope drawing attachment. Neuronal networks (158 total, 3/rat) were measured on a digitizer coupled to a microcomputer. Dendritic spines along terminal branches (474 branches, 3/network) were counted and measured microscopically for calculations of spine density.[7] Values in the tables below are group means determined from individual means. The Mann-Whitney nonparametric test evaluated differences in mean spine density. An unpaired t-test evaluated differences in mean numbers and mean lengths of terminal branches. Levels of significance are given in each table.

Age-related changes (TABLE 1) were more severe in neurons of 18-month rats than in neurons of 28-month rats, suggesting loss of the most severely affected neurons and/or animals during the intervening period. These age-related changes were susceptible to some dietary modulation, since they were not apparent in networks of 18-month rats given the 0% EDC liquid diet (TABLE 2).

Ethanol treatment was associated with a significant decrease in mean length per terminal branch only in networks of the oldest rats (TABLE 2), a structural change not associated with age alone (TABLE 1). The data suggest also that in the oldest rats other metric parameters were adversely affected by ethanol, though their decline was not statistically significant. BALs were lower in these rats than in younger rats. Under identical experimental conditions, the mean BAL for 4-month rats (N = 8) was 143.1 ± 27.8 mg/dl (unpublished data). Longer periods of chronic treatment with the 35% EDC diet may be required for a statistical demonstration of adverse effects of ethanol on other network parameters.

[a]Supported by National Institute on Alcohol Abuse and Alcoholism Grant 1 RO1 AA05592.

TABLE 1. Effects of Age on Purkinje Networks: Spine Densities and Metric Parameters[a]

Animal Age	Spines/μm (Terminal Branches)	Number of Terminals/Network	Length per Terminal Branch	Total Length of Terminal Branches/Network	Total Spines on Terminals/Network
10 Mo. (N = 8)	4.2 ± 0.4	562 ± 97	12.6 ± 1.0	7148 ± 1642	30,183
18 Mo. (N = 8)	3.3 ± 0.5[b]	399 ± 85[c]	11.7 ± 1.0	4619 ± 964[c]	15,530
28 Mo. (N = 8)	3.3 ± 0.6[b]	494 ± 54[d]	12.1 ± 1.2	5922 ± 781	19,680

[a]Mean ± SD.
[b]$2p < 0.01$.
[c]$2p \leq 0.003$.
[d]$p = 0.05$.

TABLE 2. Effect of Ethanol on Purkinje Networks: Spine Densities and Metric Parameters[a]

Animal Age and % EDC[b]	Spines/μm (Terminal Branches)	Number of Terminals/Network	Length per Terminal Branch	Total Length of Terminal Branches/Network	Total Spines on Terminals/Network
18 Mo.					
0% EDC (N = 8)	4.1 ± 0.6	508 ± 53	13.2 ± 1.0	6650 ± 749	27,036
35% EDC (N = 8)	4.1 ± 0.4	502 ± 63	12.6 ± 1.0	6335 ± 758	25,715
28 Mo.					
0% EDC (N = 6)	3.7 ± 0.6	488 ± 78	13.7 ± 0.6	6690 ± 982	24,435
35% EDC (N = 6)	3.7 ± 0.5	475 ± 48	12.6 ± 1.1[c]	5936 ± 908	21,857

[a] Mean ± SD.
[b] Ethanol-derived calories.
[c] $2p = 0.04$.

REFERENCES

1. GLICK, R. & W. BONDAREFF. 1979. Loss of synapses in the cerebellar cortex of the senescent rat. J. Geront. **34:** 818–822.
2. MERVIS, R. F. 1981. Cytomorphological alterations in the aging animal brain with emphasis on Golgi studies. *In* Aging and Cell Structure. J. E. Johnson, Jr., Ed. 143–186. Plenum Press. New York, NY.
3. NOSAL, G. 1979. Neuronal involution during aging. Ultrastructural study in the rat cerebellum. Mech. Ageing Dev. **10:** 295–314.
4. ROGERS, J., M. A. SILVER, W. J. SHOEMAKER & F. E. BLOOM. 1980. Senescent changes in a neurobiological model system: cerebellar Purkinje cell electrophysiology and correlative anatomy. Neurobiol. Aging **1:** 3–11.
5. ROGERS, J., S. F. ZORNETZER, F. E. BLOOM & R. E. MERVIS. 1984. Senescent microstructural changes in rat cerebellum. Brain Res. **292:** 23–32.
6. PENTNEY, R. J. 1986. Quantitative analysis of dendritic networks of Purkinje neurons during aging. Neurobiol. Aging. In press.
7. FELDMAN, M. L. & A. PETERS. 1979. A technique for estimating total spine numbers on Golgi-impregnated dendrites. J. Comp. Neurol. **188:** 527–542.

Index of Contributors

Page numbers in italics refer to Discussions of the Papers.